MW01069904

Urban Engineering for Sustainability

Sybil Derrible

The MIT Press
Cambridge, Massachusetts
London, England

This book was set in Stone Serif by Westchester Publishing Services. Printed and bound in the United States of America.

Library of Congress Cataloging-in-Publication Data is available.

ISBN: 978-0-262-04344-1

10 9 8 7 6 5 4 3 2 1

To all past, present, and future urban engineers
who strive to make the world
more sustainable.

Contents

Preface

As I write these lines it is Sunday, and unlike every other day I did not set an alarm to wake me this morning. Nonetheless, when I awoke I picked up my smartphone to look at the time. I stayed in bed a little, checked some emails, and then got up and took a shower. Afterward I made breakfast and had coffee. That tends to be my daily routine. I then sat in front of my computer and started to type these lines. In the hour it roughly took until I began writing this preface, I directly or indirectly used at least six infrastructure systems. With my smartphone I used telecommunication infrastructure and electrical infrastructure; while taking a shower I used water and wastewater infrastructure, as well as natural gas infrastructure; when making breakfast I used electrical and solid waste infrastructure. Moreover, just by being in a building I used electrical and natural gas infrastructure, and when I went out later in the day, I also used transport infrastructure. Virtually everything that we do involves some type of infrastructure in one way or another. Infrastructure is arguably the greatest physical manifestation of civilization, and it is the physical backbone of most of the activities in which we engage every day. Yet very few people actually understand how this infrastructure works and how it appears to seamlessly do everything we ask it to do, and this is exactly what this book is about.

I started writing this book in December 2014, in preparation for my spring 2015 class Cities and Sustainable Infrastructure, created for senior undergraduate and graduate students. I had taught the class in spring 2014, mostly using slides and hoping to generate a discussion. Most engineering students do not like to discuss concepts, however; they like to be given equations they can apply. After all, and for better or for worse, that is the way they have been trained since their first year. The second time I taught the class I therefore took some notes to recall some of the basic concepts of urban infrastructure, to make sure we were on the same page, and to use the equations and basic concepts to enable a discussion. Little did I know that these "notes" would bring me to writing an entire textbook.

In fall 2016 and fall 2017 I taught the class again, each time adding substantive content, polishing the language, and creating new problem sets. I also sent every single chapter to colleagues who were experts in the main topics of the chapters (they are thanked in the acknowledgments section) to ensure that there were no substantial errors and that the necessary material was covered. I also submitted a draft to MIT Press, which sent it out to three reviewers, one of whom recommended I add a chapter on solid waste management, which I did (chapter 9). In the end this book required tremendous effort and countless hours.

What I had not foreseen was how much I myself would learn by writing this book. Indeed, to complete it I had to read many textbooks, popular books, and scientific articles; watch documentaries and movies related to infrastructure; and engage in many amazing conversations with experts, whether in academia, in government agencies, or in the industry. Moreover, each time I traveled to a different city or country I found myself analyzing the infrastructure and subconsciously checking whether this book's contents applied there (e.g., by counting the number of wires on utility poles and transmission lines and making sure they added up to three for three-phase power—see chapter 5). I was so excited when I learned new concepts or found new quotes to add to the book.

From the beginning I purposefully used a casual writing style. Many textbooks adopt a dry tone, and I wanted to take a different approach, sometimes using personal anecdotes (for example, related to my hometown of Saint Pierre and Miquelon). In each chapter I also tried to relate the content to daily activities. For example, when I use the book in my class, I have the students evaluate their own greenhouse gas (GHG) emissions by asking them to pay attention to the number of kilometers they travel or the amount of waste they generate. When possible I also have the students measure some of the things we see in class; for example, I have a water pressure gauges that students can use to measure the water pressure at their taps at home. The students particularly appreciate the casual style of the book and the relation of many of the concepts to their personal lives. This has made the book more relatable by showing them firsthand that the concepts within it are not abstract but are actually happening in practice.

The main goal of this book is to introduce the reader to the world of urban infrastructure from an engineering perspective, within the context of sustainability and with an accent toward integrating urban infrastructure. Urban infrastructure here is plural, as in urban infrastructure systems. Specifically, five infrastructure systems are introduced: electricity, water, transport, buildings, and solid waste. Natural gas and telecommunication are missing, although they are partly discussed in the chapters on electricity (chapter 5), buildings (chapter 8), and urban metabolism and infrastructure integration (chapter 10). Moreover, one chapter is devoted to defining sustainability, and another focuses on population forecasting (as the demand for infrastructure). Finally, chapter 11 focuses on the Science of Cities and Machine Learning as novel approaches likely to take a predominant role in urban engineering. Except for the introduction (chapter 1) and the conclusion (chapter 12), most chapters are written in such a way that they can be read independently of one another. Nonetheless, I would recommend that the sustainability chapter (chapter 2) be covered toward the beginning of the class. Moreover, most students who took my class particularly appreciated the urban planning chapter (chapter 4)—especially since most were engineers, and this chapter offers "refreshing" content.

The book does not cater to only one type of engineer, or only to engineers for that matter. I am a strong proponent of interdisciplinarity, and I have never been fond of traditional disciplinary boundaries—civil engineering, computer science, mechanical engineering, electrical engineering, building science, urban planning, and so on. Why can we not integrate relevant elements from each discipline? At least, this is what I tried to do in this book, so no individual discipline can claim it. Instead, I call for a new kind of engineering: urban engineering that neatly combines the "urban" of urban planning with the "engineering" of many engineering disciplines.

As a textbook this work is written for students, but it is also written for practitioners as a general resource on urban infrastructure—for example, to recall a concept or to quickly learn important ideas about a specific field. The book can also be used to study for the Fundamentals of Engineering (FE) exam and for the Principles and Practice of Engineering (PE) exam—I know many of my students told me they used the book to study for the FE exam. Moreover, although the book was written mostly for senior undergraduate and graduate students, elements of it can be used by virtually anyone. For example, elements of chapter 2, on sustainability, can even be used in high school classes.

Finally, the data used in the book focus mostly on the United States, partly because I wrote this book as a faculty member in the United States and partly because an amazing amount of data is available for the United States. In general, the concepts, findings, and lessons developed in the book should be applicable to most cities throughout the world. It does not mean that U.S. consumption patterns can be applied everywhere—the United States tends to be one of the highest consumers in almost every category—but that many engineering systems in the world work in similar ways, and therefore the concepts covered should still be relevant. I can then only recommend that international readers obtain their own data to investigate consumption patterns in their own countries.

For instructors, I use the book in my class as follows. First, I essentially "flip" the classroom by asking all students to read a selected chapter or part of a chapter and by giving them a quiz at the beginning of every class. Quizzes are short—no more than ten minutes—and consist of two parts: (1) the students must answer ten True or False questions and (2) the students must either draw a figure, explain variables in an equation, or solve a simple numerical question from the chapter covered that day. The only exception is when we cover urban planning, for which I ask the students to draw their own city (see chapter 4). So that students do not have to focus only on the grade, I only count the ten best quizzes out of all the ones they have to take (usually between seventeen and twenty quizzes). Having to read each chapter before class forces the students to learn by themselves. As mentioned, I used the book to teach senior undergraduate and graduate students, many of whom will never take another class. I therefore try to instill in them a sense that they can read a textbook on their own and understand the content. Also, the book is fairly long, and I do not have time to cover every aspect in detail. This means that when we discuss a chapter in class I can omit some content.

The main assessment for the class is a group design project. I find one empty plot of land somewhere in the city and have the students design a livable and sustainable neighborhood: every building, road, bus line, electrical system, water system, wastewater system, and so on must be designed and accounted for in the neighborhood's final carbon footprint. Emphasis is placed on making the neighborhood livable, which forces the class to think and to be creative, as opposed to simply applying an equation from the book. Many students thrive during this exercise and get well-deserved As in the class, even though they may get Bs in other classes because they do not do well on exams.

Moreover, for every chapter I assign homework, and all homework must be submitted. If a student fails to submit even one homework assignment, I give that student a zero on all homework. I tell the students that because homework serves as *formative* assessment (as opposed to *summative* assessment, as is the case for exams), failing to turn in a homework assignment is similar to skipping class. In addition, I also give a technology assignment, generally due in the middle of the semester, for which every student must find one "sustainable" technology and report it to the entire class. Each student has to write a one-page summary that includes GHG-saving numbers and give a one-minute presentation on this technology. This assignment is essentially a means to crowdsource technologies that students might find relevant for their neighborhood design.

Finally, I assign one midterm exam that covers everything from the beginning of the semester to the midterm (generally, chapters 1 through 6) and then one more exam the week before finals that covers everything from the midterm to the end of the semester (generally, chapters 7 through 12—but I rarely cover chapter 11 because of time constraints). During finals week I have the students present their final neighborhood designs to the class, make a poster for their designs, and write a five-to-seven-page leaflet that captures the essence of what their designs are about. In the past I have compiled all these leaflets into "books" that are available on the teaching page of my website. The website also contains the syllabi of all the classes I have taught.

Overall, writing this book has been an amazing journey. I certainly never expected to write a textbook—especially since I started writing it as an untenured faculty member—and the whole process definitely took more time than I could have imagined. Nonetheless, despite the hours, writing this book has been one of the most wonderful experiences of my life, and I hope you will enjoy reading it.

Acknowledgments

Writing a textbook is never a lonely affair, especially when the textbook is highly interdisciplinary, as is this one. I therefore have many people to thank, and I am sure I will omit a few. If your name should be listed below and it is not, please accept my sincere apology.

The people who deserve to be acknowledged first are the students at the University of Illinois at Chicago who took CME494, which then became CME440, Cities and Sustainable Infrastructure, in spring 2015, fall 2016, fall 2017, and fall 2018. I wrote this book for them, and their feedback genuinely helped to shape its contents. While I cannot name them all here, I am sincerely grateful to them. Moreover, I need to thank all the students from my research group, the Complex and Sustainable Urban Networks Laboratory, who not only had to take the class themselves but had to provide ideas regarding the book and the class.

Second, I must thank all the experts who spent many hours reviewing different versions of all the chapters and providing feedback. Their contributions were instrumental to this book. In particular, I would like to thank Lynette Cheah, Nadine Ibrahim, Christopher Kennedy, Robert Kooij, Neda Mohammadi, Eugene Mohareb, John Taylor, and Thomas Theis, who took the time to read the entire book and provide comments or even use a draft version of the book in their course. They are therefore not named for the individual chapters below. Focusing on these individual chapters, I would like to acknowledge the contributions of Pierre Hélène for chapter 1; David Bristow and Ming Xu for chapter 2; Caitlin Cottrill, Kate Lowe, and Moira Zellner for chapter 4; Mohamed Badhrudeen, David Bristow, Vahe Caliskan, Sudip Mazumder, and Nicolas Stauff for chapter 5; Christopher Chini, Charlotte Frei, Amid Khodadoust, David Klawitter, Benjamin O'Connor, Matthew Reeder, Karl Rockne, Joseph Schulenberg, Ashlyn Stillwell, Mason Throneburg, and Sean Vitousek for chapter 6; Charlotte Frei, Dongwoo Lee, David Levinson, Jane Lin, Kouros Mohammadian, Daniel Work, and

Bo Zou for chapter 7; David Bristow, William Ryan, and Brent Stephens for chapter 8; Christophe Caignard, Oliver Heidrich, Daniel Hoornweg, and John Mulrow for chapter 9; and Sk Nasir Ahmad, Marc Barthelemy, Michael Batty, Patricia Bordin, Shauhrat Chopra, Matthieu Cristelli, and Francisco Câmara Pereira for chapter 11. In total, forty-eight names are listed here, and as mentioned, I may have omitted a few people as well. Moreover, I am not naming the people working for government agencies and industries who provided data for this book or answered questions about the nature of the data. Whenever applicable, the agencies and industries are directly cited in the text.

In addition, I must thank Beth Clevenger and Anthony Zannino at the MIT Press, who believed in this book from the beginning and who handled it through the lengthy editorial process. A big thank you goes to John Donohue and Wendy Lawrence at Westchester Publishing Services as well for doing an amazing work at copyediting the book. I must also thank the three anonymous readers who reviewed this book and provided great feedback that further improved it.

Finally, writing a textbook indubitably takes a significant amount of time, which led me to spend many long hours in my office, more often than not in the evenings and on weekends, away from my family. I will be forever grateful to my wife, Marie-Agathe Simonetti, who has been patient and accepting of these hours away from her while encouraging and believing in me every step of the way. To anyone aspiring to write a textbook, never underestimate the importance of a significant other in helping you achieve your goal, and I can only wish that you receive as much support as I received from my wife. Marie-Agathe: Merci.

I Urban Contexts and Sustainability

1 Introduction

We cannot solve problems by using the same kind of thinking we used when we created them.
—Albert Einstein

This quote from Einstein is one of my favorites, and it embodies many of the challenges that our society faces at the moment. Simply put, it is time to change the way we plan, design, engineer, and operate cities.

Going all the way back to the Neolithic era (about 10,000 BCE), the general fields of planning, engineering, and science have evolved tremendously, leading to the generation of knowledge, methods, and processes that have had significant impacts on the world. Over time they have allowed us to construct, operate, and monitor buildings in real time, to bring water to and from these buildings, to treat the solid waste generated in these buildings, and to travel from building to building via a large interconnected network of roads and rails.

Along with great progress, however, this development has also created quite a few problems, which should not be surprising. A solution to something almost indubitably creates some problems as well. Many engineering solutions developed during the Renaissance, for instance, allowed us to build larger cities and create bigger economies, which then offered the perfect venue for germs and disease to spread widely. As an example, in 1854 an outbreak of cholera killed 600 to 700 people in London (Johnson 2006). It was not until a man by the name of Dr. John Snow collected data about the people who had contracted the disease that the source of the outbreak, a contaminated public water well, was discovered.[1] Ways to treat water before it is distributed and to treat wastewater before it is rereleased into the environment had to be invented to keep the problem from reoccurring. As a more contemporary example, within the world of software engineering, fixing a bug in a code is known to easily create more bugs.

Therefore, all these advances that have helped us progress so much have naturally brought their own problems. Following the words of Einstein, we therefore need to think differently and come up with new sciences and practices that can solve the problems that we face, which we will broadly fit within the term *sustainability*—knowing that the solutions we come up with may very well create new problems that will need solving in the future.

1.1 On the Path to Scenario B

One great way to look at this evolution of problems is from the point of view of Joseph Tainter (1988) in his seminal book *The Collapse of Complex Societies*. In figure 1.1 the x-axis represents the addition of new "things" in a society that tend to make it more complex, like the addition of new infrastructure. Staying with our water illustration, we first get water from the most accessible source, such as a nearby river or a lake. To collect, treat, and distribute this water, we need to build the necessary infrastructure, and the complexity of our system goes from x_0 to x_1. Thanks to this new system, we get some benefits that we can measure as $y_1 - y_0$. Now, say our city grows and we need to get water from a new source. This new source will not be as accessible as the first one,

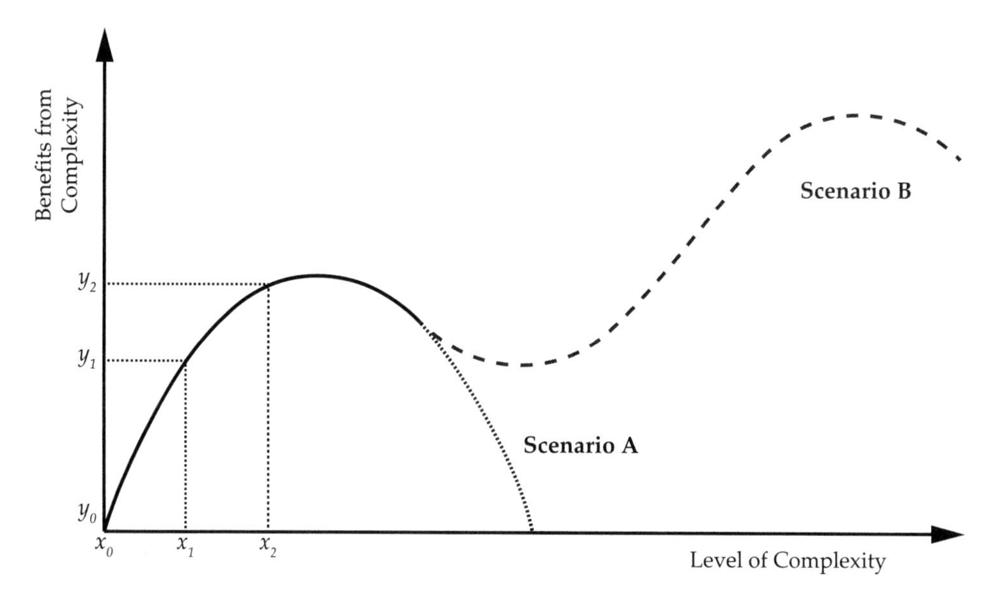

Figure 1.1
The marginal returns of increasing complexity.

and the infrastructure needed will be more costly. We may therefore have to add the same amount of complexity to our city as we did the first time around—going from x_1 to x_2—but the benefits $y_2 - y_1$ will be smaller. If we go on planning the same way, the third time, the benefits will be even smaller. This is what we call *diminishing marginal returns*—that is, the returns for every addition decrease (we will use this concept again in chapter 2). The lower marginal returns may be acceptable for a while, but at some point we will need to figure out whether we want to take Scenario A or Scenario B. Scenario A is very common and has happened many times in the past; Tainter (1988) documents a number of examples, including the Roman Empire and the Mayan civilization. Scenario B is a lot tougher since we need to come up with new solutions, therefore echoing the quote from Einstein.

The Industrial Revolution is a good example of Scenario B. Indeed, with the advances from the Renaissance and the eighteenth century, cities had grown and new technologies were needed to cope with this growth. For example, it was not before the end of the 1800s that drinking water could be pumped by large steam engines in large pipe networks. Similarly, the Industrial Revolution has created new challenges that must be faced. The difference, however, may be the scale of the problem, which has become global, and for the first time in the history of humanity, all societies in the world must work together to tackle these problems. Moreover, we cannot simply solve our problems by seeking new energy sources and consuming more resources. This is why it is called *sustainability*. As engineers, planners, and scientists, our goal is to create new solutions to make sure we follow Scenario B once again. This is a very challenging endeavor, and we need to revisit some of our current practices. Otherwise, we are bound for Scenario A.

1.2 Objective: Integrate Infrastructure Networks

Throughout this book we will go over many concepts, frameworks, methods, tools, and techniques that can help us toward our goal. We will, notably, see that defining and measuring sustainability is often not possible in practice, and in the absence of a systematic way to measure whether a design is sustainable or not, we will need to follow *sustainability principles* and hope for the best. *By far the main message of this book is the need to view infrastructure systems as integrated networks*. Indeed, as the scale of the problem is global, using more energy to fix our problem is not an option. We therefore need to improve the way infrastructure systems work, and we will see that they can certainly work much better if they are better integrated. Despite the fact that the issue is known, the state of the practice has too often been to view infrastructure as independent silos. At the time of this writing, this was why the various municipal, regional, and even national departments / ministries of transport, water, buildings, and energy rarely communicated and coordinated with one another. This practice must change, and we need to start by learning about each individual system ourselves, becoming demographers, urban planners, electrical engineers, water resources engineers, transport engineers, building engineers and sanitary engineers to determine how we can better integrate these systems.[2]

This point is particularly important because infrastructure systems are highly interdependent. Dupuy (1988) noted this point: "Although only one network is capable of having an effect on urbanization, it is the totality, the combination of several different networks, that corresponds to the new forms of spatial development, redefining the territory and bringing about the transformation of local society" (p. 295). This point was obvious for me when growing up in Saint Pierre and Miquelon. With roughly 6,000 inhabitants,[3] Saint Pierre and Miquelon is a small overseas collectivity of France located thirteen kilometers south of the Canadian shores of Newfoundland.[4] Despite being close to the North American continent, it is administratively entirely French and follows European regulations. This means that people on the island work in the metric system and use the 220-volt electrical standard. The archipelago needs to treat and distribute its own water (that is entirely gravity fed), generate its own electricity (using six diesel motors at the time of this writing), and manage its own solid waste (although a lot of it is recycled in France and in Canada). Growing up in Saint Pierre and Miquelon, I was quickly reminded of how everything is interrelated. Figure 1.2 shows a street that was opened up in summer 2017 to install a district heating system and to change a water main. The figure shows at least eight different types of infrastructure systems: buildings, transport (roads), electricity (both medium voltage and low voltage), water, wastewater, telecommunications, and the district heating system (that recuperates the heat being generated by the diesel motors that produce electricity). Most of these infrastructure systems need one another to function, and they are also amazingly collocated. So why were these infrastructure systems not planned together? Does it not seem obvious that they should have been planned together?

Moreover, most interdependencies that exist between infrastructure systems were not designed for in the first place. Instead, as individual infrastructure systems became more complex, they also became more interdependent. For example, early water distribution systems did not depend on telecommunications infrastructure, but starting in the 1960s, Supervisory Control and Data Acquisition (SCADA) systems have been installed to monitor and control water distribution systems remotely. What we need is to control for these interdependencies and design them in the ways we want

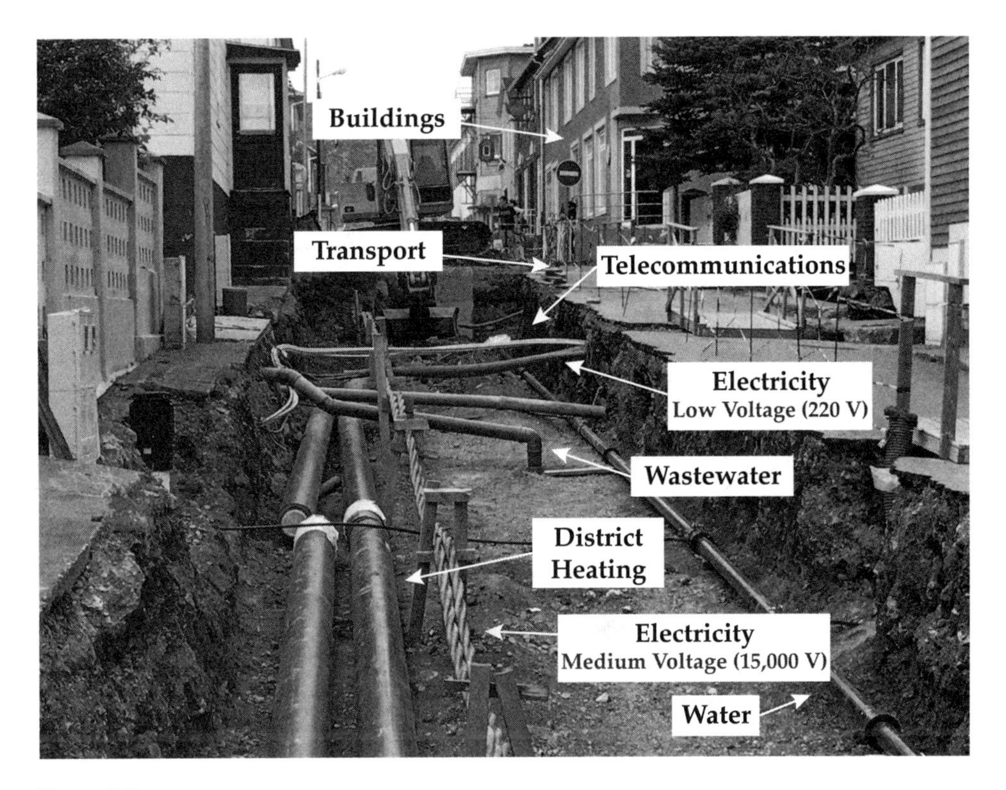

Figure 1.2
Infrastructure networks in Saint Pierre and Miquelon.

infrastructure systems to be connected. In this book, we will learn many strategies to do so.

Essentially, we need to become *urban engineers*. Another term for infrastructure integration is *infrastructure ecology*. Figure 1.3 shows how the various infrastructure systems are interrelated. Instead of land use, we will focus on buildings, but otherwise, we will learn about the systems shown in the figure as well as about solid waste management. In addition to understanding how each system functions, we will also learn and sometimes quantify these flows (that capture the interdependencies just mentioned). One good example is that streets are often designed to accommodate a certain traffic flow. But as we experience each time it rains heavily, streets also need to accommodate flows of stormwater runoff.

To partially address flooding concerns, the government of metropolitan Seoul in South Korea was quite creative. In 1968, the Cheonggyecheon River was covered with a double-story elevated expressway that was constantly congested and extremely

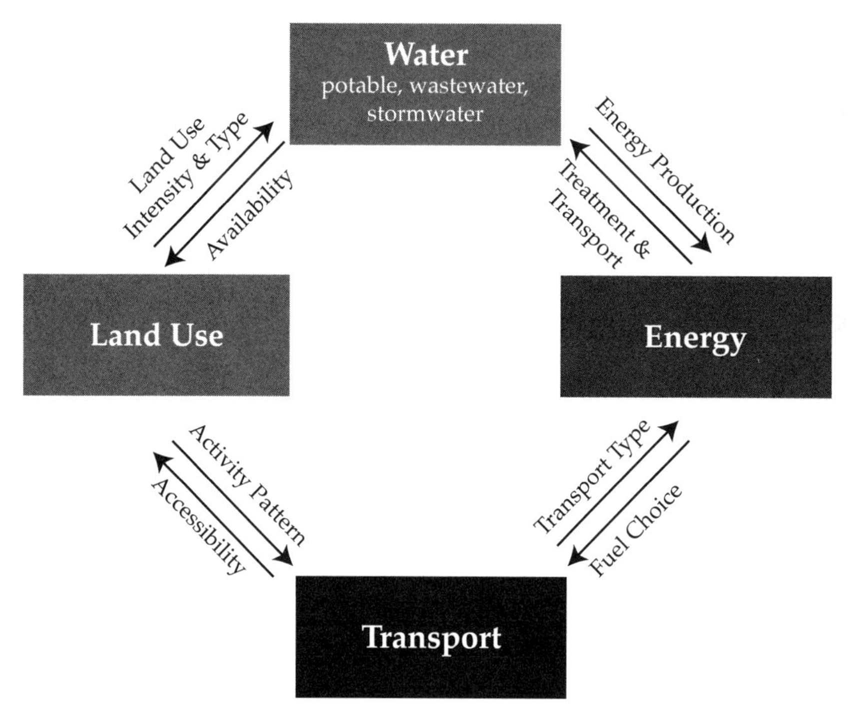

Figure 1.3
Diagram of infrastructure ecology. Adapted from Pandit et al. (2011).

polluting (figure 1.4a). After much debate, the river was restored in 2005, along with a scenic river walk (figure 1.4b), and the traffic was diverted to other areas of the city. The best part is that the river is also a stormwater channel that can be flooded during heavy rains. In this case, a transport problem became a water solution, and the interdependencies that came with the project were specifically designed for and therefore desirable. Put differently, the interdependencies are no longer constraining but enabling.

1.3 Why Cities?

Now that we have identified our ultimate goal—that we need to create new solutions to better integrate infrastructure systems—why are *cities* important? Cities are certainly not new; people have lived in cities for millennia. In fact, people lived in cities before they became farmers (i.e., before the agricultural revolution).[5] Yet cities are critical to our argument for two main reasons. First, since 2008, more people in the world live in cities than in rural areas for the first time in the history of humanity. This may not

a) Before (Courtesy of Seoul Metropolitan Government)

b) After

Figure 1.4
Cheonggyecheon River, Seoul, South Korea.

seem important to people who have lived in cities their whole lives, but it is actually quite radical. Since we consume energy and resources, the way people are distributed in space has important impacts. If we see this consumption as *demand*, then this demand for energy and resources is shifting from being distributed over large areas to being highly concentrated in small areas. At the same time, where we get the *supply* to meet this demand does not change much; for example, water resources are not getting more concentrated. This therefore means that we need to revisit the way energy and resources are supplied to cities as the cities themselves get larger and larger. Moreover, this supply is also being stretched since more people are populating the planet, therefore requiring more land and more resources. We will use this supply and demand framework extensively in chapter 2 and learn about concepts of planetary boundaries as well.

The second reason, although not definite at the time of this writing (2018), is that city residents may actually consume less than people living in rural areas. This is especially true in cities and neighborhoods where a lot of people live in a small area (i.e., high population densities). In these cities, people tend to live in smaller houses that require less heating, cooling, and lighting. Moreover, people in these cities tend to either live closer to their workplace, to grocery stores, or to restaurants or can travel to these places using less energy (i.e., walking, cycling, public transport). We will go through all these points quite exhaustively in this book. There are also more people now living on Earth than ever before, surpassing the 7 billion mark in 2011 (for which *National Geographic* magazine [2011] dedicated a full and insightful issue, including a powerful video[6]). Most of these people live in cities, and this trend will continue in the foreseeable future to 10 or 12 billion people.

Our goal is therefore to create new solutions to provide people with the infrastructure services that they need, avoiding these diminishing marginal returns while knowing that the way people are consuming "things" has changed. This goal cannot be achieved without numbers, and we will need to become energy and carbon "numerate" to compare patterns and assess different planning and design strategies.

Engineers must also change their mentalities. The engineering practice has become overly conservative in the second half of the twentieth century. Sometimes because of fear of litigation (i.e., being sued), engineers often resign themselves to following standards, codes, and their traditional tool kit of equations as opposed to coming up with better and more creative solutions. The designer Don Norman (2013) captured it well in his seminal book *The Design of Everyday Things* when he wrote: "Engineers and businesspeople are trained to solve problems. Designers are trained to discover problems. A brilliant solution to the wrong problem can be worse than no solution at all: solve the correct problem" (p. 218).[7] Engineers must first discover the fundamental problem they are trying to solve instead of providing a standard solution to an apparent problem that may not be the real issue. This is partly why we will spend an entire chapter on urban planning in order to understand how it used to be and how we got here. The most famous historian of cities is probably Lewis Mumford (1961), who wrote an amazing historical account of cities in *The City in History: Its Origins, Its Transformations, and Its Prospects*. One great quote from this book is a critique of baroque town planning: "The city was sacrificed to the traffic in the new plan: the street, not the neighborhood or quarter, became the unit of planning" (p. 391). Engineers therefore need to go back to the fundamentals of why we build cities in the first place: for people. The two basic questions therefore become: What kind of services do people require from infrastructure? And how can we best design infrastructure systems to provide them?

1.4 Civitas

Some of the concepts that we will see are quite easy to understand, while others will be a little trickier. To facilitate the process, we will systematically take the example of the city of Civitas. Civitas is a fictional city that we will use to illustrate some of the main concepts covered in each chapter. The word *civitas* comes from ancient Rome, and it represents the entity formed by all Roman citizens. It can be loosely translated to "city-state," but for us it really encapsulates the residents of a city, who make a city what it is, as opposed to the infrastructure systems that are really there to service the residents. In *Democracy in the Politics*, Aristotle wrote, "The city-state comes into being for the sake of living, but it exists for the sake of living well." This quote is particularly relevant since as human beings

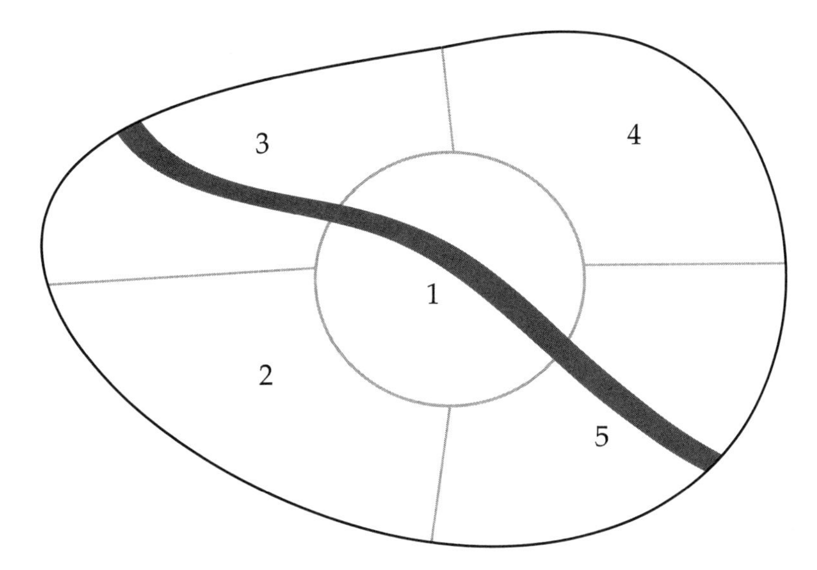

Figure 1.5
Civitas divided into five zones.

we form societies because we tend to live better together than alone. We should therefore make sure we plan, design, engineer, and operate cities for everyone as well. Jane Jacobs has a perfect quote on this particular aspect, which we will see in chapter 4.

In the 2010 census, our fictional city of Civitas was home to 60,000 people, Civitians, who went about their daily routines just as any resident of a city. Out of 60,000 people there were a total of 24,000 households, placing the average number of people per household at 2.5. The workforce of Civitas was composed of 40,000 people, who were all employed. Figure 1.5 shows a sketch of Civitas with its main river, Vita, which translates as "life" from Latin.[8]

Civitas is divided into five *zones*, akin to census tracts or traffic analysis zones. These zones will be our statistical units, and we will use them quite heavily throughout the chapters, as is done in practice. In particular, we will see that this is especially relevant in chapter 7 because we will calculate travel demand and transport emissions.

As is fairly common around the world, we will assume that zone 1 is the Central Business District (CBD), where most of the jobs are. Therefore, many people commute from zones 2 to 5 to zone 1 every day to go to work. This does not mean that zones 2–5 are suburbs. Instead, they contain a good mix of houses, schools, offices, restaurants, and any other types of buildings necessary for a livable community. We will see why having a mixture of building uses is important in chapter 4.

1.5 Book Outline

The book is divided into three main parts. In part I, "Urban Contexts and Sustainability," we will define sustainability in urban engineering (chapter 2), and we will learn how to forecast population (chapter 3). After all, people represent our main consumers of energy, goods, and resources, and forecasting future population is paramount. We will use Civitas as an example of a typical city. We will then learn about urban planning (chapter 4), going all the way back to the Neolithic era. Importantly, in chapter 4 we will learn about our limits as people and how we perceive cities as opposed to how they actually are.

In part II, "Urban Engineering and Infrastructure Systems," we will look at each individual urban infrastructure system one by one: electricity (chapter 5), water (chapter 6), transport (chapter 7), buildings (chapter 8), and solid waste (chapter 9), becoming almost experts in each. We will describe the practice at the time of this writing, study techniques to estimate energy use, and go over ways to apply the sustainability principles defined in chapter 2. This is where we will become energy and carbon numerate and where we will be able to understand electricity use in megawatt-hours [MWh], water consumption in liters per day [L/day], transport emission in grams of carbon dioxide equivalent per kilometer [g CO_2e/km], building energy use in watts per square meter [W/m^2], and solid waste in tons [t]. Here again, Civitas will constantly be used to illustrate the concepts and methods that we cover.

In part III, "Urban Metabolism and Novel Approaches," we will learn about urban metabolism (chapter 10), which integrates the various urban systems and determines the total energy and carbon footprint of cities. We will discuss to what extent infrastructure systems are interdependent and detail some of the flows shown in figures 1.2 and 1.3. In our effort to be more creative, we will then discover the Science of Cities (chapter 11), also sometimes known as Urban Science. This field emerged primarily in the 2000s, thanks to the current profusion of data and advances in Complexity Theory. Moreover, processing new data sets and data streams needed for the Science of Cities can be cumbersome, thus we will turn our attention to Machine Learning that can be incredibly useful for processing large and complex data sets. Although the application of the Science of Cities and Machine Learning to the design of infrastructure systems was uncommon at the time of this writing, they have the potential to transform urban engineering. In particular, they can help us change our way of thinking.

Finally, we will conclude by going through two important exercises. The first will determine some of the most important changes that may completely shift the current paradigm on how cities are planned, designed, engineered, and operated. Specifically, we will identify three of these changes, two of which are purely technological, while one is organizational. Then we will elaborate a simple and short four-point urban infrastructure design (UID) process that can be used for any infrastructure project, regardless of the infrastructure type. All urban engineers can use this process as a simple guideline, and as we might expect, a higher integration of infrastructure systems is part of this process.

The appendix contains important tables, diagrams, and conversion factors that can be referred to quickly. Notably, it contains a copy of the Moody diagram and the level-of-service diagram, as well as an equation list. Moreover, by U.S. state, it contains a list of the latest power grid emission factors and the per capita consumption values of electricity, natural gas, and water available at the time of this writing.

Throughout this book, we will learn, analyze, and try to develop new and more sustainable designs for the various infrastructure systems that populate our cities, focusing on five major systems: electricity, water, transport, buildings, and solid waste. For this, we will look at the people of Civitas, the Civitians, to see how they live and how their choices affect their energy and carbon footprints. By the end, everyone should be able to understand and identify the issues that need to be solved when planning and designing urban infrastructure. More importantly, everyone should be able to come up with solutions to these issues by taking an integrated and systems-thinking approach.

1.6 Measures and Units

As we learn about the various infrastructure systems, we will be using different units, which we will generally put in square brackets [ ], as is conventional. Some of these units are quite easy to understand, like liters of water per day [L/day], while others may be more difficult, like watts per meter kelvin [W/(m·K)]. From personal experience, I have found that many students have a hard time understanding power in watts [W] and energy in watt-hours [Wh]. After all, in physics classes, we typically learn that energy is in joules [J]. Power is a rate of energy production or consumption, expressed in joules per second [J/s], but it was given the unit [W] (i.e., 1 W = 1 J/s) after the Scottish engineer James Watt.[9] Then, instead of multiplying back the number of seconds produced or consumed to get energy, we like to multiply it by the number of hours. We now have watt times hours [Wh] for energy instead of [J], and 1 Wh is simply 3,600 J (i.e., 1 Wh = 1 J/s × 1 h = 1 J/s × 3,600 seconds = 3,600 J). Too often, the unit [W/h] (as in watts *per* hour) appears, although it does not exist. Many sources still use [J], however, and it may be easier to understand. Here, we prefer to use [Wh] consistently for energy since it is more common in practice.

Table 1.1 shows a list of units we will use in this book. By far, the use of the metric system is recommended since it is much easier to use and widely adopted around the globe. Although we may see gallons here and there in chapter 6, British imperial units will not be used. Another common unit that we will not use is the Btu, which stands for British thermal unit. Because we often find data in British units, however, some helpful conversion factors are shown in table 1.2. These two tables are also available in the appendix.

Moreover, we note that similar units do not necessarily have the same timescale. In water, for instance, we tend to use the day as the unit of time. In transport, we use the number of trips per day, but we also use total kilometers traveled per year. In electricity, we tend to use either monthly or yearly values in [MWh] since our electricity consumption can vary considerably depending on the season. In solid waste, we use metric tons per day, month, or year in general. For GHG emissions, we tend to prefer yearly values as well since our energy and resource consumption varies by season.

GHG emissions tend to be expressed as a mass, whether pounds or grams. But since we prefer the metric system, we will express GHG emissions in grams of CO_2 equivalent, or simply [g CO_2e]. The term *equivalent* is needed here because there are many GHGs. The most important are carbon dioxide (CO_2), methane (CH_4), and nitrous oxide (N_2O). Because they do not have the same global warming potential (GWP), the Intergovernmental Panel on Climate Change (IPCC) has come up with factors to estimate the equivalent impact of the various GHGs relative to CO_2 over time. The IPCC typically uses the 100-year factors 28 for CH_4 and 265 for N_2O. This means that CH_4 and N_2O have 28 and 265 times more GWP than CO_2, which is obviously significant.[10]

Finally, we also need to discuss conversion factors. From nano (10^{-9}) to tera (10^{12}), we often prefer to add a prefix to the unit we are using as opposed to adding a power of ten. Most of us know these conversion factors already, but here is a list:

- nano: 10^{-9}
- micro: 10^{-6}
- milli: 10^{-3}
- kilo: 10^{3}
- mega: 10^{6} (or simply ton for mass)
- giga: 10^{9} (or simply kt for mass)
- tera: 10^{12} (or simply Mt for mass)

Moreover, remember than one metric ton is 1,000 kilograms [kg] and not one megagram [Mg].[11] This is actually quite important. Although GHG emissions are usually

Table 1.1
Standard urban engineering units

Name	Unit	Symbol
General		
Time	Second	s
	Minute	min
	Hour	h
Temperature	Kelvin	K
	Celsius	C
Population		
People	Person	pers
Electricity		
Power	Watt	W
Energy	Watt-hour	Wh
Voltage	Volt	V
Current	Amp	A
Resistance	Ohm	Ω
Water		
Volume	Liter	L
Velocity	Meters per second	m/s
Flow rate	Cubic meters per second	m^3/s
Pressure	Pascal	Pa
	Meter	m
Transport		
Vehicles	Vehicle	veh
Trip	Trip	trip
Distance	Vehicle kilometers traveled	VKT
	Passenger kilometers traveled	PKT
Buildings		
Heat transfer	Watt	W
Heat transfer per unit area	Watt per square meter	W/m^2
Heat transfer per unit length	Watt per meter kelvin	$W/(m \cdot K)$
Solid Waste		
Weight	Metric ton	ton

Table 1.2
British units to metric system conversion factors

Category	Conversation factor	British units
Mass		
	1 lb = 0.453 kg	lb: pound
	1 short ton = 0.907184 ton	short ton: US short ton
	1 long ton = 1.016047 ton	long ton: US long ton
Temperature		
	0 K = 273.15°C	
	F = 1.8°C + 32	F: Fahrenheit
Distance		
	1 in = 25.04 mm	in: inch
	1 ft = 0.3048 m	ft: feet
	1 yd = 0.9144 m	yd: yard
	1 mi = 1.609 km	mi: mile
Volume		
	1 m^3 = 1,000 L	
	1 gal = 3.78541 L	gal: gallon
	1 ft^3 = 0.02832 m^3	ft^3: cubic feet
	1 yd^3 = 0.76455 m^3	yd^3: yard
Pressure		
	1 psi = 6894.76 Pa	psi: pound per square inch
Flow Rates		
	1 gpm = 0.063 L/s	gpm: gallon per minute
Density		
	1 lb/ft^3 = 16.0184 kg/m^3	ft^3: cubic feet
	1 lb/yd^3 = 0.593276 kg/m^3	yd^3: cubic yard
	1 lb/gal = 119.826 kg/m^3	gal: gallon
Energy		
	1 cal = 1 kcal	cal: calorie
	1 cal = 4.187 kJ	
	1 cal = 1.16222 Wh	
	1 Btu = 1,055.056 J	Btu: British thermal unit
	1 Btu = 0.293071 Wh	
Energy Density		
	1 Btu/lb = 0.64611 Wh/kg	
Power		
	1 hp = 745.7 W	hp: horsepower
Fuel Economy		
	1 mpg = 0.4251 km/L	mpg: mile per gallon

expressed as a mass in grams, values of GHG emissions tend to be large and are expressed in [kg], [t], or even in kilotons [kt] or megatons [Mt]. So 1 kt is the same as 1 billion grams, but we do not write gigagram. Although it may appear a little confusing at the moment, it should be natural when we get to it.

1.7 Missing Topics

Finally, one of the personal objectives for this book was that it not be too long. This means that several important topics that fit well with urban engineering were purposely omitted.

The first missing topic is life-cycle assessment (LCA). LCA is about taking into account all emissions related to the life of a product, from the extraction of the elements to make the product to the disposal of the product itself. LCA is typically used in sustainability, and in fact it is included in most books on the topic. Moreover, it requires a lot of data and a good knowledge of all the processes involved in the life of a product. Many amazing researchers spend their entire careers computing the LCA of products and adding new energy and emissions values of certain processes. Despite the relevance, we will only briefly discuss it in chapters 5 and 10. More information on LCA can easily be found on the web, and to compute LCA, the open-source open LCA platform is recommended.[12]

In terms of infrastructure, this book does not include chapters on natural gas and on telecommunications, although both are discussed a little in chapter 10. Natural gas is also discussed in chapter 8 because it is often used to heat buildings. Telecommunications will become increasingly important because it is often used to monitor the performance of the other infrastructure systems, and it has a predominant role in the emergence of *Smart Cities* (which we will discuss briefly throughout the book and specifically in the conclusion). Future editions of this book might therefore have a chapter dedicated to it.

Food production and consumption is also absent from this book. In particular, emissions related to food production are substantial (especially meat products), and there is a great deal of research going on to better understand the relationships between food, energy, and water. Food may take a larger role in future editions of the book, but at the same time, it is not an engineering system as we have defined them here.

In addition, one important concept that is gaining significant momentum and that is often associated with sustainability is *resilience*. One of the goals is, notably, to move away from fail-safe designs and instead focus on designing systems that are safe to fail (Ahern 2011), which is absolutely relevant to infrastructure design and can be done by adopting a framework of sensing, anticipating, adapting, and learning, for example (Park et al. 2013). Although we will not define or cover resilience in an independent chapter, we will refer to it at times. The omission of resilience is chiefly due to the fact that at the time of this writing (2018), the concept was still relatively new in urban planning and engineering, and the community had yet to reach a consensus about how it could best be applied. Most often, matters of resilience are handled with optimization and operations research, often missing the central point of resilience thinking—that is, by optimizing for one aspect, a design is automatically more vulnerable to other aspects that are sometimes colloquially called *unknown unknowns*. A full chapter on resilience will surely be included in the future, or resilience will be better integrated in each individual chapter, but for now the reader is referred to two insightful readings: the scientific article by Woods (2015) that discusses four aspects of resilience and the critical book *Antifragile* by Taleb (2012).

1.8 Conclusion

In this short introduction, we were able to acquire a conceptual understanding of the main problem we face. In particular, Tainter's *diminishing marginal returns* concept provides a simple and illustrative framework around which we can work. Put simply, we should focus not on efficiency but instead on providing completely new solutions to be on the path to Scenario B. As we quickly covered here, and as we will see repeatedly throughout the book, integrating urban infrastructure systems offers such a solution (think of the Cheonggyecheon River in Seoul). After all, infrastructure systems are naturally interdependent. It therefore seems logical to want to control these interdependencies, and this will be part of our role as urban engineers.

In the introduction we also covered why cities are the right places to integrate infrastructure. If we must remember one reason why this is so, we should remind ourselves that the world is increasingly urban. This does not mean we should focus only on "large" cities—in fact, we may or may not want to focus on "small" cities first since it may be relatively easier to get things done at smaller scales—but it means that we should focus on urban infrastructure (as opposed to rural infrastructure or even intercity infrastructure).

We also became acquainted with Civitas, the fictional city that we will follow throughout the book. Civitas has a population of 60,000 people, and we will soon learn more about Civitas's infrastructure and about the habits of Civitians. From sustainability, population, and transport to electricity, water, building energy, and solid waste and even to urban metabolism, by the end of the book, we will have studied Civitas from all angles of urban engineering.

Additionally, we covered some of the measures and units that will be used in this book, and we learned about the concept of GWP used by the IPCC. Moreover, although British imperial units are still widely used in the United States, we will solely use the metric system in this book.

Unfortunately, a few topics are missing from this book. Where possible, relevant resources are recommended to learn more about a topic. We must also recognize that we still have a lot to learn when it comes to urban engineering. In particular, concepts of resilience elude most of us, and this will have to change.

After this brief introduction, it is time to start diving into the main content of the book and define what we mean by *sustainability*.

Problem Set

1.1 Select an online calculator of your choice to estimate your own personal year carbon footprint (i.e., personal GHG emissions). Make sure to select a calculator that accounts and reports multiple sectors—for example, transport, building energy use, electricity use, water consumption,[a] and more, but do not include air travel in your total. Results must be reported for an entire year and in kilograms [kg] of CO_2e.[b]

1.2 Based on the results from problem 1.1, determine three ways that you can reduce your environmental footprint and estimate the savings with the calculator used.

1.3 In your own words, describe the concept of diminishing marginal returns by Joseph Tainter as applied to urban engineering.

1.4 In a fashion similar to figure 1.3 (infrastructure ecology), illustrate and report examples of how electricity, water, transport, and building systems depend on one another.

1.5 Similar to the Cheonggyecheon River project in Seoul, describe how one piece of infrastructure in a city of your choice could be replaced by a different type of infrastructure.

1.6 Based on your own knowledge, describe why cities play an important role in the global effort to become more sustainable.

---

a. If your calculator does not include water consumption, use the values given in section 2.4 ("The IPAT Equation and the Kaya Identity") in chapter 2.

b. These values can be compared with your emissions calculated for individual infrastructure sectors in chapters 5–9.

1.7 Using the web, describe what LCA is and how it can help lower the energy used to make a product.

1.8 By searching the web, report the GHG emission factors of various food types. These types may include beef, lamb, poultry, pork, seafood, milk, yogurt, eggs, cheese, fruits, vegetables, grains, and starches. Document the sources used.

1.9 Using the GHG emission factors from problem 1.8, calculate the GHG emissions of three different meals and discuss the results.

1.10 By searching the web, find four ways to dispose of waste and discuss their respective environmental impacts.

1.11 In your own words, describe the concept of resilience and why it is important to take into account for most engineering projects.

Notes

1. Removing the handle of the well then prevented people from collecting water, which stopped the spread of the disease.

2. To learn about the history and how all these systems operate in New York City, *The Works: Anatomy of a City* by Ascher and Marech (2007) is highly recommended.

3. The surface area of the entire archipelago is 242 km^2 (93 mi^2), but most people live in Saint Pierre, which has a surface area of 25 km^2 (10 mi^2).

4. Newfoundland is a large island located on the east coast of Canada. Saint Pierre and Miquelon is south of it.

5. Jane Jacobs (1970) discusses this quite thoroughly in her book *The Economy of Cities*. We will learn a lot from Jacobs in chapter 4. Although he is not an urban planner, Yuval Noah Harari (2015) also discusses this fact quite extensively in his excellent book *Sapiens: A Brief History of Humankind*.

6. Available at http://video.nationalgeographic.com/video/news/7-billion/ngm-7billion (accessed August 13, 2018).

7. The same book has another great quote: "We have to accept human behavior the way it is, not the way we wish it should be" (Norman 2013, p. 6). Designing the "perfect" system is often futile if we do not take into account how people will use the system, which is often what we do in engineering.

8. I like the following quote about water from Antoine de Saint-Exupéry (1939, p. 182): "You are not necessary to life; you are life." He said this upon getting to drink water after being stranded in the Sahara desert for several days.

9. Anecdotally, when I mention power to my students—say, a ten-watt lightbulb—many ask me, "Watts per what?" We are too used to rates being *per* something, like kilometers per hour, liters per day. This is simply not the case for power expressed in watts.

10. We should mention that these factors do not take into account possible feedback linked with the amount of water vapor in the atmosphere (which is actually the primary GHG). To learn more about this feedback and for a full list of factors and GHGs, see Myhre et al. (2013, table 8.A.1, chap. 8); and Intergovernmental Panel on Climate Change (2013).

11. As I have seen too many times.

12. OpenLCA, http://www.openlca.org/ (accessed March 14, 2019).

References

Ahern, J. 2011. "From Fail-Safe to Safe-to-Fail: Sustainability and Resilience in the New Urban World." *Landscape and Urban Planning* 100(4): 341–343.

Ascher, K., and W. Marech. 2007. *The Works: Anatomy of a City*. New York: Penguin.

Dupuy, Gabriel. 1988. "Utility Networks and Territory in the Paris Region: The Case of Andresy." In *Technology and the Rise of the Networked City in Europe and America*, edited by Joel A. Tarr and Gabriel Dupuy, 295–306. Philadelphia: Temple University Press.

Harari, Yuval N. 2015. *Sapiens: A Brief History of Humankind*. New York: HarperCollins.

Intergovernmental Panel on Climate Change. 2013. *Climate Change 2013: The Physical Science Basis; Contribution of Working Group I to the Fifth Assessment Report of the Intergovernmental Panel on Climate Change*. Edited by T. F. Stocker, D. Qin, G.-K. Plattner, M. Tignor, S. K. Allen, J. Boschung, A. Nauels, Y. Xia, V. Bex, and P. M. Midgley. Cambridge: Cambridge University Press.

Jacobs, Jane. 1970. *The Economy of Cities*. New York: Vintage Books.

Johnson, Steven. 2006. *The Ghost Map: The Story of London's Most Terrifying Epidemic—and How It Changed Science, Cities, and the Modern World*. New York: Riverhead Books.

Mumford, Lewis. 1961. *The City in History: Its Origins, Its Transformations, and Its Prospects*. New York: Harcourt, Brace & World.

Myhre, G., D. Shindell, F.-M. Bréon, W. Collins, J. Fuglestvedt, J. Huang, D. Koch, et al. 2013. "Anthropogenic and Natural Radiative Forcing." In *Climate Change 2013: The Physical Science Basis; Contribution of Working Group I to the Fifth Assessment Report of the Intergovernmental Panel on Climate Change*, edited by T. F. Stocker, D. Qin, G.-K. Plattner, M. Tignor, S. K. Allen, J. Boschung, A. Nauels, Y. Xia, V. Bex, and P. M. Midgley, 659–740. Cambridge: Cambridge University Press.

National Geographic. 2011. "Population 7 Billion." January. https://www.nationalgeographic.com/magazine/2011/01/7-billion-population/.

Norman, D. 2013. *The Design of Everyday Things: Revised and Expanded Edition*. New York: Basic Books.

Pandit, A., H. Jeong, J. C. Crittenden, and M. Xu. 2011. "An Infrastructure Ecology Approach for Urban Infrastructure Sustainability and Resiliency." In *2011 IEEE/PES Power Systems Conference and Exposition*, 1–2. New York: Institute of Electronics and Electrical Engineers.

Park, J., T. P. Seager, P. S. C. Rao, M. Convertino, and I. Linkov. 2013. "Integrating Risk and Resilience Approaches to Catastrophe Management in Engineering Systems." *Risk Analysis* 33(3): 356–367.

Saint-Exupéry, Antoine de. 1939. *Terre Des Hommes*. Paris: Gallimard.

Tainter, Joseph A. 1988. *The Collapse of Complex Societies*. New York: Cambridge University Press.

Taleb, N. 2012. *Antifragile: Things That Gain from Disorder*. New York: Random House.

Woods, David D. 2015. "Four Concepts for Resilience and the Implications for the Future of Resilience Engineering." *Reliability Engineering & System Safety* 141 (September): 5–9. https://doi.org/10.1016/j.ress.2015.03.018.

2 Sustainability

Sustainable development is development that meets the needs of the present without compromising the ability of future generations to meet their own needs.
—World Commission on Environment and Development 1987

What is *sustainability*? Whenever I pose this question to high schoolers or freshmen, many have not even heard the term before. If you have seen or heard the term before, chances are you have some understanding of what it means, but you might be unable to define it properly. If you have actively worked on sustainability problems or read books on the topic, then you probably have a good conceptual understanding of what it is, but can you really define it in a way that everyone understands perfectly? In reality, the term has become mainstream, and if you pay close attention, you will hear it constantly in the media in one form or another. For many people it is linked to climate change, although it really is not, or at least not directly. Many people also associate sustainability with the term *green* or with the environment, but again, this is only partly true (as we will see in section 2.3). We also often hear about extreme weather events, like massive floods, droughts, typhoons, and hurricanes, that are linked with climate change and that we should be *greener* or more sustainable. What do all of these things have in common and how do they relate to sustainability?

Starting with the 1969 National Environmental Policy Act, in this chapter we will learn about one formal and well-adopted definition of sustainability. After all, sustainability is at the core of this book, and we should really all be on the same page. The definition is in fact so clear that we will even be able to get an equation out of it that will be illustrated through an example; for those who have heard the term *peak oil* already, the equation will perfectly explain what it is and why we should be concerned about it. Unfortunately, we will also see that using this equation is not always possible for practical applications. To remediate this problem, we will have to define and use two sustainability principles that can be systematically applied to achieve greater sustainability. Note the use of the term *greater*, even though in theory it really does not make sense; either something is sustainable or it is not. That being said, at times, by fixing some problems, we may be less unsustainable.[1] At other times we may simply not be able to tell whether we are being sustainable, so we just try to aim in the right direction. We will also discuss some limitations of the principles and introduce concepts of rebound effect and interdependencies. Some of the content for this section was also published in my 2019 article "An Approach to Designing Sustainable Urban Infrastructure," which also includes elements of chapter 10 and the conclusion (chapter 12) (Derrible 2019) and which can be referred to for a short conceptual summary of some of the elements of the book.

Furthermore, we will learn about the triple bottom line of sustainability, also called the Three Pillars of Sustainability, that is commonly used and that consists of people, planet, and prosperity. In particular, we will learn that no project can be sustainable if it does not contribute to the Three Pillars, and we will illustrate this point through several examples.

Subsequently, we will learn about the IPAT equation and the Kaya identity that enable us to fairly easily quantify energy consumption and greenhouse gas (GHG) emissions from just about any activity. They will also provide an excellent conceptual support to understand what can be done to lower our consumption of energy and resources.

Finally, we will learn about the concept of planetary boundaries and nonlinearities that Rockström and Steffen initially defined in 2009 and that is sometimes used in sustainability. Although not directly related to urban engineering, it offers an excellent conceptual framework to understand how the impact of a linear increase in one thing does not necessarily lead to a linear increase in something else. For example, a linear increase in carbon emissions does not have a linear impact on the climate.

Overall, by the end of the chapter, we will have acquired a solid grasp of what sustainability means and why it is relevant for urban engineering, and we will be well equipped to use the lessons from this chapter throughout the book.

Naturally, we first need to start with a definition of sustainability.

2.1 Defining Sustainability

2.1.1 Formal Definition of Sustainability

The concept of sustainability is relatively well established, and the main concept was already captured in the National Environmental Policy Act of 1969. In fact, the following paragraph taken directly from the law offers a great insight into what sustainability is:

> The Congress, recognizing the profound impact of man's activity on the interrelations of all components of the natural environment, particularly the profound influences of population growth, high-density urbanization, industrial expansion, resource exploitation, and new expanding technological advances and recognizing further the critical importance of restoring and maintaining environmental quality to the overall welfare and development of man, declares that it is the continuing policy of the Federal Government, in cooperation with State and local governments, and other concerned public and private organizations, to use all practicable means and measures, including financial and technical assistance, in a manner calculated to foster and promote the general welfare, to create and maintain conditions under which man and nature can exist in productive harmony, and fulfill the social, economic, and other requirements of present and future generations of Americans. (Jackson 1969, p. 1)

Notably, from this paragraph we can read elements of people as a society along with the environment and the economy, but we will discuss those later. We can also see that our activities can have an impact on the environment, which in turn might affect ourselves in the future. Therefore, we (collectively) need to do something about it.

To take a broader perspective, we will adopt the definition given in the quote from the beginning of the chapter: "Sustainable development is development that meets the needs of the present without compromising the ability of future generations to meet their own needs." This definition comes from the World Commission on Environment and Development, more commonly called the *Brundtland Report*, from the chairman of the commission, G. H. Brundtland.[2] In fact, this definition has become one of the preferred definitions of *sustainability* and *sustainable development*. It highlights the fact that we have *needs* in order to survive, such as food, shelter, water, and energy, and that these *needs* have to be *met*, often by collecting energy and resources from the environment. From an engineering and economic perspective, these needs can be seen as a *demand* for something, and meeting these needs can be seen as the *supply* of something, similar to what we discussed briefly in chapter 1.

The definition therefore tells us that we should be able to produce the supply to meet our demand. We can thus assimilate the supply to a production P and the demand to a consumption C. The equation that we are then looking for is:

$$P - C \geq 0 \tag{1}$$

We are missing something, however, the second part of the definition: "Without compromising the ability of future generations to meet their own needs." Future generations inevitably include an element of time. We are therefore not dealing with P and

C but with $P(t)$ and $C(t)$. Yet, this is still not enough because $P-C\geq 0$ needs to apply to all times t. As a result, instead of considering production and consumption, we need to focus on the *rates* of production and consumption—as the great global health and statistician expert Hans Rosling wrote: "Rates are often more meaningful [than amounts]" (Rosling, Rönnlund, and Rosling 2018, p. 138). Therefore, as long as the rate at which we can produce something is larger than the rate at which we consume it, we are fine. Equation 1 therefore becomes:

$$\frac{dP}{dt}-\frac{dC}{dt}\geq 0 \qquad (2)$$

This is it. If we can respect equation (2) indefinitely, we should be sustainable, as in we can "sustain" our activity in the future. The term *indefinitely* is important here. It means we need to take the limit when $t\rightarrow\infty$, and this is far from obvious. Specifically, our current society is based on growth (i.e., growth in consumption to drive economic growth), which means that $C(t)$ is supposed to increase indefinitely, which is simply impractical; in fact, $C(t)$ has remained constant through most of human history. Equation (2) is therefore extremely simple mathematically, but it represents an important conundrum in practice.

Moreover, we see that the equation is not directly related to the environment, although the environment is often the main component of the *supply*, and it is therefore an important limiting factor because $P(t)$ cannot increase indefinitely either.[3] In fact, many resources that we consume—like water, for example—cannot grow. Equation (2) therefore has limits, which we will discuss a little later, but it offers a good initial conceptual framework.

To illustrate equation (2), let us consider a simple example. Say we can harvest a fixed 10 km^2 cornfield that can yield 835 kg of corn per year for every 1 km^2; therefore $P(t)=10\times 835=8{,}350$ kg per year. At the same time, we have a population of 835 people. In the first scenario, the population eats 10 kg of corn per year per person, but the population never increases, thus $C_1(t)=10\times 835=8{,}350$. In the second scenario, the population only eats 5 kg of corn per year per person but grows by 10 people every year, thus $C_2(t)=5\times 10t+5\times 835=50t+4{,}175$. How do we determine which scenario is more sustainable? Following equation (2), let us differentiate the two equations and then investigate when $t\rightarrow\infty$. In the first scenario, the population requires a lot more initially, but it is fixed, in contrast with the second scenario. Looking at the derivatives, $dP/dt=0$, $dC_1/dt=0$, $dC_2/dt=50$, and therefore:

$$\frac{dP}{dt}-\frac{dC_1}{dt}=0 \quad \frac{dP}{dt}-\frac{dC_2}{dt}=-50$$

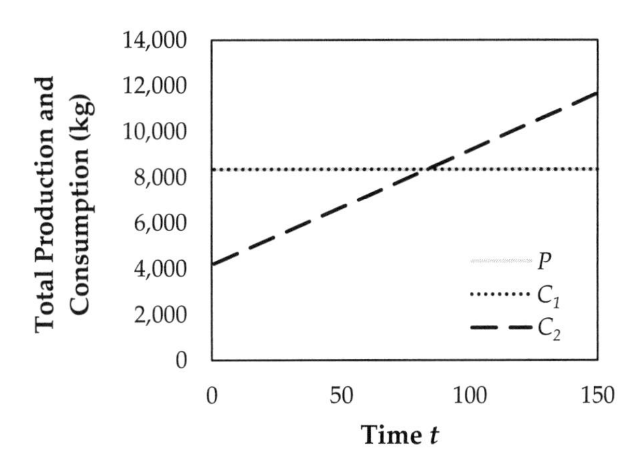

Figure 2.1
An example of production and consumption over time.

Although the population consumes more initially in the first instance, it is sustainable because the consumption does not grow in contrast with the second scenario. Figure 2.1 shows this information graphically. As we can see, the lines for P and C_1 are overlapping. We therefore see that although C_2 is initially much lower than C_1, it catches up at one point ($t = 85$), and the current production simply cannot meet the demand. According to our definition, as long as we can supply future generations we are being sustainable, and therefore the second instance is not sustainable.

While this example is simple, demand is rarely constant in real life, and the second instance (i.e., C_2) is quite common. While it is not sustainable, it is "tolerable" until a solution is found to avoid becoming unsustainable. To better illustrate this last point, figure 2.2 shows a sketch of the three scenarios that we can encounter. On the left, consumption is systematically lower than production, so it is sustainable. In the middle, consumption is currently lower than production (looking at the curves, we can see that $dC/dt > dP/dt$), but it will soon be higher than production. The situation is therefore not sustainable, but it is tolerable in the short to medium term. Finally, on the right, consumption is already higher than production, and thus the situation is unsustainable.

Instances of the three scenarios exist in real life, and we will discuss more on the topic when we learn about planetary boundaries in section 2.5, but we can list some examples in urban engineering. For instance, getting access to enough water is not a problem in many cities around the world,[4] and water consumption is

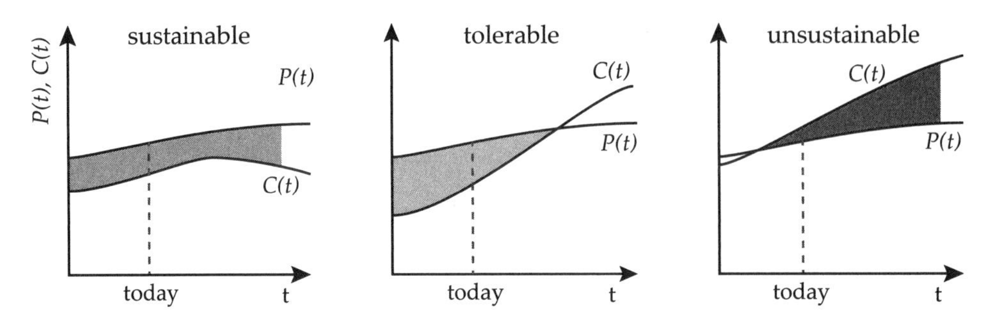

Figure 2.2
Three scenarios of production and consumption.

therefore generally sustainable. In contrast, most cities depend on power plants that run on fossil fuels to generate their electricity, which may be tolerable (for now) but not sustainable. Finally, one of the best examples of an unsustainable urban engineering system is transport since many cities must deal with serious traffic congestion, exhibiting signs that the system is over capacity (although this does not mean we need to build more roads—more on this later in this chapter and in chapter 7).

Despite the simplicity of equation (2) and figure 2.2, determining whether something is sustainable or not is often much more complicated. First, we simply cannot quantify everything that we produce and consume with a simple equation. In our example we assumed constant consumption in scenario 1 and constant growth in scenario 2, but this does not happen in practice. Sometimes consumption increases and other times it decreases. The general field of *forecasting* that tries to predict whether something is likely to increase or decrease in the future, and in which fashion, is far from trivial, and we are nowhere near to solving this problem to be able to come up with equations for every single good being produced.

Second, the consumption of a good often has an impact on the consumption of other goods. This is perhaps best illustrated by looking at figure 2.3, which shows data from the U.S. Energy Information Administration (2018). Specifically, the figure shows the evolution of total energy consumption by source in the United States from 1775 to 2017. We can see that the share of energy from biomass—that is, mostly wood burning—initially increased until the 1850s and then stayed relatively constant. This does not mean that total energy consumption remained constant—it actually increased exponentially after 1850—but it means that biomass was replaced as the main source of energy. The addition of new technologies, which cannot be imagined

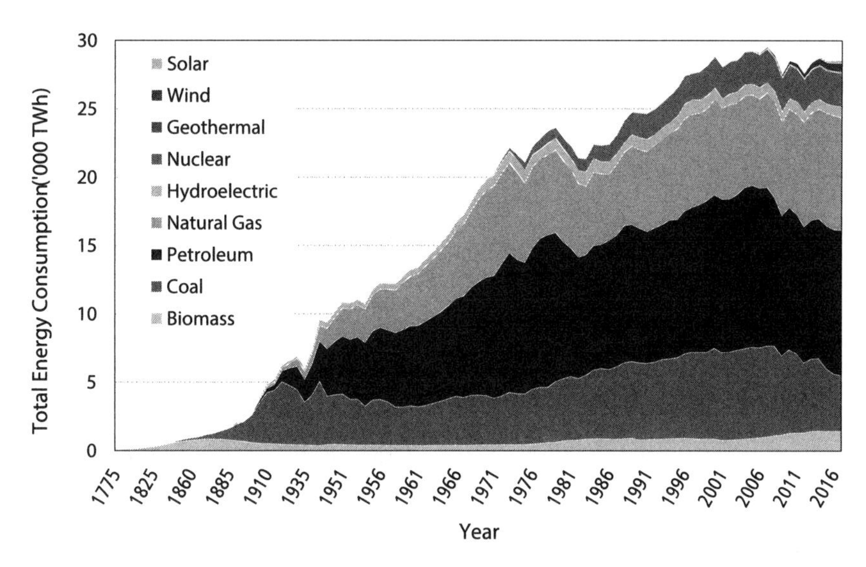

Figure 2.3
U.S. energy consumption from 1775 to 2017, by source.

before they happen, can therefore have significant impacts on how much of a good is supplied. As another example, many people also have much hope for the potential of nuclear energy, but we can see that its share has stayed fairly small even though it has been used since the 1970s. We can also see that renewable sources of energy make up a small portion of the energy consumed—we cannot even see geothermal at all—but this may change substantially. We will talk more on the topic of energy consumption in chapter 5.

Moreover, on the other side we need to be able to predict the demand to "prepare" the supply, which on figure 2.3 is the total energy consumed. In the case of energy, the consumption has increased substantially because the population has increased, but our way of life has also changed, which has had tremendous impacts on how much energy we consume. If we come back to our rates, then *sustainability is achieved when the forecasted rate of production is larger than or equal to the forecasted rate of consumption.* Now, knowing all of this, can you try to predict how the consumption of each of these energy sources will evolve to 2050 or 2100? Can you find an equation and differentiate it? The answer is obviously no. Remember that we consume millions of goods, some of which have an impact on the consumption of one another, not even mentioning the new goods that have not been invented yet.

2.1.2 Peak Oil, and Why Fossil Fuels Are Unsustainable

Staying on this topic, petroleum, coal, and natural gas, generally called fossil fuels, are fairly problematic for two reasons. First, burning these energy sources produces a lot of GHGs, including CO_2, and adding a lot of GHGs into the atmosphere actually changes our climate.[5] Although we will not dwell on the matter long, let us quickly look at the link between sustainability and climate change. When you change the climate, you essentially change how many things currently function, which especially affects the *supply* of things. For instance, crops will not grow correctly, or some important animals in the food chain will not be able to survive in their environment, which will have many impacts on the global food chain. Strong evidence also suggests that the higher frequency of extreme weather events (floods, hurricanes, extreme heat/cold, ice storms, etc.) is due to climate change (Herring et al. 2015). Overall, there is a lot of uncertainty here about what the impacts of climate change will be, but they are certain to be negative (especially because of the nonlinear nature of climate change, which we will learn more about in section 2.5), and we should at least stop releasing even more GHGs into the atmosphere. More closely related to our definition of sustainability, forecasting becomes even harder, and our ability to correctly predict and supply resources will only decrease, affecting future generations.

The second reason, which is much more closely related to our definition of sustainability, is that fossil fuels come from the processing of organic matter that has taken hundreds of millions of years to produce (hence the term *fossil*). Reserves of fossil fuels are therefore finite, and once we run out, we cannot extract more. How do we know when we will run out? We need to look at how many new reserves we have found in the past and how much fossil fuel we have extracted. Figure 2.4 shows worldwide petroleum production and consumption trends from 1980.

Naturally, both curves match almost perfectly, as supply and demand curves should. We also see that both curves currently have an increasing trend; in fact, we can fit a straight line through the points, which means that this increase is *linear*. We can also calculate the average yearly increase in consumption, and since 1984 it has been 1.14 million barrels per day. This means that on an average day we consume 1.14 million barrels of petroleum more than we did the year before. Therefore, every year, we need to be able to pump out an extra 1.14 million barrels of oil per day compared to the year before. This represents an extra 416 million barrels of petroleum per year that we need to find every year. Does that seem sustainable, considering that fossil fuels are available in finite resources?

If we look back to our rates, we will not be able to sustain this growth until we run out. In other words, at some point there will be a peak, and the rate of production will

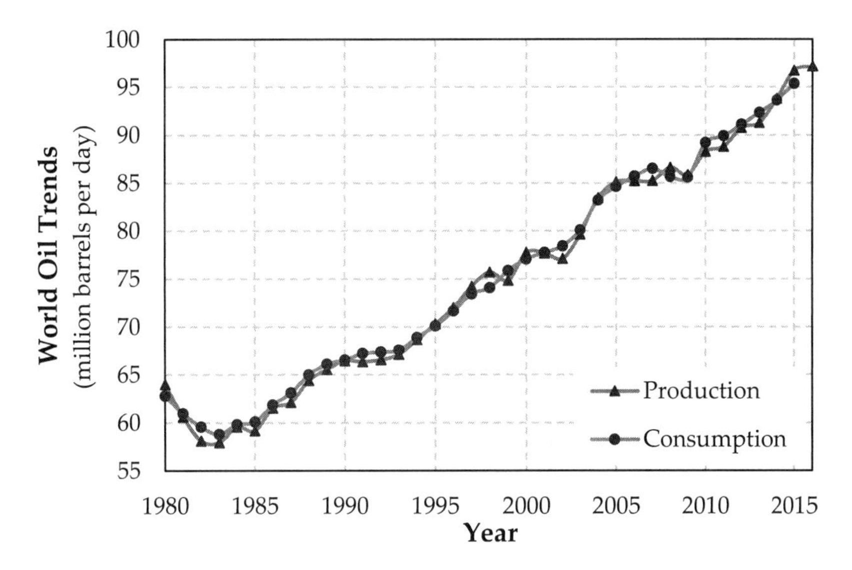

Figure 2.4
World petroleum production and consumption.
Source: U.S. Energy Information Administration. There are many relevant statistics available, and the following U.S. Energy Information Administration website offers great resources: http://www.eia.gov/beta/international/ (accessed September 3, 2018).

then decrease. Try to imagine a mirror image of the current trends along a vertical axis, a little like a mountain. This is called *peak oil*, and it is simply due to the fact that the rate of extraction (i.e., oil extracted from the ground) will have to decrease. When that happens oil prices will suddenly increase, which is likely to have severe impacts on the entire world (keeping in mind that we have already had many wars because of oil).[6] When is peak oil going to happen? There are a lot of uncertainties about it, as the largest producers of oil do not publish how much they have left in their reserves. Moreover, large efforts are being put into finding new oil resources. Regardless of when it happens, however, using oil as our main source of energy is simply unsustainable because oil is a finite resource, and we should make our transition to an oil-free society rapidly to ensure as smooth a transition as possible.

Overall, because of the absence of accurate equations for the production and consumption of everything we use, we formally cannot determine whether we are sustainable or not. There is simply no way to know the rate of something (i.e., calculate the derivatives) if we cannot forecast how something will be used. There is still hope, however, as we can rely on following two simple principles.

Example 2.1

Electricity Production and Consumption in Civitas

Civitas did not always have 60,000 inhabitants. It first started with a small settlement that grew over the years. The table below shows the evolution of population in Civitas and the evolution of per capita peak power consumption during a year (the power capacity must be known to systematically be able to meet the demand). In 1960, the Civitas power plant could produce a maximum of 4 MW, and the budget allows for a maximum additional power production of 4 MW every ten years. Based on this information, determine whether the current trend is sustainable or not.

Year	Population	Per capita peak power consumption (W)
1960	20,000	145
1970	25,000	171
1980	31,000	209
1990	38,500	244
2000	48,000	291
2010	60,000	342

Solution

We need to calculate the total power consumption, plot it against power production, and determine whether the points follow a pattern. From this, we can perform regression analyses to get an equation from the patterns, which we can then differentiate. The values are shown in the table below, as is the graph, from which we can observe two clear patterns.

Year	Population	Peak power (W)	Total (MW)	Supply (MW)
1960	20,000	145	3	4
1970	25,000	171	4	8
1980	31,000	209	6	12
1990	38,500	244	9	16
2000	48,000	291	14	20
2010	60,000	342	21	24

Example 2.1 (continued)

From the graph, we see that production follows a linear pattern, while consumption follows an exponential pattern (we will learn more about different population trends in the next chapter). Through a standard regression analysis, we extract an equation from these patterns,[7] and we then differentiate it, giving us:

$$\frac{dE_p}{dt} = 0.4 \quad \frac{dE_c}{dt} = 0.11e^{0.0392t}$$

Despite the fact that $E_p > E_c$ for smaller values of t, the electricity system is not *sustainable* since $dE_c/dt > dE_p/dt$ for larger values of t, although the situation is *tolerable* at the moment. Civitas had better change something, otherwise its consumption will soon exceed its production capacity.

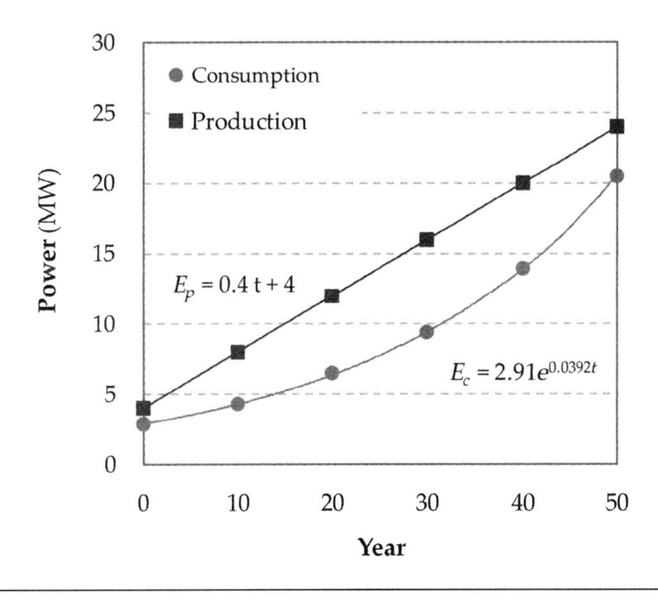

2.2 Sustainability Principles

2.2.1 Two Principles of Sustainability

While the equation that we learned in the last section is quite easy to understand and we should at least remember the main concept behind it, we also saw that it is practically impossible to determine for everything we use in reality. This becomes even harder for things that are difficult to quantify in the first place.

For instance, how do we match the quantity of "transport" used with the quantity supplied (e.g., how do we match kilometers driven with the number of kilometers of roads supplied?). Then, how do we go on and forecast something we have a hard time quantifying? This also applies the other way. We may be able to estimate how much drinking water will be needed in the future, but we may not be sure how much water will be available since the climate is changing—for example, changing rainfall patterns affect the recharge of lakes and aquifers, and the melting of natural snow and ice reservoirs used in cities like Tokyo affect how much water is available in the long term.

There is therefore a good deal of uncertainty, and while we may understand what sustainability and sustainable development means, we have substantial trouble measuring it. Our goal needs to shift from determining whether or not we are sustainable to something more practical: Are we on the right path? This is what was meant before by *greater* sustainability. Consequently, we need to come up with sustainability principles that, if followed, would ensure that we are at least going in the right direction. Looking at the two elements from our equation, we can in fact define two sustainability principles:

1. Control the demand
2. Increase the supply within reason

From our sustainability equation, we basically need to make sure that our rate of consumption is as small as possible and that our rate of production is high enough, as long as it does not affect the rates of production of other goods. Electricity offers a good example of these principles. Again linked with the fact that the number of people with access to electricity is growing and that we tend to use more electricity in our daily lives (which, arguably, is good and desirable for economic growth and social welfare), we need to control the demand so it does not increase proportionally to the population—for instance, by having more efficient appliances. Nonetheless, the demand will still increase, and we therefore need to increase the supply as well, which we can do by producing electricity from renewable resources (e.g., solar and wind—more on this in chapter 5) that are not finite like fossil fuels.

Moreover, we can make the argument that the first principle is especially important here. In fact, it is the first thing we should think about when we design a new system or add new components to an existing one. Indeed, from the beginning of humanity, we have devoted much attention to the second principle by exploiting more land, reaching new sources of water, and building new roads and structures ever bigger and farther, oftentimes without reason. Our challenge in the twenty-first century will be to successfully apply the first principle while making sure it is not impeding our development.

Controlling the demand for something is challenging, but it opens up many opportunities to create new solutions and make sure we follow Scenario B, as we saw in chapter 1.

This is especially important to remember when we talk about new technologies. While future technologies will be critical in our quest to become sustainable, not all technologies actually contribute to the sustainability principles. Every once in a while, a student comes to see me about a new technology that will revolutionize the way we live and that will help us become sustainable. Most often, these new technologies actually make things worse. If we go back to Tainter's diminishing marginal returns and if we introduce a new technology in a system, then the marginal benefits from adding this new technology should increase, and in our case they should help us reduce the demand. For example, adding new transport measures to reduce traffic congestion is great as long as the total traffic does not increase overall. This is in fact very difficult to achieve and to predict. Transport is actually a good example. To reduce traffic congestion in the 1950s, 1960s, and 1970s, many U.S. cities started building urban expressways and adding new lanes. What was the result? More traffic, more cars, and more congestion. Even now I often hear people complaining about traffic and that the solution is to add more lanes. But this is most often wrong. Traffic will not only get worse, but giving more space to transport means giving less space to buildings, parks, and other nontransport infrastructure, and if we have less nontransport infrastructure, then we may not have a reason to go wherever we wanted to go in the first place. In transport, "space" is the major resource. This phenomenon is well known, and it has resulted in the "deaths" of many cities. We will talk more about it in chapters 4 and 7.

2.2.2 Limitations and Further Considerations

Naturally, the two principles of sustainability that we defined have some limitations that should be taken into account. In particular, we can think of two items that should be discussed: the rebound effect and controlling interdependencies.

2.2.2.1 The Rebound Effect Within the same technological context discussed in the previous section, we should also be careful about claims to improve efficiency significantly. The chances are if something is less costly and more efficient, we will use more of it. This is called the *rebound effect* or *Jevons paradox* (after the English economist William Stanley Jevons). The typical example is that by buying a more fuel-efficient car, we are bound to drive more. For example, say you drive 400 kilometers per week, and it costs you about sixty dollars. You then buy a more fuel-efficient car, and it costs only forty dollars to drive the 400 kilometers. It has been observed that in many cases you may actually use the car more, such as by now driving 500 kilometers for fifty dollars. You still save money but not as much. This is a very important problem that clearly shows that efficiency is only a partial solution. Calculating or estimating this rebound effect is also a significant challenge (Gillingham, Rapson, and Wagner 2016). Although we will not dwell on the subject here, we should at least remember the rebound effect when we devise solutions to control demand, and we need to realize that working on improving technological efficiency is simply not enough.

2.2.2.2 Controlling Interdependencies Moreover, we need to emphasize here again that infrastructure systems are interdependent. By adding a new technology or improving efficiency, are we enabling or constraining interdependencies? This is especially relevant since one of our goals is to better integrate infrastructure systems.

One good example is the *solar roadway*—putting ultrathin solar panels on roads. At first the idea of using space that we do not necessarily need to produce electricity might appear sensible. But are we so short of space that the only place to put solar panels is on roads? In the end, solar roadways would be adding a significant amount of complexity and interdependence between two systems (mainly physical), but the benefits do not look that great. Surely, there must be better ways to use these new ultrathin solar panels and to integrate infrastructure systems. We will discuss infrastructure interdependencies more extensively in chapter 10.

2.3 The Triple Bottom Line of Sustainability

Naturally, we also cannot blindly apply the two principles of sustainability that we defined. Beyond enacting certain policies (e.g., parking price), we cannot simply force someone to reduce his/her consumption. Similarly, we cannot simply increase the supply without thinking about the long-term consequences, especially on the environment. Our goal will therefore be to apply the two principles by taking into account three dimensions: people, planet, and prosperity (figure 2.5).

We actually already saw some of this, back in the paragraph from the NEPA of 1969. This trio is commonly known as the triple bottom line, also called the Three Pillars of Sustainability, and is sometimes shown in the form: society/equity, environment, and economy. We can only apply the two sustainability principles successfully when all three dimensions benefit. Although including two of the three dimensions often outputs positive results (e.g., bearable, viable, equitable), it is really when the three are combined that we know we have come up with a *more* sustainable solution.

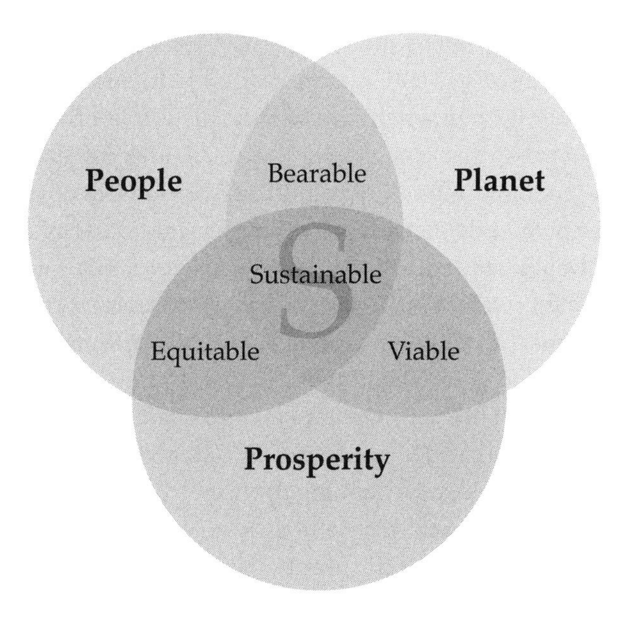

Figure 2.5
The triple bottom line: people, planet, prosperity.

Many people often assume that the benefits of the environment (i.e., the planet) outplay the other two parties in sustainability, but this is false. A truly sustainable design must benefit all three. The confusion is understandable, however. The preservation of the planet aligns more closely with the future generations from the *Brundtland Report* quote presented earlier. Plus, without a robust planet, the society and the economy cannot thrive. Seen another way, the planet can be considered a proxy for the supply of future demand (our second principles). But since we have little certainty about both what the future demand and the future supply will be, it is difficult to visualize its role, and we easily overplay or underplay it.

In complete contrast, many people argue that in dire economies and during economic crises, the environment can be omitted to focus on people and the economy. But again this is mistaken. In fact, it may only be during dire times that we can seriously rethink the current situation and be more creative. This behavior has been seen repeatedly, and Thomas Homer-Dixon (2006) calls it *catagenesis*, associating the words *catastrophe* and *genesis* (i.e., generating from catastrophes).

We can think of examples to illustrate the triple bottom line. An example of a *bearable* solution would be to ban cars; it would be desirable for the people (for health) and the planet, but prosperity would be affected. Another example would be to force the use of renewables without paying attention to market forces; again desirable for the people and the planet but not for prosperity. A *viable* example would be to not only produce electricity from renewables but also increase the price of electricity significantly; this may have positive economic impacts for manufacturers, as well as obvious environmental benefits, but it would likely hurt people without the means to pay their electric bills. Another example would be to force construction companies to insulate new buildings with the best and latest products; the vendors would gain tremendously, and our GHG emissions related to space conditioning would be reduced, but such action would come at a great cost to the people since housing prices would increase significantly. An *equitable* example would be to build only natural gas power plants—natural gas was cheap at the time of writing (in 2018), and it may be desirable for prosperity and the people—but more GHGs would be emitted, thus harming the planet. Another example would be to dump untreated wastewater back into natural streams; again more economical, thus potentially good for prosperity and the people, but certainly harmful to the planet.

Truly sustainable solutions therefore possess a mix of these strategies—for instance, by prioritizing renewables and offering incentives to reduce our energy consumption or by providing better and affordable public transport alternatives while making it a little more difficult to drive (e.g., increase gas price and parking prices). Once again we need to be creative to ensure we embark on Scenario B.

Another, and perhaps even better, way to see the triple bottom line is to look at it as a network. In a great journal article, Fiksel et al. (2014) discussed and illustrated the triple value (3V) model, shown in figure 2.6. The key here is to think about the entire system—in fact, we typically call it *systems thinking*. Changing something either for people, the planet, or prosperity likely has an impact on the other two. It is therefore important to see it through before making a decision.

Example 2.2

Triple Bottom Line in Civitas

Considering the problem seen in example 2.1, and using the triple bottom line, determine how the supply and demand of electricity can become more sustainable.

Solution

We should control the demand to lower per capita energy use as much as possible, better forecast population and electricity use (chapter 3), and develop a strategy to increase the supply of electricity if needed from renewable sources, such as solar and wind (chapter 5).

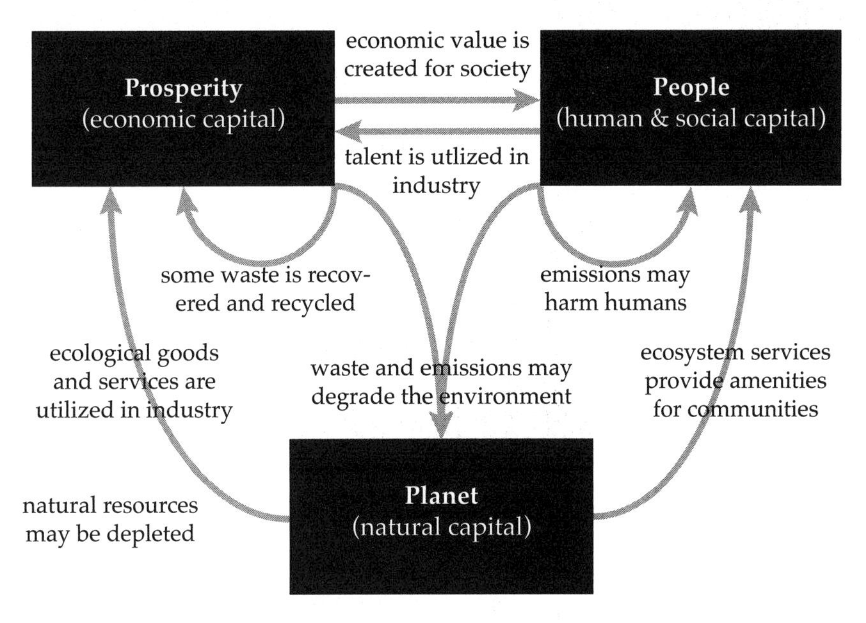

Figure 2.6
The 3V model. Adapted from Fiksel et al. (2014).

Moreover, this does not have to be only a conceptual exercise. In many cases we should be able to determine relevant metrics and estimate how these metrics would evolve by changing something. Recycling rates, air pollution in parts per million, dollars earned/lost, liters of wastewater treated, area of solar panels installed, reduction in stormwater runoff, percentage of educated/skilled population, increase in gross domestic product, contaminants released in an aquifer, and so on and so forth; all of these are variables that can be measured. Even if we do not know the exact relationship, we should be able to model it (i.e., estimate it), and if we really cannot, we should at least go through the conceptual exercise.

Two simple ways to model/measure these flows and determine whether we are aiming in the right direction are the IPAT equation and the Kaya identity, and they are the focus of the next section.

2.4 The IPAT Equation and the Kaya Identity

The IPAT equation offers a clever way to perform back-of-the-envelope types of calculations and estimate the impact I of a strategy, project, or policy on the environment. The impact I can be quantified in various ways depending on what we are trying to measure, but in our case we will mostly quantify *I* in units of energy in watt-hours [Wh] or megawatt-hours [MWh] (see units in chapter 1 or in the appendix). In some cases, we will also quantify *I* in units of emissions directly, whether in grams or tons of CO_2 equivalent [g CO_2e] or [t CO_2e]. The equation literally spells out *IPAT*:

$$I = P \cdot A \cdot T \quad (3)$$

where *P* is the population considered, *A* is the affluence or, more accurately for us, the level of consumption per person, and *T* is the impact of technology that we express in terms of energy/emissions per unit consumption. The IPAT equation was initially developed by Ehrlich and Holdren (1971), and it has had an amazing history (Chertow 2000).

The Kaya identity can be seen as an extension of the IPAT equation by systematically adding a conversion factor to estimate *I* in units of GHG emissions. The Kaya identity can take the form:

$$E = P \cdot A \cdot T \cdot \varepsilon \quad (4)$$

where *E* is essentially the impacts *I* but is expressed as emissions in grams or metric tons of CO_2e, and ε is the conversion factor to convert *T* (energy use per unit consumption) to GHG emissions; for example, ε converts electricity use to GHG emissions. In this book we will have to be flexible with the definition of the IPAT equation and the Kaya identity depending on what we are measuring. To quantify the emissions related to nonelectric car use, for example, the *T* will already be expressed in units of emissions per kilometer driven, and we will not need to use ε from the Kaya identity. The most important thing to remember here is that the units need to cancel out on the right-hand side to ensure we quantify either energy use or GHG emissions on the left-hand side.

To illustrate the two equations within the context of this book, let us take an example with water consumption. On average in the United States in 2015, people consumed 520 liters of water per day, the United States had a population of about 320 million people, and each liter of water required about 1 Wh of energy—that is, 1 Wh/L, for the collection, treatment, and distribution of drinking water and for the treatment of wastewater—we will learn more about this value in chapter 6. Then, the total energy use is:

$$I_{water} = 320 \times 10^6 \times 520 \times 1 = 166.4 \times 10^9 \text{ Wh} = 166.4 \text{ GWh daily}$$

To get the yearly impact, we simply multiply by 365:

$$I_{water} = 166.4 \times 365 = 60{,}736 \text{ GWh}$$

To reduce the impact *I*, we can try to reduce the first component of the demand, *P* (i.e., reduce population), but this really is not an option. Therefore, we turn to trying to control the second component of the demand, A (i.e., reduce the per capita consumption), for instance, by decreasing the need for something or substituting it with something else. At the same time, we can seek new supplies (or, more accurately, new technologies) that reduce the energy needed per unit consumption T (although we have to be aware of the rebound effect).

To go one step further and using the Kaya identity, if we assume that all of this energy is provided by electricity,[8] and knowing that the 2016 average power grid emission factor ε in the United States was 455 g CO_2e/kWh (or 455 t CO_2e/GWh), then the GHG emissions E_{water} attributed for the collection, treatment, and distribution of drinking water and for the treatment of wastewater is:

$$E_{water} = 455 \times 60{,}736 = 27{,}634{,}880 \text{ t } CO_2e \approx 27.6 \text{ Mt } CO_2e$$

where Mt stands for megaton. Naturally, many more factors need to be taken into account. For instance, all of the energy used does not come from electricity. Moreover, we need to account for regional differences since power grid emission factors vary greatly by region, but this simple example serves well to illustrate the IPAT equation and the Kaya identity for now.

Considering that many goods can be quantified in a per capita basis and the fact that population most often drives the demand, we can see that the IPAT equation and

Example 2.3

Emissions from Electricity Consumption in Civitas

Considering a power grid emission factor of 120 g CO_2e/kWh (kg CO_2e/MWh), use the IPAT equation and the Kaya identity to calculate the total emissions from Civitians in 2010 from electricity consumption. Assume the average power consumption in Civitas is 60% of the peak power demand.

Solution

From example 2.1, we saw that the per capita peak power consumption in 2010 was 342 W. Considering 24 hours in a day, 365 days in a year, and a 60% average-to-peak demand ratio, this means that the average per capita energy consumption A (for affluence) in 2010 was:

$$A = 0.6 \times 342 \times 24 \times 365 = 1.80 \times 10^6 \text{ Wh} = 1.80 \text{ MWh}$$

For a population of 60,000, and considering the fact that for this application the technology T is the same as the conversion factor ε, the Kaya identity gives us:

$$E = 60{,}000 \times 1.80 \times 120 = 12.96 \times 10^6 \text{ kg } CO_2e = 12{,}960 \text{ t } CO_2e$$

With a population of 60,000, this represents 12,960 / 60,000 = 0.216 t CO_2e per capita. Is this a lot? After reading chapter 5, you will know.

the Kaya identity give us a quick way to estimate the impacts of possible solutions that we can think of. We will therefore use this method quite extensively in this book and apply it specifically to the different urban systems of chapters 5 through 9. Moreover, we will develop it a little more comprehensively in chapter 10.

2.5 Planetary Boundaries and Nonlinearities

Before we move on to population forecasting, we should discuss one last concept of sustainability that is quite impactful. In 2009 a team of researchers led by Johan Rockström from the Stockholm Resilience Centre and Will Steffen from the Australian National University came up with the concept of planetary boundaries (Rockström et al. 2009a, 2009b). The concept is relatively simple. It is to identify major categories of human activities that have a worldwide impact and measure whether these activities are sustainable. The rationale is that if the activities cannot be sustained, we should stop them. The authors of these studies came up with a total of nine categories:

- Climate change
- Novel entities (chemical pollution); new substances that we introduce and for which we are uncertain of their impact (e.g., pesticides in water)
- Stratospheric ozone depletion
- Atmospheric aerosol loading
- Ocean acidification
- Biochemical flows (interference with nitrogen and phosphorous cycles that are essential for agriculture)
- Freshwater use
- Land-system change
- Biosphere integrity (includes both functional and genetic diversity)

Figure 2.7 shows whether these nine activities can be sustained or not. Activities within the innermost circle are below the planetary boundaries and can therefore be sustained at their current rate; these include stratospheric ozone depletion, ocean acidification, and freshwater use. Activities in the middle circle exceed the planetary boundaries, but some uncertainties remain about their impacts; these include climate change and land-system change. Activities that reach the edge of the circles have achieved dangerous levels, and something must be done immediately; these include

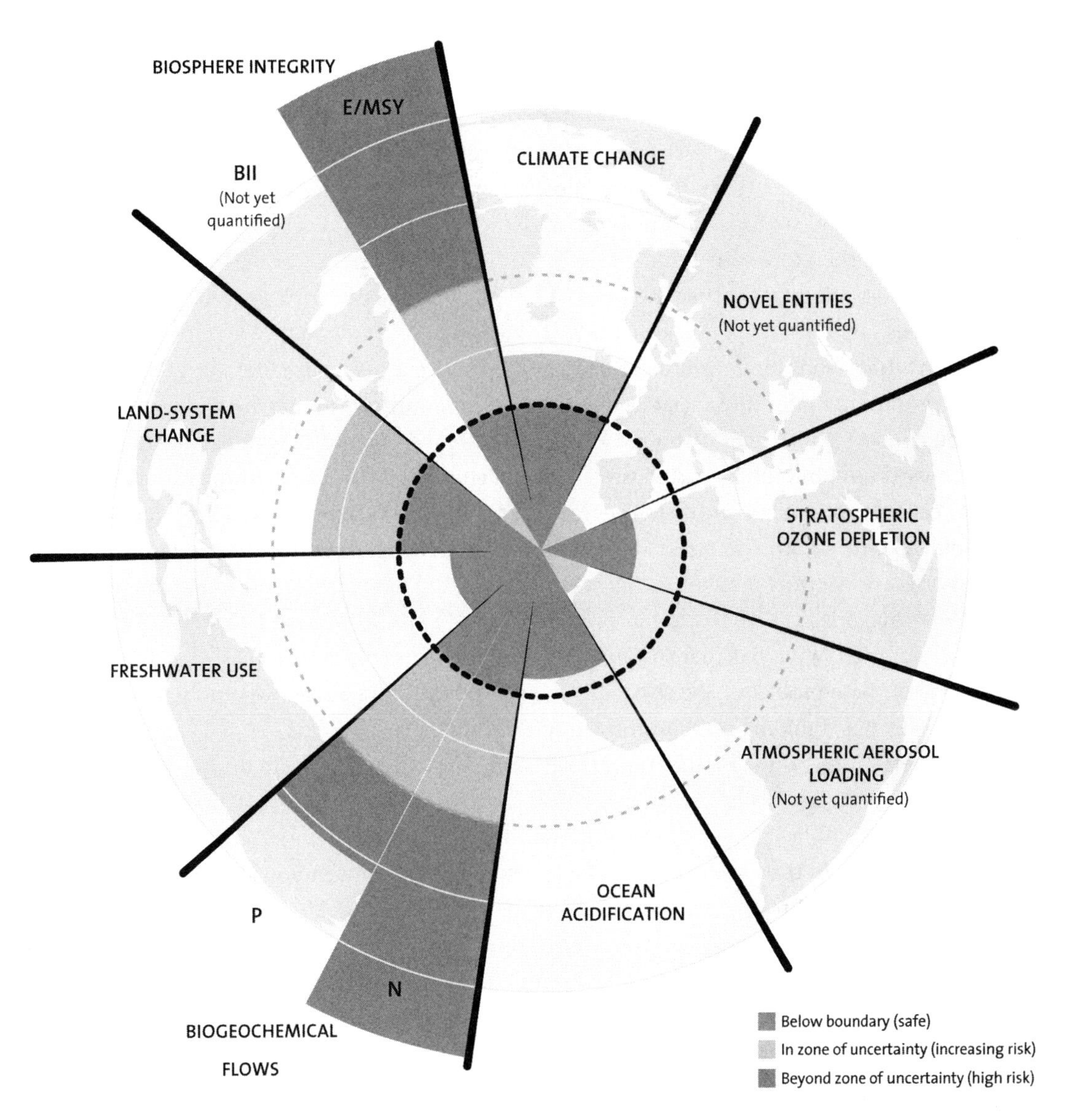

Figure 2.7
Planetary boundaries.
Source: Azote image for Stockholm Resilience Centre.

biosphere integrity (e.g., too many species going extinct) and biochemical flows (nitrogen and phosphorus). Finally, two categories remain completely unknown—the novel entities (chemical pollution) and atmospheric aerosol loading—and one subcategory is also unknown—functional diversity within biodiversity integrity.

From the figure we see that for most categories we have reached the planetary boundaries and something must be done. The debates and discussions over sustainability are therefore well founded, and they transcend the focus on the burning of fossil fuels. We also see that the nine categories deal mostly with the environmental sciences.

So what is in it for us as urban engineers?[9] First, we must realize that out of the nine categories, the ones most related to cities and urban infrastructure are freshwater use and land-system change (although here, land system deals mostly with agriculture, not the land used in cities). As discussed, climate change has an impact on us, including by causing an increase in the frequency and magnitude of extreme weather events. To some extent we are also affected by biochemical flows since both nitrogen and phosphorus are essential for the fertilizers that are used in agriculture. In general, many of these categories are related to our consumption of stuff, from food to appliances, and therefore following our first sustainability principle directly contributes to addressing some of these problems. We also need to ensure that we increase the supply in a resilient way (i.e., building resilient infrastructure).

Interestingly, these boundaries are set based on where they were during the Holocene era. The Holocene is the geological era that began roughly 10,000 years ago (after the last glacial period, around 8,000 BCE) and that provided ideal conditions for humanity to thrive. Most scientists agree that we are now entering the next geological era, the Anthropocene. The term *anthropo* essentially means "humans." In other words the current geological era is changing because of human activity—notably, because of the GHG emissions linked with the burning of fossil fuels that is causing climate change. The planetary boundaries are therefore set based on what is "good" for humans, but human activity is changing these boundaries, thus eventually harming humanity.

Just as importantly, we should pay attention to what happens once we cross these boundaries. Figure 2.8 illustrates how a "response variable" y responds to a change in variable x. Put simply, the x-axis represents the level of an activity, and the y-axis represents how a variable responds to this activity. For example, the x-axis can represent the burning of fossil fuel, and the y-axis can represent the size of the ice caps. We often assume that response variables respond linearly to a change, but this is rarely the case. The response variable y in figure 2.8 shows that we can increase the level of an

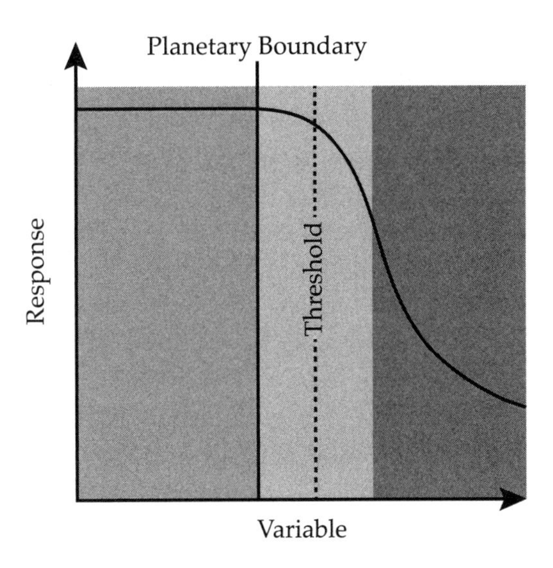

Figure 2.8
Nonlinearities in planetary boundaries. Adapted from Steffen et al. (2015).

activity *x* with little to no change up to a certain point, after which the response variable decreases rapidly. The response is therefore highly nonlinear. This observation is important. Most activities/processes in the world are nonlinear, but we often only see the linear part until it is too "late." Even the simple stress-strain relationship of any material shows nonlinearities after a point.

In urban engineering, most systems we encounter have nonlinear responses. For example, adding a new road does not increase traffic in a linear way. Changing the diameter of a pipe does not affect water pressure in a linear way. Adding a layer of insulation in a house does not change the heating/cooling energy use in a linear way. Even adding new people in a house does not increase electricity or water consumption in a linear way. This is both good and bad news. It is bad news because it makes it hard to predict what is going to happen in the longer term; that is, we do not know what will result from many nonlinear responses affecting one another. On the other hand, it is good news because it means we can do a few things to make positive nonlinear impacts. We will see some examples in the chapters to come. In particular, capturing nonlinearities is one of the main goals of the Science of Cities, but we will have to wait until chapter 11 for that. In the meantime, we should keep the concept of nonlinearities in mind and try to leverage positive nonlinearities.

2.6 Conclusion

In this chapter, we first introduced a formal definition of sustainability and sustainable development from the *Brundtland Report*, which states that "sustainable development is development that meets the needs of the present without compromising the ability of future generations to meet their own needs." We even managed to come up with an equation to measure whether we are sustainable or not. We saw that it depends heavily on the ability to forecast future demand and that sustainability can be achieved when the forecasted rate of production is larger or equal than the forecasted rate of consumption. We also looked at the example of peak oil. Nevertheless, we saw that in practice, we most often cannot determine this equation. In line with the sustainability statement, we therefore developed two principles that can be systematically applied, with the caveat that they should be framed with the traditional triple bottom line of people, planet, and prosperity. Then, we learned about the IPAT equation and the Kaya identity that give us a quick way to estimate the energy and environmental impact of a strategy, project, or policy; therefore, the two equations can help us determine whether we are respecting our two principles or not. Finally, we learned about concepts of planetary boundaries and nonlinearities, which we should keep in mind as urban engineers since a change in something (e.g., the addition of a road) often generates a nonlinear response in something else (e.g., traffic congestion).

We must realize, however, that sustainability extends beyond what we have covered in this chapter. There is no one single definition of *sustainability* that applies universally. In fact, sustainability has become more of a general *concept* than a process or goal. Although sustainability is typically related to energy use, many fields do not even consider energy. In agriculture, sustainability is partly about being able to keep growing crops in the long term. In fishing and hunting, sustainability is about avoiding overfishing/overhunting so that populations can reproduce themselves year after year at an adequate rate. In biology, sustainability can be related to diversity. Having more plants and insects tends to help an ecosystem in the long term; similarly, healthier people also tend to have a more diverse gut flora.[10] The one common trait of most uses of the term *sustainability* is the ability of a system to remain stable over time. As urban engineers, we will strive to put effort into following our two sustainability principles.

It is now time to move on to the most important aspect of cities: people. After all and in the end, it is people who consume energy and resources, and this is also the first component on the right-hand side of the IPAT equation. The first thing we must do to estimate the demand for something is to know how many people there will be in the future. In other words, we need to forecast population. Fortunately, there are several relatively easy methods that we can apply both in the short term and in the long term. We will, notably, learn about the famous S curve that dictates the long-term growth of not only populations but technologies as well. That will set us on the way to learning more about cities themselves and their various infrastructure systems.

Problem Set

2.1 In your own words, describe what *sustainability* is.

2.2 Explain in words and illustrate with an example the equation

$$\frac{dP}{dt} - \frac{dC}{dt} \geq 0$$

2.3 In your own words, describe the phenomenon of *peak oil*.

2.4 In your own words, describe the link between *sustainability* and *climate change*.

2.5 In your own words, describe how the two sustainability principles contribute to sustainability.

2.6 For each of these five categories—electricity, water, transport, building heating/cooling, and solid waste—give an example for each sustainability principle (i.e., five examples for principle 1 and five examples for principle 2).

2.7 In your own words, describe the phenomenon of *rebound effects*.

2.8 In your own words, describe why it is important to control interdependencies between infrastructure systems.

2.9 Within the context of the triple bottom line of sustainability and regardless of the infrastructure systems selected, give an example of solutions that are (1) bearable, (2) viable, (3) equitable, and (4) sustainable.

2.10 In your own words, describe the main differences between the IPAT equation and the Kaya identity.

2.11 Within the context of the planetary boundaries by Rockström and Steffen, select and describe one of the nine categories of human activities.

2.12 In your own words, explain the concept of nonlinearity in engineering.

2.13 The following production and consumption profiles have been determined for various processes. By differentiating the equations, determine whether these processes are sustainable or not.

Case	Production $P(t)$	Consumption $C(t)$
1	$36t^2$	$124t+76$
2	$2t^3+3t^2$	$5t^3$
3	$9t^{(2/3)}$	$3t^{(3/2)}$
4	$10t \cdot \ln(t)$	$10t \cdot e^t$
5	$2\left(t^2 - \frac{1}{t}\right)$	$2t^2$

2.14 The following production and consumption profiles have been determined for various processes. By differentiating the equations, determine whether these processes are sustainable or not.

Case	Production $P(t)$	Consumption $C(t)$
1	$123t^3$	$56t^2$
2	$45t\sqrt{t}$	$75t^2$
3	$26t \cdot \ln(t)$	$53t^3+45t^2$
4	$987e^{23t}$	$243e^{21t}$
5	$100t^{100}$	$e^t \cdot t$

2.15 The following production and consumption profiles have been determined for various processes. Determine whether these processes are sustainable or not.

Case	Production $P(t)$	Consumption $C(t)$
1	$50t+5$	$20t+100$
2	$2t^4+5t^3+4t^2+6t+8$	$1.23e^{0.5t}$
3	$37t^2 \cdot \ln(t)$	$4e^{0.3t}/t$
4	$36t\sqrt{t}$	$24/t^2$
5	$e^t \cdot \ln(t)$	$e^{\ln(t)}$

2.16 The table below shows the evolution of population and per capita water consumption in a city in liters per capita per day (L/day). Considering the water treatment plant can process up to 350,000 m^3 of water per day, determine when the treatment plant will not be sufficient for the city (*Note: assume the relationship is linear*).

Year	Population	Water consumption (L/day)
1980	816,000	360
1990	878,000	350
2000	954,000	340
2010	1,001,482	330

2.17 The table below shows the evolution of population and per capita natural gas in a city in cubic meters per capita per year (m^3/year). The city has a current natural gas capacity of 100,000,000 m^3/year, but a new regulation was set up to stop all natural gas use by the year 2035. Determine whether the current capacity is sufficient or whether the city will have to increase its capacity before 2035 (*Note: assume the relationship is linear*).

Year	Population	Gas consumption (m^3/year)
1980	35,258	1,462
1990	49,147	1,184
2000	66,271	956
2010	86,728	801

2.18 Current forecasts estimate a future population growth of 20%. Using the IPAT equation, estimate the percentage drop needed in consumption so that the overall impact stays the same (assume the impact of technology is unchanged).

2.19 Current forecasts estimate a future population growth of 20% and a future decline in the impact of technology of 30%. Using the IPAT equation, estimate the percentage drop needed in consumption so that the overall impact decreases by 50%.

2.20 A community of 30,000 people follows the consumption trends shown in the table below. Based on this information, calculate the yearly per capita GHG emissions in kilograms of CO_2e for: (1) electricity, (2) water (assume the electricity intensity factor to convert energy to GHG emissions), (3) transport, and (4) gas. (5) Calculate the per capita and total GHG emissions of the community in t (metric tons) CO_2e. (6) Suggest one bearable, viable, equitable, and sustainable way to lower the community GHG emissions.

Category	Consumption	Intensity factor
Electricity	4 MWh/year	500 g CO_2e/kWh
Water	300 L/day	1 Wh/L
Transport	300 km/month	300 g CO_2e/km
Gas	5 MWh/year	172 g CO_2e/kWh

2.21 From the U.S. Energy Information Administration website, download the number of residential electricity consumers by state and the total residential electricity consumption for the latest year available.[a] From the Emissions & Generation Resource Integrated Database of the Environmental Protection Agency website, download the power grid CO_2 emission factor of each state.[b] Based on these data and using the IPAT equation and the Kaya identity, determine the states with (1) the smallest, (2) the largest, and (3) the median per consumer CO_2 emissions from electricity. Taking into account the number of consumers, determine the states with (1) the smallest, (2) the largest, and (3) the median total CO_2 emissions from electricity.

Notes

1. Like substituting coal with natural gas, as we will learn in chapter 5.

2. Find more information in Theis and Tomkin (2012, chap. 1).

3. This may actually be one of the decisive factors in the end, since $dC/dt > dP/dt$ would drive prices up, which would lower dC/dt—but at what cost? We will see a practical example later with peak oil.

4. Although some cities, such as Beijing, face serious problems.

5. Find more information in Theis and Tomkin (2012, chap. 3).

6. This is a case where $dC/dt > dP/dt$.

a. URL: https://www.eia.gov/electricity/data/state/ (accessed September 3, 2018).

b. URL: https://www.epa.gov/energy/emissions-generation-resource-integrated-database-egrid (accessed September 3, 2018). If not available in kg CO_2e/MWh, find the CO_2e emission factor in lb/MWh and convert yourself. Note: make sure to download the CO_2 equivalent as opposed to CO_2.

7. Statistical software packages are really good at this. Even Microsoft Excel has an easy-to-use trend line function.

8. But it's not. Many of the pumps that ensure all water pipes have enough pressure are powered by natural gas, which is not necessarily undesirable from an emissions point of view because burning gas directly is much better than using electricity that came from a coal-fired power plant. There is more to come on this topic in chapter 5.

9. Hoornweg et al. (2016) apply the planetary boundaries to the urban scale, and they come up with new interesting socioeconomic boundaries, but we will focus solely on urban engineering here.

10. All of us have an impressive number of bacteria in our gut to help us digest the food we eat, and having more types of bacteria is generally beneficial. This is partly why it is better to have a diverse diet.

References

Chertow, Marian R. 2000. "The IPAT Equation and Its Variants." *Journal of Industrial Ecology* 4(4): 13–29.

Derrible, Sybil 2019. "An Approach to Designing Sustainable Urban Infrastructure." *MRS Energy & Sustainability* 5: E15.

Ehrlich, Paul R., and John P. Holdren. 1971. "Impact of Population Growth." *Science* 171(3977): 1212–1217.

Fiksel, Joseph, Randy Bruins, Annette Gatchett, Alice Gilliland, and Marilyn ten Brink. 2014. "The Triple Value Model: A Systems Approach to Sustainable Solutions." *Clean Technologies and Environmental Policy* 16(4): 691–702.

Gillingham, Kenneth, David Rapson, and Gernot Wagner. 2016. "The Rebound Effect and Energy Efficiency Policy." *Review of Environmental Economics and Policy* 10(1): 68–88.

Herring, Stephanie C., Martin P. Hoerling, James P. Kossin, Thomas C. Peterson, and Peter A. Stott. 2015. "Explaining Extreme Events of 2014 from a Climate Perspective." *Bulletin of the American Meteorological Society* 96(12): S1–S172.

Homer-Dixon, Thomas F. 2006. *The Upside of Down: Catastrophe, Creativity, and the Renewal of Civilization*. Washington, DC: Island Press.

Hoornweg, Daniel, Mehdi Hosseini, Christopher Kennedy, and Azin Behdadi. 2016. "An Urban Approach to Planetary Boundaries." *Ambio* 45(5): 567–580.

Jackson, Henry M. 1969. National Environmental Policy Act of 1969. 42 U.S.C. 4321 et seq.

Rockström, Johan, Will Steffen, Kevin Noone, Asa Persson, F. S. Chapin III, Eric Lambin, and Timothy M. Lenton et al. 2009a. "Planetary Boundaries: Exploring the Safe Operating Space for Humanity." *Ecology and Society* 14(2): article 32.

———. 2009b. "A Safe Operating Space for Humanity." *Nature* 461(7263): 472–475.

Rosling, Hans, with Ola Rosling and Anna Rosling Rönnlund. 2018. *Factfulness: Ten Reasons We're Wrong about the World—and Why Things Are Better than You Think*. New York: Flatiron Books.

Steffen, Will, Katherine Richardson, Johan Rockström, Sarah E. Cornell, Ingo Fetzer, Elena M. Bennett, and Reinette Biggs et al. 2015. "Planetary Boundaries: Guiding Human Development on a Changing Planet." *Science* 347(6223). https://doi.org/10.1126/science.1259855.

Theis, Tom, and Jonathan Tomkin, eds. 2012. *Sustainability: A Comprehensive Foundation*. Houston: Connexions.

U.S. Energy Information Administration. 2018. "Monthly Energy Review." https://www.eia.gov/totalenergy/data/monthly/.

World Commission on Environment and Development. 1987. *Our Common Future*. Oxford: Oxford University Press.

3 Population

> The power of population is indefinitely greater than the power in the earth to produce subsistence for man.
>
> —Malthus 1798, p. 4

Behind all the challenges that we face, there is one big driver that complicates things quite a bit: us, the people, who all together represent this *demand* that we discussed in chapter 2. If there were fewer people on Earth, our needs would be smaller, and the resources needed to meet these needs would be smaller as well. But this is not what is happening. Figure 3.1 shows the historical evolution of population from the year 1800 to three projections for the year 2100 (low, medium, and high growth); projected population scenarios start after 2015. We see that the world population evolved fairly slowly from 1800 to about 1950—although it did increase by a factor of 2.5 in that period of time—after which it suddenly increased. It may stabilize around the year 2100. We also see that we will likely reach the 10 billion mark sometime around 2050, according to the median scenario.

Naturally, this means that the population will not only *not* decrease but will in fact increase quite significantly. This causes quite a big problem. If we already consume too much and we need to decrease the current demand, how can we deal with an increasing population?

Figure 3.1 shows another interesting feature. By 2100 the world population will likely stabilize, and it may even decline after 2100. This is interesting because the world population has likely never decreased in the past. This means that most of the efforts that will be put into building sustainable cities to cater for an increasing population—that is, by not consuming more energy and resources—should also account for the fact that cities are ever changing, with populations that may eventually decline. This is another reason to stop building bigger and farther and instead think strategically about how we

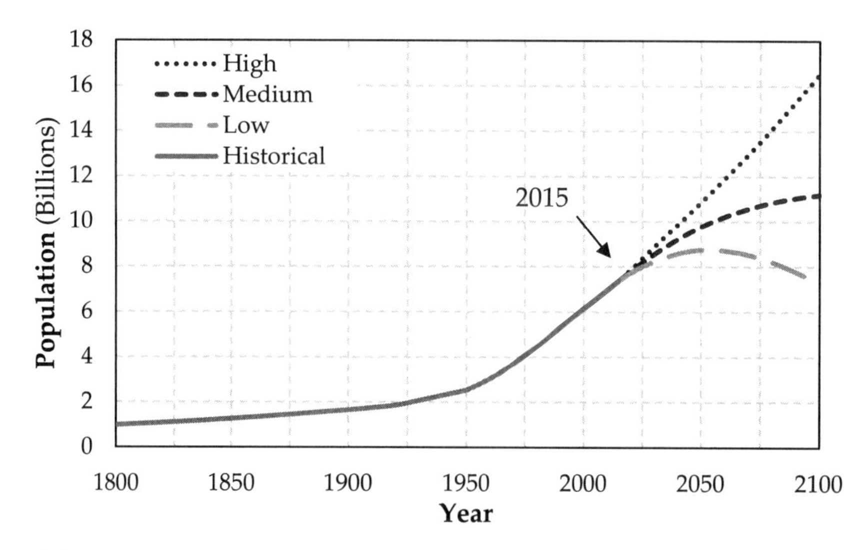

Figure 3.1
Historical and projected world population.
Source: United Nations. Historical data to 1950 come from United Nations (1999); data from 1950 onward come from United Nations (2017).

can build cities catering to the population that lives in them now and to the generations that will live in them in the future. This aspect is already relevant in many rural areas throughout the world that are witnessing a decrease in population and that are left with overbuilt infrastructure and no revenue to maintain it.

Forecasting population is therefore paramount. In this chapter we will first see that these fears of overpopulation are not new. In fact, Thomas Malthus (1798)—quoted at the beginning of the chapter—first pointed out the problem of overpopulation, and we will recall his findings, defining what geometric and arithmetic growth series are and why they are relevant when it comes to forecasting population.

Next, we will learn several models that are used in practice to forecast the population of a neighborhood, municipality, county, state, country, and even the world. Again, this is relevant because people represent the demand for the energy and resources that we consume. These models will then be able to ensure that the infrastructure systems we design serve the right number of people while respecting our two sustainability principles. More specifically, we will learn about two main families of methods: (1) the cohort-change and (2) cohort-survival methods. Cohorts are essentially groups of people, and for population forecasting, cohorts are often defined by age—for example, 0 to 4 years old, 5 to 9 years old, and so on.

Cohort-change methods are used for both short- and long-term forecasting. In particular, we will define and use the famous S curve for these methods, but we will not actually deal with age cohorts. The one cohort-survival method that we will see is a little more sophisticated than the cohort-change methods, and it requires birth, death, and migration rates. Naturally, many more techniques exist, and chapter 6 of Bauer (2010) offers a brief and descriptive introduction of several of these techniques within the context of engineering.

But first, let us see what Malthus taught us.

3.1 Malthus and an Essay on the Principle of Population

Thomas Malthus was a British political economist who was born in 1766 in Surrey, England, and who died in 1834 at the age of 68 in Bath, England; his portrait is shown in figure 3.2. His findings and opinions were quite controversial during his time, and in fact they have remained controversial ever since. He wrote his most famous book *An Essay on the Principle of Population* in 1798, which was around the time the world population reached the 1 billion mark and about twenty years before the start of the Industrial Revolution.

This essay is famous for comparing population growth with how much food can be grown naturally. Although the quote shown at the beginning of the chapter may be interpreted to have a generally positive tone, it is taken out of context—the meaning of the "power" is actually not the one that we commonly use every day. Instead, we are dealing with the mathematical sense of the term *power*. If we introduce the full context, we get a different picture:

> I think I may fairly make two postulata. First, that food is necessary to the existence of man. Secondly, that the passion between the sexes is necessary and will remain nearly in its present state. . . .
>
> Assuming then my postulata as granted, I say, that the power of population is indefinitely greater than the power in the earth to produce subsistence for man.
>
> Population, when unchecked, increases in a geometrical ratio. Subsistence increases only in an arithmetical ratio. A slight acquaintance with numbers will shew the immensity of the first power in comparison of the second.[1]

This last part is especially important. Malthus is essentially saying that the population increases faster than food production or, put differently, that the rate of population growth is higher than the rate of food production that we need to feed this population. Does that ring a bell? Yes, this relates perfectly to our definition of sustainability.

Figure 3.2
Portrait of Thomas Malthus.

If we dive a little more into what Malthus is saying, then population grows geometrically, while food production grows arithmetically. Recalling some fundamentals of mathematics, geometric series are sequences of numbers that are multiplied by one another. Typical geometric series are powers r^n (in the sense meant by Malthus), and the nth term of a series is

$$p_n = p_{n-1} r = p_0 r^n \tag{1}$$

where p_0 is an initial value, and r is a constant. We are often more familiar with the sum S_n^P of n terms in the series:[2]

$$S_n^p = \frac{p_0(1-r^{n+1})}{1-r} \tag{2}$$

In contrast, arithmetic series are sequences of numbers that are added to one another. Mathematically, starting with a_0 and adding a fixed value d (similar to p_0 and r for geometric series but changing symbols to avoid confusion), the nth term of a series is

$$a_n = a_{n-1} + d = a_0 + n \cdot d \tag{3}$$

The sum S_n^a of n terms in the series here becomes[3]

$$S_n^a = \frac{(n+1)}{2}(a_0 + a_n) \tag{4}$$

To facilitate the understanding of these two series, let us apply them to a problem. The best problem I can think of for students is free food! Figure 3.3 will help us illustrate the problem.

Say we have two friends who find a venue with free food, and each of them calls two friends, who then call two friends each, and so on and so forth. How many people do we have after three sets of calls? Our initial population p_0 is 2, and the growth rate is 2 people per person. We therefore start with 2 people, who bring 4 new people, who in turn bring 8 people, who in turn bring 16 people, and so on. Applying equation (2) to determine the total number of people after 3 sets of calls, we have $S_3^p = 2(1-2^4)/(1-2) = 30$ people (i.e., $2+4+8+16=30$ people). We can see this in figure 3.3 by counting all the dots for $n=0$, 1, 2, and 3 for the geometric series shown on the left-hand side.

At the same time, how many servings of food can we supply? If we start with an initial quantity a_0 of 2 for two people and the servers see that more people are coming and each time are able to bring more food for two additional people (i.e., $d=2$), then

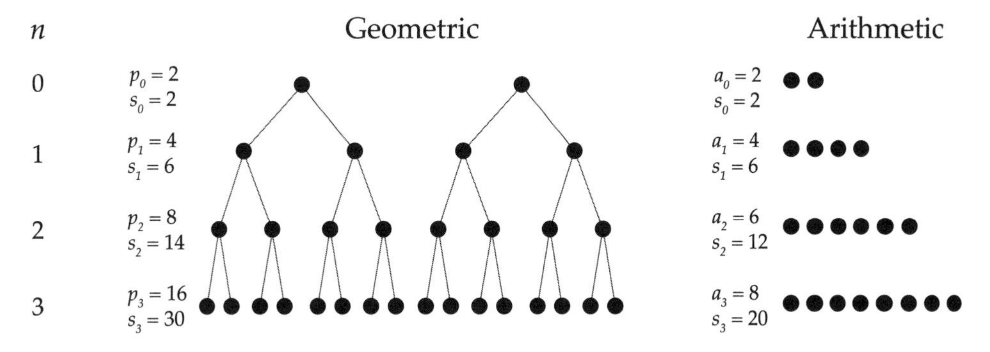

Figure 3.3
Example of geometric and arithmetic series.

they can bring food for 4 people the second time around, and then 6, and then 8. After three steps, we have a total number of servings of $S_3^a = \frac{4}{2}(2+8) = 20$ (i.e., $2+4+6+8=20$ servings). Again, we can see this in figure 3.3 on the right-hand side by counting all the dots. In simply three steps, we therefore went from having two servings of food for 2 people, to 20 servings of food for 30 people. After ten sets of calls, we have 110 servings of food for 2,046 people! This indeed represents a problem.

To view these results graphically, figure 3.4 shows the evolution of p_n and a_n as a function of n on the left-hand side and the evolution of S_n^p and S_n^a as a function of n on the right-hand side. Both figures have the x-axis on a normal scale and the y-axis on a log scale. Notably, we see that the p_n curve on the left-hand side is straight. If both axes had log scales and we had a straight line, then we would have had a power law of the form $y=a\cdot x^b$, where a and b are constants.[4] Since only one of the two axes is on a log scale and we have a straight line, then we have an exponential function of the form $y=a\cdot e^{bx}$, where a and b are constants here again.[5] Essentially, this straight line suggests that our p_n grows exponentially. This is in fact a very important property of geometric series.

If Malthus is right, then we are in big trouble since the rate of food production is much slower than the rate of population growth. This was back in 1798, however, and since then the world population has increased substantially. How did we do it? Although Malthus advocated for strong regulations to control the population (a little like the one-child policy in China), this is not what happened. Put simply, instead we

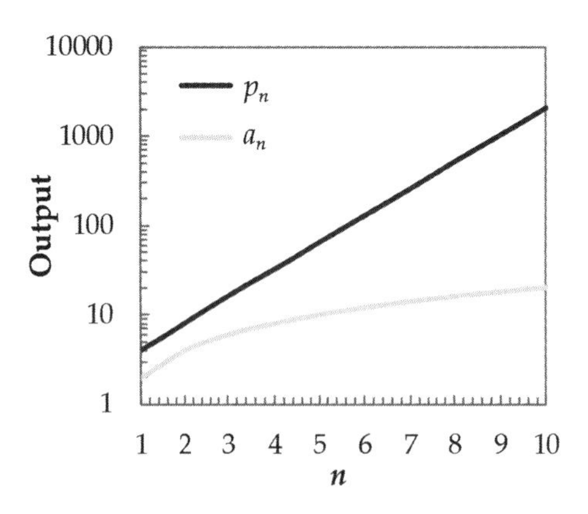

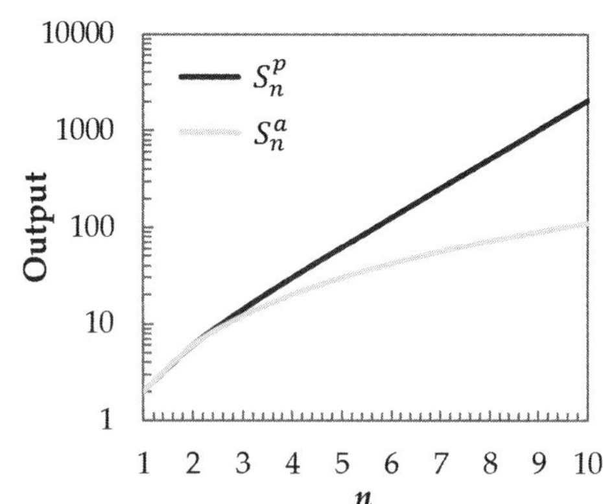

Figure 3.4
Output of p_n, a_n, and S_n^p, S_n^a as a function of n.

embarked on Scenario B. New agriculture methods were invented, and new transport modes and techniques were created to harvest all this food and bring it to urban centers (sometimes from the other side of the Earth). It is not that Malthus was completely wrong in his calculations; in fact he was partly right and partly wrong. He was partly right because the population has grown exponentially since the 1800s, and he was partly wrong because we have been intensifying land productivity (i.e., crop yield).

This is well captured by a theory initially developed by Ester Boserup (1981) and expanded and described in the model of intensification by Ellis et al. (2013). The model follows three phases: *intensification* (productive technologies enable yield faster than population growth), *involution* (peak in productivity is reached), and *crisis* (technologies cannot provide yield faster than population growth). The *crisis* then leads to efforts to find new methods and technologies that can further increase yield. In other words, we discover new ways to ensure that the rate of food production matches the rate of population growth. Going back to sustainability principles, we manage to increase the supply and embark on Scenario B.

Fears of overpopulation did not end with Malthus, however. In the 1970s a famous book titled *The Limits to Growth: A Report for the Club of Rome's Project on the Predicament of Mankind* (Meadows et al. 1972) was published by a serious group of scientists warning us about the alarming rate of population growth. For their study they calculated the rate of growth in population, and they found an exponential growth. They forgot that growth tends to follow an S curve, but that will be the topic of the next section.

Overall, we need to remember that the situation is different this time because we simply cannot expand and use more resources. We will need to be able to feed more than 10 billion people while using fewer resources than we used for 7 billion people. Again, we will need to come up with Scenario B.

The ability to forecast population is therefore paramount, which is why it is central to this chapter. Let us get started and make short-term forecasts.

3.2 Short-Term Population Predictions

Our ultimate goal is to be able to make a reasonable estimate of the future population of whichever neighborhood, municipality, county, state, or even country that we are studying. As illustrated in figure 3.5, many phenomena in real life evolve in a three-step process: (1) the growth is initially very rapid and follows a geometric series, (2) the growth stabilizes and follows an arithmetic series, and (3) the growth finally declines until the population reaches a saturation level.

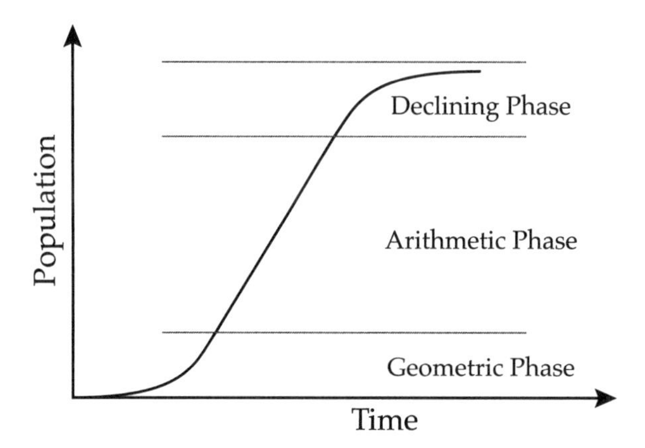

Figure 3.5
Three phases for short-term projections.

This process follows what is formally called a *logistic curve* and is sometimes termed an *S curve*. We will see the equation for this entire three-step process in the next section, for long-term projections, but for now we will concentrate on each step separately.

3.2.1 Geometric Growth Phase

This geometric phase is closely related to the one we saw in the previous section. The main argument here is that the population growth is linked to the current population, and therefore the more people are alive, the higher the population will be—similar to our friends inviting two more friends themselves. Instead of looking at the governing equation, it is easier to first approach the problem by looking at the rate of growth, which fits well with what we have been doing so far. In this case, the growth rate is

$$\frac{dP}{dt} = k_1 P \tag{5}$$

where P is the population, and k_1 is the actual growth rate. The population growth dP/dt is therefore related to P itself. This is a first-order differential equation. To solve it we simply need to put the P in the denominator on the left-hand side and the dt in the numerator on the right-hand side and integrate on both sides.

$$\int \frac{dP}{P} = k_1 \int dt$$

$$\ln(P) = k_1 t + c \tag{6}$$

where c is a constant. We then take the exponential on both sides to get P:

$$\exp(\ln(P)) = \exp(k_1 t + c)$$

$$P = c' e^{k_1 t} \tag{7}$$

where c' is simply the constant $\exp(c)$, which is the initiator (i.e., intercept) that will be our initial population at t_0 that we will call P_0, which finally gives us

$$P = P_0 e^{k_1 t} \tag{8}$$

Equation (8) is an exponential function, and naturally it grows exponentially in the same fashion as our example did in the previous section. To find the growth rate k_1, we can use equation (5) with two data points, switching the derivative d with the difference Δ. We also typically use the average of the two populations that we have to determine k_1 (see example 3.1).

The next short-term forecasting tool is the arithmetic growth phase.

Example 3.1

Short-Term Population Forecast under Geometric Growth

Considering that the population of Civitas in 1990 and 2000 was, respectively, 38,500 and 48,000 people, forecast the population of Civitas in 2010 assuming a geometric growth.

Solution

In this case t_0 is 1990, which means that $P_0 = 38{,}500$. The first step is to calibrate the constant k_1. Using equation (5):

$$\frac{\Delta P}{\Delta t} = k_1 P$$

$$k_1 = \frac{\Delta P}{\Delta t} \times \frac{1}{P} \tag{9}$$

In this case $\Delta P = 48{,}000 - 38{,}500 = 9{,}500$, $\Delta t = 2000 - 1990 = 10$, and for P we take the average of the populations from 1990 and 2000, $P = ½\,(38{,}500 + 48{,}000) = 43{,}250$. Therefore:

$$k_1 = \frac{48{,}000 - 38{,}500}{2000 - 1990} \times \frac{1}{\frac{1}{2}(48{,}000 + 38{,}500)} = 0.022$$

As our goal is to find the population for the year 2010 and since our year 0 is 1990, we need to look at the population for the year $2010 - 1990 = 20$, therefore,

$$P_{2010} = 38{,}500 e^{0.022 \times 20} = 59{,}780$$

The answer is therefore quite close to the real population of 60,000 in 2010 (see chapter 1 for details on Civitas).

3.2.2 Arithmetic Growth Phase

Again, we have seen this type of growth from the previous section. Malthus argues that food production growth follows an arithmetic progression, but sometimes population does as well. The main argument here is that the population growth is constant, and we therefore add the same number of people to our city, regardless of whether there are currently 1,000 or 1 million inhabitants. Starting again with the rate, we define an arithmetic growth as

$$\frac{dP}{dt} = k_2 \tag{10}$$

where P is the population and k_2 is a constant growth number. From this equation we can obviously see that the population growth dP/dt is not related to P, unlike the previous model. This is an even easier differential equation to solve, putting the dt in the numerator on the right-hand side and integrating on both sides:

$$\int dP = k_2 \int dt$$

$$P = k_2 t + c \tag{11}$$

where c is our initiator (i.e., intercept), or P_0, such that the model takes its final form:

$$P = k_2 t + P_0 \tag{12}$$

Equation (12) is an obvious linear equation, and its grows linearly by adding k_2 people every year. To find this constant k_2, we follow equation (10), again switching the derivative d with the difference Δ.

Our final short-term forecasting tool is the declining growth phase.

3.2.3 Declining Growth Phase

This is a new type of growth phase that Malthus did not discuss. The main argument here is that the growth rate is decreasing so that the proportion of people who are added to the population declines with time. In other words, we are still adding people every year but fewer and fewer each time. The population growth is therefore related to P, but it needs to decrease with t. This is most easily done by determining a certain maximum population that we call P_{sat}, for saturated. This saturated population can be determined based on the limits set by our infrastructure or by how much land can still be developed in a community, for instance. The growth takes the form

$$\frac{dP}{dt} = k_3(P_{sat} - P) \tag{14}$$

where P is the population, and k_3 is the growth rate that will end up having a negative sign in front. Naturally, as time passes and as the population P increases, the factor

Example 3.2

Short-Term Population Forecast under Arithmetic Growth

Considering that the population of Civitas in 1990 and 2000 was, respectively, 38,500 and 48,000 people, forecast the population of Civitas in 2010 assuming an arithmetic growth.

Solution

Similar to example 3.1, t_0 is 1990, and $P_0 = 38{,}500$. Using equation (10), our constant growth factor is

$$\frac{\Delta P}{\Delta t} = k_2$$

$$k_2 = \frac{\Delta P}{\Delta t} \tag{13}$$

Considering $\Delta P = 48{,}000 - 38{,}500 = 9{,}500$ and $\Delta t = 2000 - 1990 = 10$, we get

$$k_2 = \frac{48{,}000 - 38{,}500}{2000 - 1990} = 950$$

This essentially means that every year, there are 950 new people in Civitas, regardless of the current population.

As our goal is to find the population for the year 2010 and since our year 0 is 1990, we need to look at the population for the year $2010 - 1990 = 20$, therefore:

$$P_{2010} = 950 \times 20 + 38{,}500 = 57{,}500$$

The answer is therefore smaller than in example 3.1, which is expected since our growth rate is slower.

$P_{sat} - P$ will decrease, which will in turn decrease the growth rate dP/dt. Again, we need to integrate on both sides. This one is a little longer to write, however, and we will therefore move directly to the solution:

$$P = P_{sat} - (P_{sat} - P_0)e^{-k_3 t} \tag{15}$$

Equation (15) is also an exponential function, but this time the growth rate decreases exponentially, as opposed to increasing exponentially, because of the negative sign in front of k_3. The maximum population is systematically P_{sat}, and we then subtract a certain quantity that increases with t. If we look in the short term and therefore t is small, then the component $\exp(-k_3 t)$ will be large, but it will diminish as t increases.

Let us now look in the longer term, beyond fifteen years, and integrate the three types of growth phases that we saw in this section into one single model.

Example 3.3

Short-Term Population Forecast under Declining Growth

Considering that the population of Civitas in 1990 and 2000 was, respectively, 38,500 and 48,000 people and assuming a saturated population P_{sat} of 65,000 people, forecast the population of Civitas in 2010 assuming a declining growth.

Solution

Similar to examples 3.1 and 3.2, t_0 is 1990, and $P_0=38,500$, and in this case we also have value for the saturated population. Using equation (14), our constant growth factor is

$$\frac{\Delta P}{\Delta t} = k_3(P_{sat} - P)$$

$$k_3 = \frac{\Delta P}{\Delta t} \times \frac{1}{(P_{sat} - P)} \tag{16}$$

With $\Delta P = 48,000 - 38,500 = 9,500$, $\Delta t = 2000 - 1990 = 10$, for P we again take the average of the 1990 and 2000 populations, $P = ½\,(38,500 + 48,000) = 43,250$. Therefore:

$$k_3 = \frac{48,000 - 38,500}{2000 - 1990} \times \frac{1}{65,000 - \frac{1}{2}(48,000 + 38,500)} = 0.044$$

Again, as our goal is to find the population for the year 2010 and since our year 0 is 1990, we need to look at the population for the year $2010 - 1990 = 20$, therefore:

$$P_{2010} = 65,000 - (65,000 - 38,500)e^{-0.044\times 20} = 54,008$$

The population we calculated is smaller than both other models in examples 3.1 and 3.2, which is expected since we assumed a declining growth phase, which is slower than both a geometric and an arithmetic growth phase.

These three growth models can be practical and easy to implement when the amount of information we have relies on past population data solely and when our forecast is in the short term. Here, short term generally means between five and fifteen years. Moreover, from the three examples we see that the population of Civitas in 2010 was best estimated by the geometric growth phase model. If we wanted to forecast the 2020 population, which model would you pick? Of course, it would depend on how much activity there is in Civitas, but without any information, switching to an arithmetic growth phase model would make sense to follow the second step of the S curve.

3.3 Long-Term Population Predictions

While short-term projections are sufficient for many engineering applications, we also often want to estimate how the demand for a certain resource will evolve in the longer term, at least to establish a long-term strategic plan. The type of S curve that is apparent from figures 3.1 (world population) and 3.5 is actually quite common in many fields. In particular, we can easily relate it to the diffusion of innovations. The well-known bell curve that consists of the innovators, the early adopters, the early and late majority, and the laggards is actually related to this S curve, where the S curve is the cumulative (i.e., sum) number of people who have bought or used a technology over time.

This type of S curve thus spans far beyond population growth. In fact, it can be found for many, if not most, types of technologies as well. Figure 3.6, for instance, shows how some of the most common infrastructure systems have evolved in the United States from 1780 to 2000. Further analyses, as well as similar figures for other systems and countries, are available in Grübler (1990).

The x-axis shows the time, and the y-axis shows the percentage of the full system that had been built at the time. For instance, for canals, about half of the total 6,500 kilometers of infrastructure was built before 1840, and for roads, about half of the total 490,000 kilometers of infrastructure was built before 1955. The S shape of the curve is quite apparent. S curves are in fact ubiquitous when forecasting the use of a technology,

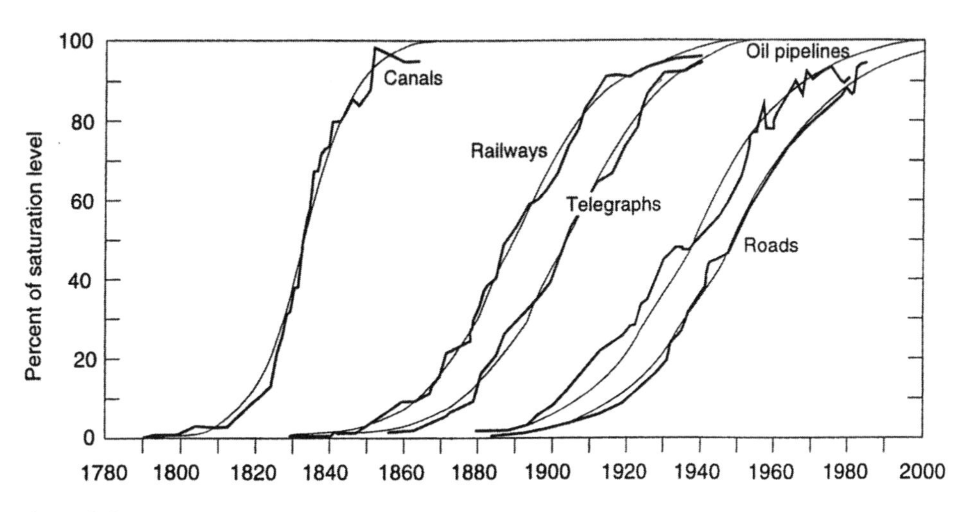

Figure 3.6
Evolution of infrastructure in the United States. Courtesy of Arnulf Grübler.

and they have even been used to model the impact of technology diffusion on the carbon footprint of cities (Mohareb and Kennedy 2012).

This type of curve is not called an S curve only because of the S shape. The *S* actually stands for *sigmoid*, and sigmoid curves belong to the family of logistic functions (hence the name *logistic* curves). Typically attributed to Belgian mathematician Pierre François Verhulst, the growth rate initially increases and then decreases after some point of inflection. The rate of change here can be expressed by

$$\frac{dP}{dt} = k_4 P\left(1 - \frac{P}{K}\right) \tag{17}$$

where k_4 is a positive growth rate, and K is a "carrying capacity," which for us will be related to the saturated population P_{sat}. If we expand the right-hand side of equation (17), we find a first element k_4P, similar to a geometric growth, and a second element k_4P^2/K. Because $K>P$ for all t (by definition since K is the saturated population), the first element is always larger than the second and dP/dt is positive. That being said, because P increases, the difference between the two elements gets smaller and smaller, which means that dP/dt gets smaller with t until it becomes 0 when $P=K$. In other words, when P is small this second element is small, and the rate is controlled by the geometric growth and is therefore high, and as P increases so does the second element, and we end up with a smaller growth rate, hence the declining growth.

Solving this differential equation,[6] changing the symbol K for our saturated population P_{sat}, the logistic function is generally expressed as

$$P(t) = \frac{P_{sat}}{1 + a \cdot e^{bt}} \tag{18}$$

where a and b are constants; in fact, $b=-k_4$, which we can see during the derivation, but we will keep b because equation (21) will already give us a negative value. Moreover, and in contrast to the declining growth phase, there is a way we can determine the value of P_{sat} analytically, based on previous data, but this time we need three previous data points, P_0, P_1, and P_2. The values that we obtain for P_{sat}, a, and b are not as accurate as they would be if we had multiple data points, but they are technically sufficient. The three parameters P_{sat}, a, and b are then determined by

$$P_{sat} = \frac{2P_0P_1P_2 - P_1^2(P_0 + P_2)}{P_0P_2 - P_1^2} \tag{19}$$

$$a = \frac{P_{sat} - P_0}{P_0} \tag{20}$$

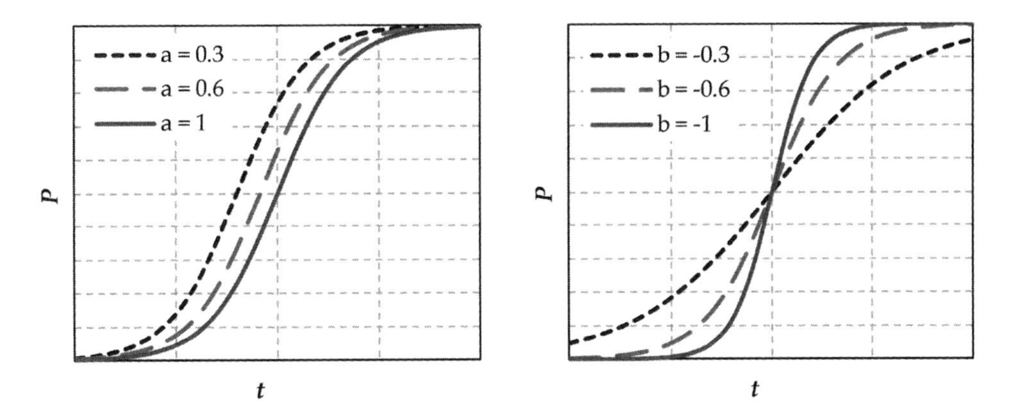

Figure 3.7
Impacts of the two constants in the logistic function.

$$b = \frac{1}{\Delta t} \ln\left[\frac{P_0(P_{sat} - P_1)}{P_1(P_{sat} - P_0)}\right] \tag{21}$$

where Δt is the time interval between each of the three data points. This means that the time intervals between the points need to be equal; for instance, we need data for 1980, 1990, and 2000, which are separated by ten years, or for 1900, 1950, and 2000, which are separated by fifty years.

In some cases the value of P_{sat} calculated can be completely off or sometimes even negative. We will therefore need to use our engineering judgment to determine whether or not the value of P_{sat} is reasonable. Selecting values for P_0, P_1, and P_2 is also not evident, and depending on the values we select, we may get completely different results, so again we need to use our engineering judgment to assess whether the final results make sense or not.

To better understand the impacts of the constants a and b, the two graphs in figure 3.7 plot the logistic function with different parameter values. On the left-hand side, the parameter a changes, and the higher it is, the longer it takes for the initial geometric phase to begin—because P_0 is small compared to P_{sat} in equation (20), and it takes "time" for the population to grow. On the right-hand side, the growth rate b changes, and the higher it is, the steeper the distribution becomes—that is, the population grows faster due to the exponential nature of equation (21) that is in the denominator with a negative b.

In total, the three short-term and the long-term modeling techniques are simple to apply, and they require few data points. Although we will not cover the full spectrum of all modeling techniques to forecast population, there is one more that is often used that is based on birth, death, and migration rates as opposed to past population values.

Example 3.4

Long-Term Population Forecast Using a Logistic Function

Considering that the population of Civitas in 1960, 1970, and 1980 was, respectively, 20,000, 25,000, and 31,000 people, forecast the population of Civitas in 2010 assuming the population trends follow a logistic function.

Solution

First, we need to assign our three data points and ensure that they are separated by equal intervals. Here, $\Delta t_1 = 1970 - 1960 = 10$ and $\Delta t_2 = 1980 - 1970 = 10$ also, which means we can adopt the logistic model. For P_0, P_1, and P_2, we assign the values 20,000, 25,000, and 31,000, respectively.

We can first calculate our P_{sat} value:

$$P_{sat} = \frac{2 \times 20{,}000 \times 25{,}000 \times 31{,}000 - 25{,}000^2(20{,}000 + 31{,}000)}{20{,}000 \times 31{,}000 - 25{,}000^2} = 175{,}000$$

This value of P_{sat} is high, but it may be reasonable for very long-term projections. We will therefore keep it for this modeling exercise. We can now compute the constants a and b:

$$a = \frac{175{,}000 - 20{,}000}{20{,}000} = 7.75$$

$$b = \frac{1}{10}\ln\left[\frac{20{,}000\,(175{,}000 - 25{,}000)}{25{,}000\,(175{,}000 - 20{,}000)}\right] = -0.026$$

The value of a is again quite high, and we may expect a lower value. That being said, because it is high the denominator of our logistic function (equation [18]) will also be high, which should offset our large P_{sat}. Applying equation (18), with a time $t = 2010 - 1960 = 50$, we get

$$P_{2010} = \frac{175{,}000}{1 + 7.75 \times e^{-0.026 \times 50}} = 56{,}232$$

The result is lower than our actual population of 60,000 in 2010 but not that far off considering our latest data point was in 1980. Out of all the methods we have used so far, the geometric growth model was the closest to the actual value. This means that the population growth in Civitas has been higher than what we would expect. This scenario is in fact typical for a city that is "young" because the number of newborns and migrants is significant. It is also typical of a city that is doing well economically, attracting many people to move there.

3.4 The Cohort-Survival Method

So far, we have looked purely at past population data, and we assume that we can forecast the general trend that will occur in the future. In contrast, the cohort-survival method is based on the likelihood of the survival of an age cohort. An age cohort means that the total population is divided into age groups, often five years apart; for instance, the number of people between ages 0 and 4, 5 and 9, 10 and 14, and so on. *Survival* means the probability that someone will survive in an age cohort; for instance, 99% of all children between 10 and 14 survive, which therefore accounts for death rates per age cohort. To apply the method, we also need to add the number of new births per age cohort, using fertility rates, and the number of migrants (i.e., people moving to the community and leaving the community). Watkins (2015) offers great examples if you are interested in learning more.

At the most fundamental level, and for the entire population P at time t and $t+1$, we are basically trying to solve this equation:

$$P(t+1)=P(t)+B(t+1)-D(t+1)+M(t+1) \tag{22}$$

where B is the number of births, D is the number of deaths, and M is the net number of migrants (i.e., immigrants minus emigrants). P here is the entire population, but because different cohorts go through different dynamics, we will actually divide the population into cohorts.

One requirement of this method is that we can only forecast in as many years as the size of our age cohorts. For instance, if we use 2000 census data and if every age cohort is divided into five-year intervals, then the closest year we can forecast is 2005, when everyone from a certain age cohort will move to the next. To forecast population for any years in between, we can simply interpolate between the two years that we have; for instance, we can interpolate the values between 2000 and 2005 to get the population in 2003.

We will go through the cohort-survival method step by step. Table 3.1 shows how the initial survival procedure works. At time t, given an age cohort i other than the first age cohort (i.e., the babies), a total population per cohort P_i, and a survival rate S_i, the population P_i at time $t+1$ is given by

$$P_i(t+1)=P_{i-1}(t)S_{i-1}(t) \tag{23}$$

Equation (23) is applied to all cohorts except the first cohort since it is only populated by new births and migrants (i.e., there are no "survivors" below 0 years old).

Table 3.1
Cohort-survival method—step 1: survival

Age cohort	Population year t	Survival rate	Population year $t+1$
0–4	P_1	S_1	
5–9	P_2	S_2	P_1S_1
10–14	P_3	S_3	P_2S_2
…	…	…	…
80–84	P_{17}	S_{17}	$P_{16}S_{16}$
85+	P_{18}	S_{18}	$P_{17}S_{17}+P_{18}S_{18}$

Moreover, it should not be applied to the last cohort n since it is not in a five-year interval. For the last cohort, the equation to use is

$$P_n(t+1)=P_{n-1}(t)S_{n-1}(t)+P_n(t)S_n(t) \tag{24}$$

Essentially, we are just adding people from the previous cohort to the last cohort, but because survival rates at n are low, the number of people in this cohort tends to remain at the same level.

More generally, the final "surviving" population for year $t+1$ for all cohorts except the first and the last can be calculated as

$$P(t+1)=\sum_{i=2}^{n-1} P_{i-1}(t)S_{i-1}(t) \tag{25}$$

Most often, men and women are separated and are given different survival rates since the survival rates of the men tend to be lower than the women's.

Subsequently, we need to add the number of newborns in the past five years. Similar to the survival rate, we use a fertility rate per age cohort, which is the number of births in a year divided by the total number of women in a cohort. This naturally just applies to women, hence the need to separate men and women. In the event that men and women cannot be separated, the cohort fertility rates need to be adjusted by an estimation of the proportion of women in the cohort.

Because fertility rates are computed by year, we need to estimate the number of births in five years. For this, we can simply calculate the number of newborns for one year and multiply it by five. Although this is valid, it poses one problem. Women at the edge of the cohort in the first year will move to the next cohort in the next year. For instance, a 24-year-old woman will join the next cohort in one year, not in five years. To remediate this issue, we take the average fertility rates of the current cohort and the one after, and we then multiply it by the size of the cohort (e.g., five years). Table 3.2 shows the procedure.

Table 3.2
Cohort-survival method—step 1: births

Age cohort	Women year t	Fertility rate	Births per cohort $t+1$
0–4	P_1	0	0
5–9	P_2	0	$P_2 \times 5 \times ½ \times (0+F_3)$
10–14	P_3	F_3	$P_3 \times 5 \times ½ \times (F_3+F_4)$
15–19	P_4	F_4	$P_4 \times 5 \times ½ \times (F_4+F_5)$
...	...	...	...
80–84	P_{17}	F_{17}	0
85+	P_{18}	F_{18}	0

Naturally, the fertility rates of the prepubescent and postmenopausal cohorts are 0, but in general the number of births B in year $t+1$ is calculated as

$$B(t+1) = \sum_{i=1}^{n-1} P_i(t) \times \Delta t \times \frac{1}{2}(F_i(t) + F_{i+1}(t)) \tag{26}$$

where P can either be only the population of women in this case or the various fertility rates needed to account for the presence of the male population. In addition, Δt is the size of the age cohort—for example, 5. The only problem with equation (26) is that the sum needs to stop at $n-1$, otherwise we would need F_{n+1}, which does not exist. If the last cohort n has a nonzero birth rate (e.g., a cohort of forty plus), then F_{n+1} needs to be created and given a value of 0.

One more item that needs to be discussed in the number of births is the sex ratio. If we differentiate between sexes, then some of the new babies are added to the female population, and the others are added to the male population. Although we may be tempted to simply split the number in half, most statistics show that there are systematically more baby boys than baby girls. There are no systemic explanations for this,[7] but a good estimation is to take the ratio of boys to girls to be 1.05, which means that out of all newborns about 1.05 / (1.05 + 1) = 0.5122 = 51.22% are male and therefore 48.78% are female.

Once all the new births are calculated, we then need to sum them and place them in the first age cohort at $t+1$. Because of the way births are calculated, we are tempted to add births to specific age cohorts. For example, we are tempted to add the births from the 20–24 age cohorts at time t to the 20–24 age cohorts at time $t+1$, but no one was ever born 20 years old—this is a classic mistake that I have seen in many a homework. All new births must be placed in the first age cohorts at time $t+1$.

The last feature to add to our model is the number of migrants per age cohort, accounting for all the people who have left the community (emigrants) and all the people who have joined the community (immigrants). The rate is basically the number of immigrants minus the number of emigrants divided by the population per age cohort. This sort of information can be a little more difficult to acquire, especially by age cohort. Local offices may have relevant data (i.e., local tax offices, schools, etc.), but typically, we resort to census data to do some estimation based on exact population counts from previous counts. Taking equation (22) and changing the dependent variable to $M(t+1)$, we get

$$M(t+1)=P(t+1)-P(t)-B(t+1)+D(t+1) \tag{27}$$

Therefore, if we know $P(t+1)$ and $P(t)$, for instance, by using decennial census data (e.g., using 2000 and 2010 census data to forecast the 2015 population), then we can find previous migration rates to apply to the future. Naturally, migration depends on many criteria that can change over time (e.g., the local economy), but in the absence of any current data sets, this method gives us a valid approximation.

Overall, the cohort-survival method is more detailed than the four methods we saw earlier, and as a result it can often be more accurate, especially in the short term (e.g., in the next five to ten years). Thanks to the additional information, we can also tailor policies to make sure they target the right populations. In a "young" city, we might favor the construction of new schools. In a city where the demography is older, we might favor the construction of leisure areas and medical services. Naturally, the cohort-survival method is also more data-intensive and therefore not always an option. It remains, however, a preferred method if available.

3.5 Conclusion

This concludes our chapter on population. This chapter is arguably the most important of the book since population represents the demand for energy and resources, and being able to estimate how demand will evolve in the future is paramount. In fact, this aspect of population is especially relevant since we expect the global population to increase to 10 billion by 2050.

In this chapter we saw that worrying about being able to match population growth with our ability to provide for this population is not new. Although Thomas Malthus is famous for being one of the first scholars to write about it, surely it must have been a concern much earlier, at least in Alexandria (Egypt) and Rome (Italy), which are supposedly the first cities to ever reach the 1-million-inhabitants mark,

Example 3.5

Cohort-Survival Method

Considering the information presented below, which represents Civitas in 2000 as well as previously calculated net migration rates by cohort, estimate the number of men and women by age cohort in Civitas in 2010.

Demographic statistics for Civitas

			Survival rates			Net migration rates	
Age cohort	Men	Women	Men	Women	Fertility rates	Men	Women
0–9	3,201	3,087	0.996	0.998	0	0.141	0.138
10–19	3,152	3,001	0.998	0.999	0.041	0.161	0.163
20–29	3,078	2,980	0.994	0.995	0.201	0.178	0.179
30–39	2,981	2,958	0.988	0.989	0.072	0.166	0.171
40–49	2,756	2,801	0.984	0.985	0.006	0.153	0.156
50–59	2,583	2,686	0.961	0.965	0	0.135	0.141
60–69	2,342	2,451	0.914	0.934	0	0.121	0.129
70–79	2,168	2,213	0.681	0.701	0	0.078	0.087
80+	1,739	1,823	0.324	0.363	0	0.054	0.061

Solution

Starting with survival rates and applying equations (23) and (24), we get the number of people who are still alive in 2010, shifting from their initial cohort to one cohort older. The results for this step are shown in table A, with the 2010 values shown in the last two columns.

Second, we apply equation (26) to the fertility rates and the number of women in 2000, ensuring that we multiply by ten and that we take the average of two fertility rates. For the 10–19 cohort, for example, the number of new babies is $3{,}001 \times 10 \times ½ \times (0.041 + 0.201) =$ 3,631. The results for this step are shown in table B. We then take the sum and add it to the 0–9 cohort for 2010.

Third, we simply multiply the population cohort by their net migration rates to get the number of new migrants, noting that in this case we simply stay on the same row (i.e., no need to shift down one row to estimate the net new number of people). The results for this step are shown in table C.

Finally, the population forecast for each cohort in 2010 is presented in table D. Note that because of rounding issues, small differences may appear. Here, the rounded values were taken from the results in the spreadsheet software package used for this example.

(continued)

Example 3.5 (continued)

Table A

2010 survival numbers by age cohort in Civitas

			Survival rates		2010 Survival numbers	
Age cohort	Men	Women	Men	Women	Men	Women
0–9	3,201	3,087	0.996	0.998		
10–19	3,152	3,001	0.998	0.999	3,188	3,081
20–29	3,078	2,980	0.994	0.995	3,146	2,998
30–39	2,981	2,958	0.988	0.989	3,060	2,965
40–49	2,756	2,801	0.984	0.985	2,945	2,925
50–59	2,583	2,686	0.961	0.965	2,712	2,759
60–69	2,342	2,451	0.914	0.934	2,482	2,592
70–79	2,168	2,213	0.681	0.701	2,141	2,289
80+	1,739	1,823	0.324	0.363	2,040	2,213

Table B

Births between 2000 and 2010 by age cohort in Civitas

Age cohort	Women	Fertility rates	Births
0–9	3,087	0	633
10–19	3,001	0.041	3,631
20–29	2,980	0.201	4,068
30–39	2,958	0.072	1,154
40–49	2,801	0.006	84
50–59	2,686	0	0
60–69	2,451	0	0
70–79	2,213	0	0
80+	1,823	0	0
Total			9,570

Example 3.5 (continued)

Table C

2010 Migrant numbers by age cohort in Civitas

			Net migration rates		Net migration	
Age cohort	Men	Women	Men	Women	Men	Women
0–9	3,201	3,087	0.141	0.138	451	426
10–19	3,152	3,001	0.161	0.163	507	489
20–29	3,078	2,980	0.178	0.179	548	533
30–39	2,981	2,958	0.166	0.171	495	506
40–49	2,756	2,801	0.153	0.156	422	437
50–59	2,583	2,686	0.135	0.141	349	379
60–69	2,342	2,451	0.121	0.129	283	316
70–79	2,168	2,213	0.078	0.087	169	193
80+	1,739	1,823	0.054	0.061	94	111

Table D

Population of Civitas in 2010

Age cohort	Men (2010)	Women (2010)	Total (2010)
0–9	5,353	5,094	10,447
10–19	3,696	3,570	7,266
20–29	3,694	3,531	7,225
30–39	3,554	3,471	7,025
40–49	3,367	3,362	6,729
50–59	3,061	3,138	6,198
60–69	2,766	2,908	5,674
70–79	2,310	2,482	4,791
80+	2,134	2,324	4,458
Total	29,935	29,880	59,813

We can see that this methodology gives us a final population of 59,813 people. Although it is below the actual 60,000 people recorded, the accuracy remains significant. In particular, we see that the number of new births is significant, which suggests that Civitas is relatively "young," as we suggested before. The high migration rate also suggests it is an attractive place since many people are moving to Civitas. This type of information is quite telling about an area, and we cannot get the same insights from the other forecasting models that we have seen.

between 100 BCE and 100 CE, and in Chang'an (China), Hangzhou (China), Nanjing, (China), and Baghdad (Iraq), which supposedly reached that stage between 500 CE and 1500 CE.

Providing for a fast-growing population in a sustainable way is a colossal challenge. This is especially relevant in the twenty-first century since we need to be able to provide for this growing population while using less energy and fewer resources. But a colossal challenge can be tackled if we are smart about it. Our first task was therefore to be able to predict our future demand, and we learned about five different techniques.

First, we saw three techniques that can be used to estimate the evolution of population in the short term. These three techniques assume three very different trends in population growth. The geometric growth phase evolves rapidly and increases exponentially. The arithmetic growth phase is slower and increases linearly. The declining growth phase is much slower and in fact tapers off to a saturated population.

Second, we combined the three previous techniques into one, giving us an S curve or, more formally, a logistic curve, which is most applicable to long-term forecasts. Logistic curves depend not only on a saturated population but also on two other parameters that can have a large impact on the forecasts. We also detailed how we could calibrate these three parameters using three past data points, although relying on more data is preferable.

Finally, we saw one more method that requires more data. It is called the cohort-survival method because the population is divided into different age cohorts. Survival rates as well as birth rates and migration rates are needed to run this model. The estimations tend to be more accurate, but it is not always possible to use if the data are not available. It is, however, likely the most popular method used to forecast population. For us, it might also be relevant since different age cohorts consume resources in different ways, and we may be able to better forecast how energy and resource consumption will evolve in the future. We will not go into this level of detail, however.

Now that we have learned how to estimate our future consumption, we will spend time on urban planning. We will get to start designing infrastructure systems soon, but first we need to understand how a city functions as a whole. More importantly, we will learn about how human beings perceive cities. After all, we do not experience a city as it is. We have only our perception of it, and we should make sure that we design infrastructure systems with that aspect in mind. Just as importantly, we will learn that the way our brains work when we design cities is fundamentally flawed.

Problem Set

3.1 In your own words, explain why population forecasting is important in urban engineering.

3.2 In your own words, recall the main argument of Thomas Malthus and describe why it was and still is controversial.

3.3 In your own words, explain the difference between geometric and arithmetic series in the context of population and food growth. The use of a practical example is encouraged.

3.4 In your own words, explain how food yields have historically been able to match population growth and why the previous solutions will have to be different in the future.

3.5 In your own words, describe the three stages of the logistic curve (i.e., S curve). You are welcome to make use of equations and illustrations.

3.6 By searching the web, find and report an example of any process other than those seen in the chapter that has followed an S curve. Make sure to include a graph of the process and fully cite your sources.

3.7 In your own words, explain the main elements and limits of the cohort-survival method.

3.8 The spread of a contaminant in a water reservoir was found to grow in a geometric fashion. Initially, 1 m^3 of water was contaminated, but more water gets contaminated every hour with a power of 2. Calculate (1) the new volume of water that gets contaminated after six hours and (2) the total volume of water contaminated after six hours.

3.9 New roads are being built in a new city. The growth rate follows an arithmetic series. Initially, there was only one road, but the growth rate increases by two roads every year. Calculate (1) the number of new roads added in year ten and (2) the total number of roads in year ten.

3.10 The population in Seattle, Washington, from 1990 to 2010 is shown in the table below. From the 1990 and 2010 values, estimate the 2014 population in Seattle, assuming (1) an arithmetic growth and (2) a geometric growth. Using a logistic growth model, (3) estimate the 2014 population in Seattle (if P_{sat} is negative, use 700,000). (4) The real population in 2014 was 668,342. Comment on the results based on your judgment of long versus short-term population projections.

Year	Population
1990	516,259
2000	563,374
2010	608,660

3.11 The population of Detroit, Michigan, in the early twentieth century is shown in the table below. From the 1900 and 1910 values, estimate the population in Detroit for the years 1950 and 2000, assuming: (1) a geometric growth, (2) an arithmetic growth, and (3) a declining growth (use a P_{sat} of 3,000,000). From the three values, (4) estimate the population in Detroit for the years 1950 and 2000, assuming a logistic growth. (5) The real populations in 1950 and 2000 were 1,849,568 and 951,270, respectively. Comment on the most accurate method and on the results.

Year	Population
1900	285,704
1910	465,766
1920	993,768

3.12 Download the historical population of any city other than the ones seen in this chapter. From historical values, forecast the population in the latest census year, assuming (1) an arithmetic growth, (2) a geometric growth, (3) a declining growth, and (4) a logistic growth. (5) Comment on your result, comparing your forecasts with the actual population. (6) Select the most accurate method to forecast the population of your city in thirty years and comment on the results.

3.13 The table below shows the population of Cook County by decade from 1900 to 1960. Estimate the population in Cook County for the year 2000 using: (1) graphical extension (*hint: use the trendline function in Microsoft Excel*). From 1940 and 1960 values, estimate the 2000 population, assuming (2) a geometric growth, (3) an arithmetic growth, and (4) a declining growth (assume a P_{sat} of 6,000,000). (5) Using 1900, 1930, and 1960 as your P_0, P_1, and P_2, estimate the 2000 population using a logistic curve. The actual population of Cook County in 2000 was 5,367,741. (6) Comment on the most accurate method. (7) Using the most accurate method, project the population in 2050 and comment on your results.

Year	Population
1900	1,838,735
1910	2,405,233
1920	3,053,017
1930	3,982,123
1940	4,063,342
1950	4,508,792
1960	5,129,725

3.14 The table below shows the population of the Bay Area by decade from 1900 to 1980. Estimate the population of the Bay Area for the year 2010 using: (1) graphical extension (*hint: use the trendline function in Microsoft Excel*). From 1960 and 1980 values, estimate the 2010 population, assuming (2) a geometric growth, (3) an arithmetic growth, and (4) a declining growth (assume a P_{sat} of 12,000,000). (5) Using 1920, 1950, and 1980 as your P_0, P_1, and P_2, estimate the 2010 population using a logistic growth model. The actual population of the Bay Area in 2010 was 7,150,739. (6) Comment on the most accurate method. (7) Using the method that you think is the most appropriate (explain why), project the population in 2050 and comment on your results.

Year	Population
1900	658,111
1910	925,708
1920	1,182,911
1930	1,578,009
1940	1,734,308
1950	2,681,322
1960	3,638,939
1970	4,630,576
1980	5,179,784

3.15 The table below shows the 1990 population statistics for a community. Using the cohort-survival method for 2010, calculate (1) the surviving population, (2) the number of births, (3) the net number of migrants to the community, and (4) the total forecasted population for 2010.

Cohort	Population	Survival rates	Fertility rates	Net migration rates
0–19	20,081	0.992	0.008	0.055
20–39	18,213	0.979	0.121	0.047
40–59	15,724	0.874	0.065	0.002
60–79	11,054	0.352	0	–0.013
80+	6,234	0.012	0	–0.015

3.16 The table below shows the 2000 population statistics for a community. Using the cohort-survival method for 2005, calculate (1) the survival numbers, (2) the number of births, (3) the number of migrants to the community, and (4) the total forecasted population for 2005. (5) Using the same rates, estimate the population of the community in 2010. (6) Considering the population in 2010 was 310,237, comment on the accuracy of the model.

			Survival rates			Net migration rates	
Age cohort	Men	Women	Men	Women	Fertility rates	Men	Women
0–4	10,520	10,019	0.995	0.996	0	0.121	0.119
5–9	9,994	9,567	0.997	0.998	0	0.117	0.117
10–14	9,452	9,134	0.998	0.999	0.001	0.116	0.115
15–19	8,961	8,875	0.996	0.997	0.008	0.113	0.114
20–24	8,523	8,598	0.991	0.993	0.112	0.114	0.115
25–29	8,013	8,234	0.989	0.991	0.132	0.112	0.113
30–34	7,777	7,897	0.987	0.989	0.132	0.109	0.109
35–39	7,498	7,610	0.982	0.984	0.125	0.108	0.107
40–44	7,123	7,251	0.981	0.986	0.023	0.093	0.094
45–49	6,799	6,998	0.979	0.981	0.001	0.082	0.081
50–54	6,432	6,712	0.959	0.925	0	0.074	0.076
55–59	6,098	6,321	0.939	0.943	0	0.051	0.055
60–64	5,601	5,891	0.899	0.912	0	0.038	0.039
65–69	5,324	5,596	0.972	0.985	0	0.027	0.026
70–74	5,012	5,165	0.681	0.713	0	0.021	0.025
75–79	4,499	4,602	0.458	0.558	0	0.012	0.012
80–84	3,781	4,012	0.331	0.421	0	0.010	0.009
85+	2,630	3,149	0.299	0.345	0	0.009	0.009

3.17 The table below shows the 2000 population statistics for a city. Using the cohort-survival method for 2005, calculate (1) the survival numbers, (2) the number of births,

(3) the number of migrants to the community, and (4) the total forecasted population for 2005. (5) Using the same rates, estimate the population of the community in 2010. (6) Considering the population in 2010 was 3,354,381, comment on the accuracy of the model. (7) Based on the results, describe the types of policies you would recommend to the city.

			Survival rates			Net migration rates	
Age cohort	Men	Women	Men	Women	Fertility rates	Men	Women
0–4	101,987	99,947	0.994	0.995	0	0.109	0.108
5–9	97,364	97,176	0.997	0.997	0	0.098	0.099
10–14	97,234	97,123	0.998	0.999	0.002	0.097	0.097
15–19	94,119	93,401	0.995	0.996	0.007	0.012	0.009
20–24	92,236	91,319	0.991	0.992	0.109	0.023	0.032
25–29	90,392	89,916	0.988	0.991	0.131	0.048	0.056
30–34	88,584	89,147	0.986	0.988	0.129	0.093	0.101
35–39	86,812	88,380	0.982	0.984	0.116	0.102	0.101
40–44	85,076	85,477	0.981	0.983	0.018	0.087	0.091
45–49	83,374	86,366	0.978	0.981	0.001	0.078	0.081
50–54	81,707	83,641	0.955	0.961	0	0.043	0.054
55–59	80,073	81,674	0.941	0.949	0	0.029	0.035
60–64	78,471	81,009	0.899	0.912	0	0.015	0.018
65–69	76,902	77,217	0.976	0.991	0	−0.002	0.002
70–74	75,364	76,123	0.673	0.702	0	−0.011	−0.009
75–79	73,856	76,277	0.444	0.523	0	−0.101	−0.111
80–84	72,379	75,554	0.332	0.426	0	−0.122	−0.129
85+	70,932	77,004	0.296	0.333	0	−0.198	−0.187

3.18 The data for the world population from figure 3.1 is shown in the table below. (1) Fit a logistic curve to data from 1750 to 2010 using any technique desired other than the three-data-point technique seen in the chapter. Make sure to clearly state the P_{sat} calculated as well as the constants a and b and the sum of the squared difference in populations—that is, $\sum\left(P_i^{actual} - P_i^{fitted}\right)^2$. *Hint: search online "how to fit a logistic curve in Excel" and iterate many times the values given by the solver, playing with the values of P_{sat}, a, and b (look in the chapter for how these parameters impact the final results).* (2) Plot your results with the actual data, and using your model estimate the 2100 population. Comment on the accuracy of your model based on the scenarios seen in the chapter.

Year	Population (billions)
1750	0.79
1800	0.98
1850	1.26
1900	1.65
1910	1.75
1920	1.86
1930	2.07
1940	2.30
1950	2.52
1960	3.02
1970	3.70
1980	4.44
1990	5.27
1999	5.98
2000	6.06
2010	6.79

Notes

1. Malthus 1798, p. 4.

2. The derivation for this sum is actually easy to do and well explained in Wikipedia at http://en.wikipedia.org/wiki/Geometric_progression#Derivation (accessed August 14, 2018).

3. Here again, the derivation for the sum is quite easy to do and well explained in Wikipedia at http://en.wikipedia.org/wiki/Arithmetic_progression#Derivation (accessed August 14, 2018).

4. Just take the log on both sides, and you will easily see that the equation becomes linear.

5. Technically, we are using r as opposed to e (Euler's number) for our exponential function, but the message is the same. Moreover, here, take the log on the right-hand side only, and you will easily see that it becomes linear.

6. This is relatively easy to do and someone posted a YouTube video on it at https://www.youtube.com/watch?v=vsYWMEmNmZo (accessed August 14, 2018).

7. And there is a lot of speculation. One of the best explanations I have seen is offered by Watkins (2015) and relates to the difference in weight between the longer and therefore slightly heavier female X chromosome compared to the male Y chromosome.

References

Bauer, Kurt W. 2010. *City Planning for Civil Engineers, Environmental Engineers, and Surveyors*. Boca Raton, FL: CRC Press.

Boserup, Ester. 1981. *Population and Technological Change: A Study of Long-Term Trends*. Chicago: University of Chicago Press.

Ellis, Erle C., Jed O. Kaplan, Dorian Q. Fuller, Steve Vavrus, Kees Klein Goldewijk, and Peter H. Verburg. 2013. "Used Planet: A Global History." *Proceedings of the National Academy of Sciences* 110(20): 7978–7985.

Grübler, A. 1990. *The Rise and Fall of Infrastructures: Dynamics of Evolution and Technological Change in Transport*. Heidelberg: Physica-Verlag.

Malthus, T. R. 1798. *An Essay on the Principle of Population: As It Affects the Future Improvement of Society*. London: Printed for J. Johnson, in St. Paul's Church-Yard. http://www.esp.org/books/malthus/population/malthus.pdf.

Meadows, D. H., D. L. Meadows, J. Randers, and W. W. Behrens III. 1972. *The Limits to Growth: A Report for the Club of Rome's Project on the Predicament of Mankind*. New York: Universe Books.

Mohareb, E., and C. Kennedy. 2012. "Greenhouse Gas Emission Scenario Modeling for Cities Using the PURGE Model." *Journal of Industrial Ecology* 16(6): 875–888.

United Nations. 1999. "The World at Six Billion." ESA/P/WP.154. http://mysite.du.edu/~rkuhn/ints4465/world-at-six-billion.pdf.

———. 2017. *World Population Prospects, the 2017 Revision*. https://esa.un.org/unpd/wpp/.

Watkins, T. 2015. "The Cohort Survival Projection Method." Department of Economics, San José State University. http://www.sjsu.edu/faculty/watkins/cohort.htm.

4 Urban Planning

> Cities have the capability of providing something for everybody, only because, and only when, they are created by everybody.
>
> —Jacobs 1961, p. 238

When we think about designing cities, we quickly think about urban planning. Although the planning of cities has existed for a long time, the modern definition of urban planning is relatively new, and it developed at the same time cities became prominent economic engines in the nineteenth century. Sometimes also called city planning or regional planning, these interchangeable terms simply reflect the fact that cities are getting larger, and their areas of influence are so great that we have institutions to look at region-wide issues.[1] In the United States, these institutions are called Metropolitan Planning Organizations (MPOs).[2]

The role of planners is a difficult one since they have to consider nearly every possible aspect and feature of a city in their plans, from land use and police stations to schools and fire stations, systematically considering the population they are serving. This directly contrasts with the engineering realm, in which we tend to look at individual systems simply as separate entities. In fact, as engineers we are often given design specifications from planning authorities for our designs. If we are to successfully integrate urban infrastructure systems together, however, engineers need to become more like planners, calling for a new breed of engineers: *urban engineers*, as discussed in chapter 1. This means that transport engineers must realize that streets are also relevant to water resources engineers, and any street design must accommodate both transport and water concerns. Similarly, electrical engineers must understand that the demand for electricity follows similar patterns to the demand for water, and both can potentially be better integrated to work at the neighborhood or district scale. At the time of this writing, only the transport system and the building stock (in terms of land use, not energy demand) were starting to be seen as integrated entities—as we will see in chapter 7—and there is therefore much room for improvement.

Interestingly, the roles of city planner and engineer were the same throughout most of human history. Lewis Mumford (1961), whom we briefly saw in chapter 1, has a great quote about this: "From Hippodam[u]s to Haussmann: regimenters of human functions and urban space" (p. 172). In the next section we will learn both about Hippodamus, the father of urban planning who lived in ancient Greece, and Baron Haussmann, who in the nineteenth century made the Paris that we know today. Moreover, in this chapter we will pay particular attention to the "urban space" feature of the quote. The selection of the word *regimenter* is fitting for our goal since *regimenting*, like *engineering*, implies a sense of order and design.

Beyond engineering, however, earlier city planners or chief architects (prior to the nineteenth century) not only had to possess a strong mathematical and technical background but also needed to have an incredible appreciation for the humanities (i.e., culture) to take into account all aspects of a city. The same person had to deal with transport problems and water problems. At the same time, this person needed to have an intricate understanding of social issues to include plazas, public venues, and markets in the urban fabric. Through the twentieth century this aspect has been largely lost in the engineering profession. Instead, engineers are required to be ever more specialized. It is not even enough to concentrate on engineering; engineers are asked to pick a field from transport and water to electricity and structural engineering. Even within a field, engineers are required to label themselves—for example, as a traffic engineer or a pavement engineer. Although the engineering profession may have become more "efficient," this constant specialization has also generated many negative impacts.

Formerly, at the center of the entire planning process was the citizen, the human, the worker, the shopper, who generally had to walk everywhere, thus the "pedestrian" was at the nexus of the planning process. In fact, it is partly thanks to its complex, hierarchical, and regional transport system, designed in part for the "pedestrian," that the city of Rome managed to house 1 million people sometime between 0 and 100 CE. This was an incredible feat for the time. On the one hand, people had to be able to walk everywhere and reach their destination within a relatively small and congested area. On the other hand, enough food and products had to be imported into the city to provide for 1 million people, thus necessitating an extensive regional transport network (both on land and using rivers).[3] Several cities reached the 1-million-people mark before the Industrial Revolution, and they had to deal with similar issues; other cities that reached 1 million people include Alexandria (Egypt), Chang'an (China), Hangzhou (China), Nanjing, (China), and Baghdad (Iraq), and interestingly, they did not necessarily come up with the same solutions.[4]

As engineers, we will adopt the viewpoint of Bauer (2010) who wrote a book on urban planning for civil engineers. He views the city as a physical plant that "comprises the buildings and structures that house the individual land uses; the utility facilities that serve those buildings and structures with power, light, heat, communications, sewerage, water supply, and drainage; and the street and highway, transit, and transport facilities that provide for the movement of people and goods between two buildings and structures, and that connect the city to the rest of the nation and world" (Bauer 2010, p. 13).

In this chapter we will learn some of the rudiments of urban planning. But first, we will review the evolution of urban planning. This section will unfortunately be short, but it should help us understand why cities took on the form they did, and therefore we will better understand the origin of some of the contemporary problems we face. In particular, we will learn that cities used to be unsanitary and heavily polluted, and the initial push toward suburbanization was an initially desirable response to these problems before effective solid waste management strategies were developed, as we will learn in chapter 9.

Second, and in line with other chapters, we will see some essentials of urban planning. In contrast with the other chapters, these essentials are not laws of physics, but they do set certain limits on what can and cannot be done—hence the term *essentials* as opposed to *fundamentals*, although it is a technical detail. In fact, we will almost take a scientific approach to the topic, in particular by considering the extensive work of Kevin Lynch.

Finally, we will see some desirable traits in city designs. One of the lessons from the second section will be that the way we design urban environments, the logical process that we adopt as human beings, is inherently flawed. This applies particularly to engineers in our quest to optimize systems and omit the fact that systems (and demand) evolve over time. We therefore need some guidelines to steer our mental processes toward more desirable outcomes. Notably, we will learn four conditions for diversity from one of the best urban thinkers of all time: Jane Jacobs.

All the knowledge that we will acquire will be useful when we start to integrate infrastructure systems, to ensure our designs are not only sustainable but capable of providing livable communities.

But first, let us go back in time.

4.1 A Brief History of Urban Planning

In this section we will go through a brief evolution of urban planning. Although we will go back to the Neolithic era, we will put more emphasis on the various movements and practices that emerged in the nineteenth and twentieth centuries, since this is when the urban planning profession flourished. This will be more relevant for us to better understand past practices and ensure we embark on Scenario B.

For each urban planning principle that we will see, we must remember that planners respond to the problems of their time. As Sir Peter Hall, one of the leading scholars on the study of cities, said, "Historical actors do perform in response to the world they find themselves in, and in particular to the problems that they confront in that world" (Hall 2002, p. 4). In fact, cities often change to respond to a problem. In other words, as I wrote in one of my publications: "Cities are often shaped by the challenges they need to face" (Derrible 2017, p. 2). We are now in a similar situation, and we need to shape future cities to respond to the sustainability problems we have been discussing.

4.1.1 The Neolithic Era

The Neolithic era corresponds to the end of the Stone Age, running from roughly 10,000 BCE to 2,000 BCE. This is the era when agriculture and pottery were invented across the world. The *Encyclopedia Britannica* (2015) defines it best:

> [The] Neolithic Period, also called New Stone Age, final stage of cultural evolution or technological development among prehistoric humans. It was characterized by stone tools shaped by polishing or grinding, dependence on domesticated plants or animals, settlement in permanent villages, and the appearance of such crafts as pottery and weaving. The Neolithic followed the Paleolithic Period, or age of chipped-stone tools, and preceded the Bronze Age, or early period of metal tools.

Naturally, what matters for us is the "settlement in permanent villages," which essentially corresponds to the creation of cities. In fact, despite the lack of metal tools (since we were before the Bronze Age), cities managed to form, grow, and thrive. Mumford explains the process incredibly well:

> The great many diverse elements of the community hitherto scattered over a great valley system and occasionally into regions far beyond, were mobilized and packed together under pressure, behind massive walls of the city. Even the gigantic forces of nature were brought under conscious human direction: tens of thousands of men move into action as one machine under centralized command, building irrigation ditches, canals, urban mounds, ziggurats, temples, palaces, pyramids, on a scale hitherto unconceivable. (Mumford 1961, p. 34)

The key term in Mumford's quote is *centralized command*. It was only when we came together, often under one ruler,[5] that we were able to achieve these things. It directly echoes the quote from Aristotle that we saw in chapter 1, about the fact that city-states exist for the sake of living well. Moreover, Mumford's regimenters also represent this central command, as they often acted for a king or a queen.

The Neolithic era essentially represents a social revolution in how we, as human beings, started to live together as a society. As cities grew in size and population, the need to better plan cities rapidly emerged, and the birth of urban planning was imminent.

4.1.2 Ancient Greece and Rome

Hippodamus is often recognized as the father of urban planning. He designed the city of Miletus (now in the south of Turkey) in 470 BCE. The city of Miletus had been captured and destroyed, and when it was taken back, instead of rebuilding it as a small city like it had been, Hippodamus planned the entire city to make it a major commercial and military center (Reader 2004).

Hippodamus's plan for Miletus is shown in figure 4.1. The city could accommodate a population of about 50,000, although it housed about 100,000 people 500 years later. Within the fortifications, the city followed a typical grid plan. The main activities of the city were located in the geographical center, which includes the agora, or main plaza, in Greek cities for public gatherings. The grid orientation was designed to conveniently follow the geography of the peninsula and to take advantage of the wind in the summer to cool down the city, and yet it also favored exposure to the sun to keep warm in the winter. Utilizing the wind in the best way possible was also important for health issues, to avoid the spread of disease. At the same time, the grid pattern defined particular social neighborhoods, therefore linking engineering with the humanities, as mentioned previously.

The Romans evolved at roughly the same time but were much less developed initially. First, for what were agrarian people, the foundational element of planning in Rome was the law of property. Having ownership of a piece of land was considered extremely important, and the authorities made sure to keep a record of land ownership and punish individuals who trespassed boundaries. All land in a city was zoned based on its type that included:

- Res communis (owned by all)
- Res publicae (owned by the government)
- Res nullius (owned by no one, abandoned land that could be claimed)
- Res privatae (owned by individuals)
- Res universitatis (owned by a public group; e.g., university, hospital)
- Res divini juris (owned by the gods)

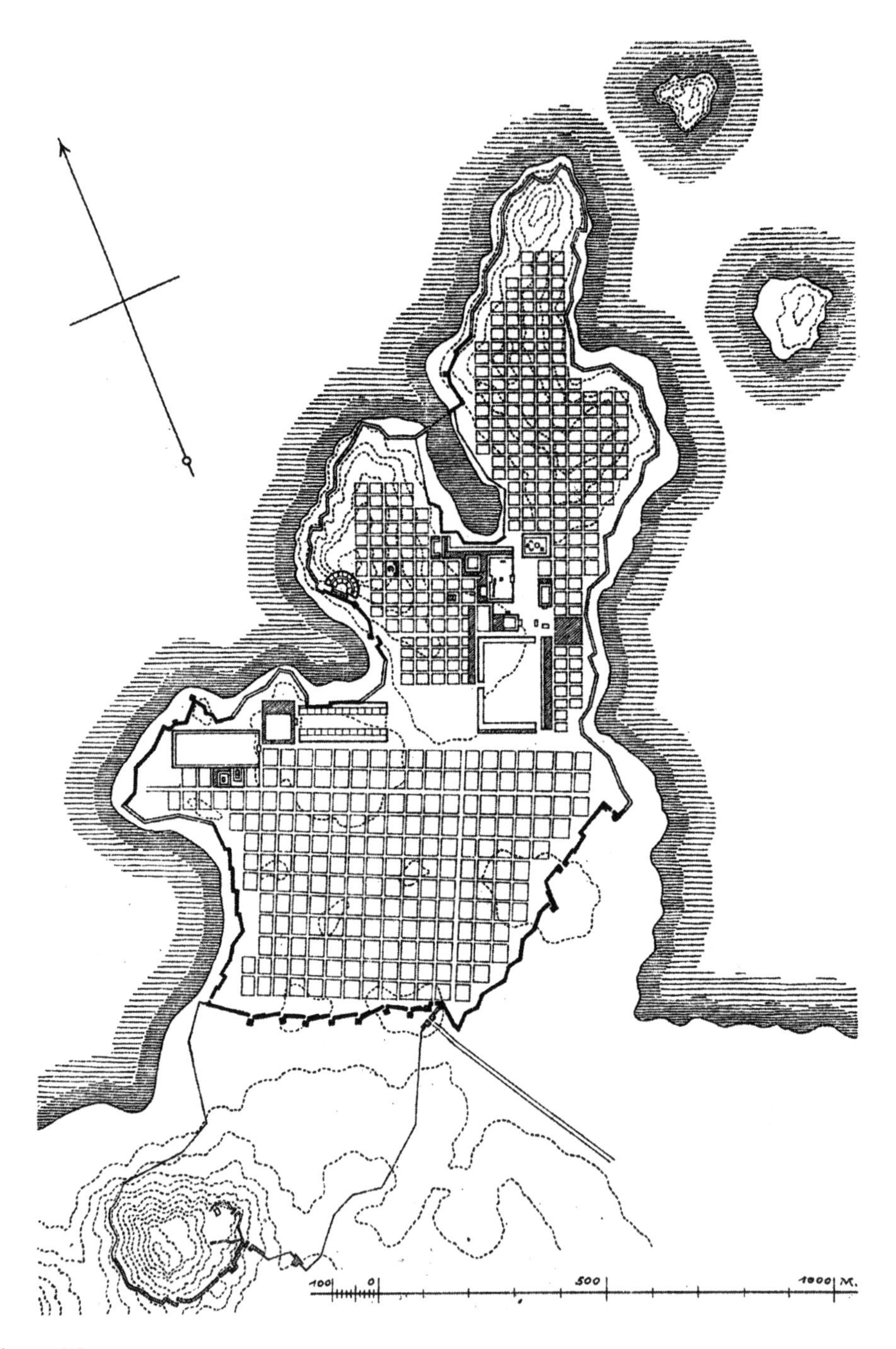

Figure 4.1
The grid structure of the street network of Miletus.

Keeping a strict cadastral survey and obeying the law of property helped significantly as the Roman kingdom turned into a republic and then into an empire. Eventually, land planning was somewhat codified by Vitruvius in 15 BCE in his famous ten books on architecture (Pollio 1999).

Roman cities had two main thoroughfares, along the west–east and north–south axes. These two thoroughfares were paved, and they received most of the traffic. A grid matrix with smaller streets was then laid out around these two axes. Here again, the orientation of the grid depended on the main winds. Moreover, akin to the Greek agora, the center of larger cities included a forum for public gatherings, especially when the cities were inland. In the case of port cities, public venues were generally located closer to the harbors.

For large cities, local wells could not provide enough freshwater, and the Romans therefore had to rapidly develop their water resources engineering skills. In fact, they largely pioneered water resources engineering, described as Water 1.0 by Sedlak (2014). The residents of the city of Siena, for instance, located on top of a hill, had to walk down the hill to get water from a well and walk back up. The residents cleverly designed a system that used gravity to create enough pressure to bring the water up the hill naturally. Water fountains could then be stationed at several locations in a city. In addition, the Romans pioneered underground wastewater systems. The best example is Cloaca Maxima, which translates to "Greatest Sewer" and is located in Rome. It was first built in 6 BCE as an open drain and likely covered around 3 BCE. It was further expanded in 1 CE. Considering the system is still in use, it is arguably one of the most profitable urban infrastructure projects ever built.

Cities also hosted major monuments that became the foci of these cities. In particular, Rome and Constantinople (now Istanbul) used their engineering ingenuity to build monuments that dwarfed absolutely all other buildings in a city. The current structure of Hagia Sophia (or Ayasofya) in Istanbul, for instance, was built in 537 CE after less than six years of construction, and it still has one of the largest domes currently standing. How immense it must have been compared to the rest of the city at the time!

In the end the Greek and Roman cities were designed for pedestrians.[6] The grid system served well to guide winds while making it easy to travel through a city. The grid was accompanied by public squares with shops and water wells to avoid unnecessary walking. Despite the many advantages of the grid, however, it is not ideal for pedestrians in large cities, as we will see later in this section.

4.1.3 Medieval Towns and the Renaissance

After the fall of the Roman Empire, towns in Europe remained relatively small. The primary purpose of formal planning in the Middle Ages was for military purposes. The small cities therefore tended to be located on hills and were surrounded by fortifications. Figure 4.2a shows a map of the city of Charleville (France) in 1625. We can see that the fortifications are fairly intricate. The street patterns kept the common grid shape, and at the center of the city lay the main square.

In the Middle Ages, however, most cities were not "formally" planned. Instead, they grew "organically," as illustrated by the city of Pérouges (France) in figure 4.2b.[7] Organic planning is in complete contrast with formal planning. The objectives of organic planning are simply different. Instead of being straight, streets followed the paths made by nature itself; for example, if a tree was in the way, it was easier to go around it than to remove it. This was not a problem since streets were not adapted for wheeled traffic. One decision therefore had an impact on another decision that then had an impact on another decision, and so on and so forth, together creating an amazing cohesiveness in the city. To some extent, organic planning is an algorithm. A specific set of rules was followed each time without necessarily thinking of the long-term implications. This is also a great example of a "bottom up" approach.

Here again, cities were designed for easy accessibility, and everything was located within walking distance. The emphasis was not on throughput but on being able to access a destination.

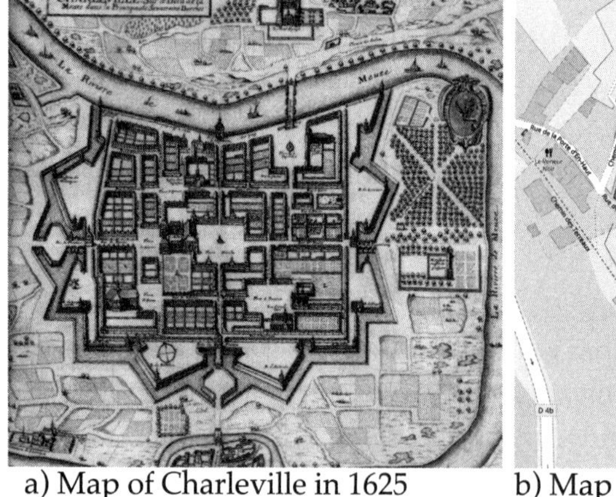

a) Map of Charleville in 1625

b) Map of Pérouges (France)
(© OpenStreetMap contributors)

Figure 4.2
Medieval towns.

Although this did not happen in Europe or during the Middle Ages, we can also see remnants of organic planning in New York City. The world-famous Broadway Avenue is oddly diagonal simply because it originated as a natural path followed by Native Americans.[8]

Once again, Mumford defines it best as follows:

> Organic planning does not begin with a preconceived goal: it moves from need to need, from opportunity to opportunity, in a series of adaptations that themselves become increasingly coherent and purposeful, so that they generate a complex, final design, hardly less unified than a pre-formed geometric pattern....
>
> Those who dismiss organic plans as unworthy of the name of plan confuse mere formalism and regularity with purposefulness, and irregularity with intellectual confusion or technical incompetence. The towns of the Middle Ages confute the formalistic illusion. For all their variety, they embody a universal pattern; and their very departures and irregularities are usually not merely sound, but often subtle, in their blending of practical need and esthetic insight. (Mumford 1961, p. 302)

This does not mean that cities could not grow. For example, the city of Bruges in Belgium became a significant economic power with 200,000 people in 1500, but none could rival the importance that Rome had acquired a millennium before.

There has been a lot of misconception about organic planning over the years. Now, many argue that we should include more organic planning in neighborhood plans, to make cities less "mechanical" and more humane. This statement is absolutely valid, but we also have to find a way to combine both practices properly, and this is where infrastructure integration can play an important role.

By the end of the Middle Ages, as kings became more powerful and could assure the security of the land (for their lords), cities began to expand. During the Renaissance, cities became very active culturally. Monuments and palaces began to replace fortified buildings. These structures then began to have the same impact that monuments previously had during the Roman times. The Cathedral of Florence with its famous Duomo offers a great example.

This was the start of a new era for cities, constituting the start of *baroque* planning.

4.1.4 Baroque Planning, the Expansion of Cities, and the Pedshed

With relative peace and order, accompanied by a rich cultural heritage from the Renaissance, cities started to expand and change significantly. Sometimes these changes were "aided" by major disasters, such as the Great Fire of London in 1666, which forced the city to be rebuilt—although this time building codes required the wood structures to be replaced with masonry.

Figure 4.3
Plan for Washington, DC, by L'Enfant in 1791.

The spread of monuments as foci of neighborhoods also dominated the new urban planning practice. In line with these changes, street plans started to shift from the common grid with small streets to radial patterns with large avenues. One of the great examples is Washington, DC, with its original design by L'Enfant in 1791 (figure 4.3). These radial patterns are particularly conducive to walking, offering a straight line to a point. In fact, these designs are sometimes referred to as *pedestrian sheds*, or *pedsheds*, as a direct analogy to watersheds. They are much more manageable for large cities in contrast to the pure grid system. Moreover, these pedsheds make it easier for pedestrians to navigate in cities.

Pedsheds also offer great intermodal locations—that is, locations where people change transport mode. For instance, one may walk to a pedshed to hop on a horsecar or, in a more contemporary note, to hop on a bus or to enter a metro/subway station.

Paris is another great example of a city with many pedsheds. Baron Haussmann largely rebuilt Paris in the second half of the nineteenth century, as commanded by Napoleon III, who wanted a modern city akin to London after the Great Fire of 1666. The modernization of Paris enabled the construction of an extensive sewer system and the planning of large avenues, such as the famous Rue de Rivoli that runs by many of the most famous monuments of Paris, including the Louvre. These large avenues and their sewer tunnels also greatly helped to clean up the city.[9] Several parks were also built throughout the city to provide green space.

Barcelona also went through substantial changes and adopted a similar street layout with pedsheds, designed by the Spanish urban planner Ildefons Cerdà. Cerdà developed the concept of *vialidad* that possesses a strict hierarchy, from major foci to small intersections.

These practices represent applications of baroque planning. In direct contrast to organic planning, baroque planning seeks order at a larger scale, with well-defined and straight streets, uniform buildings, and strategically located administrative offices and monuments. This all sounds good, right? This is actually not necessarily the case. Often, these systems were built not to satisfy the actual needs of the city but to fulfill an overarching vision of one individual. Efficiency also started to overplay basic human needs. This means that cities were not built for their residents anymore. Naturally, most of us would agree that many of these changes were necessary and beneficial. A fundamental element of planning was lost, however, and this is the element that we need to add back to our practice as urban engineers.

4.1.5 The City Beautiful, the Garden City, and the Radiant City

By the mid-nineteenth century, and as Scenario B, the Western world saw the onset of the Industrial Revolution thanks to the birth of the steam engine. Steam engines, however, run on coal, and coal is extremely polluting. Cities around the Western world therefore became extremely polluted.[10] More importantly, as walking was the main mode of travel, workers lived close to manufacturing facilities in areas that we can qualify as slums. This combination of pollution and high social inequality became extremely problematic and called for a new city form.

One response to crowded tenements was the rise of a new type of urban planning, called the *City Beautiful* movement, that originated in Chicago, specifically during the World's Columbian Exposition of 1893. Inspired by the grand monuments of European cities (e.g., from baroque planning, as discussed earlier), the premise of the movement was to build large structures that would attract businesses and improve the economy.

The model was adopted in many places, initially in North America, and then even back in Europe.

Another response was to let people live away from industrial areas in neighborhoods supposedly closer to nature. This practice, developed in England in 1898 by Ebenezer Howard, is called the *Garden City* movement. The premise was to separate the dirty industrial activities from residential areas. Figure 4.4 shows the initial plan from Howard. We can see that the walkable grid pattern was replaced by completely secluded neighborhoods. The title on the map of figure 4.4 reads "Group of Slumless Smokeless Cities." While the plans certainly solve the problem of having people living too close to industry and to some extent put the human back at the center of the planning practice, they brought significant problems in terms of access to industries. The same steam engine that benefited the Industrial Revolution now had to be used to run trains to take people to their jobs, further polluting cities. To do Howard justice, it is worth mentioning, however, that his initial idea was to develop livable communities next to large urban centers (like small villages), which is different from suburbanization as we know it. In the end, suburbanization was in part a response to certain economic incentives on both the developer and homeowner sides.

What the City Beautiful and the Garden City movements have in common is the separation of residential and industrial land uses, which led to the modern zoning practice. As a result, many cities started to assign a zone type to every single plot of land, from residential and commercial to industrial and agricultural. By the twenty-first century, zoning practices have actually become somewhat of a problem, as we will see, since cities seldom have large polluting industries. Moreover, zoning practices in certain neighborhoods have been exploited to purposefully exclude certain populations.[11] Zoning practices must therefore be rethought and instead a mixed use of the land should be encouraged.

One architect pushed this idea of separating people from the noise of the city to the extreme. In 1932, with his Broadacre city, Frank Lloyd Wright wanted everyone to have a large house, a large plot of land, and a car, which could be caricatured as a "suburbia on steroids." Although Wright is recognized as a brilliant architect, it does not take long to realize that this would be a terrible idea for a city. After all, we live in cities to connect with people, and we often go somewhere because there is a lot of human activity.[12] In fact, in many of Wright's drawings we see groups of people together, so why separate them in the first place? Moreover, to travel around the city, Wright imagined flying machines that were a little like helicopters. Of course, he only added a few of these machines in his drawings, whereas the reality would have produced massive congestion.

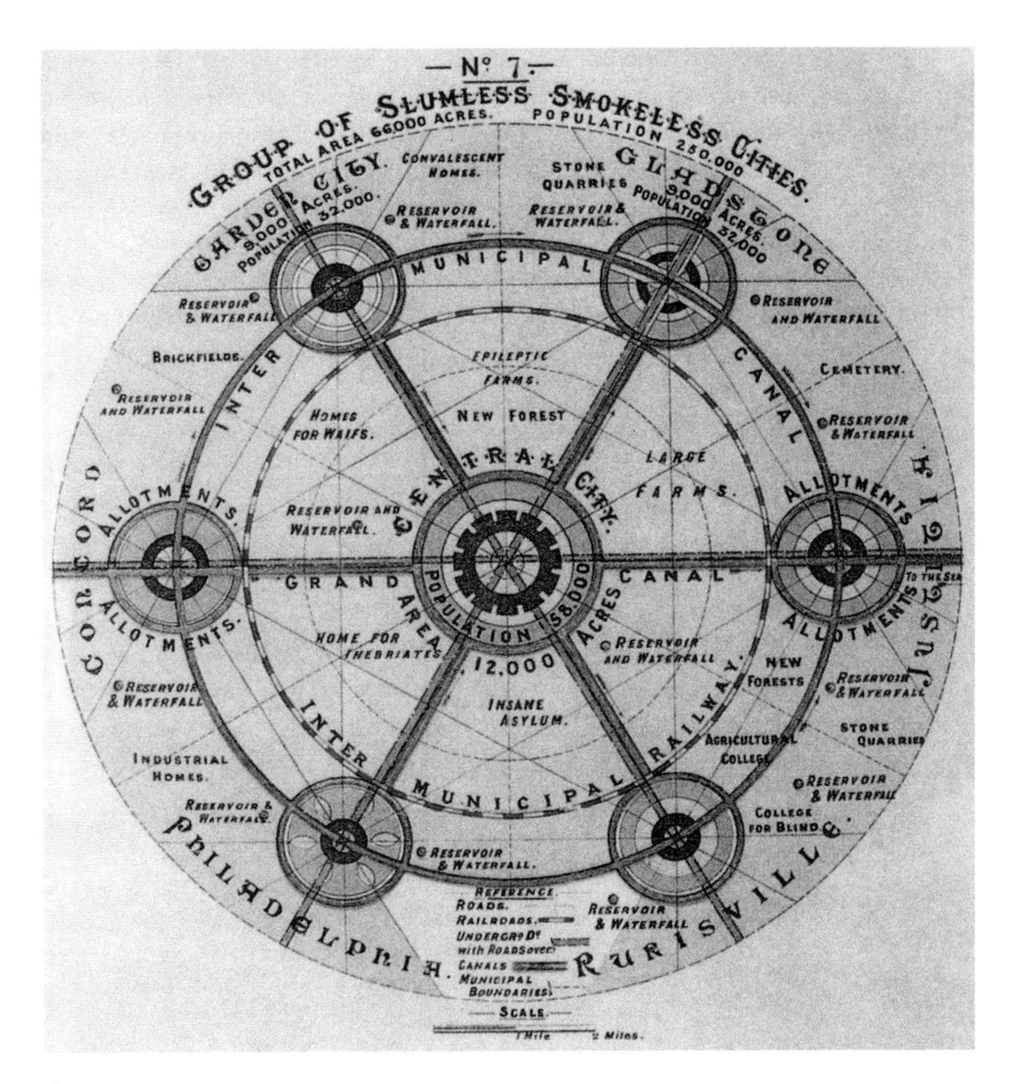

Figure 4.4
The 1902 Garden City concept.

More importantly for us, a critical paradigm shift occurred in planning at that time. Essentially, the "pedestrian" was replaced by the "mobile" as the center of attention, and this mobile could now use new modes of transport. This meant that people could live farther away and take the train, the streetcar, and later the private car to go to work. At the beginning, train and streetcar lines were built and operated by private companies that would sometimes also finance mortgages for people to live farther away, therefore requiring them to take the streetcar every day. Interestingly, we now see the private automobile as the cause of urban sprawl, but it really started much earlier because of streetcars. The "walking city" had lost its place to the "tracked city" as defined by Schaeffer and Sclar (1975).

Following these movements, the deindustrialization of urban centers, and the rise of the private automobile, another type of urban form came about called the *Radiant City*.[13] Created and championed by the famous architect Le Corbusier, the premise of the Radiant City is to host all daily activities in skyscrapers and let the rest of the land be green space. Unlike Wright's vision, here, people would be concentrated in apartments within one tower. Moreover, these towers would also have schools, grocery stores, city halls, and so on. Figure 4.5 shows examples of plans laid out by Le Corbusier; figure 4.5a shows a sketch of the Radiant City that comes directly from Le Corbusier's work, and figure 4.5b shows a plan for Paris.

This movement had a substantial influence in the mid-twentieth century. One of the most famous adherents of this urban planning practice was Robert Moses, who had

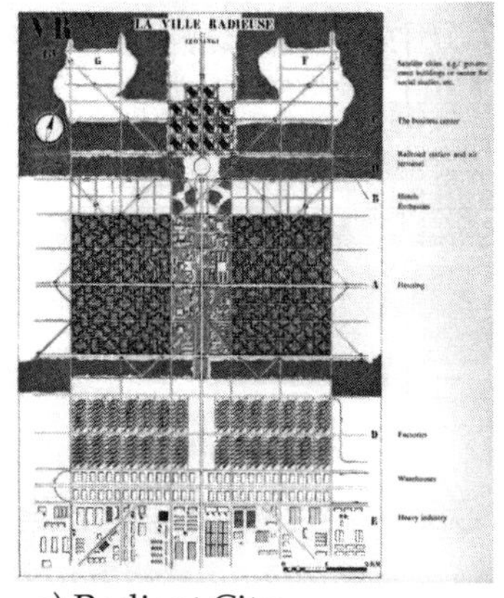

a) Radiant City

b) Plan for Paris

Figure 4.5
The Radiant City by Le Corbusier.

a great impact on New York City—notably, by building expressways along the Hudson and East Rivers, thus depriving pedestrians of the riverfronts. Moreover, the capital of Brazil, Brasilia, was built in 1960 following these principles. Many, if not most, projects that followed this philosophy ended up as massive failures because they tend to put "efficiency" at the center of focus, without considering how humans perceive cities, as we will discuss in the next section.

While the Radiant City philosophy was initially popular, it most often contributed to high crime rates and social unrest, and it completely lost its appeal in the 1970s and 1980s. Indeed, this type of planning was often adopted in low-income neighborhoods, and the poorly lit green spaces in the evening quickly turned into ideal locations for crime.

One of the most notable failures of a Radiant City–type project is the Pruitt-Igoe project in Saint Louis (figure 4.6), designed by Minoru Yamasaki. At the design phase, the project was highly supported by local stakeholders as well as by many architects and planners. Completed in 1956, the neighborhood rapidly became a hot spot for crime. The number of vacant apartments rose rapidly and many buildings closed. In the mid-1970s, barely twenty years after its construction, the whole complex was demolished.

Figure 4.6
The Pruitt-Igoe Housing Project in Saint Louis.
Source: U.S. Geological Survey.

The dramatic consequences of the decisions to follow Le Corbusier's plans remain present in many parts of the world. For example, in the close suburbs of Paris, as well as in most large cities in France, many housing projects similar to the one in Saint Louis were built, and many bear similar signs of failures.

In contrast to the Radiant City, the City Beautiful movement lost its appeal even earlier, partly due to the Great Depression. The Garden City, however, kept thriving, especially in the *City of Highways*.

4.1.6 Greenbelt Towns and the City of Highways

After the Second World War, millions of soldiers came back to the United States and benefited from the G.I. Bill. The bill allowed the soldiers to get an education and also to partly finance the purchase of property. With the advent of the automobile, the American suburb took on a new meaning from its streetcar era, and cities across the United States grew at an incredible rate. All land at the periphery of cities started to host greenbelt towns, consisting of rows of houses with front yards and backyards.

To be able to travel from these suburbs to downtown jobs, massive expressways were built, which resulted in a further flight to the periphery and increased congestion.[14] Figure 4.7 shows the spread of the American city from 1890 to 2000. In the nineteenth century, most travel was done by walking or by horsecar. Initial suburbanization then followed with streetcars in the early twentieth century, which was quickly succeeded by the recreational auto up to 1945. The auto then became the dominant mode of transport, and cities started to further grow around expressways, giving rise to what Schaeffer and Sclar (1975) describe as the "rubber city."

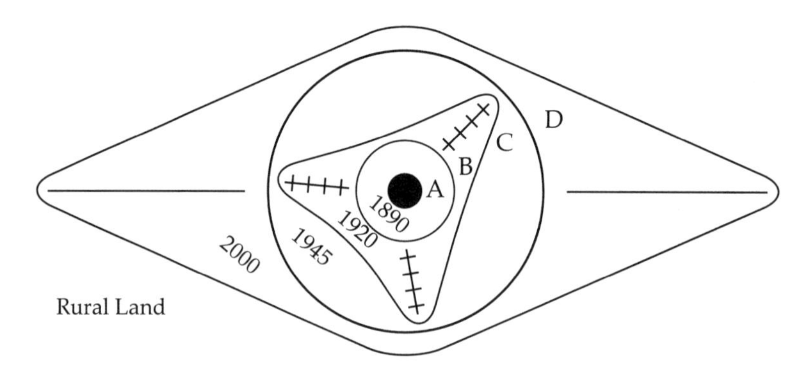

Figure 4.7
Evolution of intraurban transport. *A*, walking and horsecar; *B*, electric streetcar; *C*, recreational auto; *D*, freeway. Adapted from Muller (2004).

A possible exception is the Stockholm region in Sweden. Instead of letting the city grow, the government created satellite towns a little farther away. These towns were supposed to have all the amenities needed for any city. Traveling between these satellite towns and the Stockholm city center could then be achieved with the then newly built Stockholm T-Bana (i.e., the metro), which really acts more like a regional rail system. As is often the case, residents rarely live the way planners anticipated, and Stockholm rapidly suffered from severe congestion, forcing it to establish traffic congestion pricing. The book *The Transit Metropolis: A Global Inquiry* by Cervero (1998) offers an excellent resource to learn more about the evolution of transit practices around the world.

Overall, this Garden City–inspired movement dominated the urban planning practice in the second half of the twentieth century. The street pattern also changed. Neither a grid nor a radial pattern, streets in the suburbs are most often curvilinear, supposedly emulating countryside paths, and making roads unnecessarily long for walking.

This massive suburbanization has been named *urban sprawl*, and it came at a great cost, adding severe traffic congestion and requiring enormous amounts of energy; it is therefore partly responsible for the current environmental crisis. In fact, there is now overwhelming evidence that urban sprawl has significant negative impacts, not only environmentally but also socially and economically. What is worrying is that this practice is exporting itself well, even to places where conditions are simply not adequate for this type of land use. Figure 4.8 shows a typical American suburb—except that it is not in the United States but in the middle of a desert, in Dubai. Similar types of neighborhoods can be found in China as well, often destined for high-income populations in search of emulating the American lifestyle.

On the positive side, the trend has been changing in the United States since the beginning of the 2000s. Many city centers have experienced an increase in population, and more people have decided to trade the autocentric neighborhoods for those that offer higher accessibility. Many suburbs, also, are getting denser, and they try to emulate a number of the features of city centers. The design focus has largely shifted from the "mobile" to the "accessor," thus emphasizing accessibility. Urban planning practices not only increase accessibility by providing necessary amenities nearby, for the "pedestrian," but also allow longer trips to be made with all the amenities and opportunities available. We will more rigorously quantify accessibility in chapter 7. One of the pioneers of this new urban planning way of thinking was Jane Jacobs, and as we will see in section 4.3, she has come up with four conditions that can serve as great guidelines.

In addition, the concepts of *Smart Growth* and *New Urbanism* have emerged as particularly relevant to improve livability in cities. The goal is to put humans within their societal context back at the center of planning. It is about designing human-scale and

Figure 4.8
American urban sprawl in Dubai.
Source: Google Earth.

walkable neighborhoods where people can both live and work. It is about offering multiple modes of transport to travel to a destination. In America the car has long been the symbol of freedom, which has led us to become prisoners of the car. As a response, the ability not to take a car has now become a stronger symbol of freedom. Again, the focus has shifted to the accessor, and we will learn more about it in section 4.3. For a quick recap of the different urban planning stages we have covered in this section, have a look at the conclusion, which shows a drawing from a student who took a course based on this book.

As one last note in this section, beyond urban planners, we have not talked about the role of developers in shaping cities, and their role is incredibly important since, in most countries, it is they who seek the capital to actually build buildings. While urban planners work mostly for public agencies and develop plans that feed into zoning regulations, real estate developers finance the construction of buildings. Developers are therefore the ones making the decisions to build suburban lots, condominium towers, or mixed residential and commercial buildings. In some instances (especially in city centers), developers work both with the public sector and with the local communities.

Moreover, developers often make decisions based on recommendations from architects and engineers. Overall, many actors are involved in the shaping of cities, including us, urban engineers, and we need to do a great job at it.

Let us now move on and learn some essentials of urban planning.

4.2 Essentials of Urban Planning

Although urban planning is not a science, we can list some essentials that we should keep in mind when designing urban systems. There are two types of essentials that we can discuss.

First, we have the essential responsibilities of an urban planner. These essentials include the definition of the central role of the planner and the legal limitations, as well as the need for planners to consider all aspects of cities, including housing, schools, fire districts, police stations, zoning, and so on. Put simply, these essentials include what a Planning 101 course would include. We will not discuss these essentials, however. Instead, we will pay closer attention to the second type of essentials.

Second, we have the biological limitations, which are closer to our limitations as human beings. How does a person perceive/see/sense a city? Is our logical process flawed, and are there any processes that we should be careful about when designing? As Dehaene (1999) puts it in his excellent book *The Number Sense: How the Mind Creates Mathematics*: "The brain is not a logical, universal, and optimal machine" (p. 252). This logical conundrum is actually a big problem, which may be at the origin of our inability to better integrate urban infrastructure systems.

Understanding how cities are perceived should be paramount for engineers. In the architecture profession, architects design rooms and buildings while taking into account how humans use the space. Why can't engineers do the same? Can we design cities and engineering systems for how people perceive and live in cities?

In this section we will first discuss this issue of logic and design by recalling the seminal work of Christopher Alexander. We will then discuss not how cities are designed but how they are perceived, learning from Kevin Lynch's *The Image of a City*. Finally, we will quickly consider one lesson from Jane Jacobs about the importance of having eyes on the street.

4.2.1 A City Is Not a Tree

Here, we will actually spend quite some time on one single article published in 1965 by the architect and mathematician Christopher Alexander. The article, titled "A City Is Not a Tree" (Alexander 1965), was published in *Architectural Forum* magazine, and

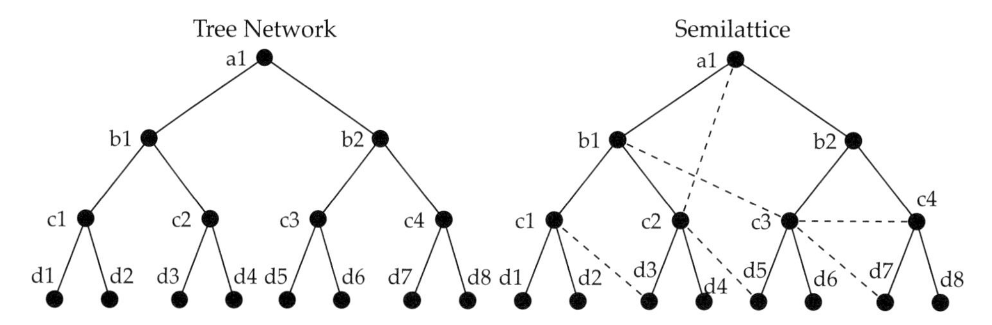

Figure 4.9
Examples of tree and semilattice networks using a network representation.

the lessons of article are incredibly important for urban engineers. Namely, the article firmly criticized the designs of the Radiant City and the Garden City since they separated urban elements as opposed to integrating them—does that ring a bell?

By *tree*, Alexander does not refer to the plant but to a mathematical structure. More particularly, he refers to a type of network: tree networks. Consider a network G composed of a set of nodes N and links L, such that $G=\{N, L\}$. Figure 4.9 (*left*) shows an example of a tree network where all nodes are labeled. Essentially, it is a network with no "loops" (i.e., overlapping links), and it necessarily contains $L-1$ nodes. In contrast to tree networks, Alexander defines semilattices. These semilattices are also networks, except that they can have loops and overlapping links, as shown in figure 4.9 (*right*); these "extra" links are shown as dotted lines. What Alexander argues is that human beings think and design in terms of tree networks, while cities are semilattices.

To initially explain the concept, Alexander gives the example of a newsstand located on a sidewalk close to an intersection. When the traffic light turns red, people are more prone to reading the headlines of the newspapers. Despite being individual elements, the traffic light and the newsstand are interconnected.

Going back to figure 4.9, let us imagine that the nodes at the bottom (i.e., d-level) are houses and that the next level up (i.e., c-level) represents grocery stores. The tree network tells us that we design cities so that every house can only go to one grocery store. In contrast, the semilattice tells us that some houses may actually have access to several grocery stores. To take a practical example, at the time of this writing I lived in a neighborhood in Chicago called Lakeview, but I bought my groceries in another neighborhood called Lincoln Park. This does not mean that Lakeview did not have grocery stores—because it did—but considering all my options are at equal distances, I preferred the grocery store in Lincoln Park. In fact, the dividing line between Lakeview and Lincoln Park was artificial and purely used for administrative purposes. Human beings, however, systematically "containerize" activities within space, and had one single individual been in charge of deciding exactly how many grocery stores should be planned, the store I used would have likely not even existed.

We can redraw figure 4.9 using Venn diagrams as opposed to networks. The tree network in figure 4.10 looks much more orderly. As human beings, we like symmetry because we can make more sense of systems that show some symmetry, and unsurprisingly, we put a lot of symmetry in our urban designs.

Cities that have evolved over time, however, are not like that. Look at a map of Chicago, Paris, Cairo, or Tokyo from above. Can you find the symmetry? All the seeming "disorder" that we see on the semilattice in figure 4.10 that makes the diagram look much more confusing and inefficient is actually a much better representation of a city. In fact, the symmetry from master plans may be more pleasing to see from a top-down view, but walk in these neighborhoods and you can easily get lost since everything looks the same. Back to figure 4.10, Alexander is quick to point out that this overlap does not actually mean the city is "disorderly." In fact, this overlap that seems confusing to us actually possesses some order that we just cannot see. The structure is simply more complex and subtle. The Science of Cities tries to capture this order from the apparent disorder, and we will touch briefly upon it in chapter 11.

A good network analogy is the structure of social networks. In a group of friends, these types of overlaps of one or many people belonging to several groups are common. Even in companies that follow strict tree organograms (i.e., strict hierarchies), informal relationships form that break this tree structure.

While most famous urban planners of the twentieth century argued that the total separation of city elements was preferable, with pedestrians on one side and cars on the

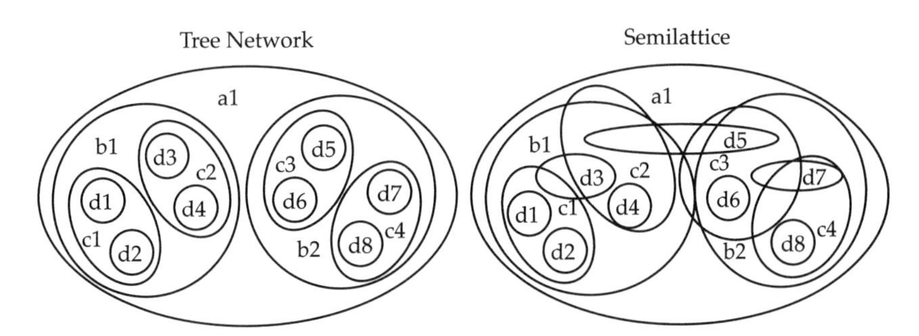

Figure 4.10
Examples of tree and semilattice networks using a Venn representation.

other, and houses in one area and businesses in another area, this is actually not how cities evolve organically (back to our definition from section 4.1).

Does this separation of elements remind you of anything? How about infrastructure systems? We know that infrastructure systems are interdependent, and yet we do not design them together. This is obviously partly due to the fact that we, as engineers, are too specialized. In fact, I wrote a scientific publication on the parallels between infrastructure systems planning and the ideas of Christopher Alexander. Perhaps unsurprisingly, the title of the publication is "Urban Infrastructure Is Not a Tree" (Derrible 2016), and we will discuss it in chapter 10.

Back to Alexander, he goes on to explain how this process of forming trees occurs in our brain. He gives the simple example of remembering four objects: an orange, a watermelon, an American football, and a tennis ball. How can we keep them in mind? We will necessarily group them to make it easier, either based on function (orange and watermelon as fruits and football and tennis ball as sports balls) or on shape (orange and tennis ball as round and watermelon and football as egg-shaped), as illustrated in figure 4.11, *left*. Although the structure of a semilattice here is evident, we simply do not think in terms of semilattices that look much more complex (figure 4.11, *right*).

The main lesson is that we may very well be incapable of designing semilattices, and this is obviously a problem. What designs do we choose if we cannot design successful cities? For starters, we should not design entire cities. Although we need master plans to track the evolution of a city and provide major goals and guidelines, the actual design should be done in a much more piecemeal manner, and most importantly, it should allow room for improvement and natural evolution. That being said, there are desirable traits that we can emulate, and this will be the topic of section 4.3.

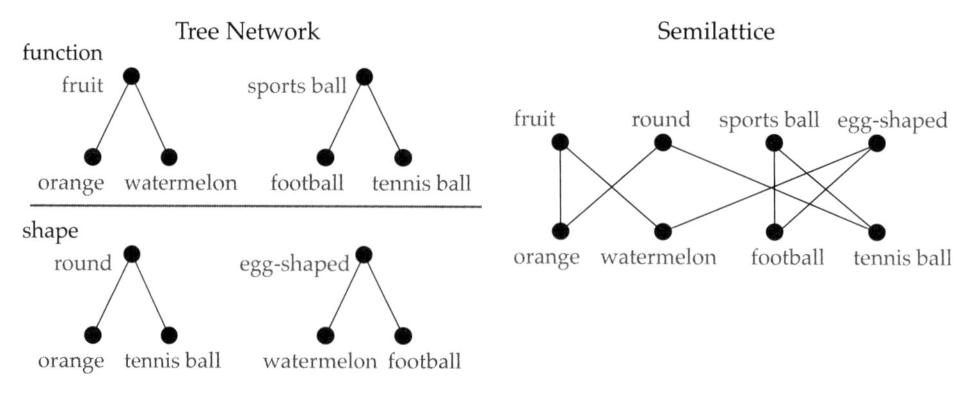

Figure 4.11
Mental grouping of items.

For now, this is all we are going to discuss about human beings producing designs, and we will move on to learn about how human beings perceive cities.

4.2.2 The Image of the City

The Image of the City is the title of a seminal book written by Kevin Lynch in 1960. In this book, Lynch compared how people from Boston, Jersey City, and Los Angeles perceived their city, based on personal interviews. Right at the beginning of the book, echoing what we have just discussed, Lynch wrote: "Not only is the city an object which is perceived (and perhaps enjoyed) by millions of people of widely diverse class and character, but it is the product of many builders who are constantly modifying the structure for reasons of their own" (Lynch 1960, p. 2). Cities are not the products of one single person; instead, they consist of layers after layers of modifications, giving them their semilattice structures, from which we establish our own order.

Here, we will not discuss the three cities Lynch used. Instead, we will focus on the main elements that stood out about how cities are perceived, the apparent order within the inherently complex structure of cities.

Before we go on, I would encourage you to take a moment and think about the city where you live. Draw a freehand map—that is, an aerial view—of your city on a blank sheet of paper. Do not look at any maps, even if you forget street names. What matters is your perception of your city, not the actual city.

Take your time. Make sure to go through the exercise.

Done? Let us analyze your map.

Drawing our own map of our city to illustrate how we perceive/remember a city is exactly what Lynch asked from his participants. From the sixty people he interviewed, he identified the emergence of five elements that we perceive in cities, which we will briefly discuss:

- Paths: streets, sidewalks, and other corridors that we travel on
- Edges: physical boundaries, whether geographic (e.g., lake) or man-made (e.g., border of a park)
- Districts: large sections of cities that are perceived as belonging together
- Nodes: points and strategic spots, more related to transport, to enter a city
- Landmarks: easily identifiable objects, such as a tall building

Paths are exactly what they sound like. Paths are the streets and thoroughfares that we either use daily or that we know well. In Chicago, for instance, most of us will have the expressways in mind, along with our own streets that we use daily. We will also remember the streets that have a lot of activity, like Broadway Avenue in New York City and Market Street in San Francisco. Paths also include transit paths, such as metro lines, as well as bike lanes or pedestrian paths. We can never remember all the streets that we use, but chances are you will remember several. In cities with streets that go off at angles or that turn, like in older European cities, we will tend to exaggerate by how much they go off since this is an important feature for us.

Edges are the general boundaries of a city. In fact, for the map-drawing exercise, chances are you started drawing the city by sketching the city boundaries, especially in the presence of large geographic constraints like lakes, rivers, and mountains. Boundaries also capture abrupt changes in land use, whether because of a park or the presence of a railroad track, for instance. Edges also include highways, and most people drawing Paris will certainly start by drawing the *périphérique* (ring road delimiting the city of Paris), the river Seine, and the two islands in the river.

Districts are the neighborhoods that are not necessarily physically separated but that express a different character. As mentioned in the previous section, the boundary between Lakeview and Lincoln Park in Chicago was not distinct, and I therefore did not consider these two neighborhoods as separate; instead, I would put the Lakeview boundary more north, halfway through Lakeview. In contrast, the Loop area (Chicago's central business district) has a decidedly different character from its surroundings, and I would consider it to be a district. In fact, districts sometimes emerge because of the main activity within them, and although we may give the same name to the district as the neighborhood, they tend to be quite different. Moreover, districts sometimes follow paths; think of the main commercial corridors that go through several neighborhoods, and yet we may put them in one single district (e.g., Broadway Avenue in New York City, Milwaukee Avenue in Chicago, or Yonge Street in Toronto all go through several neighborhoods).

Nodes are localized points that represent entrances or exits in a city or a district. A large and noticeable intersection, a train station, or the end of an expressway can be a node. Nodes can also have a linear shape, like a delimiting point, very much like an edge. In fact, a street can be both an edge and a node, but the street edge may be longer than the street node. Here, the emphasis is on transport, and we therefore need to think about personal, "symbolic" entryways. In Seoul, Seoul Station is the major train station, and it therefore acts as a node for the thousands of people arriving and departing from the station every day. In Stockholm, the central station of the T-Bana can be seen as a node since it acts as the main hub for residents of the satellite towns. In New York City, Times Square can be seen as a node, not only for auto traffic but also for pedestrians and transit users. In London, Piccadilly Circus is a typical node; in fact, there is a small fountain, which I have never seen functioning in the square, that is one of the preferred meeting points in the entire city.

Landmarks are exactly what they sound like. They often help us orient ourselves in a city—by comparing our location to a certain tower, for instance. They can be just about anything from fountains and statues to skyscrapers and entire small areas. Go back to your drawing and see what landmarks you chose to draw. Some of the most famous buildings in the world are landmarks, from the Eiffel Tower in Paris, the Hagia Sophia in Istanbul, and the Marina Bay Sands in Singapore to the Fernsehturm (television tower) in Berlin or the Tokyo Tower in Tokyo. These landmarks also help give character to an entire district.

Overall, these five elements can be common to many people living in the same city, but they are also highly subjective.[15] Some people will think of certain paths and nodes because they use them often, but other people will think of other paths and nodes. Similarly, we all have certain landmarks that are not well known but with which we have developed a certain emotional connection. Moreover, singular physical locations and structures can belong to multiple elements. A park can very well be an edge, as well as a landmark, as well as a path. This is simply the way that we perceive cities, putting our own order and simple hierarchy on a complex system.

This is the end of our section on how we perceive cities. Ultimately, it is an incredibly intimate experience, and a designer cannot possibly design a city so that everyone perceives it the same way. This is why it is important to let a city grow and form, gain character, and allow thousands of people to perceive it in thousands of different ways.

We must discuss one more essential before moving on to desirable traits, and it deals with safety.

4.2.3 Eyes on the Street

Most of the urban planners who we have talked about in the evolution of urban planning were highly intellectual individuals who tried to place the *human*, as a rational person, at the center of the urban context. What we just discussed, however, is that we are all different, and we all have different relationships with our cities. Moreover, we live in a society, and it is the *society*, not the *human*, that should be at the center of our designs. In the end, the best ideas in urban planning did not come from one of the famous and influential urban planners but from someone with no training at all in urban planning: Jane Jacobs.

Jane Jacobs became an extraordinary figure in urban planning. Born in 1916, she studied at Columbia University, taking courses, notably in economics and political science. She was later hired as a writer for an architecture magazine, giving her access to the works of the most influential urban planners who we have discussed. Jacobs rapidly became a staunch critic of the massive urban planning projects that adopted a top-down approach (such as Le Corbusier). Instead, she placed the community first, advocating for bottom-up approaches (and she greatly inspired Smart Growth and New Urbanism). In 1961 she published one of the most famous books on urban planning, titled *The Death and Life of Great American Cities* and quoted at the beginning of this chapter (Jacobs 1961). In the field of urban planning, this book has remained one of the most influential (or even the most influential) guidebooks ever written.[16] Although we will discuss the lessons of the book later, one of the main messages it carries is the importance of having eyes on the street.

Eyes on the street is a basic societal phenomenon that makes the places we visit safer, and often more lively and enjoyable, when more people are there. Jacobs gives a great example of her own neighborhood, Greenwich Village in New York City, composed of rows of medium-rise buildings with commercial activity on the first floor. The butcher, barber, postal worker, and so on were all sharing the streets and therefore providing eyes that made them safer. Funnily, eyes on the street was not a criterion we had to worry about during the pedestrian era since streets were constantly full of people.

In contrast, Le Corbusier advocated for large towers in which people carry on with all of their activities (thus indoors) surrounded by large, empty green spaces with no "eyes." One of the adherents of Le Corbusier, Robert Moses, was actually working hard in New York City to make these ideas come to reality. Moses had already changed New York City dramatically by 1960. His next project was to tear down Greenwich Village and replace it with Le Corbusier–style towers. Many protests were started against the plans, and Jacobs quickly took a leadership role. In the end, the plans were abandoned, and Greenwich Village has become one of the most sought-after neighborhoods in New York City.

Later, when she moved to Toronto, Canada, Jacobs again had to protest. Plans were in motion to tear down a lively portion of the downtown to make room for an urban expressway. Here again, the plans were abandoned, and the area that was saved has become one of the most desirable neighborhoods in Toronto as well. Unfortunately, many cities did not have a Jane Jacobs to protest similar plans, leading to the loss of great neighborhoods. While trying to give access to downtowns, many planners took away land for business activities, leading to the "death" of many U.S. cities, as we discussed a bit when learning about the two sustainability principles.

Although the eyes-on-the-street criterion is first about safety, it adds a cornucopia of benefits. Eyes on the street also relates to a more lively economy and a more active social scene. Importantly for us, these shops with the "eyes" also provide amenities at close proximity, thus relieving us from needing motorized transport for basic trips.

We can even relate eyes on the street with the lessons of Alexander, who advocated against the separation of space by use. Indeed, eyes on the street tells us that the street "belongs" to everyone and that having pedestrians, consumers, workers, visitors, cyclists, and even automobile drivers around is more desirable than segregating space into discrete corridors.

In the end, and to conclude this section on essentials, it is practically impossible for us as human beings to design cities in ways that are enjoyable to everyone since we are all different. The life span of a city is also much longer than the average life span of a human being, and to grow organically, cities need to become the processes of many builders who add their contributions over time. That being said, there are several traits that are more desirable than others, some of which we have just discussed and others that we will learn now.

4.3 Urban Design and Desirable Traits

Although it is impossible to provide the recipe to design a "perfect" city—and, in fact, the term *perfect* is moot for urban design—there are certain properties and features that we can keep in mind. We can refer to these features as *traits*, partly because of the etymology of the word *trait*—a *stroke* on a drawing—echoing the process of drawing when designing.

Funnily, what we will essentially do is to characterize cities and neighborhoods based on few properties. This is extremely reductive since cities are inherently complex, as we have seen, and we are therefore simply fitting cities within a tree network. But we are after all human beings, and this is how we understand and learn more easily. We therefore need to be cautious when designing a city not to simply maximize or minimize a certain objective. We should try to think of the city as a whole, and most importantly, we need to "leave room" for future construction.

Here again, we will learn from two people whom we just saw, Kevin Lynch and Jane Jacobs. With Kevin Lynch, we will see five dimensions as well as two metacriteria of urban form that will help us understand various axes/levers we need to consider and that will partially shape how we will perceive the city. With Jane Jacobs, we will discover four conditions for diversity that are important to design a lively and pleasant neighborhood. These four conditions will actually be much easier to apply. They relate to only one type of urban design practice and may therefore not please everyone, but they will be extremely useful to be able to design for a community as opposed to for an individual.

4.3.1 Lynch's Five Dimensions and Two Metacriteria

The same Kevin Lynch who came up with the five elements of cities that we perceive most (section 4.2.2) also came up with five performance dimensions we can use to assess a city. On top of the five dimensions, he added two important metacriteria that can help us better gauge the five dimensions. These five dimensions were published in another seminal book from Lynch (1984) simply titled *Good City Form*. The five dimensions are:

- vitality: related to the basic biological needs of human beings, such as food and safety;
- sense: related to orientation—how we can situate ourselves in a city and create a mental map of a city;
- fit: related to the way the various elements of a city fit together in a pleasing way;
- access: related to the ability to enter and move within a city; and
- control: related to a personal level of "security" of having control of a household while being able to share public space.

The two metacriteria are:

- efficiency: related to the cost of operating and maintaining a city; and
- justice: related to the way the benefits are distributed not to favor one population over another.

In particular for the two metacriteria, we will enumerate the different trade-offs between the five dimensions, ensuring there is no imbalance toward one dimension over another and that efficiency does not come at the cost of justice. But first, let us go over the five dimensions sequentially.

Vitality is purely about the biological, or "health," needs of human beings. It relates to how a city provides for our survival needs, not only in terms of food and safety but also in terms of clean air and water, waste disposal, and fire protection. This is purely an anthropocentric condition. Lynch discusses three principal features for vitality: sustenance, safety, and consonance. The first two are self-explanatory; the third is a little more complicated. Consonance relates mostly to the spatial environment, which needs to be adapted to our needs, whether in terms of temperature or how items are sized (e.g., steps and doors) for human use. A city that cannot provide these basic needs cannot be successful. In urban design, vitality is either inherently planned (e.g., when sizing urban elements) or completely omitted as a negative externality (e.g., designing streets for traffic without thinking about pollution).

Sense is directly related to how we perceive a city, ensuring that buildings and streets have a certain identity and are not laid out in a way that is confusing. This is where modern-type planning à la Le Corbusier fails since it tends to be very symmetrical, as discussed. Similarly, suburbs with cookie-cutter houses simply cannot provide an element of sense. In older towns, the main clock offers a landmark for sense, for instance. We can be brought to any place within the city, and most likely we will know where we are. In the suburbs, however, we may feel lost one kilometer (0.6 mile) away from our house. Certain elements of a city must be present for us to acquire a mental map, which we will draw as paths, edges, districts, nodes, and landmarks when asked to. In urban design, plans for urban settings will tend to include elements needed for sense. In suburban settings, the meandering streets and the homogeneity of every house cannot contribute to sense.

Fit is about how the various elements of a city "fit" together and are representative of the population, both spatially and temporally. Shops should therefore be open when people can shop. Fit is also related to how "comfortable" we are in a setting or how "satisfied" we are. An adult without a driver's license may not fit in a suburb, and similarly, a person who enjoys having a large backyard may not fit in a dense urban setting. People from different cities will often argue about how their city is better, most typically by citing elements of their city that best fit them. Often, elements of fit will seem insignificant, such as the presence of an extra step on which we can sit, a doorway, a small park, or an area that is conducive to talking. In urban design, fit is very important, and in architecture, fit is crucial. Architects will go on extensively to establish a feeling of coherence in their design. We therefore need to ensure that various amenities are present for the right context (e.g., a playground in a family neighborhood), but fit also remains a highly personal dimension.

Access is about accessing a certain location. We have discussed access fairly extensively already. By the beginning of the twenty-first century, access had arguably become the most important feature of urban planning and design, and many locations are selected or designed to be accessible by all modes of transport (walking, bikes, transit, and private and shared vehicles). This problem actually offers a significant trade-off that Lynch explains well: "Many people, when asked what they imagine would be the best place to live, think of a house in a secluded garden, which is but one step from the center of a great city" (Lynch 1984, p. 193). Moreover, access is actually not only relevant for travel within a city but also between cities. One of the reasons New York grew so much is because it was strategically located on the East Coast and was well connected to the inland thanks to the construction of the Erie Canal. In urban design, access is one of the most important features that will contribute to the success or failure of a neighborhood. In Toronto, for instance, the Distillery District offers a fantastic setting to carry out many activities, but because its accessibility is poor, it gets fewer visitors than it likely deserves. We can recall here the most important modus operandi of real estate: "Location, location, location," which often comes down to access.

Control is about the way space is regulated. Control over the land that we may own is an important feature of how we enjoy cities. If we are deprived of any control, we may decide to change cities. As discussed with the law of property, control was at the heart of early Roman civilization. Quite differently, although related, we know that we do not have control of public areas, and yet we need to feel that we can use a public area and therefore exercise some kind of control. Funnily, we set some invisible boundaries of control in a public space. For example, we will choose a public bench that is vacant if we can, and if someone sits next to us, we may feel like we are losing some control; after all, human beings are territorial animals. In urban design, control is both a regulatory element to ensure property owners can exercise control of their property and a way to instill a feeling of "belonging" to a community.

Naturally, performing well in each of these five dimensions is significantly challenging, and we may feel tempted to favor some aspects over others. This is why Lynch added the two metacriteria of efficiency and justice. Efficiency is about the overall cost of a city beyond the simple monetary cost—that is, including the environmental costs that we have discussed. Justice is about the distribution of these costs and of the benefits. In particular, Lynch identifies six interdimensional conflicts that highlight the trade-offs between efficiency and conflict:

1. Vital environment versus decentralized user control: because many vital elements are hard to monetize (i.e., clean air), they sometimes conflict with control.
2. Vital environment versus well-fitted one: vital elements may require us to make some efforts that go against fit—for example, the need to walk for health versus the desire to walk the least amount possible.
3. Sense versus adaptability of fit: an element that contributes to sense, like an old clock, may not fit well in a new, vibrant area, for instance.
4. Present and future fit: an element of a city may be a good fit now but will likely be a problem in the future; this seems to be particularly the case in architecture when a design goes out of fashion.
5. Good access for all versus local control of territory: this goes back to the quote about having access to all, including a secluded garden. Good access is one of the most litigious points, though also one of the most critical.
6. Personal access versus health: the example used by Lynch is the automobile. Despite providing access, automobile use may lead to a lack of physical activity, and non-electric automobiles significantly pollute the air we breathe.

For our purpose, we will keep in mind these five dimensions when designing urban engineering systems. We need to size our systems adequately while expecting some evolution. We need to design our streets efficiently while keeping sense in mind. We need to design our water systems efficiently and integrate low-impact development strategies in streets while thinking about fit. These dimensions are also relevant when we think about decentralization of the power grid, such as the harvesting of local resources (e.g., wind vs. solar). Even control may be affected, for example, if district heating and cooling become more common. We could very well see that neighborhood, block, and homeowner associations become more common, akin to current condominium associations. Again, here, the integration of infrastructure is key, whether visible (e.g., streets) or not (e.g., subsurface infrastructure).

We will now focus even more on specific urban design traits with Jane Jacobs's four conditions.

4.3.2 Jacobs's Four Conditions for Diversity

Jane Jacobs was not only an activist and an architecture critic but also a prolific writer. Whether her famous *The Death and Life of Great American Cities* (Jacobs 1961) or her equally excellent *The Economy of Cities* (Jacobs 1970), her writings are likely going to be relevant for a very long time. Where Jacobs succeeds compared to her peers is in her ability to put the individual back in society as opposed to by herself or himself. Although she may not have known it at first, she was one of the pioneers of adopting a complexity theory mind-set to study cities and of understanding how societies behave. She was therefore a firm promoter of bottom-up approaches and a firm critic of top-down approaches.

Relevant for us, she came up with four conditions of urban design that promote diversity and that can therefore provide the eyes on the street that we discussed. In fact, she even wrote: "The necessity for these four conditions is the most important point this book has to make" (Jacobs 1961, p. 151). These four conditions are:

1. Mixed primary uses: areas should have more than one function so that they are occupied throughout the day and on weekends.
2. Small blocks: blocks should be small so as to promote walking and make it easy to turn and explore a city.
3. Aged building: areas should themselves be heterogeneous and therefore, they should have buildings of varying ages to attract diverse people.
4. Concentration: areas should have a population density high enough to provide the critical mass of people walking the streets.

Unlike Lynch's five dimensions, these four conditions are very simple to understand but perhaps not as simple to implement. We will go over each of them sequentially.

Mixed primary uses relates to the idea that to be vibrant, an area must have multiple functions. We go somewhere because we want to carry out some type of activity. We go to work during the day, and as a result, areas with large office complexes are empty at night and on weekends. Similarly, solely residential areas are empty during the day. If we aspire to build a diverse and vibrant area, it needs to be visited throughout the day. The area therefore needs to incorporate residential, commercial, and recreational land use, as well as any other types. In zoning, areas that have mixed primary uses are simply called *mixed-use*. Mixed-use areas may therefore have some offices, but they also have stores, banks, restaurants, and other types of establishments that bring people in the evening and on weekends. They can also have some sort of recreational activity, such as museums, to bring people on weekends. The concept of a mixed-use area is strongly promoted by the Smart Growth and New Urbanism movements. Having these multiple primary uses will also provide the eyes on the street that we discussed previously.

Small blocks are more conducive to walking since we can easily turn and reach our destination. For pedestrians, they are much more manageable. City centers in older cities typically have smaller blocks, and these cities were designed for pedestrians (see the figures shown earlier in the chapter). Areas with small blocks are also inherently more complex, thus directly contributing to Alexander's semilattice argument. Smaller streets also enable us to explore an area more, and they often have more character. Smaller streets also naturally have a more human scale and hence contribute to Lynch's fit dimension. At the same time, they are a deterrent to auto use since traffic is affected by either stop signs or traffic lights. The purpose is naturally not to restrict the use of private vehicles but simply to make walking an equally attractive option. Certainly, one of my favorite cities to walk in is Venice in Italy. Venice is full of streets with small blocks that are highly walkable and that inherently possess a human scale.

Aged buildings have two direct purposes. First, a mixture of buildings of different ages will contribute to the character of the city. They will therefore directly help both the sense and fit dimensions: sense, because they will help us orient ourselves in an area—for instance, we might use a building of a certain shape and color as a reference point—and fit, because they will offer more chances for everyone to develop a feeling of "comfort." People preferring older buildings will find their fit, and so will people preferring newer buildings. Aged buildings will also directly contribute to interdimensional conflict of present versus future fit. Too many areas are developed all at once and look dated after only a few years. Second, a mixture of buildings of different ages provides a larger range of rent prices, thus providing options to a wider array of incomes.

Naturally, we cannot design an entire area with aged buildings, but we can make sure to leave room for future construction, which will both provide this mixture of building ages in the longer term and enable future builders to bring their own contribution (i.e., toward Alexander's argument).

Concentration relates to population density. Finally, we can only have eyes on the street if people are there. It is therefore important to ensure a sufficient population density for the streets to be used at all times. Jacobs recommends 100–200 units per net acre of residential land. This is not a standard unit since it does not include land for transport, commercial, and other types of uses. A rough conversion is somewhere between 10,000 and 25,000 pers/km^2 (about 25,000 to 65,000 pers/mi^2). This may sound like a lot compared to city averages, but dense places with large condominiums in large cities easily reach more than 30,000 or even 50,000 pers/km^2. As a comparison, the population density of Manhattan is about 28,000 pers/km^2, and this accounts for all land, including Central Park and all office buildings. In fact, Jane Jacobs is against tall towers and too-high population densities. Her "ideal" city is composed of midrise buildings of four to six stories that have shops on the first floor and apartments on the floors above (representing a typical European city). This type of urban form is not necessarily for everyone, but neighborhoods of 10,000 pers/km^2 can still have single-family detached houses—neighborhoods around the Gold Coast and Lincoln Park areas in Chicago have population densities higher than 10,000 pers/km^2, and they are some of the most sought-after neighborhoods of the entire region.

Overall and as discussed, these four conditions are easy to understand. Chances are if you try to think about neighborhoods that you particularly like, they will respect some, if not all, of these four conditions. Keeping in mind these four conditions can therefore significantly help when designing neighborhoods. Moreover, and incredibly relevant for us, if infrastructure systems are to become more decentralized—discussed in chapter 10—we need to know the patterns of daily demand per neighborhood. If our neighborhood is not visited during the day and becomes crowded at night, how do we make sure we can generate enough electricity and treat enough water while accommodating a fluctuating demand? Having mixed primary uses will ensure a relatively stable demand throughout the day. Jacobs's four conditions therefore also have many benefits for urban engineering systems.

4.4 Conclusion

This concludes our chapter on urban planning. Urban planning is a difficult profession since planners need to account for all the needs of all the residents. Although engineers at the moment tend to stay within one discipline, this will need to change in the future, especially if urban infrastructure systems become decentralized. We therefore need a new breed of engineers: urban engineers. This is actually not new, and engineers used to be multidisciplinary up until the end of the nineteenth century (which in fact corresponds to the birth of modern urban planning). Moreover, in many smaller cities, municipal engineers have to plan and manage multiple infrastructure systems, from roads to water pipes. With the "rise" of the neighborhood as the unit of work in engineering systems, this practice may come to large cities as well.

In this chapter, we first went through a quick review of the evolution of urban planning, from the birth of cities in the Neolithic era to the modern suburb. Since Hippodamus, in the past two and a half millennia, cities have evolved tremendously. In particular, we saw that planners first focused on planning for the "pedestrian," which shifted to planning for the "mobile" in the twentieth century. Rising traffic congestion and poor urban form, however, call for a new focus as a hybrid between the two, which we defined as the "accessor." Cities in the twenty-first century will therefore again likely go through substantial changes, especially in how urban infrastructure systems are planned. Figure 4.12 shows a wonderfully illustrative and informative quick recap of all the urban planning practices covered in this chapter. The drawing was made by Jamei Borges, who took a class based on the notes for this book in fall 2017.

Overall, Patrick Geddes—another leading scholar and pioneer of taking an evolutionary approach to cities—says it best when looking back at the evolution of cities: "A city is more than a place in space, it is a drama in time" (Geddes [1905] 2004, p. 107). To learn more about cities and their history, in addition to the works mentioned, I recommend *History of Urban Form before the Industrial Revolution* by Morris (1994); *The City Assembled: The Elements of Urban Form through History* by Kostof, Castillo, and Tobias (1999); and *The Evolution of Great World Cities: Urban Wealth and Economic Growth* by Kennedy (2011). One of my personal favorites again is *The City in History: Its origins, Its transformations, and Its Prospects* by Mumford (1961); it is not the easiest book to read, but it offers a great historical and analytical account of how cities have evolved since they first appeared.

Subsequently, we discussed three essentials of urban planning, here focusing on our "limitations" as human beings, as opposed to listing the various responsibilities of an urban planner. First, we went over the ideas of Christopher Alexander in his article "A City Is Not a Tree." Alexander argues that as human beings, we cannot design cities effectively since we are prone to designing tree networks as opposed to semilattices. It is therefore important to leave space for growth in our plans. Then, we learned about how we perceive cities thanks to the work of Kevin Lynch. Notably, we learned about the five elements of cities that we clearly differentiate: paths, edges, districts, nodes, and

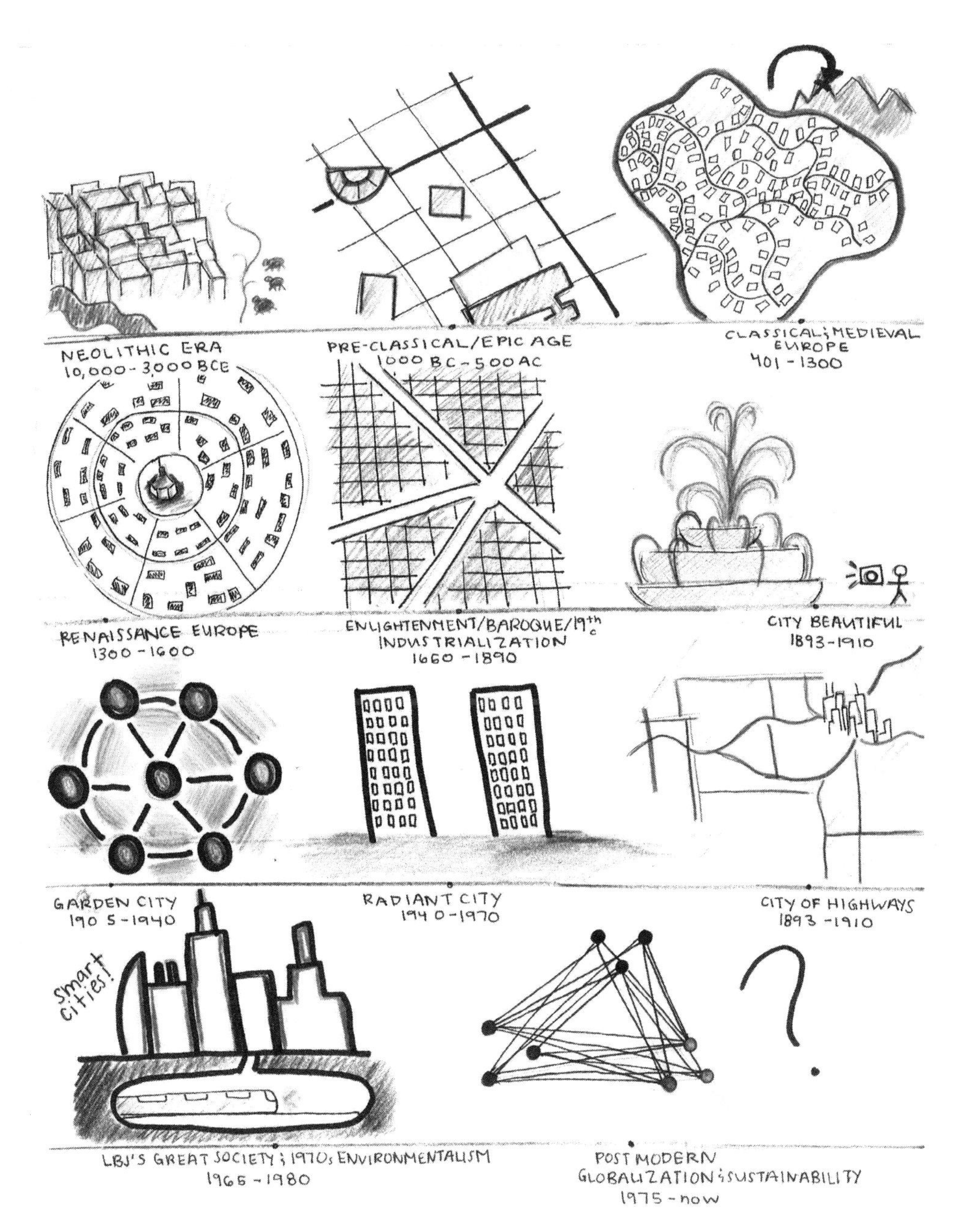

Figure 4.12
Quick recap of urban planning practices. Courtesy of Jamei Borges.

landmarks. Then, we went over Jane Jacobs's argument about the need to have eyes on the street to have a safe and vibrant area.

In the last section of the chapter, we discussed some desirable traits of cities. Although we cannot design semilattices, there are several desirable traits in cities that we can plan. First, we discussed Lynch's five performance dimensions of cities: vitality, sense, fit, access, and control. To these five dimensions, Lynch adds two metacriteria: efficiency and justice, to address several trade-offs between the dimensions. The five dimensions are informative but much harder to translate into practical designs. In contrast, we also learned Jane Jacobs's four conditions for diversity, which are incredibly clear. These four conditions are essential ingredients for the development of vibrant and diverse neighborhoods.

At the time of this writing, most engineering curricula did not include classes on urban planning, and this is a big mistake. The engineering profession tends to confine itself in discrete disciplines although these disciplines each involve the same fundamental unit: the city. These disciplines need to be better integrated. In fact, we often forget that as spatial systems, the infrastructure systems themselves are inherently integrated, and we need to plan them accordingly. It is now time to get more technical, and in the next chapter, we will cover our first major infrastructure system: electricity.

Problem Set

4.1 By performing your own research, beyond the material covered in the chapter, describe how cities evolved during the Neolithic era.

4.2 By performing your own research, beyond the material covered in the chapter, describe how Romans planned cities.

4.3 By performing your own research, beyond the material covered in the chapter, define and compare *organic planning* and *baroque planning*.

4.4 By performing your own research, beyond the material covered in the chapter, describe how the practice of "zoning" works and explain why many cities have adopted zoning practices.

4.5 By performing your own research, beyond the material covered in the chapter, describe in detail one of the following three movements: the City Beautiful, the Garden City, or the Radiant City.

4.6 By performing your own research, beyond the material covered in the chapter, describe the phenomenon of *urban sprawl*.

4.7 From the chapter and any other relevant sources, make a time line of the various urban planning practices from the Neolithic era to today. Draw the time line as an axis and list the practices along the axis. Importance should be given to the visualization of the time line as well as its accuracy.

4.8 In your own words, describe and illustrate with an example the differences between a tree network and a semilattice as Christopher Alexander defines them.

4.9 Take a blank piece of paper and draw a city of your choice without looking at any map. Next, identify the features in the map that relate to the five elements of a city as discussed by Kevin Lynch.

4.10 In your own words, describe and illustrate with examples the five elements of a city as defined by Kevin Lynch.

4.11 In your own words, describe and illustrate with personal examples why having more eyes on the street, as argued by Jane Jacobs, tends to be preferable in urban planning.

4.12 In your own words, describe and illustrate with examples Kevin Lynch's five performance dimensions and two metacriteria.

4.13 Select any neighborhood/city in the world and discuss how this neighborhood/city performs using Lynch's five dimensions and two performance criteria.

4.14 In your own words, describe and illustrate with examples Jane Jacobs's four conditions for diversity.

4.15 Select any neighborhood/city in the world and discuss how this neighborhood/city performs using Jacobs's four conditions for diversity.

Notes

1. Trying to determine the boundary of a city versus a region is trickier than it seems. Some urban planners have strict definitions of urban vs. regional planning, but we will use the term loosely here.

2. Technically, MPOs only have to look at transport issues by federal mandate, but many also look at land use, water, economic, and many other planning issues.

3. As we will see, the Romans are also well known for being incredible water resources engineers.

4. Tokyo (then called Edo) supposedly reached the 1-million-people mark in the seventeenth century. Unlike Rome, Tokyo did not have a wastewater system, but it had a formidable network

of canals and rivers. Farmers from the rural areas would come in on boats to bring food, and they would collect human waste to use as fertilizer on their fields (Sorensen 2012).

5. This might not have been the case everywhere, however. Some of the Indus Valley civilizations of the era, including the cities of Harappa and Mohenjo Daro, show order and planning, but they lack the presence of monuments, temples, and palaces that are representative of a central command. Nonetheless, this "decentralized" order remains to be formally proven, as it would diverge from just about every other civilization in the world.

6. Although wider streets were also purposefully built to accommodate horse carriages as well.

7. The figure was made with Openstreetmap, which is an amazing crowdsourced mapping platform. To learn more about Openstreetmap and copyright, see https://www.openstreetmap.org/copyright (accessed September 5, 2018).

8. In Chicago, many streets stem from old Native American trails as well, including Lincoln Avenue, Milwaukee Avenue, and Ogden Avenue.

9. These sewer tunnels (or *égouts* in French) are actually quite famous, and they can still be visited today. You can even visit an entire museum dedicated to the égouts. The attraction is actually quite popular for tourists and locals.

10. Not to mention the enormous sanitary issues linked with animal and human waste, which became increasingly problematic as cities grew in size. We will discuss this in chapter 9.

11. Notably, by preventing any affordable housing projects to be implemented. That practice is commonly called *exclusionary zoning*.

12. That is actually quite an important point that we will develop in chapter 11, when we are introduced to the Science of Cities.

13. La Ville Radieuse in its original French.

14. Funnily, these expressways were often supported by city leaders since they thought they would help downtowns become more vibrant and thrive economically. Most often, they did not.

15. When we cover this chapter in my class, I always ask students to draw Chicago on a blank piece of paper, and I have yet to find two drawings that look alike. They may all be "rational" engineers, but we all perceive our cities in different ways.

16. And her lessons travel far. During my time in Singapore, I rapidly saw that the planners there made sure to follow her recommendations.

References

Alexander, Christopher. 1965. "A City Is Not a Tree." *Architectural Forum* 122(1): 58–62.

Bauer, Kurt W. 2010. *City Planning for Civil Engineers, Environmental Engineers, and Surveyors*. Boca Raton, FL: CRC Press.

Cervero, Robert. 1998. *The Transit Metropolis: A Global Inquiry*. Washington, DC: Island Press.

Dehaene, Stanislas. 1999. *The Number Sense: How the Mind Creates Mathematics*. New York: Oxford University Press.

Derrible, Sybil. 2016. "Urban Infrastructure Is Not a Tree: Integrating and Decentralizing Urban Infrastructure Systems." *Environment and Planning B: Urban Analytics and City Science* 44(3): 553–569.

———. 2017. "Complexity in Future Cities: The Rise of Networked Infrastructure." *International Journal of Urban Sciences* 21 (supp. 1): 68–86.

Encyclopedia Britannica. 2015. s.v. "Neolithic Period." https://www.britannica.com/event/Neolithic-Period.

Geddes, Patrick. (1905) 2004. *Civics: As Applied Sociology*. EBook #13205. Online edition, Project Gutenberg, http://www.gutenberg.org/ebooks/13205.

Hall, Peter. 2002. *Cities of Tomorrow*. 3rd ed. Oxford: Blackwell.

Jacobs, Jane. 1961. *The Death and Life of Great American Cities*. New York: Random House.

———. 1970. *The Economy of Cities*. New York: Vintage Books.

Kennedy, Christopher. 2011. *The Evolution of Great World Cities: Urban Wealth and Economic Growth*. Toronto: University of Toronto Press.

Kostof, Spiro, Greg Castillo, and Richard Tobias. 1999. *The City Assembled: The Elements of Urban Form through History*. Boston: Bulfinch Press.

Lynch, Kevin. 1960. *The Image of the City*. Cambridge, MA: MIT Press.

———. 1984. *Good City Form*. Cambridge, MA: MIT Press.

Morris, A. E. J. 1994. *History of Urban Form before the Industrial Revolution*. Essex, UK: Pearson Education.

Muller, Peter O. 2004. "Transportation and Urban Form: Stages in Spatial Evolution in the American Metropolis." In *The Geography of Urban Transportation*, 3rd. ed., edited by Susan Hanson and Genevieve Giuliano, 59–85. New York: Guilford Press.

Mumford, Lewis. 1961. *The City in History: Its Origins, Its Transformations, and Its Prospects*. New York: Harcourt, Brace & World.

Pollio, Vitruvius. 1999. *Vitruvius: Ten Books on Architecture*. New York: Cambridge University Press.

Reader, J. 2004. *Cities*. New York: Grove Press.

Schaeffer, K. H., and Elliott Sclar. 1975. *Access for All: Transportation and Urban Growth*. Baltimore: Penguin Books.

Sedlak, David L. 2014. *Water 4.0: The Past, Present, and Future of the World's Most Vital Resource.* New Haven, CT: Yale University Press.

Sorensen, A. 2012. "Uneven Geographies of Vulnerability: Tokyo in the Twenty-First Century." In *Planning Asian Cities: Risks and Resilience*, edited by S. Hamnett and D. Forbes, 40–66. Abingdon, UK: Taylor & Francis.

II Urban Engineering and Infrastructure Systems

5 Electricity

> Every aspect of daily life depends on electricity. In fact, our society's prosperity and security hinges on the instantaneous availability and unwavering reliability of electricity and central to achieving both is the role of electric system operations.
>
> —Vadari 2013, p. xix

As the Vadari quote highlights, the impact that electricity and power systems have had on the world since Thomas Edison opened the first ever power plant, the Pearl Street Station in New York City, in 1882 is quite simply impossible to fully grasp. Almost everything we do nowadays involves electricity, from the obvious, like using a computer or lighting a room, to the not so obvious, like drinking water, driving a car, or even reading a physical book that required electricity to produce. It is therefore not so surprising to see that one of the largest sectors of greenhouse gas (GHG) emissions is electricity. Figure 5.1 shows the breakdown of GHGs by economic sector for the United States in 2016, and electricity is nearly tied in the lead with 28.4%.[1] Reducing the carbon footprint of the electricity sector therefore has substantial benefits.

We easily forget, however, that electricity is actually a second source of energy. Unlike energy sources such as minerals or fossil fuels, electricity does not exist as a raw quantity on Earth (or at least not in significant amounts). All the electricity that is produced comes from the conversion of a raw energy into electricity, whether from the movement of water (hydroelectricity) and air (wind power) or from the burning of fossil fuels. And yet, electricity seems to be the primary source of energy for so many things that we use. The entire challenge is therefore to find the right raw material or energy source to produce electricity while respecting the triple bottom line.

On the plus side and unlike transport, where a reduction in GHG will require a change in consumption behavior (i.e., we need to change the way we live), the electricity

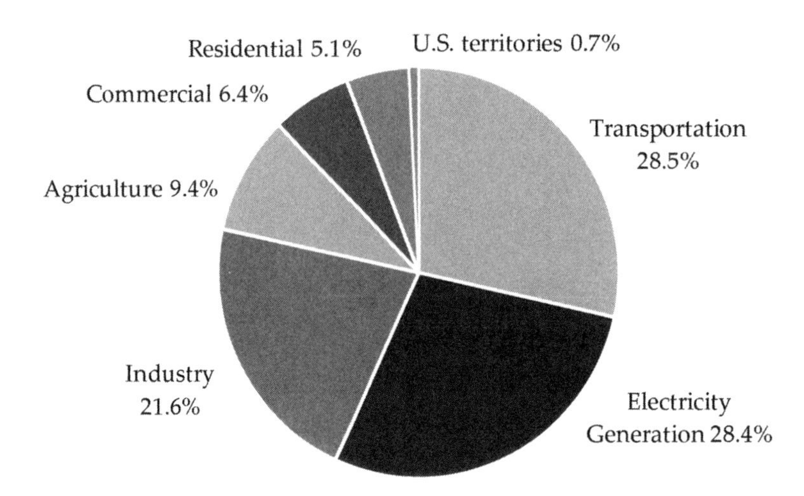

Figure 5.1
U.S. GHG emissions by economic sector in 2016.
Source: U.S. Environmental Protection Agency (2018).

sector does not directly impact people. Obviously, we should make sure that the overall demand decreases over time, but whether our electricity is completely supplied from renewable sources or not generally does not impact our consumption behavior.[2] Naturally, like everything we have seen so far, the problem is a little more complicated. The power grid that transmits all the electricity that we consume is an immense complex system that is incredibly hard to operate and maintain. Fox-Penner (2010) actually calls for a "second" revolution, after the first one that saw the rise of electricity, and the entire power sector has been undergoing significant changes. Some of these changes include the concept of a *smart grid*, where we harvest the power of information and communication technologies to better run the power grid. We will discuss smart grids and microgrids a little at the end of the chapter.

In this chapter we will first review some of the basic principles of electricity, going over the definitions of voltage, intensity, resistance, power, and energy. We will also recall Kirchhoff's two laws and learn how circuits in series and in parallel work. We will also see the differences between alternating current (AC) and direct current (DC), and we will learn about three-phase power.

As we will do with most upcoming chapters as well, we will then analyze patterns of electricity demand to understand how electricity is being used, both from a power grid perspective and in individual households. We will also learn how electricity is supplied to meet this demand instantaneously. In terms of data, we will rely heavily on the U.S.

Energy Information Administration (EIA) and the U.S. Department of Energy (DOE), which are formidable resources that we will use again in chapter 8.

Subsequently, we will address the major aspect of electricity that emits GHGs: generation. First, we will classify the different types of power plants (including coal, natural gas, nuclear, hydroelectric, solar, and wind), and we will review some of their properties—notably, their GHG emission factors.

Finally, we will briefly discuss the future of the power grid by going over some of the latest trends in the electricity sector, including electricity storage, the smart grid, and microgrids.

But first, we need to go back to our basic physics classes and recall some fundamentals of electricity.

5.1 Fundamentals of Electricity

Despite being essential to our lives, electricity tends to be a fairly abstract concept. Indeed, unlike pipes of water, congested roads, wall insulation material, and garbage that we can physically see, electricity is much tougher to visualize. In fact, a 2008 report from the DOE showed that only about 12% of a group of people interviewed could pass a basic electricity-literacy test (Sovacool 2009). To remediate this problem, in this section we will first recall a few fundamentals of electricity that we should keep in mind as we go along. Moreover, we will learn about circuits in series and in parallel, about three-phase power, and about the power grid that controls all matters of electricity from the time it is generated to the time it gets to your home.

5.1.1 Basics of Electricity

There are four essential measures in electricity, all of which we know quite well, and to this list we can add one more, energy, that is important for us:

- Voltage (also called tension) V, expressed in volts [V] = [J/C]
- Intensity (also called current) I, expressed in amps [A] = [C/s]
- Resistance R, expressed in ohms [Ω] = [V/A]
- Power P, expressed in watts [W] = [J/s]
- Energy E, expressed in watt-hours [Wh] = [3,600 J]

Although all engineers have used these measures countless times, they often remain mysterious for nonelectrical engineers. In particular, note the unit [C], which is the coulomb, the fundamental unit of electrical charge, representing the charge of 6.241×10^{18} electrons. This unit is a little like mass for physical properties.

Figure 5.2
Illustration of Ohm's law.

Moving to voltage, many people forget that it is not an absolute quantity but a difference between two electric potentials, which is why we often hear the term *voltage drop* as opposed to simply *voltage*. Moreover, voltage is expressed in joules per coulomb, and it is therefore a rate, not an amount, that expresses a difference in the energy between two points. Also a rate, intensity is a flow of electric charge, expressed in coulombs per second. With R as the resistance in your system, these two quantities are directly proportional as:

$$V = R \cdot I \tag{1}$$

Equation (1) is famously known as Ohm's law. It is probably the first law that all students learn when they are introduced to electricity. But what does it actually mean? Figure 5.2 offers a funny sketch. As an electric potential, Volt "pushes" energy as much as possible. As the intensity, Amp represents the flow of charge that gets to go through despite the presence of Ohm, a resistance. Below, we will also compare electricity with water, which will further help us understand the concept.

Ohm's law is only half the story, however. It only applies to the right-hand side of figure 5.3, when we do not have a power source V_S. In the presence of a power source (i.e., electricity generator), the voltage of the circuit is actually:

$$V_C = V_S - RI \tag{2}$$

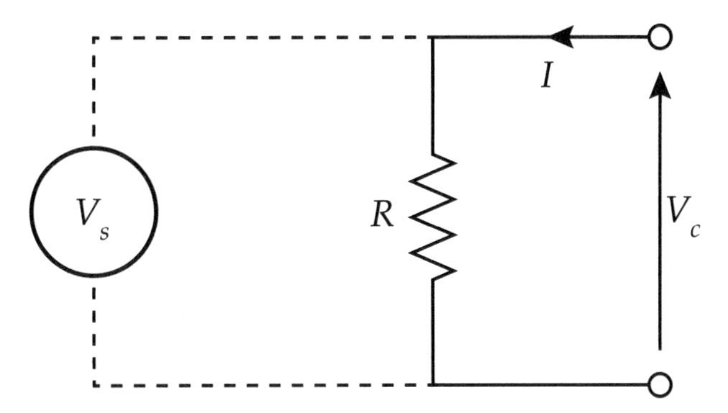

Figure 5.3
Simple circuit.

The term *RI* actually represents the losses in the system. Sometimes we want these losses—for instance, to decrease V_C or to create heat in the resistance. Other times, we want to minimize the losses, especially in electricity transmission and distribution, and we will see why in section 5.1.2.

These measures, however, are not intuitive, and we can therefore benefit from making an analogy with another system that we understand and can visualize: water. Although water will be the main subject of our next chapter, it can help us at this stage. Voltage is essentially like pressure in water systems (or, more accurately, the difference in pressure between two points[3]); intensity is like the flow rate, *Q*; and resistance is like the energy loss. Flow pressure decreases in a pipe because of friction and other losses (i.e., a constriction like a resistance *R*), similar to a voltage drop increasing with a higher resistance *R*. However, flow rate is not affected by *R* because of continuity ($Q_1 = Q_2$), only the pressure is affected—that is, with a higher resistance R, only V_C is affected, not *I*.

One of the quantities that we are most interested in is power *P*. Remember that power is a rate—that is, it is energy over time. Energy over time is the same thing as energy density times a flux (like a flow rate). Relating power to voltage and intensity, the fundamental equation is:

$$P = V \cdot I \tag{3}$$

Checking the units, [V] = [J/C] and [I] = [C/s], therefore, power is expressed in [J/s], which is the [W]. In water, power is the energy potential of water (the *energy head*) times the flow rate. Power is important because it represents the demand that needs to be met immediately by the supply.

The final amount of energy *E* that is consumed is therefore:

$$E = P \cdot t \tag{4}$$

Example 5.1

Electrical Property of a Refrigerator

The average Civitian has one refrigerator at home that requires an intensity of about 5A at any given time. Calculate (1) the equivalent resistance of the refrigerator, (2) the power drawn, and (3) the energy consumed in a year. Assume $V=120$ V.

Solution

(1) Since $V=R\times I$, the resistance can be calculated as $R=V/I$. Therefore:

$R=120\ /\ 5=24\ \Omega$.

(2) Power is defined as $P=V\times I$, and $V=120$ V and $I=5$ A, therefore:

$P=120\times 5=600$ W.

(3) Energy is $E=P\times t$, where t is the number of hours in one year, i.e., $t=24\times 365=8{,}760$ hours, therefore:

$E=600\times 8{,}760=5{,}256{,}000$ Wh $=5{,}256$ kWh $=5.256$ MWh

Considering that the average US household consumed 10.8 MWh in 2016. This means that a refrigerator consumes roughly 49% of the total consumption of an American household. This is obviously too large, which shows that although the refrigerator has an intensity rating of 5A, it is not always drawing that much intensity from the grid—in fact, with many refrigerators, we can hear when the motor turns on.

where t is time. If time is in seconds, then energy is expressed in joules [J]. In electricity systems, however, we tend not to use joules. Instead, we measure how many hours a certain power is consumed and simply multiply it by the number of hours. Energy is therefore expressed in [Wh], which is equivalent to 3,600 J. We discussed this unit briefly in chapter 1. The [Wh] is supposed to make things a little easier to understand, but I am not sure it is doing a good job as I constantly have students say "kW per h," which does not make sense; kilowatts per hour is energy divided by time squared, thus it is more like an acceleration of energy consumption, whereas what we want is an energy value.

The typical example to illustrate [Wh] is a light bulb. How much energy does a 10 W light bulb consume in one day? The answer could be $10\times 3{,}600\times 24=864{,}000$ J, but typically, it will be expressed as $10\times 24=240$ Wh. Again, it may not seem so intuitive at first, but it is actually quite helpful, and we can quickly get used to it.

A typical household in the United States consumed a total of 10.8 MWh in 2016 (U.S. Energy Information Administration 2017c) compared to 4.2 MWh in the United Kingdom in 2013 (Engineering Council UK 2014). This is therefore an energy. On average, a house in the United States therefore consumes about $10.8/365 = 0.03$ MWh $= 30$ kWh in a day. This means that the average power used at any time is about $30/24 = 1.25$ kW for a typical American household; *make note of this figure of 1.25 kW because we will use it a lot in this chapter.* Look around your house at some of the appliances that you have. They will have a power rating and/or an intensity rating, and if you plug them in all at once and go over 3 kW, chances are your circuit breaker will trip depending on the appliances used because the intensity required ($I = P/V$) will have become too great.

We can now look at what happens when we plug in multiple electric devices that all require voltage and intensity.

5.1.2 Kirchhoff's Laws and Load Types

Gustav Kirchhoff was a nineteenth-century German physicist who came up with two fundamental laws of electricity, shown in figure 5.4. Kirchhoff's current law states that the sum of currents at a node must be zero. Kirchhoff's voltage law states that the sum of the voltages in a loop must be zero. These two laws perfectly explain the complexity of the power grid. They are essentially telling us that we do not decide where the electricity goes. We cannot simply direct the current to go in one direction and to skip a power line that we do not want to use. When an electric flow gets to a junction, it chooses the path of least resistance. Put differently, think about all the millions of kilometers of power lines that are laid out across cities and regions. There is absolutely no way for us to trace the path of an electron from the power station to your home. To make an analogy with structural engineering, the same thing happens when forces are

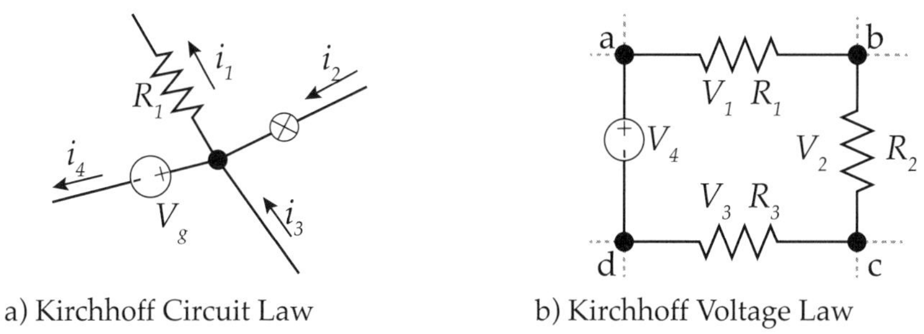

Figure 5.4
Kirchhoff's circuit and voltage laws.

Example 5.2

Voltage and Current in Small Circuit

Using Kirchhoff's laws, determine (1) V_1 (i.e., V_{ab}), (2) V_2 (i.e., V_{be}) and V_3 (i.e., V_{cd}), and (3) i_2, and i_3 for the small circuit below.

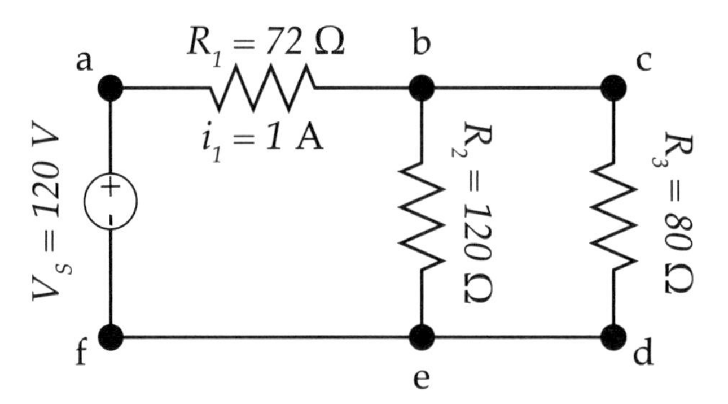

Solution

(1) Applying Ohm's law, $V_1 = R_1 \times i_1 = 72 \times 1 = 72$ V. This means that the drop in voltage between *a* and *b* is 72 V (i.e., a "pressure" loss of 72 V).

(2) Applying Kirchhoff's voltage law on the left-hand side loop: $V_S + V_1 + V_2 = 0$. Since V_S runs in the opposite direction to V_1 and V_2, we have:

$V_2 = V_S - V_1 = 120 - 72 = 48$ V.

Similarly, applying Kirchhoff's voltage law on the right-hand side loop:

$V_2 = V_3 = 48$ V.

(3) Applying Kirchhoff's current law on node *b*, we have $i_1 = i_2 + i_3$. Moreover, we have $i_2 = V_2/R_2 = 48\ /\ 120 = 0.4$ A. Therefore:

$i_3 = 1 - 0.4 = 0.6$ A

Similarly, we could have used $i_3 = V_3/R_3 = 48/80 = 0.6$ A.

being distributed in a structure; you cannot simply "expect" that a force will not be distributed in a beam.

These laws are also saying that if a current is not consumed in one place, then it will have to go somewhere else. For instance, when you turn off your TV, you are forcing more current into parallel wires. This is responsible for some of the power outages we had in the 2000s. One power line went down, and the current was redistributed, forcing some new lines to break. The current was therefore again redistributed, forcing more lines to break, and so and so forth, causing what is called a *cascading failure*. Now, lines are disconnected automatically if the current is too high.[4]

All the current flowing in power lines must therefore be distributed, and all the electrical energy being generated must therefore be consumed right away. One way to simply test this is to plug a voltmeter into any electrical outlet and record the voltage over time. First of all, you will notice that the voltage tends to be around 115 V or 117 V, as opposed to 120 V.[5] Second, if you pay attention, over time the voltage will sometimes decrease and sometimes increase a little. This directly reflects changes in the load—that is, changes in the demand for electricity. If the demand is high, the voltage will decrease a little, and if the demand is low, it will increase.

The same laws actually apply in water systems. At a node, the sum of the flow rates must be zero—that is, all the water arriving at the node must leave somewhere. Moreover, the pressure of water at the tap changes during the day. If everyone had their tap open at the same time, the water pressure would drop significantly. In water, the average pressure is around 300 kPa (~ 40 psi), which is the equivalent of the 120 V in electricity; in water, however, pressure can fluctuate more without problems (we will discuss this in the next chapter).

These two laws of electricity thus pose a fairly large problem. Electricity providers need to be able to estimate future demand so they can actually provide the electricity at about 117 V. Ideally, we would produce as much electricity as possible, even when we do not need it, then simply store it and use it later when we are not able to generate as much electricity. Unfortunately, storing electricity is not easy and quite costly. At the time of this writing, we could not depend on storage, but we will discuss it toward the end of the chapter. For now, let us learn about series and parallel circuits.

5.1.3 Series and Parallel Circuits

From large electric appliances to tiny electric systems, most devices that consume electricity consist of a combination of circuits in series and in parallel. Figure 5.5 shows an example of a circuit in series and a circuit in parallel. This is significant because a circuit in series operates differently from a circuit in parallel, exactly like water systems (we will learn about pipe networks in the next chapter).

Therefore, to be able to estimate how much current is drawn and how much power is consumed by a circuit, we like to calculate an "equivalent" resistance R_{eq} for the entire circuit. Put differently, if the circuits in figure 5.5 only had one resistance, what would it be? We will actually follow the same procedure when learning about R values in chapter 8.

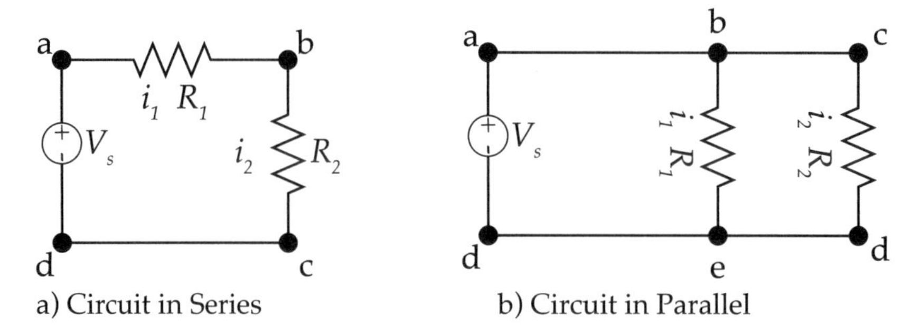

Figure 5.5
Circuits in series and in parallel.

For circuits in series, the process is actually quite simple. If we apply Kirchhoff's voltage law to the circuit in series in figure 5.5a, we have $V_S+V_1+V_2=0$, where V_1 is V_{ab}—the voltage drop (i.e., "pressure" loss) between nodes *a* and *b*—and V_2 is V_{bc}—the voltage drop between nodes *b* and *c*. Since V_1 and V_2 are in the opposite direction to V_S, we have:

$$V_S=V_1+V_2 \tag{5}$$

Applying Ohm's law, we have

$$i_S \cdot R_{eq}=i_1 \cdot R_1+i_2 \cdot R_2 \tag{6}$$

And from Kirchhoff's current law, we know that i_s must be equal to i_1 and i_2, therefore:

$$R_{eq}=R_1+R_2 \tag{7}$$

That is it. For circuits in series, we can simply sum the resistances to get an equivalent resistance. More generally, for circuits in series:

$$R_{eq}=\sum R_i \tag{8}$$

For circuits in parallel (figure 5.5b), we need to start with Kirchhoff's current law. Applying the law at node *b*, noting that i_1 and i_2 are in the opposite direction to i_s, we have:

$$i_S=i_1+i_2 \tag{9}$$

Applying Ohm's law, we have:

$$\frac{V_S}{R_{eq}}=\frac{V_1}{R_1}+\frac{V_2}{R_2} \tag{10}$$

Example 5.3

Equivalent Resistance in Small Circuit

The figure below shows the circuit from example 5.2. (1) Calculate the equivalent resistance of the circuit. (2) Verify whether i_1 really is 1 A, as given in the figure.

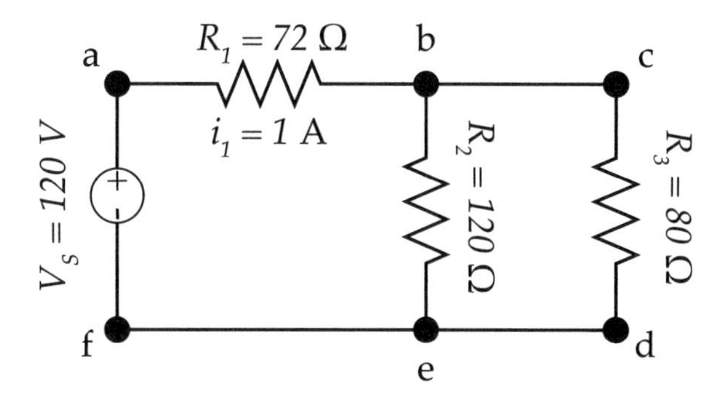

Solution

(1) Starting with the two resistances in parallel on the right-hand side (i.e., R_2 and R_3), we have:

$$R_{23} = \frac{1}{\frac{1}{R_2}+\frac{1}{R_3}} = \frac{1}{\frac{1}{120}+\frac{1}{80}} = \frac{1}{0.0083+0.0125} = 48\ \Omega$$

Adding R_1 as a resistance in series, we have:

$$R_{eq}=R_1+R_{23}=72+48=120\ \Omega$$

(2) Applying Ohm's law, the current $i_S=V_S/R_{eq}=120/120=1$ A. Moreover, we note from Kirchhoff's current law that $i_S=i_1$, therefore, $i_1=1$ A as shown in the figure.

And from Kirchhoff's voltage law, we know that $V_S=V_1$ and that $V_1=V_2$ since the parallel circuits create two loops; therefore $V_s=V_1=V_2$, giving us:

$$\frac{1}{R_{eq}} = \frac{1}{R_1}+\frac{1}{R_2}$$

$$R_{eq} = \frac{1}{\frac{1}{R_1}+\frac{1}{R_2}} \tag{11}$$

More generally, for circuits in parallel:

$$R_{eq} = \frac{1}{\sum \frac{1}{R_i}} \tag{12}$$

That is it. The equation is a little more complicated for resistances in parallel, but the process remains simple. Interestingly, one of the reasons to construct circuits in series and in parallel is that we rarely find a resistance with the exact value that we want. Instead, we need to build circuits in parallel and in series such that the R_{eq} is exactly what we want.[6]

Up until now we have discussed voltage and current as if there were only one type, but we have all heard of DC and AC. Let us see how they defer and why this is important to generate and distribute electricity.

5.1.4 Alternating Current and Direct Current

So far, we have assumed that the voltage and the current were supplied at a constant rate. We call this direct current (DC) power. It makes a lot of sense, and we use DC power in many devices. Even batteries supply DC power.

DC power is not the only type of power that is supplied, however. In fact, we have all heard of alternating current (AC). The big difference between AC and DC is that in AC the current and voltage constantly alternate between being positive and negative—that is, the electrons in the wires physically alternate direction constantly. Figure 5.6 shows how voltage changes in AC and DC systems. First, we see that the voltage in DC systems simply does not change at all.

To be more specific, figure 5.6 shows one type of AC power—a sine wave—but AC can follow a step function, a triangular function, or any other type of periodic function. The sine waveform is the one that is used in practice. Precisely, the voltage being supplied to most buildings follows the function:

$$V(t) = 170 \cdot \sin(2\pi \cdot 60 \cdot t) \qquad (13)$$

where 170 is the maximum voltage supplied, and 60 is the frequency f at which the voltage is alternating. In the United States, the frequency f is 60 Hz[7]—that is, the voltage changes sixty times per second. In Europe it is 50 Hz.

We therefore see that the voltage fluctuates constantly between −170 V and 170 V, but we generally say that the voltage is 120 V. This is because we quote the root mean square (RMS) of the voltage,[8] and when we calculate the RMS of the sinusoidal curve shown in figure 5.6, sure enough we get 118 V. This is close to the 117 V that we discussed earlier.

A pretty epic and nasty battle occurred in the late 1800s between Thomas Edison and Nikola Tesla. Edison advocated for the widespread use of DC power, while Tesla was advocating for AC power. In the end, AC power proved to be a much better choice.[9]

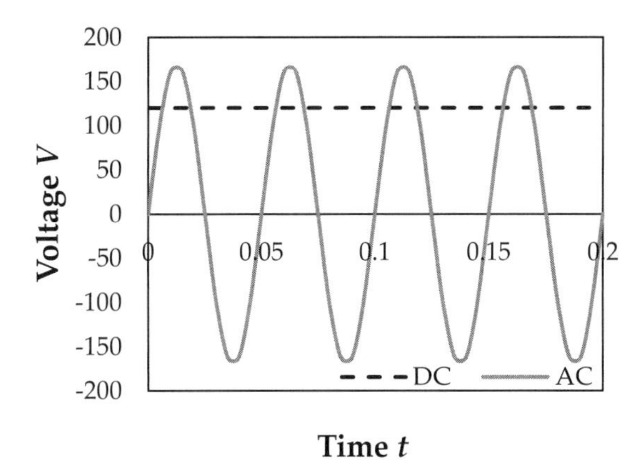

Figure 5.6
Alternating current (AC) and direct current (DC).

The reason is quite simple, and it boils down to simple economics. To minimize transmission losses, voltage should be high (we will see why in the next paragraph). At the power plant, the voltage that is generated therefore must be transformed to a high voltage, and it will be transformed to a low voltage when it reaches individual buildings. While this is fairly easy to do with AC power (we just need a basic transformer), it is pretty difficult to do with DC power. This means that if the grid used DC power, we would need power plants in every single neighborhood—that is, power plants would service buildings within roughly a 1.6 kilometer (one mile) radius, which does not make sense.

Why is high voltage better to minimize the losses? Remember equation (2): $V_C = V_S - RI$. V_S is the voltage that we supply from the power plant, and V_C is the voltage in the line. V_S is also P/I, which gives us:

$$V_C = \frac{P}{I} - RI \tag{14}$$

We know that the term *RI* represents the losses, which we want to be as low as possible when distributing electricity, and our power output *P* is fixed; here, *R* is simply a property of the wire that we use (like friction losses in water pipes). This means that *I* should be as small as possible, which directly translates into a high voltage V_c.

To be fair, DC power can be transmitted in high voltages too, and in fact, losses are smaller with DC power than for AC power. Again, the problem is the conversion process from low to high and then from high to low voltage. High-voltage direct current

(HVDC) lines do exist in the world, but they only make sense economically when they are very long.[10]

AC power is therefore a better option, but this is not the end of the story. The grid does not have only one AC line, but three.

5.1.5 Three-Phase Power

Most generators that produce alternating current are alternators. Alternators tend to consist of magnets that turn around a shaft (called a rotor) inside a hollowed cylinder with coils on its surface (called a stator). To make more efficient use of the alternator, several coils can be installed, thus generating several alternating currents. The standard is to use three sets of coils, placed at an angle of 120 degrees from one another.[11] Figure 5.7 shows a schematic of a typical *Y* or *star* connection. In practice, the three phases are called *a*, *b*, and *c* in the United States. At the center of the Y connection we can see a neutral line.

The voltage between any of the three points and the neutral is 120 V. This means that $V_{an} = V_{bn} = V_{cn} = 120$ V. The sine waves for the three voltages are shown in figure 5.8.

This is pretty significant because now we are generating three times the amount of power. This is why transmission lines will have primarily three cables, one for each phase (or six cables for double-circuit transmission). One or two cables are normally present at the very top, but they are only there to prevent any damage from lightning

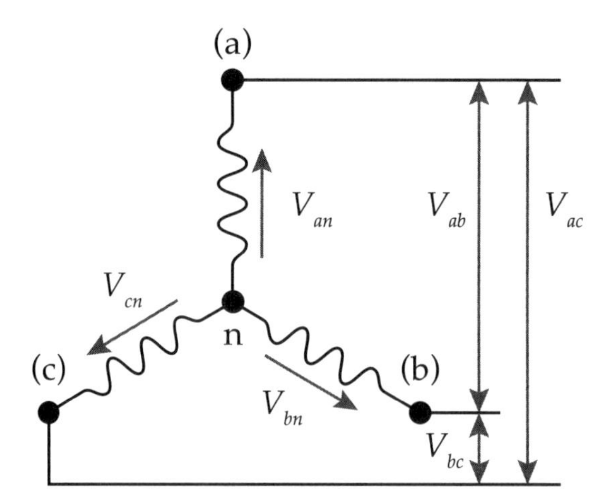

Figure 5.7
Three-phase-power Y connection.

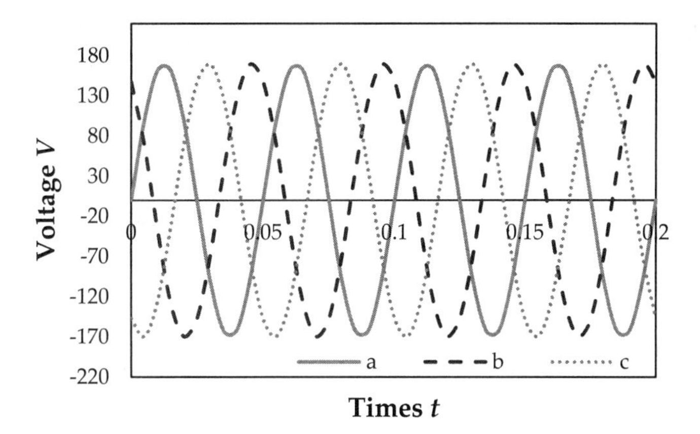

Figure 5.8
Voltages in three-phase-power systems.

strikes—they are in fact linked to another cable attached to the utility pole that runs into the ground.

For distribution lines in cities, the same three a, b, and c cables are present, as well as the static wire for lightning protection. In some cities, telecommunication cables may also be attached to utility poles (i.e., television and Internet), which is why you might see more than three or six cables (plus one or two cables sometimes for lightning protection). Moreover, telecommunication cables are generally placed lower on utility poles since they are "safer"—that is, we cannot get electrocuted from them.

In the United States, most buildings are only serviced by two of the three phases. Therefore, most buildings receive two cables—one for each of the two phases—as well as a third cable for the neutral line. The medium-voltage lines on the utility poles are first converted to low voltage (i.e., 120 V). This is done by transformers—the gray devices on top of utility poles that sort of look like garbage bins, as shown in figure 5.9 (*left*). Also notice the three wires on the top of the utility pole, one for each phase. The two phases are also shifted so that they supply alternating currents at 180 degrees from each other. This is practical because if you now measure the voltage directly between the two phases, then you get a voltage of 240 V as opposed to 120 V, and many house appliances require the extra voltage,[12] such as heating, ventilation, and air-conditioning systems (HVAC), as well as some dishwashers, stoves, ovens, electric water heaters, washers, and dryers.

To know which outlet uses which phase, simply open your electrical panel, as shown in figure 5.9 (*right*; courtesy of Dongtian Ji[13]). All circuit breakers on the left-hand side are connected to one phase, and all circuit breakers on the right-hand side

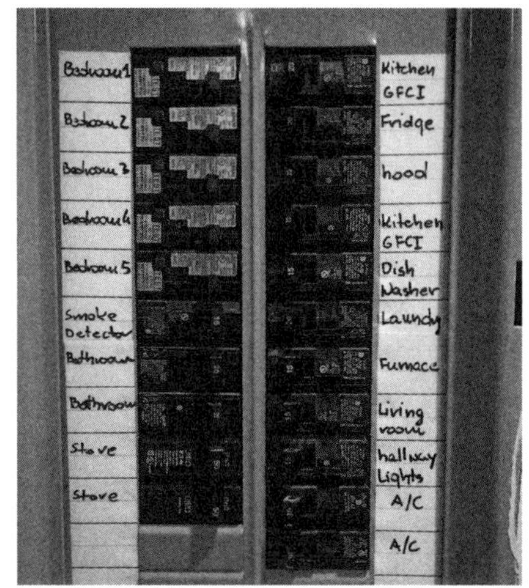

Figure 5.9
Electrical equipment: utility pole and transformer (*left*) and electrical panel (*right*). Courtesy of Dongtian Ji.

are connected to the other phase. You will also see two types of circuit breakers: (1) single-pole breakers that provide 120 V and (2) double-pole breakers that provide 240 V (as discussed). On figure 5.9 (*right*), we can see two double-pole breakers, one for the stove and one for the air-conditioning unit.

Finally, the two wires that carry the electricity should be black, the neutral line should be white, and you may see a green wire, which is the ground.

In areas that use 240 V directly, typically only one phase comes into the building. In Europe, for example, electric panels have several rows of horizontal breakers (as opposed to vertical in the United States), and the colors of the wires are different.

It is time to put all this knowledge together and learn how the entire power grid works.

5.1.6 The Power Grid

From its humble beginnings, the power grid has become this giant network of wires that run in every nook and cranny of cities. The very first power grid originated in Chicago, when Samuel Insull (who worked with Thomas Edison) bought and merged many power utility companies together in the early 1900s.[14]

Quite naturally, the power grid starts at the electricity generator and ends at the device that consumes electricity. To quote Glover (2012, p. 1), the power grid "is probably the largest and most complex industry in the world." Figure 5.10 shows a schematic of the power grid, which consists of three main elements: (1) generation, (2) transmission, and (3) distribution.

Electricity generation consists of the systems of power plants that generate the electricity that we consume. They range from significantly large power plants, generating 200 MW of electricity, to small, neighborhood-scale power plants (including small wind turbines) that generate 2–3 MW of power.

If we remember that in the United States an average household consumes 1.25 kW of electricity, a 200 MW plant can power 200×10^3 kW / 1.25 kW = 160,000 homes. Electricity generation is fundamental to our sustainability goal, and we will review most types of power plants in section 5.3. Although power plants generate electricity at voltages between 2.3 and 22 kV, it is then "stepped up" to more than 200 kV for transmission.

Electricity transmission consists of the systems of large power lines that transmit power across states or even nations, carrying electricity at more than 200 kV (i.e., 200 kV is close to 1,700 times higher than the common 120 V) and typically around 345 kV. These are the lines that we often see when driving on highways (the next time you see them, count the number of wires). Electricity distribution is similar to transmission but at much lower voltages. These are the lines that we see in streets and that bring electricity to individual households. The voltages are much lower, from 50 kV to 1 kV. The cables are mostly above ground and in back alleys in North America and Asia, but they tend to be underground in Europe.

In this chapter we will mostly focus on electricity generation and ignore, for the most part, electricity transmission and distribution. However, we should account for average transmission and distribution losses of about 6% whenever we make calculations.

The power grid is incredibly large and complex, and it is difficult to make significant changes. At the time of this writing, in the United States, the power grid is divided in three: Eastern, Western, and Texas.[15] Having a large grid offers substantial benefits since it helps balance loads. However, it can also make the entire system more vulnerable to cascading failures.

The power grid personally reminds me of a quote by the Roman emperor Tiberius, who had problems controlling his people: "We have the wolf by the ear, and we can neither hold him nor safely let him go." The quote was also used several times by Thomas Jefferson. Adapting a power grid that has evolved tremendously since the 1880s is no small feat, especially when dealing with a dynamic demand that requires a variable power generation system while keeping in mind that all electricity produced

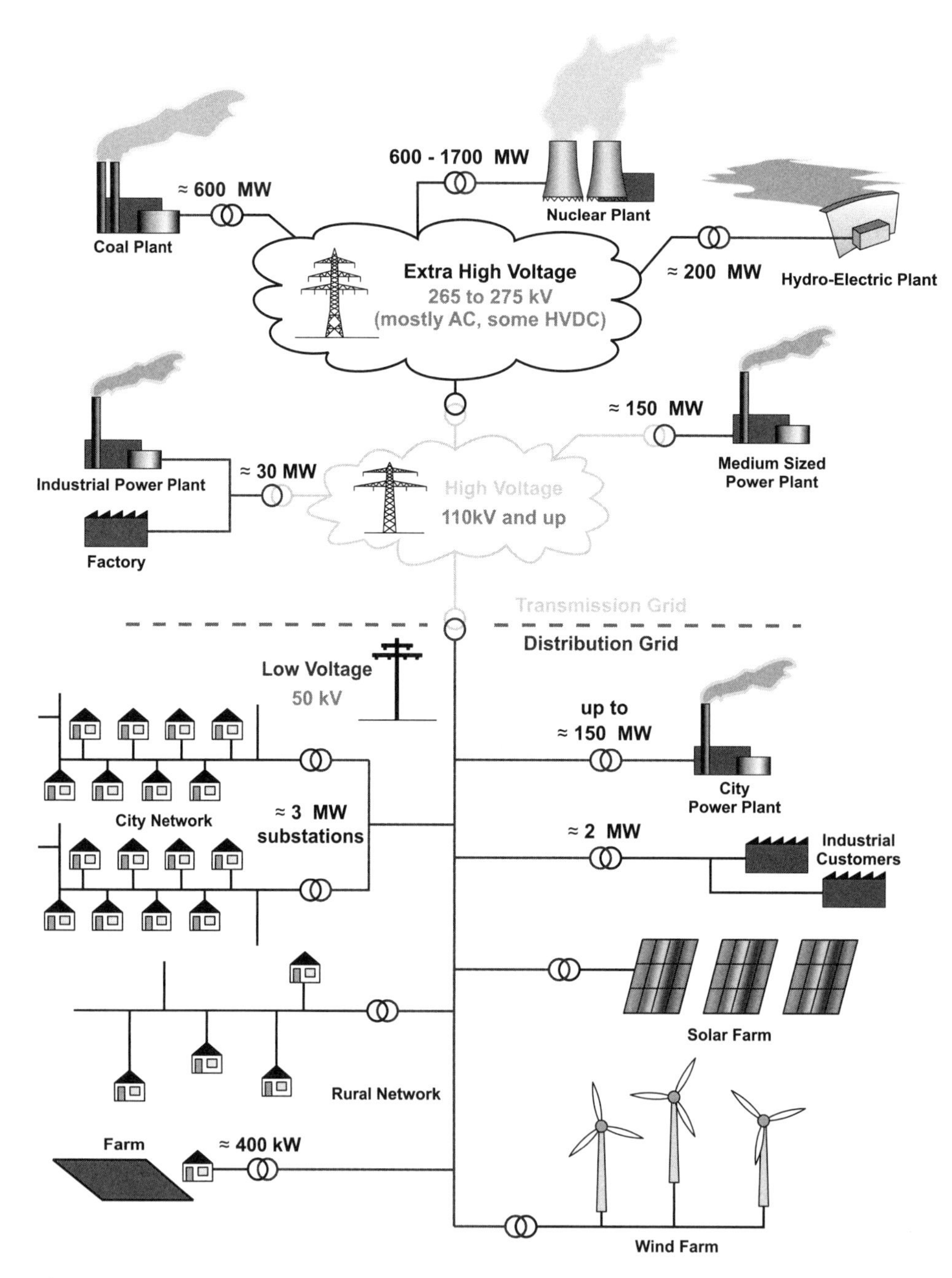

Figure 5.10
Schematic of the power grid.
Source: MBizon (CC-BY).

Example 5.4
Calculating Transmission Losses

A new 30 MW hydroelectric power plant is about to be built in Civitas. The plant is located 100 km away from the city, and the transmission line has a resistance per unit length of 4×10^{-4} Ω/m. A transformer is used to supply electricity at 200kV at the power plant. Calculate (1) the voltage drop and (2) the percentage of power lost due to line losses.

Solution

(1) The resistance of the line is $R = 4 \times 10^{-4} \times 100 \times 10^3 = 40\ \Omega$. Since $P = V \times I$, the current is $I = P/V = 30 \times 10^6 / 200 \times 10^3 = 150$ A. The voltage at the end of the line:

$V_c = V_s - RI = 200{,}000 - 40 \times 150 = 194{,}000$ V

The voltage drop is therefore 200 kV − 194 kV = 6 kV.

(2) The power that can be provided in Civitas is $P = V \times I = 194 \times 10^3 \times 150 = 29.1 \times 10^6$ W = 29.1 MW. Therefore the percentage loss in power is:

$$\text{Losses} = \frac{30 - 29.1}{30} = 0.03 = 3\%$$

Similarly, the voltage drop represents a $100 \times (6/200) = 3\%$ loss.

must be consumed almost instantly. And yet, the demand for electricity fluctuates constantly, which is the subject of the next section.

5.2 Electricity Demand

The whole reason why we produce electricity is to meet a demand. This means that we produce electricity in such a way that it follows the same characteristics as demand. Put differently, if demand increases over time, electricity generation must also increase over time. If demand is variable during the day, then electricity generation must also be variable during the day.

In this section we will cover three topics related to electricity demand. First, we will see how demand has evolved over time in the United States and how consumption varies by state. Second, we will pay close attention to how demand evolves during a typical day. In fact, we will see that demand is far from being constant throughout the day, and we must account for that if we are to find a Scenario B that works. Finally, we will investigate the power consumption of typical home appliances. After all, it is these appliances that consume the electricity, not the people, and lowering their power needs can help us decrease the overall demand.

5.2.1 Temporal and Spatial Analysis of Electricity Demand in the United States

At the beginning of this chapter, we learned that the average household in the United States consumed 10.8 MWh of electricity in a year (2016 value). This was not always the case, however. In fact, in 1950 the average yearly household consumption was around 1.6 MWh (Fox-Penner 2010).[16] Figure 5.11 shows the historical evolution of electricity demand for residential purposes in the United States from 1960 to 2014. The solid line is the per household value, which increased from 4 MWh in 1960 to about 12 MWh in 2010 and then decreased slightly. The dashed line is the total residential electricity demand, which increased from 200 million MWh (= 200 TWh) to close to 1,450 TWh in 2010. In fifty years, from 1960 to 2010, demand therefore increased by a factor of three on a per-household basis and by a factor of seven in absolute terms.

Linking these numbers back to power production, a yearly electricity demand of 1,400 TWh translates into an average power requirement of $1{,}400 \times 10^6 / (24 \times 365) \approx$ 160,000 MW. This amount of power is enormous, and the power grid and the number of active power plants must therefore be able to provide this power at all times. What is more, this is just the power required for residential purposes. When we add businesses, industries, and all other sectors along with residential, the total yearly electricity consumption in the United States came to 3,695 TWh (2012 value), which is 2.64 times higher than the residential demand. This demand required an average power of more

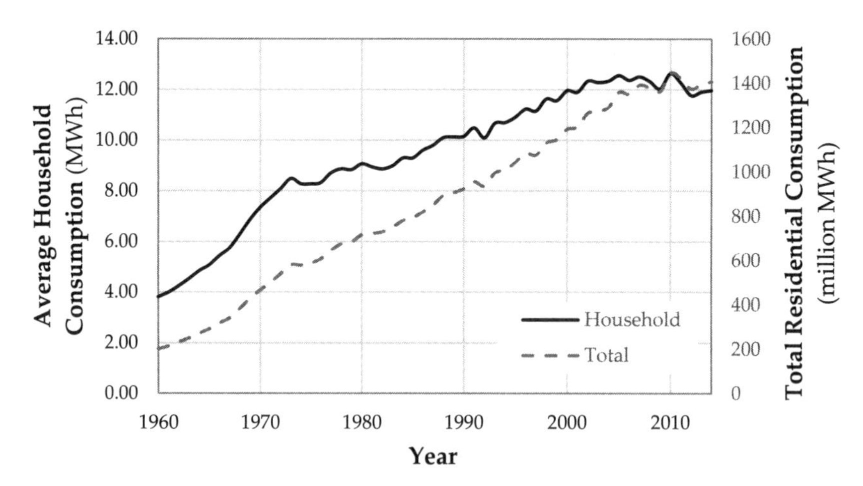

Figure 5.11
Historical evolution of residential electricity demand in the United States.
Sources: U.S. Energy Information Administration and U.S. Census Bureau.

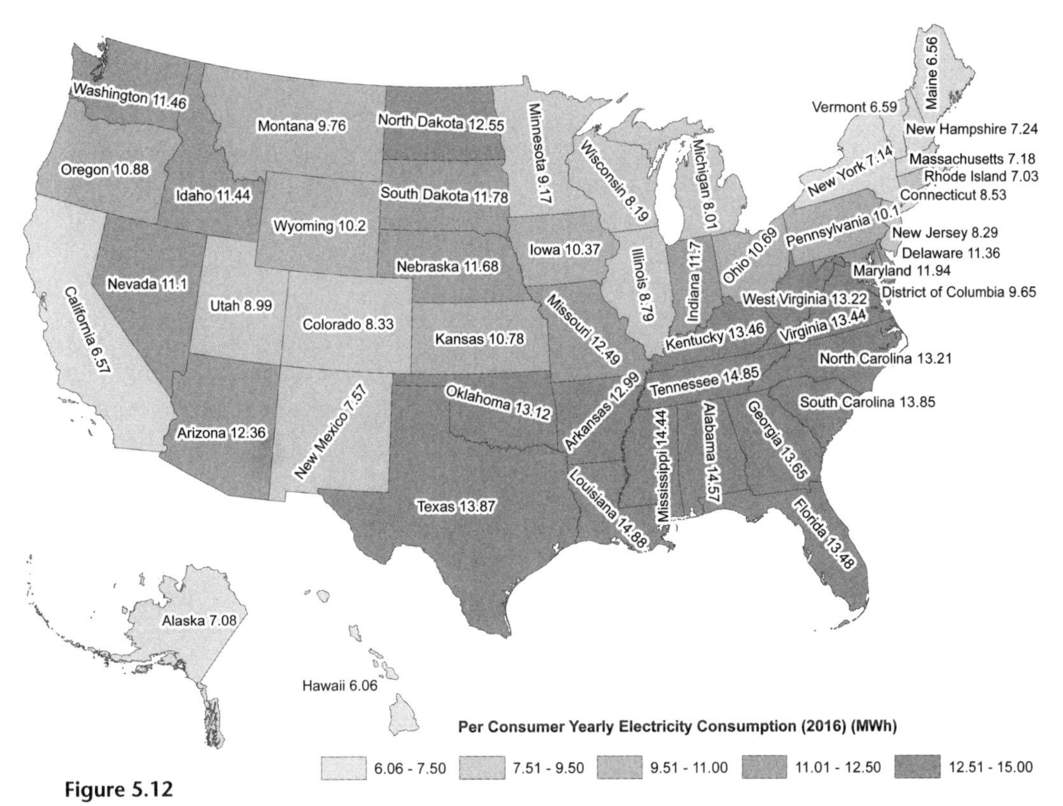

Figure 5.12
Average U.S. household electricity consumption in 2016 by state.
Source: U.S. Energy Information Administration.

than 420,000 MW. In section 5.3 we will understand why these numbers are incredibly high when we learn how much power a typical power plant can generate.

Demand also varies geographically. Warm places tend to consume more electricity because of space-cooling requirements. Figure 5.12 shows the average household yearly electricity consumption by state in the United States in 2016. As we will see, water heaters are the largest consumers of electricity in households (assuming heat is provided by natural gas), but they are closely followed by air conditioners. With the exception of New Mexico, we see that southern states stand out, which directly relates to their high consumption of electricity for space cooling (i.e., running air conditioners). Climate has a large impact on electricity consumption, and it is not surprising that more temperate states like Maine and California have lower electricity usage then Texas and Louisiana.

Out of all the northern states, we see that North Dakota stands out. At 12.55 MWh in 2016, the average electricity consumption of a household is higher than the consumption of its neighboring states—for example, Minnesota is at 9.17 MWh and Montana at 9.76 MWh. This high electricity consumption suggests that many North Dakotans may use electricity to heat their homes, as opposed to using natural gas. That would be fine if the power grid emission factor of North Dakota was low, but it was actually among the highest in the United States, as we will see in the next section. This is a substantial problem in terms of GHG emissions (burning gas directly for heating would be preferable).

Similar to population, predicting future demand is significantly challenging. This is, however, critical, especially in the electricity sector, which must respond to demand instantly. As new power plants cost millions of dollars, selecting the most effective is not an easy task. Fox-Penner (2010, p. 124) gives an account of this process for a large power provider in Missouri.

Electricity demand has therefore evolved quite a lot in the past. In addition, demand patterns differ based on geography, making it difficult to predict how demand will evolve. On top of that, we should remember that demand varies constantly throughout the day, and this is the topic of our next section.

5.2.2 Real-Time Electricity Demand

The fact that power demand varies during the day is quite intuitive. At night most people sleep and demand is small. During the day it increases, first in the morning when we wake up and turn on the light, then again when we are active during the day, and finally again in the evening, when we are at home cooking and watching TV. This variation is indeed expected, but the question is by how much.

Figure 5.13 shows the power demand profile of New England for August 14, 2018, along with the forecasted demand for the day after. August 14 was actually a typical summer day. Exactly as we predicted, the demand was lowest at night and started to increase at 4:00 a.m., initiating the morning ramp. It then increased more during the day but actually peaked in the evening, at around 6:00 p.m.—when people were cooking and perhaps still running their air conditioners—before decreasing toward the night. In the case of New England, the lowest demand was 11,835 MW at 2:25 a.m. This value represents the minimum value that must be supplied, and it is called the *base load*. This base load is generally provided by the power plants that we want to use the most, like nuclear and hydroelectric,[17] as we will see in section 5.3. The highest demand, on the other hand, was 20,289 MW, which occurred at 4:50 p.m. and which was 1.71 times higher than the base load.

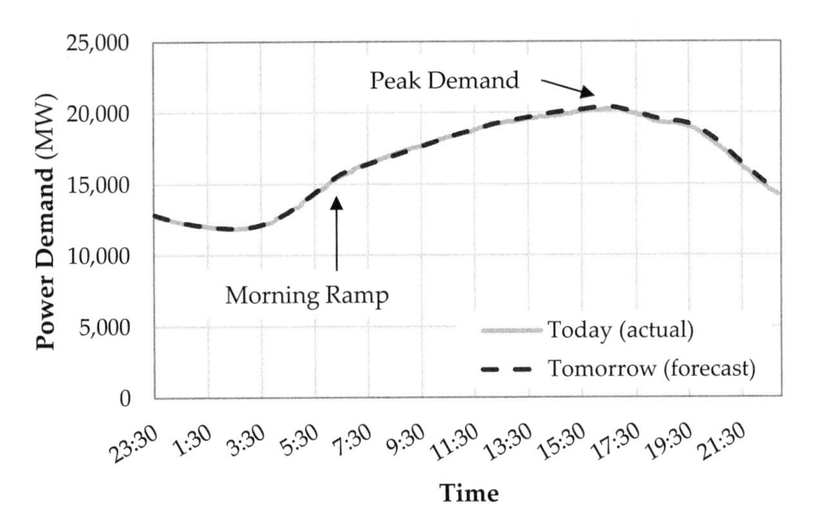

Figure 5.13
Power demand in New England for August 14, 2018.
Source: Independent System Operator New England (2015).

The difference between the base load and the peak demand is substantial. Essentially, an extra 8,500 megawatts must be provided by power plants that were not running during the night. There is therefore a need for additional power to be generated at will. Moreover, this power must be reliable, which is a characteristic that does not agree well with intermittent wind and solar power. Nevertheless, as we can see, power demand is relatively easily predictable, and it tends to follow the same curve every day. The general curve decreases in the winter—because air conditioners are not running anymore—but the shape remains generally the same. Although it depends on several factors, including local weather, predicting demand, day after day, is manageable.

To accommodate this variability in demand, a series of power plants are used every day, some of which only function for a short period of time. Figure 5.14 shows how electricity generation is typically distributed by power plants during a day.

First, a base load must be ensured, which is the minimum amount of power that must be produced at all times. This is typically supplied by the "best" power plants—the most efficient ones that we can run at 100% performance. In addition to nuclear and hydroelectric power plants, these include new coal-fired power plants and natural gas plants that are more energy-saving than older ones but still emit a significant amount of GHGs. Counterintuitively, these power plants normally include all renewables,

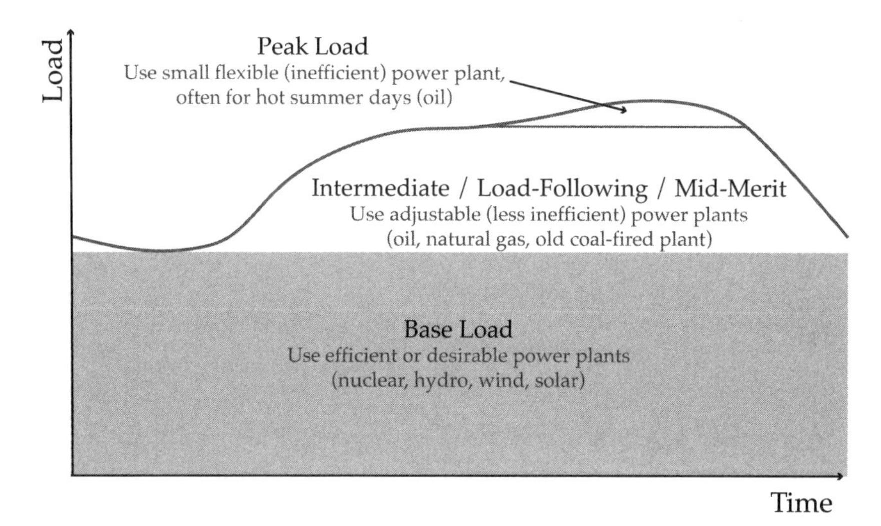

Figure 5.14
Matching power generation with load profile.

including wind and solar farms. Even though the base load is not constant in that case, we want to use renewable energies as much as possible. In fact, the base load is rarely as flat as figure 5.14 suggests—notably, because of maintenance and repairs. Renewables, including geothermal, biomass, solar thermal, wave, and so on are therefore part of the base load, as we will discover in the next section.

To supplement this base load during the day, other power plants are used. These power plants need to be adjustable. Old coal-fired plants, as well as oil and natural gas plants, are common for this purpose. Depending on the region, some areas may require additional plants to run for peak load. This is especially important on hot summer days when the demand for space cooling is high. These are also the days when the sun is the strongest and therefore when solar energy is at its best. Having a solar power plant as part of a grid can be a good strategy, especially in warm places.

A initiative called *demand response* also encourages people to lower their electricity use in summer months at peak times. Customers are asked to lower their demand during a peak, and they get credited money for the electricity they did not use. This initiative has a bright future in the age of the smart grid as we will discuss later. But first, if one of our end goals is to decrease the demand, then we need to know what drives that demand. In the next section, we quickly review typical load patterns of home appliances.

5.2.3 Typical Power Rating of Appliances

Loads are generally divided in two categories: (1) resistive loads and (2) inductive loads. Resistive loads are appliances that are actual resistances, like a light bulb or a toaster (think of figure 5.2, which illustrates Ohm's law). Appliances with inductive loads tend to require more power when they are started, and they create a small spike in the demand—this is why sometimes when you start an appliance, you might see a light bulb flicker a little. Inductive loads usually have a motor like refrigerators and dishwashers.

Table 5.1 shows a list of appliances with their typical power ratings, annual usage, and a yearly consumption estimate in kWh. The nice thing about this table is that it shows us how much power is drawn when the appliance is idle and off. TVs, for instance, consume about 97 W when they are on, but even off they consume about 4 W. The larger appliances, like refrigerators, are hard to estimate, and we therefore do not have an annual usage but an average power when the motor is running and a final energy in kWh. Other appliances also do not add up, therefore the final annual consumption is the more accurate value to consider. Overall, we see that the largest consumer of electricity is the water heater, with 4.5 kW.[18] As an informal guide, anything that heats something up will consume a lot of power.

In our quest to reduce demand, these are the appliances that must be targeted. The use of compact fluorescent light bulbs is a big success for energy conservation, and other appliances will soon reduce their power needs. Other solutions are to recuperate energy wasted during a process. Some water heaters now exist that use the heat left from the water going down the drain to preheat water. These sorts of ideas can significantly reduce the current demand. This is especially important since the population will increase, which means demand will likely increase as well.

This concludes our section on electricity demand. It is now time to move on to the main reason why electricity is responsible for such a proportion of GHG emissions. Let us learn how electricity is generated.

5.3 Electricity Generation

The main reason why electricity is crucial for sustainability is because of electricity generation. There are many different ways to generate electricity. Again, the whole objective is to find a source of energy, whatever it is, and convert it to electricity. The most obvious sources of energy, which are the ones we have always used, are those present in nature, usually in the form of mechanical power, that we can convert to electricity

Table 5.1
Typical power rating and yearly consumption of common appliances

	Power draw (W)			Annual usage (h/year)			Annual consumption (kWh/year)
	Active	Idle	Off	Active	Idle	Off	
Kitchen							
Coffee maker	1,000	70	0	38	229	8,493	58
Dishwasher	1,300			–			120
Microwave oven	1,500		3	70		8,690	131
Toaster oven	1,051			37			54
Refrigerator–freezer	180			–			660
Freezer	200			–			470
Lighting							
18 W compact fluorescent	18			1,189			20
60 W incandescent lamp	60			672			40
100 W incandescent lamp	100			672			70
Torchiere halogen lamp	300			1,460			440
Bedroom and bathroom							
Blow-dryer	710			50			40
Waterbed heater	350			3,051			1,070
Laundry room							
Clothes dryer	3,000			–			1,000
Clothes washer (exc. water heating)	500			–			110
Home electronics							
Desktop PC	75	4	2	2,990	330	5,440	237
Notebook PC	25	2	2	2,368	935	5,457	72
Desktop computer monitor	42	1	1	1,865	875	6,020	85
Stereo system	33	30	3	1,510	1,810	5,440	119
Television	97		4	1,860		6,900	222
Analog, < 40 in.	86			1,095			184
Analog, > 40 in.	156			1,825			312
Digital, ED/HD TV, < 40 in.	150			1,095			301
Digital, ED/HD TV, > 40 in.	234			1,825			455
Set-top box	20	0	20	6,450	0	2,310	178
DVD/VCR	17	13	3	170	5,150	3,430	78
Video game system	36	36	1	405	560	7,795	41

Table 5.1 (continued)

	Power draw (W)			Annual usage (h/year)			Annual consumption (kWh/year)
	Active	Idle	Off	Active	Idle	Off	
Heating and cooling							
Air conditioner (window)	1,000			–			–
Air conditioner (central)	3,500			–			–
Dehumidifier	600			1,620			970
Furnace fan	295			1,350			400
Ceiling fan (only fan motor)	35			2,310			81
Space heater	1,320	1		584			314
Water heating							
Water heater—family of 4	4,500			64			4,770
Water heater—family of 2	4,500			32			2,340
Portable spa	4,350	275		25	8,735		2,525
Miscellaneous							
Rechargeable power tool	13	4		73			38
Vacuum cleaner	542			37			55
Pool pump	1,000			792			790
Well pump	725			115			80
Lawn sprinkler	11			0			32
Aquarium equipment	24			6,534			153

Adapted from U.S. Department of Energy (2012).

(e.g., moving water for hydroelectricity). An alternative is to use a chemical reaction to create a mechanical movement, from which we can then generate electricity.

In this section we will go through twelve different ways to generate electricity. Although more ways exist, these twelve are the most common. Importantly, we will specify where the primary energy source comes from, and we will review typical efficiencies that are critical for us. We will then group all of these into one table to analyze their GHG emissions properties. This is naturally key since we want to emit as few GHGs as possible. To quantify GHGs, the common unit used g CO_2e/kWh, which is the same as kg CO_2e/MWh or t CO_2e/GWh.

There are also several ways to classify electricity generation methods. Figure 5.15 shows one such classification. Horizontally, the methods are classified by energy type: thermal, kinetic, and other. Usually, most thermal plants use the primary energy to boil water and generate steam, which then turns a turbine that produces electricity. The exceptions are oil and natural gas plants that work like jet engines by simply turning a

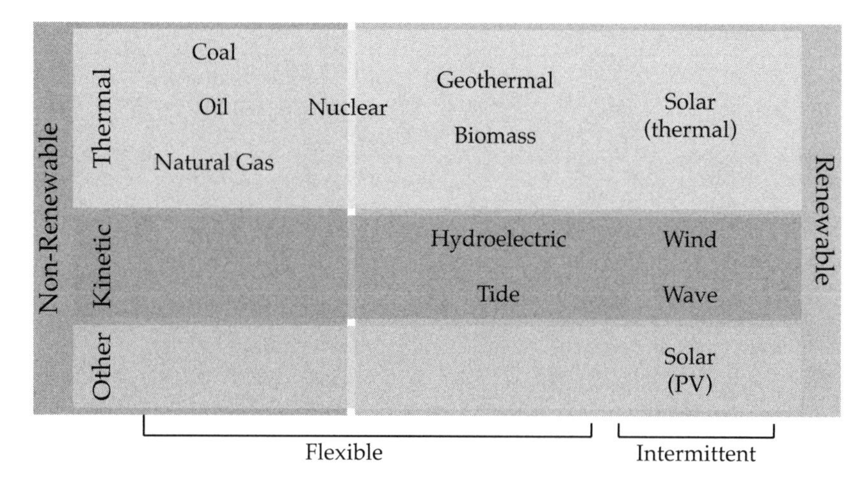

Figure 5.15
Electricity generation classification.

shaft to directly create electricity. Thermal plants are by far the most common. Kinetic methods require a primary source that is already moving—that is, possessing kinetic energy—which we can harvest directly, such as wind and hydro. Finally, although they are less common, some methods can generate electricity without involving any moving parts, and the most famous is solar photovoltaic (PV).

Vertically, the methods are classified in two ways. First, all the methods enclosed in the box on the right-hand side harvest "renewable" energies; the one exception is nuclear, which is not renewable, but because it does not emit any GHGs, we tend to group with it with the renewables. The three nonrenewable sources, coal, oil, and natural gas, belong to the family of fossil fuels. Fossil fuels, as well as nuclear fuels,[19] only exist in finite quantities on Earth and are therefore technically nonrenewables.

Finally, the methods are also classified as to whether they can generate electricity in a "flexible" manner (i.e., whether the rate of electricity production is flexible and can be predicted) or whether they are intermittent (i.e., like wind energy, which depends on weather). Intermittent technologies pose an important problem since it is difficult to ensure that the supply can meet the demand—that is, what happens on a windless day for wind energy?—although in some contexts, they have specific properties that can make them better suited to produce electricity than flexible technologies, as we will see later.

Naturally, these methods are not split evenly. Figure 5.16 shows the share of electricity generation by source in 2014 for the world (*left*) and in 2016 for the United States (*right*).

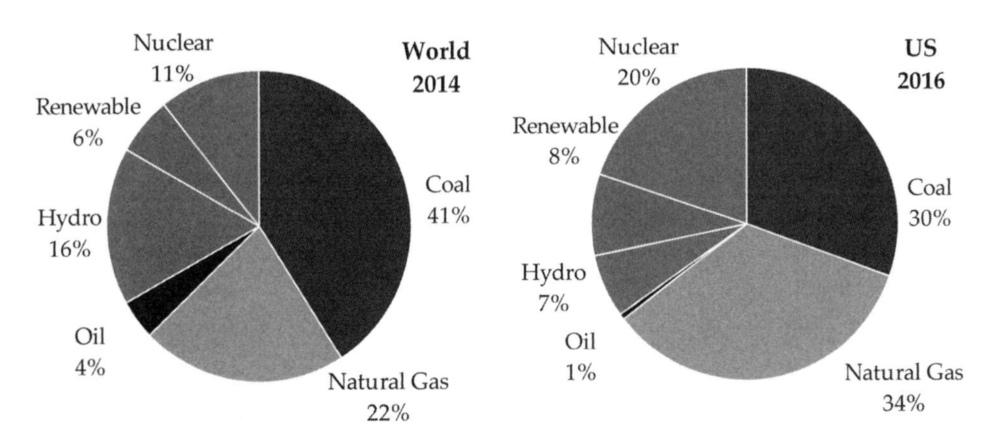

Figure 5.16
World and U.S. electricity generation by source in 2014.
Sources: World Bank and U.S. Energy Information Administration.

Although its share is slowly decreasing, coal is heavily used in the world to produce electricity. Coal used to be the largest energy source used to generate electricity in the United States as well, but the supply of cheap natural gas changed the trend, and in 2016, natural gas was the dominant source of energy.

As coal is the heaviest polluter, let us therefore start with it.

5.3.1 Coal-Fired Power Plants

Coal-fired power plants tend to be the most common type of power plant in the world. This is partly due to the abundant amounts of coal that can be collected at the surface of the earth (in contrast to oil) and easily transported. Moreover, coal-fired power plants can be shut down and restarted relatively easily, and older plants are typically used to accommodate peak demand, as we saw in the previous section.

As shown in figure 5.17, in coal-fired power plants coal is burned and pulverized to boil water, which creates super-heated steam. This steam then turns a turbine, which creates electricity (e.g., using an alternator). The process is relatively simple and common to many thermal power plants.

A typical coal power plant produces about 500 MW of power, which can serve about $500 \times 10^3 / 1.25 = 400{,}000$ homes in the United States. Small 5 MW plants exist, however, and the largest coal plant in the world produced 6,720 MW of power at the time of this writing. Coal-fired power plants have a relatively low efficiency of about 35%. This means that for an equivalent 100 Wh of coal burned, only 35 Wh of electricity is produced.

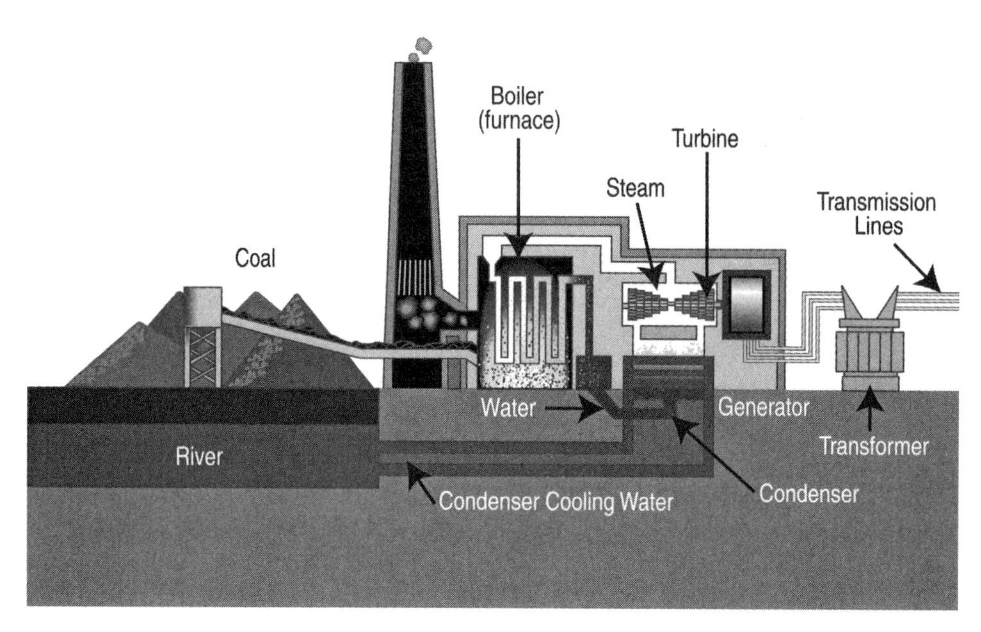

Figure 5.17
Schematic of coal-fired power plant.
Source: Tennessee Valley Authority.

Some projections estimate that we have enough coal on Earth to last roughly to 2100. This is actually a very difficult question to answer, and you will find many websites showing different numbers. On the one hand, we may find more coal reserves, but on the other hand, we may burn more coal if we run out of oil and gas first. The simple bottom line is that coal is a finite resource, and burning coal emits the GHGs that are contributing to climate change. We simply cannot rely on coal in the long term, and we need to find alternatives.

On average, a coal power plant emits 1,000 g CO_2e/kWh, placing it as the largest emitter of GHGs for electricity generation. On the plus side, new devices are being installed on coal power plants to capture almost all the CO_2 before it is emitted.[20] These systems can in fact be installed on any carbon-emitting facility, which not only includes oil and gas plants but also cement plants, steel mills, and so on. Preventing emissions is therefore highly beneficial, and emissions can decrease significantly. However, what to do with the captured carbon is an unsolved problem.[21] Moreover, the device's effectiveness remains to be proven. More importantly, capturing the carbon does not solve the nonrenewable aspect of coal. Capturing the CO_2 emitted may therefore be an acceptable strategy in the medium term, but coal will eventually need to be phased out.

5.3.2 Oil- and Natural Gas–Fired Power Plants

Oil- and natural gas–fired power plants work in similar ways, but they do not boil water to create steam. Instead, the fossil fuel is burned directly in an engine (similar to jet engines that power aircraft), which directly turns a turbine that produces electricity. In figure 5.17, the entire process left of the turbine is therefore simply replaced by one motor. These types of plants can also be used for peak demand.

Oil-fired power plants have a low efficiency of about 31%, while the efficiency of natural gas plants is closer to 42%. However, the loss of energy is mainly in terms of heat during the combustion process. This heat can therefore be recuperated through a heat exchanger and used to heat buildings (especially relevant for district heating systems, as we will see in chapter 8). These types of plants are called combined heat and power plants, and their efficiency can be as high as 85%. Some heat can also be recuperated from nuclear and coal-fired power plants.[22]

A typical natural gas power plant produces about 200 MW (i.e., 160,000 homes), but sizes can easily range from 20 MW up to 5,600 MW for the largest gas plant in the world at the time of this writing. Small-size oil-fired power plants work well in smaller communities, and they are even used to generate electricity on large ships.

Oil-fired power plants are not as common, as we can see in figure 5.16. This is partly due to the fact that we use oil for many other purposes, and this is partly good news for us. Indeed, the main difference between oil and natural gas power plants is in their GHG emission factors. Natural gas plants emit, on average, two-thirds as many GHGs as oil plants, with a typical emission factor of 500 g CO_2e/kWh, compared to 750 g CO_2e/kWh for oil. It is therefore preferable to use natural gas to produce electricity rather than oil (and coal). Here again, CO_2 can be captured in both oil- and natural gas–fired power plants.

Some project that the remaining reserves of oil and natural gas will run out by about 2060. What we just saw on coal is equally applicable here. We also need to remember peak oil from chapter 2 and all the problems associated with it (e.g., wars) before we run out of oil. In fact, *peak coal* is an important problem too.

This concludes the list of fossil fuel power plants. We will now review the alternative methods to generate electricity that are much "cleaner" and do not emit any GHG directly.

5.3.3 Nuclear Power Plants

Nuclear power plants are also thermal power plants. In fact, the way they produce electricity is similar to coal-fired power plants. Nuclear reactors simply produce heat that is used to create steam that is then used to turn a turbine. As opposed to coal, however, the fuels used are uranium or plutonium. No combustion is happening in the process, and therefore no GHGs are emitted. Nonetheless, once the energy from the fuels has been harvested, the resulting nuclear wastes are extremely radiotoxic and must be contained securely.[23] Moreover, similar to coal plants, the efficiency of currently operating nuclear power plants is about 33%, although advanced high-thermal-efficiency nuclear plants could be achieving efficiencies of 40% at the time of this writing. We will later discuss the life cycle emissions that account for all GHG emitted during the lifetime of the plant, therefore accounting the construction, operation, and disposal of all elements of the plant.

The large gray towers that we typically see in pictures of nuclear power plants are cooling towers. The white smoke coming out of the towers is simply water. This is why nuclear plants are often located close to bodies of water, like rivers, that can supply large quantities of water. This is part of something called the *energy-water nexus*. Simply put: we need electricity to treat and distribute water, and we need water to generate electricity (back to our infrastructure integration problem).

The size of a nuclear reactor is limited to a maximum of 1,600 MW of power—that is, 1.28 million homes—but several reactors can be located on the same site. As a result, a typical nuclear power plant produces about 2,000 MW of power—that is, 1.6 million homes—but nuclear plant power outputs ranged from about 500 MW up to close to 8,000 MW for the largest nuclear power plant in the world at the time of this writing.

The major advantage of nuclear plants is that they function in a similar fashion to coal plants because large facilities can be built virtually anywhere (pending there is water) and produce a great deal of electricity. Moreover, they do not emit any GHG. Although few had been built in years prior this writing, nuclear power plants will likely maintain a strong presence in the power grid, especially as it becomes possible to extract even more energy from uranium and plutonium, which means nuclear wastes are not as radiotoxic. Nonetheless, like coal, oil, and gas, uranium is a finite resource. Simply put, nuclear is better than fossil fuels because at least it does not produce CO_2 emissions, but power plants that use renewable energy are preferable.

5.3.4 Geothermal Power Plants

Geothermal power plants also use steam to turn a turbine and produce electricity. The steam used, however, is found in hot reservoirs below the ground. Several types of geothermal plants exist based on the temperature of the reservoir, some of which need to process the produced steam before it gets used.[24] Although no combustion is required, the steam that is used also carries some other gases, including GHG, but generally in negligible amounts.

Geothermal plants tend to be relatively small, however, producing on average about 100 MW of power—that is, 80,000 homes—with plant sizes ranging from 3.3 MW up to 1,500 MW for the largest geothermal plant in the world at the time of this writing. Moreover, because the steam is at a lower temperature than the other systems we have seen, the efficiency tends to be quite low, at about 12% (Zarrouk and Moon 2014). Despite this low efficiency, in the presence of a hot reservoir, geothermal remains an attractive solution.

5.3.5 Biomass Power Plants

Biomass power plants are again similar to coal power plants in the sense that steam is being generated to turn a turbine. What is burned here, however, are plants and plant-derived matter that "grow," like corn, wood, and sugarcane. Naturally, the burning produces GHG emissions. The logic is that the GHGs emitted were first captured by the plants during their growth, and "new" GHGs are therefore not being emitted.

Biomass power plants can also be used to burn organic waste (i.e., food wastes, sawdust, and even plastics), which may be better than simply putting waste in landfills, but we must wait until chapter 9 to learn more about generating energy from burning waste.

Unlike coal and nuclear power plants, however, and similar to oil power plants, biomass power plants are quite scalable, and their size can range from 0.5 MW—that is, 400 homes—up to 740 MW for the largest in the world at the time of this writing. The average biomass power plant produces about 30 MW of power (i.e., 24,000 homes).

5.3.6 Solar Thermal Power Plants

Our last thermal power plant is solar. We often associate solar energy with PV panels (that we will cover later), but solar energy can also be harvested for its heat. Rays from the sun can be concentrated on pipes that carry water, called a line concentrator system (figure 5.18, *left*). This water is then used to produce steam that turns a turbine. Alternatively, many mirrors can reflect the rays of the sun on one thermal receiver, called a dish system (figure 5.18, *right*), which uses the heat to produce steam here again.

In comparison to solar PV systems, solar thermal plants do not require silicon to produce electricity. Silicon requires a lot of energy to produce, and this is therefore a nonnegligible advantage. In contrast, solar thermal power plants are not scalable. They require a lot of space and a lot of sun. They are therefore ideal for hot and dry areas, and they have been very successful. Figure 5.19 shows a map of the direct normal solar irradiance of the United States.

Figure 5.18

Left, concentrated solar thermal system: line concentrator.
Source: U.S. Bureau of Land Management.
Right, dish system.
Source: SOLUCAR PS10 (CC-BY).

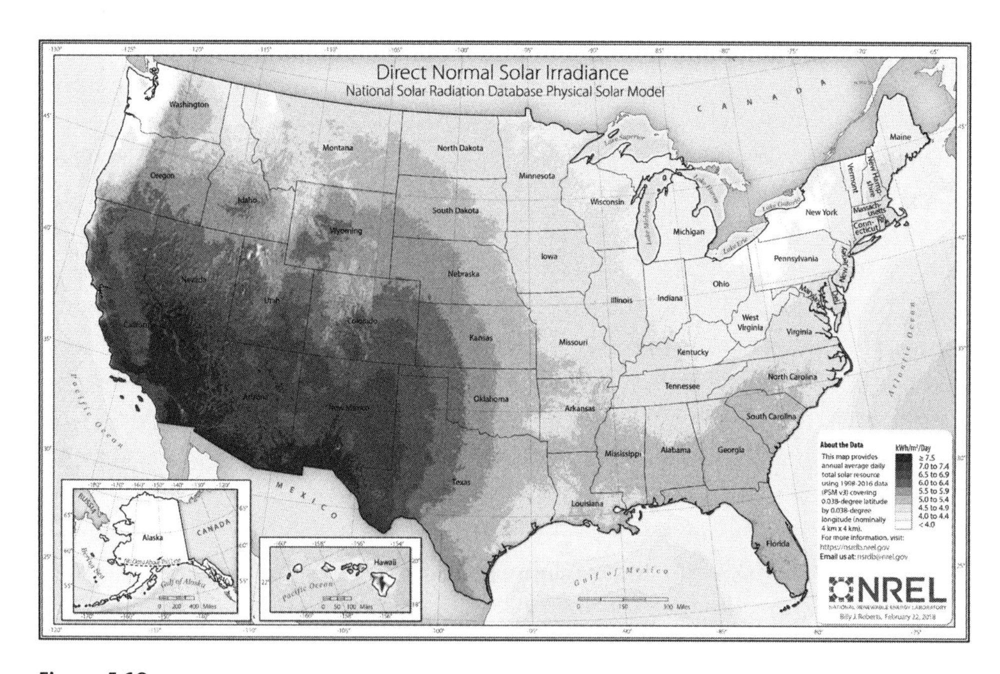

Figure 5.19

U.S. concentrating solar power map.
Source: National Renewable Energy Laboratory.

Solar thermal power plants range from as small as 0.5 MW up to nearly 510 MW of power for the largest solar thermal plant in the world at the time of this writing. It is much tougher to give an average size for solar thermal since they are not scalable, but the average seems to produce around 70 MW of power (i.e., 56,000 homes). One source cites an efficiency of 17%,[25] which is quite high considering the energy needed to create steam—remember that for geothermal, the efficiency was about 12%.

This completes the review of thermal systems. We now move on to mechanical systems.

5.3.7 Hydroelectric Power Plants

Using the mechanical power of water is nearly as old as humanity itself, and the idea to use rivers to produce electricity came almost as soon as electricity started to be generated. Of all renewable energies in the world, hydroelectricity retains the largest share of electricity generation, at 16% in 2014 (see figure 5.16). The concept is quite simple. As figure 5.20 shows, the flow of water is directly used to turn a turbine, which produces electricity.

Thermal power plants convert heat to mechanical energy, which is then converted to electrical energy. This heat-to-mechanical conversion is inefficient (as captured by the laws of thermodynamics). Here, the primary source is mechanical energy, and it is directly converted to electrical energy, resulting in efficiencies of up to 90%. Moreover, because the energy comes from the movement of water, there are absolutely no GHG emissions.

Hydroelectric power plants are therefore very attractive, and in fact, many, if not most of them, have already been built where the potential existed. This is not to say that many hydroelectric power plants will not be built in the future, but they will likely not be able to overtake the share of coal, for instance.

One positive feature of hydroelectric power plants is that they are quite scalable. In fact, microhydroelectric plants can be built to produce power low enough to provide for a house. The US Department of Energy classifies microhydropower as producing less than 100 kW. Small hydropower facilities produce between 100 kW to 30 MW, and large hydropower facilities produce more than 30 MW (U.S. Department of Energy 2015).

The power of water is so vast that the largest power plant in the world at the time of this writing was a hydroelectric power plant that could produce 22,500 MW of power (i.e., 18 million homes). In fact, the five largest power plants in the world were hydroelectric, and the sixth was a nuclear power plant with a capacity of 8,000 MW, nearly three times smaller. Considering the range of sizes, from 1 kW to 22,500 MW, we will

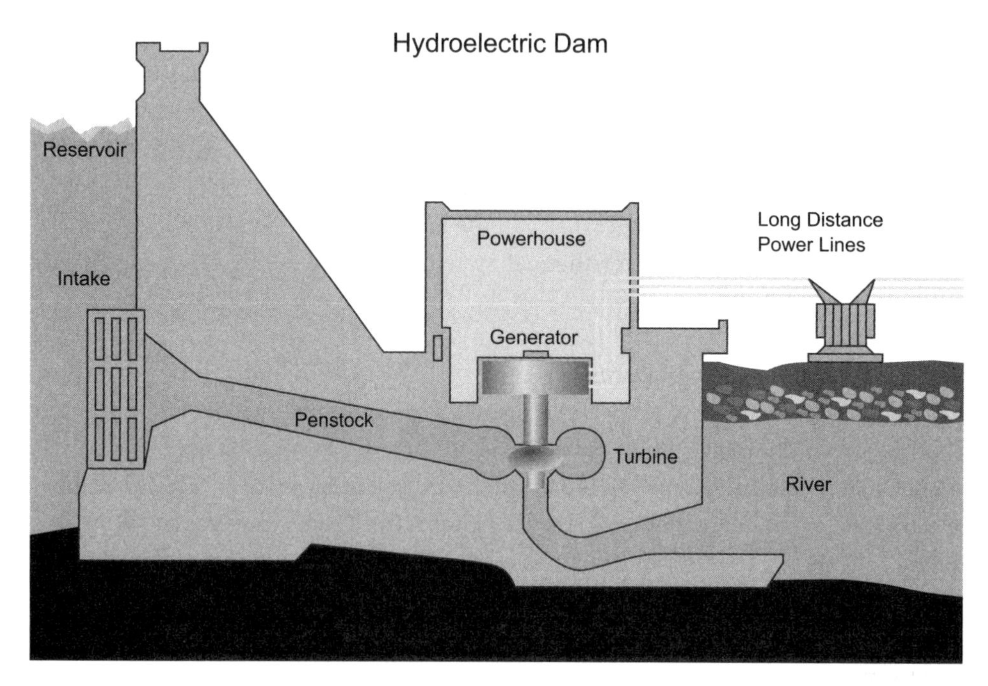

Figure 5.20
Schematic of hydroelectric power plant.
Source: Tennessee Valley Authority.

not state directly an average size here as we have done with the previous electricity generation methods.

5.3.8 Wind Farms

Wind is an energy that has gotten a lot of traction since the early 2000s. Most people are familiar with the conventional three-blade turbine. Harvesting the power of the wind is natural, and the use of wind as an energy source is very old (at least as old as the use of sails on boats). Here again, there is only one conversion of energy from mechanical to electrical. The relationship between wind speed, v, and power, P, is actually interesting and highly nonlinear:[26]

$$P = \frac{1}{2}\rho A v^3 \tag{15}$$

where ρ is the density of air at the specific height (i.e., about 1.2 kg/m^3), and A is the area formed by the blades. The relationship is therefore cubic, and the more the wind blows, the more power can be generated. The unit of P in equation (15) is in watts. To

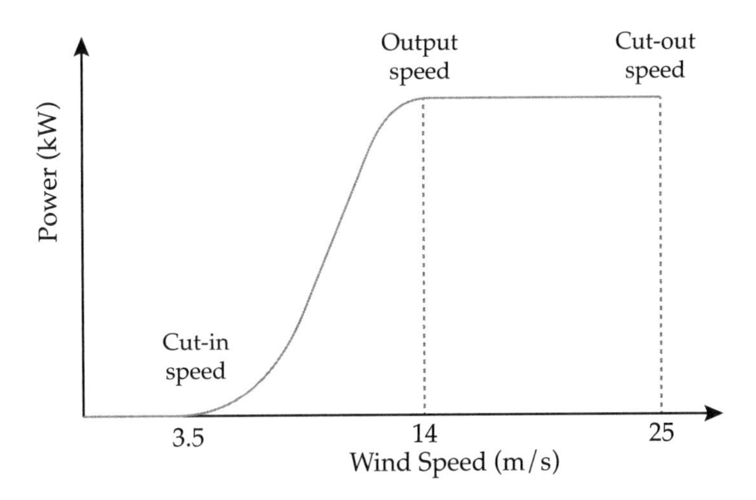

Figure 5.21
Typical power curve of a wind turbine.

calculate the energy, we therefore simply need to multiply equation (15) by the number of hours the wind turbine is operating.

We can see that because we cannot control air density and wind velocity, equation (15) tells us that to increase power output, we need to increase A (i.e., the size of the blades), which is why offshore wind turbines are so large.

Moreover, the wind is not always blowing, and a wind turbine cannot produce this power at every wind velocity. Figure 5.21 shows the power curve of a typical wind turbine. We can see that at low wind speeds, no power is produced. Power output then picks up after the cut-in speed around 3.5 m/s (i.e., 12.6 km/h or 7.83 miles per hour [mph]). It then reaches its maximum power output at its output speed, around 14 m/s (i.e., 50 km/h or 31 mph). Wind speed can then increase up to a cut-out speed around 25 m/s (i.e., 90 km/h or 56 mph), after which the electric motor in the turbine simply cannot produce any electricity and needs to be switched off.

Wind turbines vary greatly in size. Large offshore turbines have power outputs of 8 MW. Wind farms are groups of many individual wind turbines that constitute a power plant. At the time of this writing, the largest offshore wind farm could produce 660 MW, but another farm was under construction to be able to produce 1,200 MW.[27] In contrast, the largest onshore wind farm was said to be able to reach close to 8,000 MW, although it was expected to produce 20,000 MW by 2020.[28] In contrast, wind turbines can be as small as 1 kW or even 0.5 kW and are common for individual houses in rural areas to recharge batteries.

The efficiency of wind turbines to convert mechanical power to electrical power is quite high, at around 80%. Wind turbines do not function all the time, however, depending on wind conditions, and a good approximation is to assume that they generate their capacity one-third of the time in one year. So, a 1 MW turbine would produce about $\frac{1}{3} \times 1 \times 24 \times 365 = 2{,}920$ MWh in a year. Considering a 2016 average annual consumption of 10.8 MWh, this means a 1 MW turbine can provide electricity to about 270 homes. This 1 MW of power is not constant, however, and alternative methods of electricity generation must be used at the same time, as we discussed when we learned about electricity demand in section 5.2.

Wind turbines are also getting more popular in urban settings, with small turbines appearing on buildings. Like solar, wind power needs a lot of space, and the potential of wind in dense urban areas is limited. The potential in suburban areas, however, is much larger, similar to solar PV.

Figure 5.22 shows the annual average wind speed at an altitude of 30 m in the United States. All areas in green (see colored figure online) have low wind speeds and are not good candidates for wind power. Areas in yellow or darker, however, have average wind velocities of more than 5 m/s. The potential for wind power is therefore quite large in the United States.

Akin to solar thermal, the major advantage of wind over solar PV is the absence of silicon in electric motors. As we will see, wind tends to perform better than solar when considering the life cycle of these electricity generation methods.

5.3.9 Wave and Tide Power

Wave and tide power is much less common, and we will not discuss them too much here. Nonetheless, the power produced by waves and tides is considerable, and these methods of generating electricity may have a bright future.

Tide power is especially interesting because the power that can be produced is fairly predictable—that is, tides are predictable, unlike waves that depend on wind. The conditions of the sea (salt, algae, etc.), however, make it hard at the moment to judge the long-term costs and needs of tide power. Figure 5.23 shows the Annapolis Tidal Power Station located in Annapolis Royal in Nova Scotia (Canada). Annapolis Royal is located in the Bay of Fundy, which enjoys some of the largest tides in the world (up to 16 m). The tidal station was opened in 1984. It can produce up to 20 MW at peak, and it generates about 80 to 100 MWh per day.[29] Although this prototype works well, several technological issues have to be solved before tidal power plants become more common.

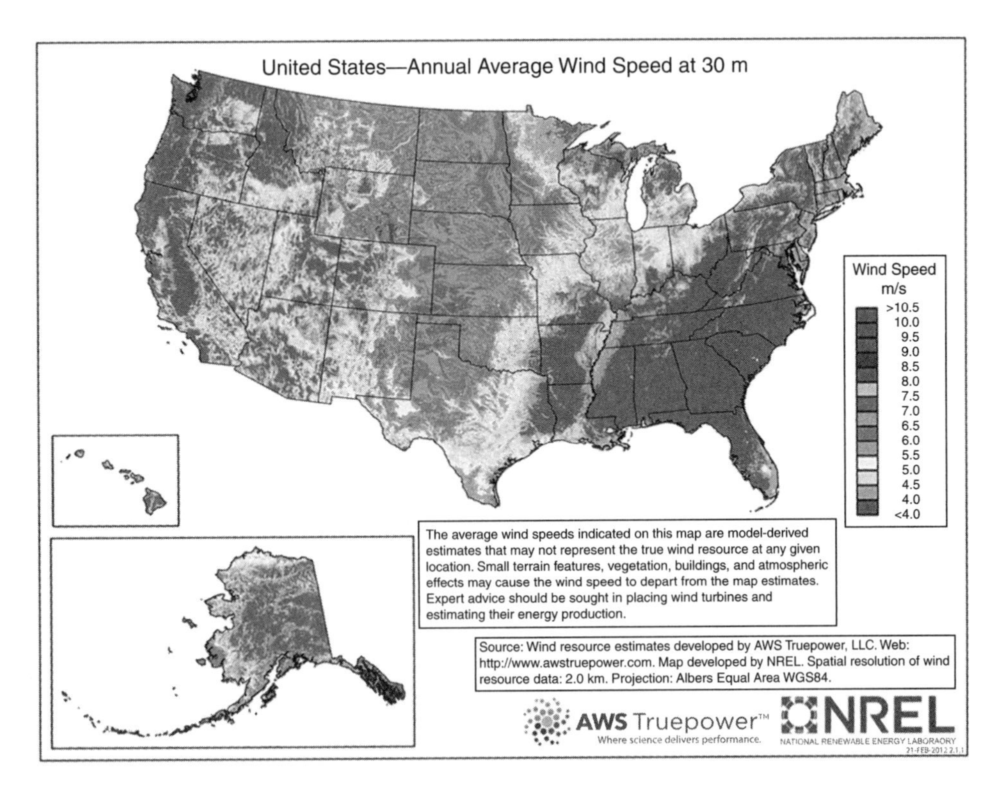

Figure 5.22
Map of U.S. annual average wind speed at thirty meters.
Source: National Renewable Energy Laboratory.

Figure 5.23
Annapolis Tidal Power Station, Nova Scotia, Canada.

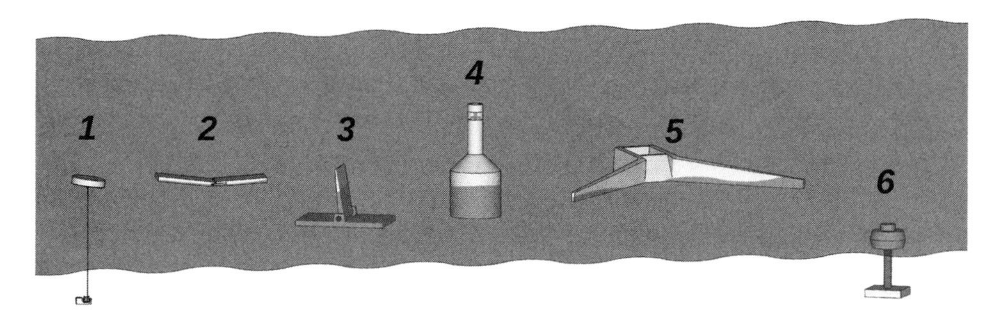

Figure 5.24
Wave power generation concepts.
Source: Ingvald Straume.

Harnessing the power of waves is also possible, with various concepts developed and attempts made, more or less successfully. Figure 5.24 shows some of the ways that wave power can be captured. At the time of this writing, one of the most well-known types of wave power generator was the pelamis wave energy converter that essentially moves like a snake—see design (*2*) in figure 5.24—and this movement is transformed into electricity. Typical values cannot be reported yet. A life-cycle analysis will be important to determine how long these systems can last in the sea compared to how much power can be produced.

For more information on wave and tide power, sections 4.6 and 4.7 of Grigsby (2012) are recommended.

This completes the review of kinetic systems.

5.3.10 Solar Photovoltaic Power Plants

The last electricity generation method that we will cover is solar PV. Along with wind, solar PV is the most famous intermittent renewable energy method. The potential to use sunlight to produce energy was first discovered by the French physicist Edmond Becquerel in 1839, but the first solar cell was not developed until 1954. PV panels have cells that are made of layers of silicon, phosphorous, and boron. The main mechanism to determine how much energy, E, can be produced by solar cells is dictated by Planck's constant, h ($=6.63\times10^{-34}$ J · s), and the frequency, υ (in hertz), of the photon, in the form:

$$E=h\upsilon \tag{16}$$

Solar panels are mostly rectangular and possess a familiar blue color. Panels can be bought individually and placed just about anywhere. Large solar farms also exist that

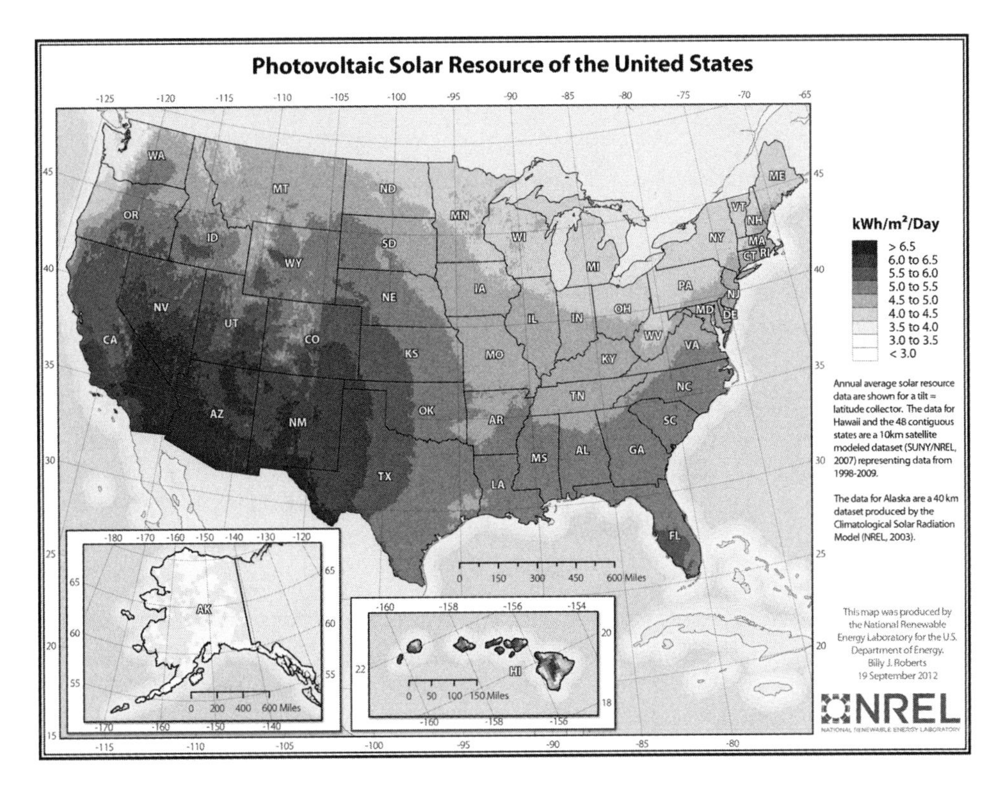

Figure 5.25
U.S. solar PV power map.
Source: National Renewable Energy Laboratory.

produce a lot of power. Because they vary so much in size, power ratings tend to be in the form of watt per square meter [W/m^2] of panel. The efficiency of solar PV is not terribly great, at about 10 to 15%. At the time of this writing, newer panels were able to have efficiencies of around 25%, with one panel going as high as 48% efficiency.[30]

Similar to the solar thermal and wind maps, the National Renewable Energy Laboratory (NREL) publishes solar potential maps for the United States. Figure 5.25 shows that again the southwest United States possesses the biggest potential for producing electricity from solar PV, which is not surprising. In areas with higher concentrations of sun, solar thermal is preferred, however, because it is more efficient, and the life cycle tends to be more favorable. Nonetheless, many places in the United States can still produce ample electricity from solar. This is particularly relevant for individual buildings.

To estimate how much energy can be produced from solar, E_p in [Wh], the map from NREL can be used. In particular, the direct energy from the sun, E_s [Wh/m^2], can simply be multiplied by a specific efficiency rating of the solar panel selected, η, and the total area of panels, A in [m^2], such that:

$$E_p = \eta \cdot E_s \cdot A \tag{17}$$

Naturally, this equation does not take the angle at which the sun hits the panel into consideration, although it has an impact. Ideally, a solar panel should be angled so that it has a maximum exposure to the sun. In the Northern Hemisphere, this means that the panel should face the south, and the opposite is true for the Southern Hemisphere. But for rough estimations, equation (17) will suffice.

Moreover, akin to wind, the power produced from solar panels is intermittent, and solar PV plants must therefore work alongside more flexible sources of power.

This completes the review of electricity generation methods. We will now compare the various methods with one another.

Example 5.5

Wind and Solar PV Power in Civitas

In Civitas, the average household consumes 1 kW, and there are 24,000 households. Assuming a solar exposure of 5 kWh/m^2/day and a solar PV efficiency of 20%, determine (1) the size of the solar PV plant needed to provide enough power to the city. Considering an average wind velocity of 7 m/s and a wind turbine blade length of 30 m, calculate (2) how much power can be produced by one wind turbine. Assuming the turbine operates one-third of the time, determine (3) the number of houses that can be supplied with one turbine. Based on the results, compute (4) the number of wind turbines needed to provide enough power to the community.

Solution

(1) The total power consumed $P_{consumed} = 1 \times 24{,}000 = 24{,}000$ kW $= 24$ MW. The total daily energy needed $E_p = P \times 24 = 576$ MWh. Applying equation (17):

$A = 576 \,/\, (0.2 \times 5 \times 10^{-3}) = 576{,}000 \text{ m}^2 = 0.576 \text{ km}^2$

(2) The total power that can be produced is $P_{max} = ½\ \rho A v^3$, where $\rho = 1.2$ kg/m^3 and $A = \pi r^2 = \pi \times 30^2 = 2827.43$ m^2. Therefore:

$P_{max} = ½ \times 1.2 \times 2827.43 \times 7^3 = 581{,}886 \text{ W} = 581.9 \text{ kW}$

(3) The actual power produced is $P_{produced} = ⅓ \times 581.9 = 194$ kW. Therefore: Number of houses $= 194 \,/\, 1 = 194$ houses.

(4) Since Civitas has 24,000 households:

Number of wind turbines $= 24{,}000/194 = 124$ turbines

5.3.11 Greenhouse Gas Emission Factors

Out of the twelve types of electricity generation methods, only four are direct GHG emitters: coal, oil, natural gas, and biomass (although part of the GHGs emitted for biomass are offset during the growth of the biomass). The eight other methods do not emit any GHG emissions at all. Nevertheless, we need to account for all the GHGs emitted, not only for the production of electricity but also for the construction, the operation, and the disposal of the power plant. For example, the cement that is used to build hydroelectric power plants requires a lot of energy. This type of measure is typically called life-cycle analysis or life-cycle assessment (LCA).

Table 5.2 displays typical emission factors for the twelve methods covered. Most of these values were extracted from ranges given in the "Energy Systems" chapter of the *Climate Change 2014: Mitigation of Climate Change* report by the IPCC (Intergovernmental Panel on Climate Change 2014). In fact, the IPCC has been extremely active in analyzing world trends in GHG emissions from all sectors and from all over the world. The "Energy Systems" chapter of the report as well as several other IPCC chapters are strongly recommended, including those related to transport and buildings.

Overall, we can see that traditional fossil fuel power plants are the worst emitters, even when considering the complete life cycle of a power plant. Moreover, although CO_2 capture decreases the impact of coal and natural gas,[31] it is still not enough to make these technologies competitive with the non-GHG emitters. In addition, we can see that biomass is also quite significant. The rest of the renewable sources vary, and the method with the most potential given the context should be selected for any location.

Table 5.2
Life-cycle GHG emissions by electricity generation method

Electricity generation method	GHG emission (g CO_2e/kWh)
Coal (with CO_2 capture)	1,000 (250)
Oil	750
Natural gas (with CO_2 capture)	500 (150)
Biomass	200
Geothermal	50
Hydroelectricity	50
Solar PV	50
Solar thermal	30
Nuclear	20
Wind	10
Wave and tide	10

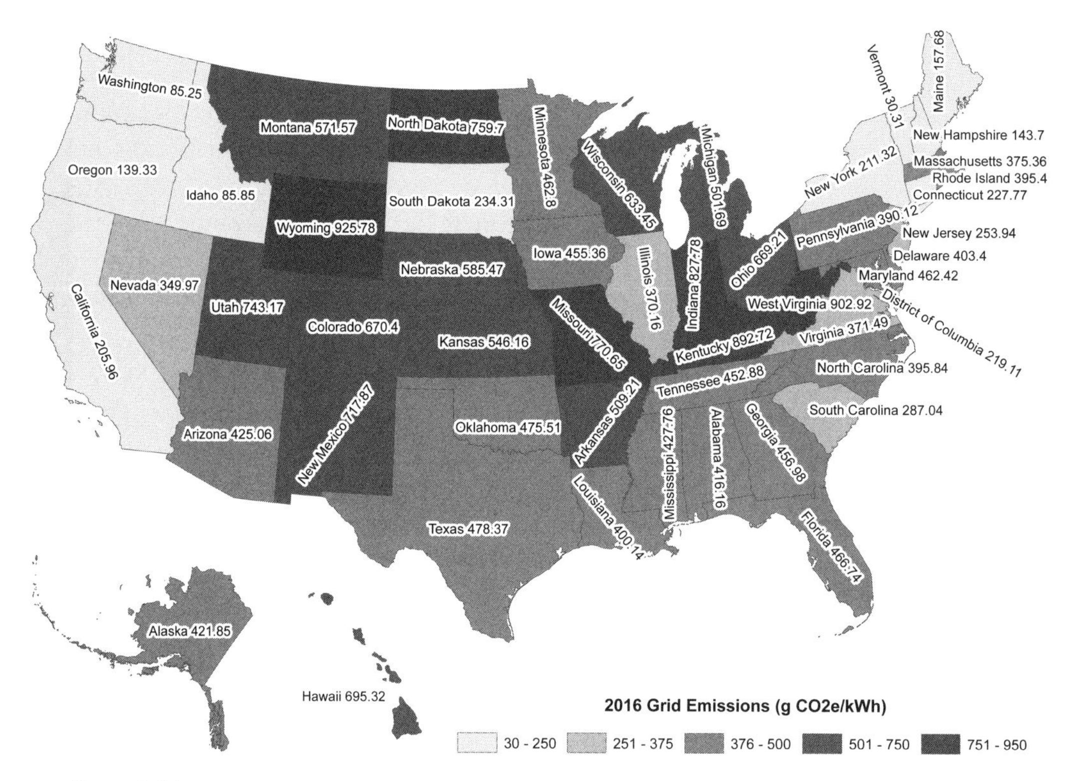

Figure 5.26
Electrical grid GHG emission factors in 2016 by U.S. state.
Source: U.S. Environmental Protection Agency.

In general, however, most places tend to have a mix of various power plants, perhaps including coal and wind power plants.[32] To calculate the carbon footprint of electricity without knowing exactly from which power plant our electricity comes from, we need to use a regional GHG emission factor. Figure 5.26 shows the grid emission factors for all U.S. states in 2016. Lighter-colored states have low emission factors. In the United States, they include most states in New England and on the West Coast, all of which enjoy large amounts of hydroelectric power, either located in the state or nearby—for instance, eastern states buy a lot of power from Canada, which possesses large hydroelectric power plants. Darker states tend to be more reliant on coal and therefore have much higher emission factors.

Looking at figure 5.26, we also now understand why using electricity for heating is not such a good idea in North Dakota. Remember that coal-fired plants have an efficiency of about 35%. In contrast, gas furnaces have an efficiency of more than 80%

(as we will see in chapter 8). Considering that coal emits more GHGs than natural gas, heating a house with electricity that comes from coal is more than five times as bad as using a natural gas furnace.[33] For states where households consume a lot of electricity, paying attention to the emission factor of their power grid is paramount.

Similarly, we can compare electricity grid emission factors at the international level. In 2016 the United States had a grid emission factor of 455 g CO_2e/kWh, which compares, for instance, to a factor of 71 g CO_2e/kWh in France (which produced most of its power from nuclear), 672 g CO_2e/kWh in Germany (which mainly used coal), and 443 g CO_2e/kWh in Japan (Brander et al. 2011).

As a means of comparing these values, Kennedy (2015) published a commentary piece arguing that all countries should aim to have electricity grid emission factors at least below 600 g CO_2e/kWh by 2020, which would be needed to keep the global temperature rise below 2°C. Using the 2011 numbers, the world average emission factor was 624 g CO_2e/kWh, and out of the 139 countries included, about 40% were above the 600 threshold. Moreover, beyond the 600 threshold, we should really be as close to 0 as possible, and most countries have much work to do to reduce the emission factors of their power grids.

This concludes our section about electricity generation. Next, we will discuss how the electricity grid might change in the future to become more flexible and more environmentally friendly.

5.4 Future Grid

Overall, the power grid as we know it needs to change. At the moment it is too inflexible and quite vulnerable (major power outages often make the news coverage). Here, we will see that electricity storage could play a major role in the future of the grid, especially if we are to generate more electricity from intermittent renewable sources. We will also discuss two more important concepts that have the potential to transform how we use electricity: smart grids and microgrids.

5.4.1 Electricity Storage

Although we will not discuss this topic in detail, at the time of this writing, the storage of electricity was one of the main issues why we were having trouble converting the power grid to more renewable sources. In the absence of a major river on which we can install a hydroelectric power plant, the remaining major renewable sources of energy—that is, solar, wind, wave, and tide—are intermittent. Even the flexible sources that we know can be difficult to adjust. The demand throughout the day, moreover, is not fixed, and other thermal systems can more easily adapt to changes in demand.

The logical answer is to store the electricity when we produce too much and use it during peak periods. The problem is that electricity in itself is very difficult to store. Super capacitors may have a future, but at the time of this writing, they were not ready. This means that electricity has to be converted to another form of energy and converted back to electricity afterward, which naturally incurs high losses—conversion from electrical energy to chemical or mechanical energy is roughly about 80 to 90% efficient.

For short-term storage (i.e., to store power for a short period of time), many power plants have a flywheel with a heavy mass. When power gets too high (i.e., the load decreases), a motor turns the wheel, therefore using electricity. Thanks to its heavy mass, when the motor stops the wheel keeps turning and can produce electricity. Another short-term method is to run a compressor to compress air in a tank, which can then be used to produce electricity when needed.

Naturally, batteries have a major role to play. Batteries store energy as chemical energy. Storing any significant amount of electricity therefore requires very large batteries that are not easy to manage. At the time of this writing, much innovative research was being carried out on batteries, and with some technological breakthroughs, their use is likely to increase significantly in the future.

As a more long-term storage strategy, some systems pump water to a higher elevation. When electricity is needed, the system then works like a hydroelectric power plant. This is actually not a new method, and it has been used at least since the 1950s in Canada, but its applicability is obviously location-dependent.

Another solution is to convert electricity into heat, which can then be used to produce electricity once again. Other research is also trying to capture CO_2 directly from the atmosphere and turn it into fuel, which would offer a substantial solution in the medium term.

Overall, a fair bit of research is needed for energy storage to become mainstream, and we have not discussed many other issues (including nontechnical issues) related to storage.[34] Yet it is likely the area that offers the most hope for electricity. Because we must harvest more renewable energy, we need to find a way to circumvent its shortcomings, and energy storage is critical.

5.4.2 Smart Grid and Microgrid

A smart grid essentially leverages the power of telecommunications and the Internet. You may not know this, but the amount of electricity that you are billed each month is often not the actual amount of electricity that you consumed. Meters are stationed at every consumer's location, and it would be much too expensive to send someone to read individual meters every month. Instead, what most electricity (and gas) providers do is make an estimate based on past consumption data. About once a year, they then send someone to read the value on your meter and adjust your bill based on your actual consumption. The smart grid essentially enables the reading of your actual consumption in real time. A device is placed on the meter that electronically sends the value of your meter every once in a while (that can range from seconds to hours, depending on the electricity provider). Figure 5.27 shows an example of a smart (electric) meter and a conventional (gas) meter. At the time of this writing, many smart electric meters were being installed.

Beyond being able to bill your actual monthly consumption, these smart meters offer vast opportunities. Indeed, it becomes easier now to detect power outages and to understand the demand for electricity at a high resolution, both in time and space. For example, we can visualize how much power is being consumed in any neighborhood at any given time. This can help electricity generators better balance their loads. Moreover, the consumers can now also visualize how much power is being consumed in real time, thus helping them save energy by switching off appliances if they wish. Smart meters also allow utility companies to track conservation efforts, especially during hot

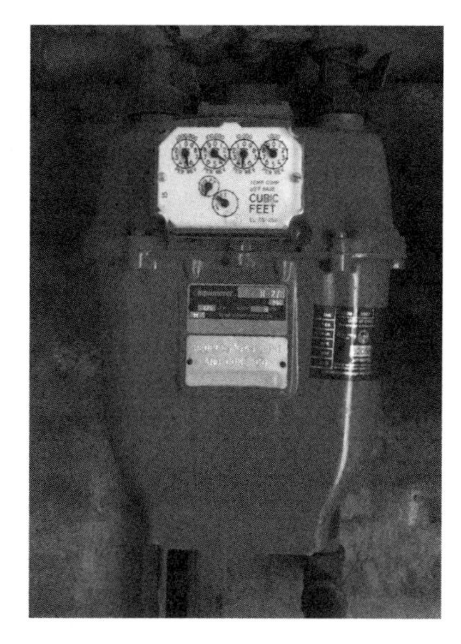

Figure 5.27
Left, smart (electric) meter; *right*, conventional (gas) meter.

days—remember the concept of *demand response* from section 5.2.2. These are only some examples of the benefits of the smart grid, but there are many more, including many that we do not know just yet.

Microgrids tackle a completely different issue. As we have seen, most power plants are large, which can be both good and bad. It can be good because large facilities may benefit from increasing economies of scale. However, this also makes the grid much more vulnerable because the failure of one power plant can affect the entire grid. What would be preferable is to have smaller power plants to service individual neighborhoods. This would essentially recreate a small grid, hence the term *microgrid*. The concept is actually not new. Small towns, college campuses, and large facilities like hospitals tend to run their own power stations, acting like a microgrid. The goal here is to make the practice wider, to individual neighborhoods. However, at the time of this writing, microgrids tended to run on natural gas, and ideally, we would rather use renewables or small, modular nuclear reactors. This is also why storage is likely to play a key role for microgrids as well. Again, because demand must be instantly met, having a "buffer" in the form of electricity storage would be a real game changer.

Another important aspect should be mentioned. Although it is not technical, it plays a big role in the resistance to change. This aspect is the conservative culture. As mentioned several times, the grid is inherently relatively vulnerable, and electricity providers tend to adopt the proverb "If it ain't broke, don't fix it." There are actually many successful stories of integrating more renewable sources of energy and of managing the grid. These stories are unconventional, however, and it will take more informed managers before we can truly change the power grid (Sovacool 2009).

5.5 Conclusion

This concludes our chapter on electricity. We first started by learning some fundamentals of electricity that often determine what can and cannot be done. We also defined the main measures that we use in electricity, including power P and energy E. Applying Ohm's law to series and parallel circuits with electricity generators taught us that it is better to transmit and distribute electricity at high voltages. Kirchhoff's laws also showed us that the power grid is complex and that we do not decide where the power goes. We then reviewed the fundamentals of AC and DC power, and we learned about three-phase power. Clearly, managing and maintaining the grid is challenging.

We then concentrated on electricity demand, studying how it has increased since 1960 and how it varies geographically. Taking into account regional variations is important for several reasons, and matching local demand characteristics with the potential of renewable energies is necessary. We then learned how the demand for power varies hourly across a typical day and how we need to accommodate our power supply to function with this demand. Afterward, we reviewed typical average power ratings for common appliances, which will need to decrease in the future in line with our first sustainability principle.

Subsequently, we reviewed twelve types of electricity generation methods. We classified these twelve types as thermal, kinetic, and other on one axis and as renewable and nonrenewable on the other. We detailed how they function to produce power. We learned that although they are popular, coal-fired power plants are also the largest emitters of GHG emissions. This is significant and their emission patterns must be reduced (at least by capturing their carbon). Natural gas plants are slightly better, but they still emit many GHGs. In contrast, other thermal methods grouped in the category of renewables can benefit from nearly the same properties while having a much smaller footprint. Apart from hydroelectric, the share of solar and wind plants is small at the time of this writing, but it will undoubtedly increase in the future. Taking into account life-cycle emissions, we then reviewed effective GHG emission factors for these twelve methods, and wind power was revealed to be the method with the smallest footprint.

Finally, we briefly discussed the future of the power grid. In particular, we discussed the role of energy storage, which will likely increase significantly in the future. We then learned about two important concepts, smart grids and microgrids, that are likely to take a predominant role in the future of the power grid, especially when electricity storage becomes a viable option. We also need to ensure that the generators used in smart grids and microgrids emit no or few GHGs.

Overall, the electricity system is incredibly complex. It is also the primary source of GHGs on the planet, and this must change. To learn more about electricity, you can refer to the Switch Energy Project (2017),[35] which also provides a useful seven-page document that recalls many of the concepts covered in this chapter.

Another utility that we consume heavily, similar to electricity, is water. Water is essential to life, but water infrastructure systems will need to change as well. In our next chapter, we will learn about water and stormwater systems, estimate how much water we use daily, and discover how much energy is required to distribute and collect this water.

Problem Set

5.1 The main goal of this question is to estimate your yearly carbon footprint from your electricity usage. Locate your latest electricity bill, which should have information about your usage for the past twelve months, or locate your last twelve bills. If you only have your usage for one month, multiply it by twelve months.[a] Otherwise, locate your meter and record the reading for a day, then multiply it by 365 and by an appropriate factor. Based on this information, calculate (1) your total household electricity usage in [MWh], (2) the average power drawn in [W], and (3) your personal electricity usage (i.e., divide the total electricity consumption by the number of people in your household). Using your regional grid emission factor, calculate (4) your personal carbon footprint from electricity in [t CO_2e].

5.2 Similar to question 5.1, locate your natural gas consumption for the past twelve months. Depending on your gas provider, the units expressed may be in [therms] or [Btu]. Calculate (1) the total amount of gas consumed in [MWh] (use the appropriate conversion factor), (2) your personal gas usage (i.e., divide the total gas consumption by the number of people in your household), and (3) your personal carbon footprint from gas in [t CO_2e] using a factor of 0.172 t CO_2e/MWh (as we will see in chapter 8).

5.3 By selecting any year other than 2016 for the United States (used to make figure 5.1) or by selecting any country of your choice, (1) plot as a pie chart the GHG emissions by economic sector and (2) comment on the results.

5.4 In your own words, explain and give the units for voltage, intensity, resistance, power, and energy in electricity. Feel free to use analogies with any field other than electricity.

5.5 From the EIA website, download the number of residential electricity consumers by state and the total residential electricity consumption for the latest year available.[b] From this data, calculate the average residential power consumption and compare the values with the value of 1.25 kW used in the chapter as the average power consumed by a typical American household.

5.6 A toaster draws 10 A of intensity when it is on. Assuming a voltage of 120 V, calculate (1) the equivalent resistance of the toaster, (2) the power drawn, and (3) the energy consumed when it is on for two minutes.

5.7 A laptop computer has a power rating of 60 W. Assuming a voltage of 120 V, calculate (1) the intensity drawn and (2) the equivalent resistance. If the computer runs for a total of 1,000 hours in a year, calculate (3) the total energy consumed. Assuming an

---

a. If appropriate, apply a factor such as 0.8 to account for lower electricity usage in the winter months.

b. URL: https://www.eia.gov/electricity/data/state/ (accessed September 4, 2017).

average electricity consumption of 10 MWh, (4) calculate the proportion of the energy used by the computer.

5.8 In your own words, explain Kirchhoff's voltage and current laws.

5.9 For the circuit given, determine (1) an equivalent resistance R_{23} for R_2 and R_3, (2) an equivalent resistance R_{123} for the entire circuit, (3) i_1, (4) V_1, V_2, and V_3, and (5) i_2 and i_3.

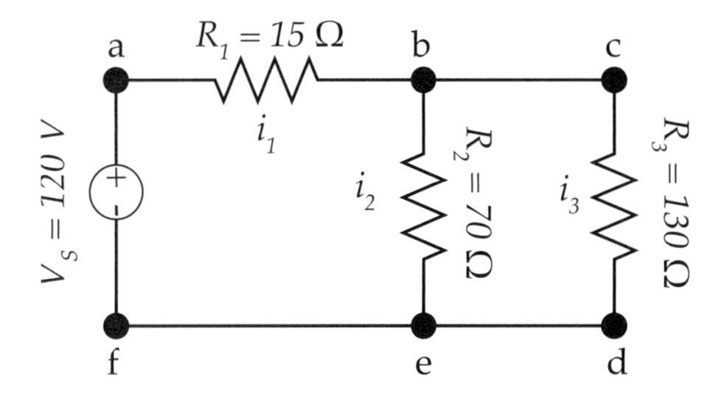

5.10 For the circuit given, determine (1) an equivalent resistance R_{345} for R_3, R_4, and R_5, (2) an equivalent resistance R_{2345} for R_2 and R_{345}, (3) an equivalent resistance R_{12345} for the entire circuit, (4) the total current i_S, (5) V_1, (6) i_1 and i_2, (7) V_2, (8) V_3, (9) V_4 and V_5, (10) i_3, and (11) i_4 and i_5.

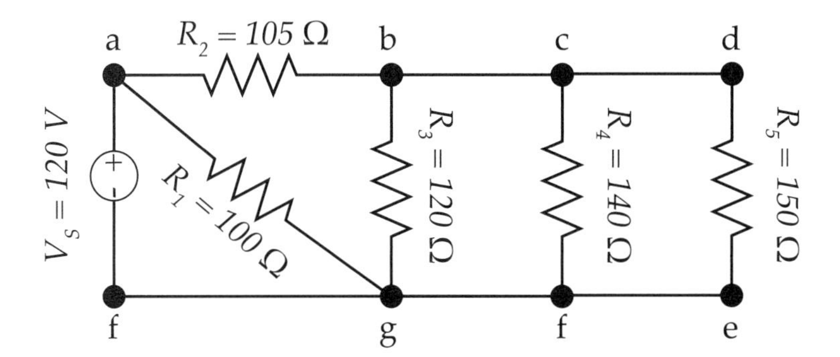

5.11 In your own words, describe: (1) why it is preferable to use high voltage to transmit and distribute electricity and (2) why AC power is usually preferable to DC power for electricity transmission and distribution.

5.12 By searching the web, describe how an alternator that produces AC power works. You are encouraged to submit drawings.

5.13 In your own words, describe what three-phase power is and why it is beneficial.

5.14 In your home, find your electrical panel and identify which appliances run on 240 V. Include a picture of your electrical panel.

5.15 In your own words, describe the three main elements of the power grid.

5.16 A 100 MW wind farm is being built 80 km from the city it will serve. Assuming the transmission line has a resistance per unit length of 4×10^{-4} Ω/m, and losses should not exceed 6%, calculate the voltage that will need to be supplied at the source.

5.17 A 15 MW hydroelectric power plant is producing power for a community located 50 km away. If the resistance in the transmission line is rated as 4×10^{-4} Ω/m, (1) determine the total effective resistance of the transmission line. If the power losses are not to exceed 3%, determine (2) the intensity (current) and (3) the voltage that needs to be supplied at the source. Based on this information, calculate (4) the voltage that will arrive at the community.

5.18 In your own words, describe how and why electricity demand varies (1) by year and (2) during a day.

5.19 In your own words, describe why households in warmer climates tend to consume more electricity.

5.20 List between ten appliances in your home and record their respective electrical properties (often given in terms of intensity or power). Assuming a voltage of 120 V, (1) determine the total power consumption of your appliances if they are all running at the same time in [W]. (2) Estimate a total usage per year in hours for each appliance and calculate the total energy consumption in [kWh]. (3) Based on the results from (4), estimate the share of these appliances on your personal carbon footprint from problem 5.1.

5.21 By searching the web, find one power plant located near where you live and describe it (i.e., type, power capacity, yearly electricity produced).

5.22 In your own words, describe how thermal power plants generate electricity.

5.23 By searching the web, find five hydroelectric power plants in the world and describe them (i.e., location, year opened, power capacity).

5.24 In your own words, describe the main disadvantages of solar and wind power to generate electricity.

5.25 The average household in Phoenix, Arizona, consumes 12 MWh of electricity per year. Provided NREL estimates that Phoenix receives about 6.5 $kWh/m^2/day$ and assuming an efficiency coefficient of 20%, (1) calculate the areas of solar panels needed to provide for one household. Provided NREL estimates that Phoenix receives an average wind speed of 5 m/s and assuming a wind turbine blade length of 30 m, (2) determine how much power could be provided by a wind turbine (use an air density value of 1.2 kg/m^3). Assuming the wind turbine operates one-third of the time, (3) compute the number of houses in Phoenix that could be supplied by one turbine.

5.26 Assume a household located in Chicago has an average yearly electricity consumption of 8 MWh. Using the NREL solar map and an efficiency coefficient of 20%, (1) estimate the area of solar panels needed to provide for this household. Using the NREL wind map and assuming a wind turbine blade length of 35 m, (2) determine how much power could be provided by a wind turbine. Assuming the wind turbine operates one-third of the time, (3) compute the number of houses in Chicago that could be supplied by one turbine.

5.27 By searching the web, expand on the content given in the chapter and describe what LCA is and how it works.

5.28 Assume a household has an average yearly electricity consumption of 10 MWh. Using life-cycle emission factors (i.e., table 5.1), determine the carbon footprint of the household if the grid is composed of power plants that only use: (1) coal (with and without CO_2 capture), (2) oil, (3) natural gas (with and without CO_2 capture), (4) biomass, (5) geothermal, (6) hydroelectricity, (7) solar PV, (8) solar thermal, (9) nuclear, (10) wind, and (11) wave and tide.

5.29 By searching the web, (1) report the electricity grid emission factor of any country or region not included in the chapter and (2) comment on the value.

5.30 In your own words, describe the main opportunities and challenges of electricity storage.

5.31 In your own words, describe the concept of a smart grid.

5.32 In your own words, describe the concept of a microgrid.

Notes

1. Electricity generation actually used to lead with more than 30%, but with the rise of natural gas power plants in the United States this percentage has decreased, and in 2016 transport became the largest U.S. GHG emission sector.

2. This may (or should) change in the future, thanks to dynamic pricing—that is, changing the price of electricity to reduce the overall demand during peak times.

3. Even in water, pressure is technically the difference between the pressure at one point and the air pressure.

4. A message from a relay is sent through a fiber optic cable that transmits the information at the speed of light (thus faster than electricity) to another relay, which disconnects the lines.

5. Yes, the voltage in North America is now 120 V, but it used to be 110 V. Similarly, in Europe the voltage is now 230 V, but it used to be 220 V. We will learn more about this in the next section.

6. As a side note, resistance ratings follow a famous color code, with colored bands drawn around resistances. Most of us will remember these colored bands from basic physics classes.

7. The unit hertz [Hz] is simply [s^{-1}].

8. For N points, the RMS is defined as $\sqrt{\frac{x_1^2 + x_2^2 + \cdots + x_N^2}{N}}$.

9. To learn more about the epic battle between Tesla and Edison, the CityLab featured a great short article by Small (2017).

10. A pretty exhaustive list of HVDC projects can be found on Wikipedia at https://en.wikipedia.org/wiki/List_of_HVDC_projects (accessed August 15, 2018).

11. Three phases offer a more stable power output for some sensitive devices, and they can also offer mechanical benefits for the generator. Nonetheless, more coils could be added to produce four, five, six, or even more alternating currents, but the benefits would only be marginal, while more phases would require more infrastructure.

12. Technically they could work with 120 V, but they would then draw more current since $V = RI$, which could cause some wires to overheat and possibly burn.

13. Dongtian Ji took a class based on the notes for this book in fall 2017.

14. To learn more about Samuel Insull and the power grid, I recommend Wasik's *The Merchant of Power* (2009).

15. National Public Radio has made available a great visualization to show the current power grid and how it could evolve in the future. I highly recommend you take a look at "Visualizing the U.S. Electric Grid," http://www.npr.org/2009/04/24/110997398/visualizing-the-u-s-electric-gridm (accessed September 26, 2017).

16. Fox-Penner (2010, p. 4) states that the average American home consumed approximately 138 kWh per month in 1950.

17. Although seasonal effects have to be taken into account.

18. In states with high grid emission factors, it is therefore preferable to use water heaters that run on natural gas. More on grid emission factors in section 5.3.

19. Nuclear fuels are actually quite abundant on Earth and could supply electricity for a very long time.

20. Sometimes referred to as *clean coal*, although it obviously does not make coal "clean."

21. This process is called *carbon capture and storage*. Ideally storage is underground, where oil and gas wells used to be. Nonetheless, we have no certainty that this CO_2 will not leak back into the atmosphere in the future.

22. As long as there is "heat," we can recuperate it.

23. We should remember that everything is "radioactive" but normally in tiny amounts. Funnily, nuclear power plants have been so effective at blocking any type of radiation that fossil fuel power plants tend to emit more radioactivity than nuclear power plants simply because coal, oil, and natural gas have nonnegligible (but not harmful) levels of radioactivity.

24. The NREL has a great learning center where you can learn not only about geothermal energy but also about all other renewable energies: https://www.nrel.gov/workingwithus/learning.html (accessed August 15, 2018).

25. Available at Solar Cell Central, "Concentrated Solar Power," http://solarcellcentral.com/csp_page.html (accessed August 15, 2018).

26. Power is energy density times flux. Here, energy density (i.e., per unit mass) is in terms of kinetic energy in the form of $\frac{1}{2}\rho v^2$ and flux is $A \times v$.

27. From Electrek, "World's Largest Offshore Wind Farm Starts Construction," https://electrek.co/2018/01/30/worlds-largest-offshore-wind-farm/ (accessed August 15, 2018).

28. From Interesting Engineering, "The 11 Biggest Wind Farms and Wind Power Constructions That Reduce Carbon Footprint," https://interestingengineering.com/the-11-biggest-wind-farms-and-wind-power-constructions-that-reduce-carbon-footprint (accessed August 15, 2018).

29. The power plant was designed to be unidirectional, so it only works about eleven hours per day—that is, two times 5.5 hours when the tide is low.

30. The National Center for Photovoltaics within the NREL keeps a close watch on these matters. https://www.nrel.gov/pv/national-center-for-photovoltaics.html (accessed August 15, 2018).

31. A CO_2 capture value was not available for oil and biomass. Although it would help reduce GHG emissions here as well, it is unlikely to beat non-GHG emitters.

32. CarbonBrief published a great article mapping all U.S. power plants in 2017 by type and by power capacity. It can be found at https://www.carbonbrief.org/mapped-how-the-us-generates-electricity (accessed August 15, 2018).

33. Burning natural gas emits 172 kg CO_2e/MWh. With a 90% efficiency, this gives us an emission factor of 191 kg CO_2e/MWh, compared to 1,000 kg CO_2e/MWh for a coal-fired power plant (assuming an efficiency of 100% with electric heating), which is 5.23 times higher. We can achieve even better values with ground-source heat pumps, but we need to wait for chapter 8 for that.

34. To learn more, see de Sisternes, Jenkins, and Botterud (2016).

35. The website is found at http://www.switchenergyproject.com/ (accessed August 15, 2018).

References

Brander, Matthew, Aman Sood, Charlotte Wylie, Amy Haughton, and Jessica Lovell. 2011. "Electricity-Specific Emission Factors for Grid Electricity." *Ecometrica*, August. https://ecometrica.com/assets/Electricity-specific-emission-factors-for-grid-electricity.pdf.

de Sisternes, Fernando J., Jesse D. Jenkins, and Audun Botterud. 2016. "The Value of Energy Storage in Decarbonizing the Electricity Sector." *Applied Energy* 175 (August): 368–379.

Engineering Council UK. 2014. "Chapter 3: Domestic Energy Consumption in the UK between 1970 and 2013." In *Energy Consumption in the UK (2014)*, vol. 18. London: Department of Energy and Climate Change. https://www.gov.uk/government/uploads/system/uploads/attachment_data/file/338662/ecuk_chapter_3_domestic_factsheet.pdf.

Fox-Penner, Peter S. 2010. *Smart Power: Climate Change, the Smart Grid, and the Future of Electric Utilities*. Washington, DC: Island Press.

Glover, J. Duncan. 2012. *Power System Analysis and Design*. 5th ed. Stamford, CT: Cengage Learning.

Grigsby, Leonard L., ed. 2012. *Electric Power Generation, Transmission, and Distribution*. 3rd ed. Electric Power Engineering Series. Boca Raton, FL: CRC Press.

Independent System Operator New England. 2015. *Real-Time Maps and Charts*. February 9. http://www.iso-ne.com/isoexpress/.

Intergovernmental Panel on Climate Change. 2014. *Climate Change 2014: Mitigation of Climate Change*. Edited by O. Edenhofer, R. Pichs-Madruga, Y. Sokona, E. Farahani, S. Kadner, K. Seyboth, and A. Adler et al. Cambridge: Cambridge University Press.

Kennedy, Christopher. 2015. "Key Threshold for Electricity Emissions." *Nature Climate Change* 5(3): 179–181.

Small, Andrew. 2017. "When Cities Went Electric." *CityLab*. March 15. https://www.citylab.com/tech/2017/03/how-the-war-of-currents-brought-power-to-cities/519402/.

Sovacool, Benjamin K. 2009. "Rejecting Renewables: The Socio-technical Impediments to Renewable Electricity in the United States." *Energy Policy* 37(11): 4500–4513. https://doi.org/10.1016/j.enpol.2009.05.073.

Switch Energy Project. 2017. *Electricity*. http://www.switchenergyproject.com/education/CurriculaPDFs/SwitchCurricula-Secondary-Electricity/SwitchCurricula-Secondary-ElectricityFactsheet.pdf.

U.S. Department of Energy. 2012. *Operating Characteristics of Electric Appliances in the Residential Sector*. http://buildingsdatabook.eren.doe.gov/TableView.aspx?table=2.1.16.

———. 2015. *Types of Hydropower Plants*. http://energy.gov/eere/water/types-hydropower-plants.

———. 2018. *Inventory of U.S. Greenhouse Gas Emissions and Sinks: 1990–2016*. Report 430-R-18-003. Washington, DC: U.S. Environmental Protection Agency.

Vadari, Mani. 2013. *Electric System Operations: Evolving to the Modern Grid*. Boston: Artech House.

Wasik, John F. 2009. *The Merchant of Power: Sam Insull, Thomas Edison, and the Creation of the Modern Metropolis*. New York: Palgrave Macmillan.

Zarrouk, Sadiq J., and Hyungsul Moon. 2014. "Efficiency of Geothermal Power Plants: A Worldwide Review." *Geothermics* 51(July): 142–153. https://doi.org/10.1016/j.geothermics.2013.11.001.

6 Water

> If water is the essential ingredient to life, then water supply is the essential ingredient to civilization.
>
> —Sedlak 2014, p. 1

The critical nature of water to life is evident. In fact, the presence of water is the one determining factor that is required for life to exist. What is harder to appreciate, however, is the importance of engineering systems to collect, treat, distribute, and dispose of the water that we need in our daily lives, and it is very well captured by the quote above. Much like electricity, we expect water to flow whenever we open the tap. In most high-income countries, we also expect water to be safe to drink. This was not the case so long ago, however. In fact, it evolved very much at the same time as the power grid, and both infrastructure systems are highly interdependent, as we will see in this chapter and again in chapter 10.

To give a bit of context, we often associate the development of water resources engineering systems to the Romans, who were able to bring in water from rural areas to urban communities and even right to the middle of Rome, with its population of a million people, through a clever system of pipes, channels, and aqueducts. This is what Sedlak (2014) calls Water 1.0. These systems were incredibly ingenious, and in fact they only required gravity to run; no pumping of any sort was used. It then took nearly 1,800 years to see substantial innovation in how cities use their water. Water 2.0 is the active treatment of water, which we discussed briefly in chapter 1 as an example of Scenario B with the story of John Snow in London. Water 3.0 came about half a century later with the much-needed treatment of sewage. Faced with new challenges, Sedlak argues that we need to come up with further innovation and get to Water 4.0. Indeed, cities are getting larger, while water sources are not changing; at the same time, the frequency of extreme drought and extreme rainfall that lead to flooding is increasing,

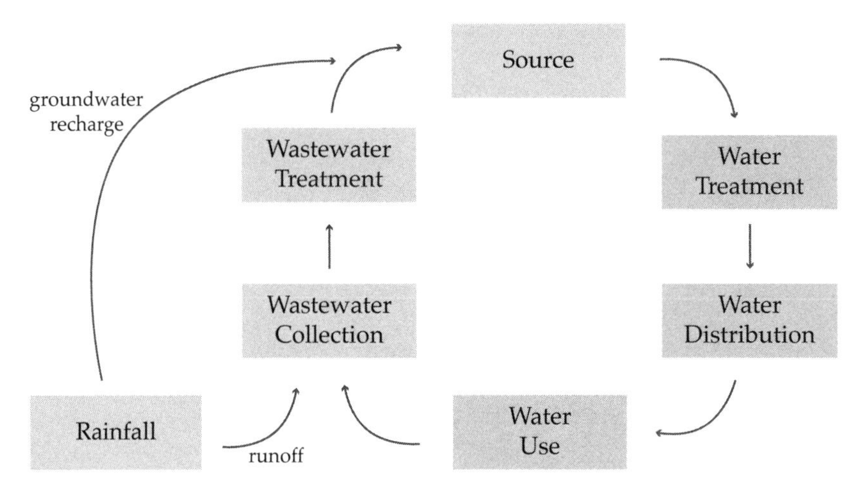

Figure 6.1
The urban water cycle.

and we desperately need to reduce the amount of energy that we use to operate this water system.

In this chapter we will largely follow the urban water cycle (figure 6.1) that starts at the source, whether the source is a river or a lake, or even whether we are tapping the water located underground (i.e., groundwater). We will therefore start with some hydrology principles, in part to learn about watersheds and hydrographs and to be able to understand and quantify what a 100-year rain means. We will then review the fundamentals of flow in closed conduits and open channels and discuss the foundations of groundwater engineering.

Similar to the previous chapter, we will then study patterns in the demand/consumption of water, across states and during a typical day. More specifically, we will look at past trends and identify how much water we consumed by the beginning of the twenty-first century, how it was used, and how water consumption changed with household size (i.e., how many people live in a household). We will also go through a typical water and wastewater treatment process. Although we will not cover it, techniques also exist that use natural environments to treat both drinking water (e.g., using sand dunes) and wastewater (e.g., using wetlands). These treatment techniques will likely be important for the future to decrease our need for chemicals—thus decreasing the energy used in the process.

Subsequently, and quite critically for this chapter, we will address the problem of stormwater management (Water 4.0) as urban flooding is becoming a predominant issue in cities around the world, and ideally, rain should replenish the groundwater as opposed to flooding streets or ending up in the sewers. Notably, we will learn about two methods to estimate runoff from different types of rainfall events. Finally, we will look at how much energy is used for each of these processes and identify how it can be reduced.

6.1 Fundamentals of Water Resources Engineering

Water resources engineering is a broad and relatively well-established discipline with many fundamentals. In this section we do not have time to recall all the fundamentals, but four concepts are relevant for us. First, we will go over some fundamentals of surface hydrology, recalling the concept of watersheds, intensity-duration-frequency (IDF) curves, and hydrographs. We will then learn about some fundamentals of flows in closed conduits—focusing on pipes—that deliver water to the tap and flow in open channels like rivers and stormwater channels, as well as sewers. Finally, we will briefly discuss the rudiments of groundwater engineering since 30% of all freshwater on Earth in actually located underground.

6.1.1 Surface Water Hydrology

Surface water hydrology is a large field, and while we will not be able to cover all relevant topics, three are particularly important for urban engineering. First of all, we will recall basic watershed hydrology principles. Second, we will briefly get introduced to hyetographs and hydrographs. Finally, we will learn about the IDF curves that are used in practice to "quantify" different types of rain events.

6.1.1.1 Watershed When rain falls and hits the ground, it inevitably tries to flow from a higher elevation to a lower elevation because of gravity. After all, the basic hydrologic water cycle—as opposed to the urban water cycle from figure 6.1—clearly shows a constant flux of water from the ocean to evaporating in the air, to falling to the ground (or being intercepted and perhaps evaporated), to slowly (e.g., thousands of years) or rapidly (e.g., hours, minutes, or even seconds) flowing "down" to the ocean again, whether it is on the surface, in rivers, through the ground, or a combination of all. This view is a bit simplistic and incomplete, but basically, water flows, and because this flow is often governed by gravity,[1] in a given area this water will tend to flow to the same outlet. In hydrology, the area that water falls over is called a *watershed*. More formally using the definition of the U.S. Geological Survey (USGS), a watershed "is an area of land that drains all the streams and rainfall to a common outlet such as the

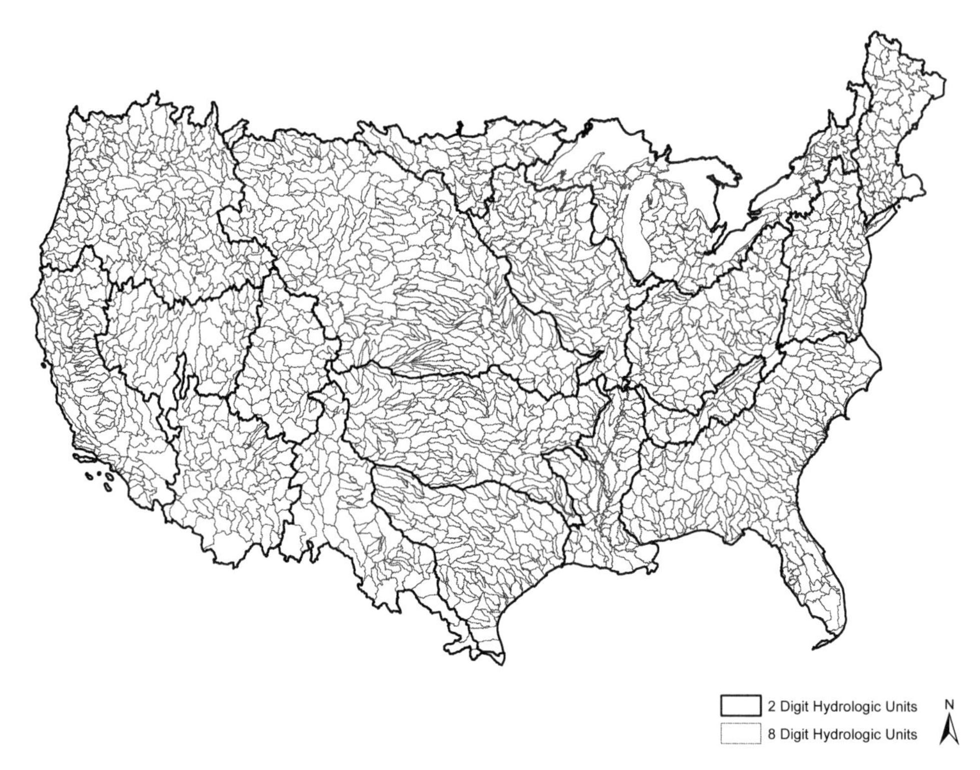

Figure 6.2
Eight-digit hydrologic units (watersheds) in the United States.
Source: U.S. Geological Survey.

outflow of a reservoir, mouth of a bay, or any point along a stream channel." The USGS has another and simpler definition: "A watershed is a precipitation collector" (U.S. Geological Survey 2016).

Watersheds are also called *drainage basins* or *catchments*. They can be tiny or enormous, simply because the outlet of a watershed feeds into another watershed. Figure 6.2 shows two "scales" of watersheds in the United States from the USGS. The larger scale bounded by thick lines shows the *regions*, technically called two-digit hydrologic units. The United States has twenty-one regions, including eighteen in the continental United States. These are the largest watersheds defined by the USGS. The smaller scale bounded by thin lines shows the *subbasins*, technically called eight-digit hydrologic units. The United States has about 2,200 subbasins. The ten-digit hydrologic units (called *watersheds*) are not shown in the figure, and there are about 22,000 of them. Even smaller-scale watersheds exist, the twelve-digit hydrologic units (called *subwatersheds*) that amount to about 160,000.

The concept of watersheds is important for urban engineers because of urban flooding problems. Whenever rain falls in cities, the water has to go somewhere. This is why streets are slightly arced,[2] for the rain to flow to the gutters and then to the sewers. An area served by a single storm drain (i.e., sewer grate) essentially acts as a watershed. Accordingly, sewers need to be sized correctly, otherwise they can overflow. Once stormwater enters the sewer, we then talk about *sewersheds*—that is, essentially, sewer pipes feeding into each other, often in the form of a tree network (as in chapter 4) until the final pipe reaches a wastewater treatment plant. Several sewer pipes therefore feed into one pipe, hence the term *sewershed*. We will discuss sewers and wastewater treatment later.

Watersheds are also important from an urban planning perspective because multiple municipalities and counties can share a watershed, and an overconsumption of water from one municipality can affect its neighboring municipalities.

One watershed concept that we will use is *time of concentration* T_c. Basically, time of concentration is the time it takes for water at the farthest point of the watershed to reach the outlet. Time of concentration is site specific, and therefore it needs to be estimated for individual projects (e.g., using a version of the Manning equation that we will learn in section 6.1.3). This is important because, as we will soon see, water flows in a nonlinear fashion, and peak flow typically happens at the time of concentration—that is, when the water that fell at the farthest part of the watershed reaches the outlet. To learn more about how rainfall and flow rate evolve with time, we turn to hyetographs and hydrographs.

6.1.1.2 Hyetographs and Hydrographs The terms *hyeto* and *hydro* both come from the Greek and mean "rain" and "water," respectively. Although the terms may look a little complicated, hyetographs and hydrographs essentially show how rainfall intensity and flow rate evolve over a period of time.

Hyetographs are collected at weather stations. The traditional way to measure rainfall is with a tipping bucket that collects water and counts how many times and when it tips.[3] Tipping buckets are still in use to get precise data at specific locations, but radar stations are used to cover large areas. To collect rainfall intensity, rainfall (e.g., in [mm]) is recorded during a given time interval (e.g., fifteen minutes) and then turned into intensity (e.g., in [mm/h]), which is why most hyetographs are displayed as bar charts.

Hydrographs measure the *discharge* at any location, including streams, detention ponds, and individual areas—the term *discharge* is synonymous with *flow rate* and is more common in stream hydrology.[4] The hydrograph of a river, for example, shows how discharge evolves over time. Similarly, hydrographs are measured for stormwater

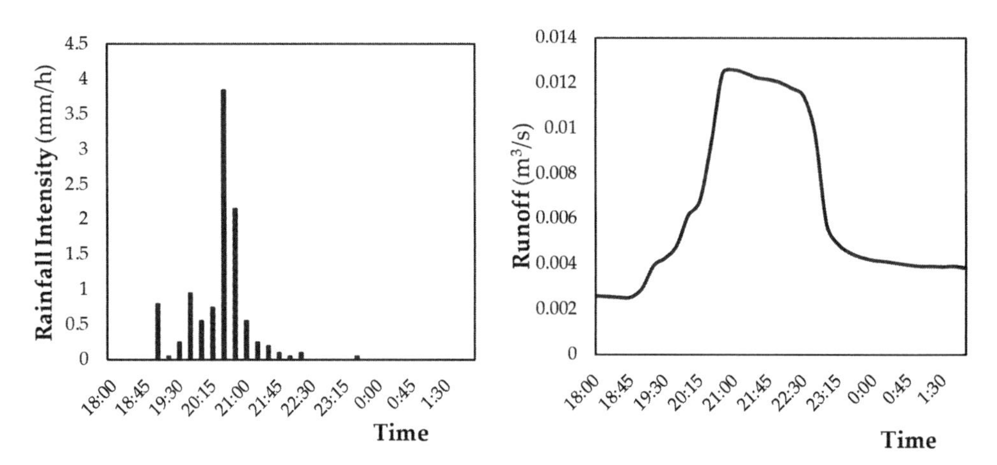

Figure 6.3
Hyetograph and hydrograph of permeable concrete in Chicago on May 10–11, 2017.

runoff on streets. Figure 6.3 shows an example of a hyetograph (*left*) and a hydrograph (*right*) for a parking lot on the University of Illinois at Chicago campus. The data shows the records for May 10–11, 2017, from 18:00 (i.e., 6:00 p.m.) to 5:00 (i.e., 5:00 a.m.).

From the hyetograph, we can see that rainfall started at 18:45, the peak occurred at 20:30, and rainfall mostly stopped a bit before 22:30. From the hydrograph, we first can see a base flow of 0.003 m^3/s. The runoff discharge then started to increase immediately when the rainfall started. It peaked when the rainfall peaked as well, but it then stayed relatively high until 22:30, when the rain stopped. This relatively flat peak occurs because the detention basin where the measurements were taken is equipped with a restrictor plate, which slows down the flow of water to the sewers (we will learn more about detention basins in section 6.4).

Although every stream is unique, figure 6.4 shows the anatomy of a typical hydrograph. First, we can see that a hydrograph is often accompanied by the hyetograph on the top, but reversed. Second, most streams tend to have a *baseflow* (like a typical river flow), which is shown by the fact that the discharge never goes to zero on the figure and is separated from the *stormflow* by the dotted line. When rainfall starts, the initial rise in discharge is called the *rising limb*, which ends at *peak flow* and is followed by the *falling limb*.

A special kind of hydrograph is the *unit hydrograph* (UH). The UH is a hypothetical hydrograph of a stream for a constant unit input of rainfall. For example, a UH can be plotted for a constant rainfall intensity of 10 mm/h. It is useful since once we have the

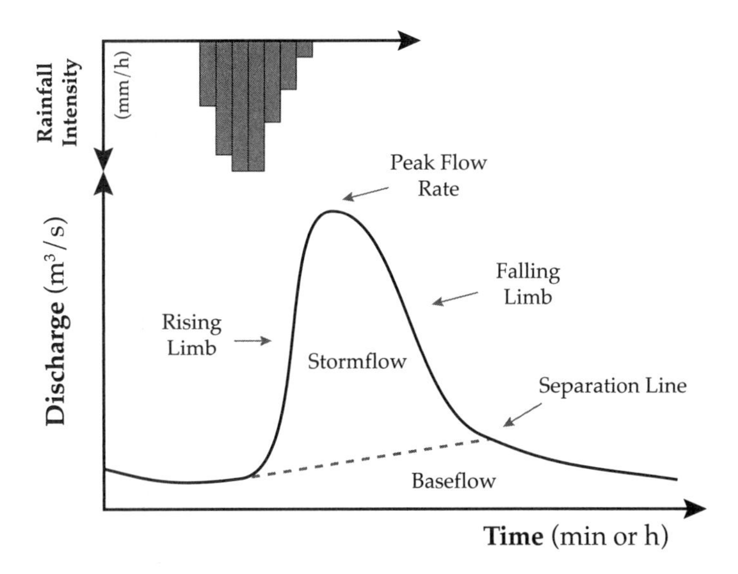

Figure 6.4
Anatomy of a hydrograph.

unit hydrograph, we can model the discharge of any rainfall, but it is not within the scope of this chapter.

Naturally, the hydrograph of a stream depends both on the intensity of the rainfall and on the physical characteristics of the stream. Focusing on the latter, figure 6.5 shows schematics of possible runoff hydrographs for channels with various characteristics. In the figure, three main characteristics are evaluated: slope, surface roughness, and soil porosity (i.e., infiltration potential). Overall, we can see that smaller slopes, rougher surfaces, and high infiltration potentials are preferred if we want to avoid high peaks that occur rapidly.

Hydrographs become handy, especially when we model large storm events, like a 100-year rain. How do we determine what a 100-year rain is? This is the topic of the next section.

6.1.1.3 Intensity-Duration-Frequency Curves With climate change, it seems that 100-year-rain events are frequently reported by the media, while they are supposed to occur roughly once in every 100 years.[5] But how do we determine what a 100-year rain is in the first place? We need to start with the definition of an IDF curve.

IDF curves inform us about the duration of rainfall for a certain intensity and a given return period. The return period is this 100-year rain or 50-year rain, or whatever period

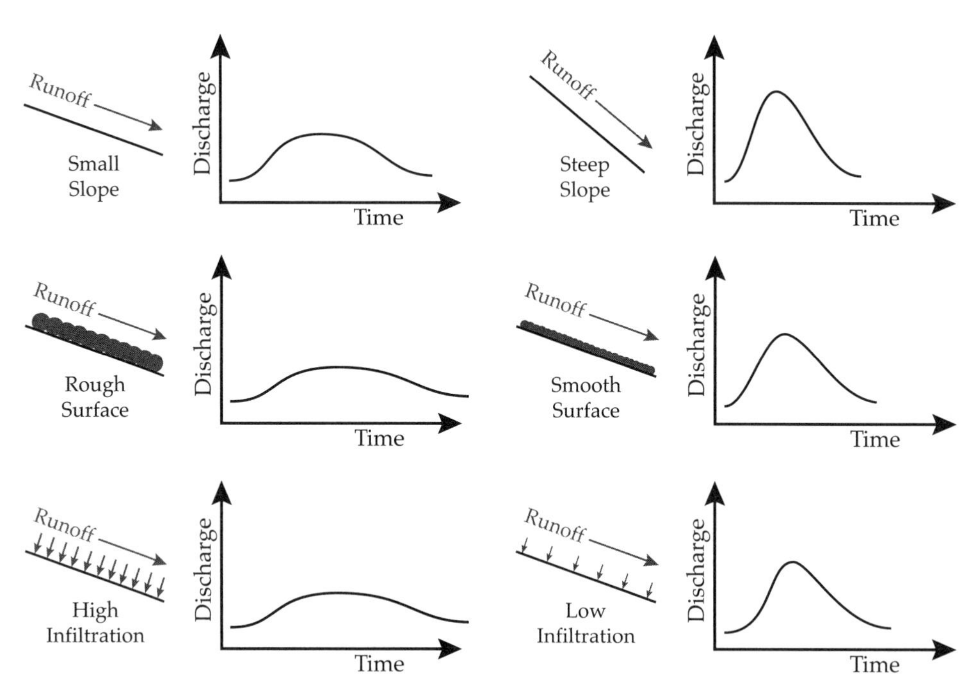

Figure 6.5
Schematics of a runoff hydrograph based on physical characteristics.

we wish to consider. The intensity is the amount of rain, typically in millimeters or inches, per minute or per hour (e.g., [mm/h]). The duration is the length of time for a given intensity, like an hour, or more typically twenty-four hours. We therefore expect very high intensities for short durations and low intensities for longer durations. Figure 6.6 shows the IDF curves of Chicago for seven different return periods in [mm/h]. Note that the axes are in log scales, and the curves are nearly straight.

To produce these IDF curves, we first need some data about the area that we are considering. This data typically comes in the form of rain depths per time interval; for instance, it rained 30 mm in 30 min, and we have this data for n number of years. Then, we simply rank the depths from highest to lowest. The return period for each rain depth is generally determined by using the Weibull formula:

$$T = \frac{n+1}{m} \tag{1}$$

where n is the number of years and m is the rank. If we have data for fifty-three years, then the maximum rain depth becomes the rain depth for a fifty-four-year return period, and the second-largest rain depth becomes the rain depth for a 54/2=27-year return period.

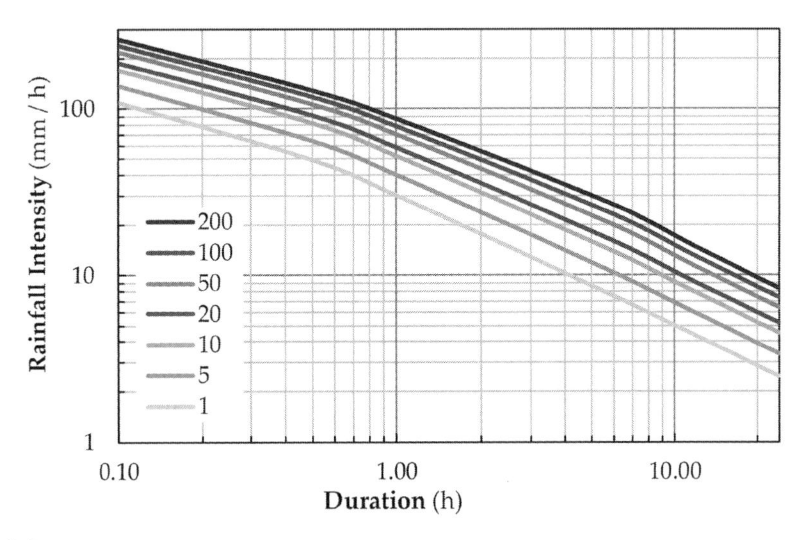

Figure 6.6
IDF curve for Chicago.
Source: National Oceanic and Atmospheric Administration.

This ranking feature is nearly identical to something called Zipf's law, which we will learn in chapter 11. We therefore have these values for any intervals with which we work—for example, all the rain depths for 5-min intervals, 10-min intervals, 30-min intervals, and so on. If we look for other return periods within two current data points, we can perform a simple linear interpolation—although this is not completely accurate, it is often sufficient. Finally, we simply need to turn these depths into intensities by dividing the values by the time intervals. For instance, 30 mm of rain in 30 min represents an intensity of 30/30 = 1 mm/min or 30/0.5 hours = 60 mm/h. In our designs, we then refer to these IDF curves to estimate how much rain we can expect to get in a certain period of time.

The resulting line is nearly straight on a log-log plot, but it is not totally a power law. There are various equations that can be fitted to the lines we see in figure 6.6, including a simple power law.[6] Commonly, we associate the lines to the Sherman equation defined as:

$$i = \frac{a}{(b+t)^c} \tag{2}$$

where i is the intensity in [mm/h]; t is the time interval selected (whether in [min] or [h]); and a, b, and c are constants. In particular, a is related to the return period that we are considering. Moreover, b does not tend to be very large compared to t (up to 15 min/0.25 h perhaps). For large values of t, b therefore becomes small, and the equation becomes a simple power law. This equation can be calibrated for any point in the United States using a U.S. Department of Commerce report titled *Rainfall Frequency Atlas of the United States*, which was published in 1961. The report is typically called TP40 because it is Technical Report 40 from the U.S. Weather Bureau. Most places, however, usually refer to the National Oceanic and Atmospheric Administration (NOAA) Atlas 14 that collects data nationwide or use local reports—in Illinois, it was the *IL Bulletin 70* at the time of this writing.

These IDF curves are essential for the design of many systems, including stormwater sewers. In practice, we talk about two main kinds of events: minor and major. Minor events include ten- to thirty-year rains, while major events include 100-plus-year events. Similarly, in cities, minor stormwater systems include underground sewer systems that are generally designed to handle minor events. Major stormwater systems include roadside ditches and other types of open channels that are designed to handle major events. We will focus on stormwater management in section 6.4.

The problem, however, is that the IDF curves that we calibrated historically are no longer valid because of climate change (especially the curves calibrated with TP40). We tend to have both more extreme events (i.e., more 100-year-rain events) than we used to, and the rain events themselves have become worse. Recalibrating these IDF curves offers one solution, but we may need to recalibrate them every year, which is obviously not a solution when we design a system for the next 50 to 100 years. A better solution is to increase our standard to a 500- or 1,000-year rain while carefully weighing the risks and the costs and by exercising our engineering judgment not to overbuild. Arguably, the even better solution is to build adaptable infrastructure that can cope with even bigger rain events, like 5,000-year events, without causing significant damages due to flooding. This type of design is at the heart of *resilience*, as discussed in chapter 1, but no perfect solutions exist yet. Nonetheless, some strategies can help. This will be a topic for later.

Enough about IDF curves—to learn more, I recommend the book *Environmental Hydrology* by Ward et al. (2015). For now, let us focus on flow in pipes and learn why we need pumps to distribute water across cities.

6.1.2 Flow in Closed Conduits

Closed conduits include all the pipes and confined pieces of infrastructure that are full and therefore allow for pressure to build up—unlike open channels, as we will see in section 6.1.3. There are four main fundamentals of closed conduits that we will cover in this section: conservation of energy, friction losses, pumps, and pipe networks (i.e., pipes in series and in parallel).

Example 6.1

Plotting IDF Curve in Civitas

The table below shows the five largest rain depths in millimeters by time interval for the past 99 years in Civitas. Using this data, (1) draw the IDF curves of Civitas.

	Rain depths time interval (mm)				
Rank	5 min	30 min	60 min	12 h	24 h
1	21.59	58.67	78.23	154.69	176.53
2	19.81	53.09	70.10	134.62	153.92
3	18.43	49.46	64.40	121.67	139.93
4	17.78	47.75	61.72	115.57	133.35
5	17.02	45.55	58.59	108.63	125.22

Solution

Using the Weibull formula for $n=99$ years, the largest return period is $T=(99+1)$ / $1=100$ years, the second largest is $100/2=50$ years, and so on and so forth. In terms of return periods, the table above therefore becomes

	Time interval				
Return period T	5 min	30 min	60 min	12 h	24 h
100 years	21.59	58.67	78.23	154.69	176.53
50 years	19.81	53.09	70.10	134.62	153.92
33 years	18.43	49.46	64.40	121.67	139.93
25 years	17.78	47.75	61.72	115.57	133.35
20 years	17.02	45.55	58.59	108.63	125.22

Switching rain depths to rain intensities in [mm/h]—for example, the 21.59-mm 5-min rain has an intensity of $21.59/5=4.318$ mm/min $\times 60=259.08$ mm/h—we get:

(continued)

Example 6.1 (continued)

	Time interval				
Return period T	5 min	30 min	60 min	12 h	24 h
100 years	259.08	117.35	78.23	12.89	7.36
50 years	237.74	106.17	70.10	11.22	6.41
33 years	221.16	98.92	64.40	10.14	5.83
25 years	213.36	95.50	61.72	9.63	5.56
20 years	204.22	91.10	58.59	9.05	5.22

Finally, plotting these values on a log-log scale, we obtain:

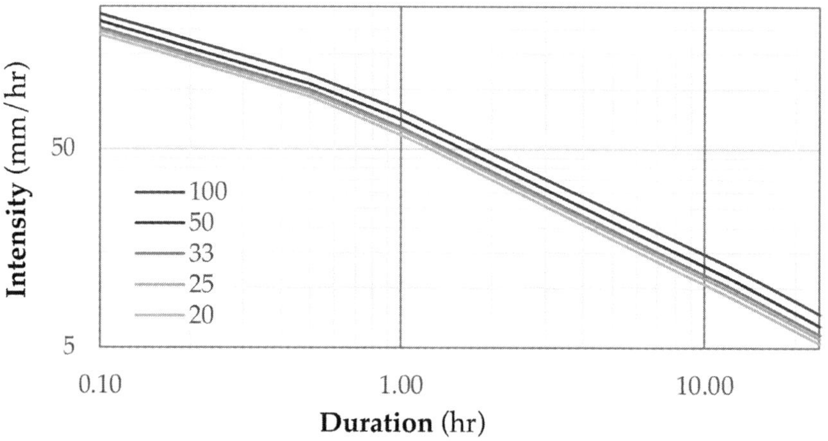

Intensity (mm/hr)
50
5
100
50
33
25
20
0.10
1.00
10.00
Duration (hr)

6.1.2.1 Conservation of Energy When we think about water distribution, we often think about the water that flows in the hundreds of kilometers of pipes located underneath the streets. To flow, a force must be exerted on the water that not only overcomes the weight of the water but also the losses that constantly try to stop the flow. Instead of talking about forces, however, we speak in term of energy in water resources engineering, and the energy of water at any given point is the sum of three different energies:

$$total\ energy = hydraulic + kinetic + potential \quad (3)$$

Hydraulic energy is the energy given by the pressure of water, kinetic is the energy given by the velocity, and potential is the energy given by the elevation of the pipe.

These terms should be familiar, and the first time you saw them may have been in an introductory fluid mechanics class, when you learned the Bernoulli equation:

$$P+\frac{1}{2}\rho V^2+\rho gz=\text{constant} \tag{4}$$

where P is the flow pressure, V is the flow velocity, z is the elevation, ρ is the fluid density, and g is the gravitational constant. We therefore have the three elements of equation (3) in the Bernoulli equation.

You may also recall that the Bernoulli equation can only be applied to flows that are steady and laminar and to fluids that are incompressible (i.e., constant density like water). Naturally, pipe flows in water distribution systems are not laminar, and the Bernoulli equation does not apply. Or more accurately, the definition of energy—that is, the left-hand side of equation (4)—is still valid, but the total energy at two points is not constant because of friction and minor losses. But before we get to losses, we can change the way we consider energy a little. Again, you likely remember energy in joules, but there is actually an easy way to manipulate the energy equation to have it expressed in meters. By dividing the three elements of energy by the specific weight of water, $\gamma=\rho g$, the total energy, which will now be called head H, takes the form:

$$H=\frac{P}{\gamma}+\frac{1}{2}\frac{V^2}{g}+z \tag{5}$$

In water distribution systems, P should be around 300 kPa (~40 psi).[7] The specific weight of water, γ, is equal to 9.79 kN/m^3; it is derived from $\gamma=\rho g$, where ρ is the density of water—about 998.2 kg/m^3 at 20°C—and g is the gravitational acceleration—9.81 m/s^2. The flow velocity V is generally expressed in meters per second, and finally, the elevation z of the pipe is already in meters. Because the three elements are in meters, it becomes easy to compare the contribution of each of the three "heads" (i.e., energies) to the total head.

Potential energy is purely based on the elevation of the pipe. Changes in potential energy therefore only happen if the pipe is on an upward or downward slope.

Kinetic energy is purely based on the flow rate and the dimension of the pipe. In fluid mechanics, continuity (i.e., conservation of mass) states that the flow rate $Q=VA$ must be constant in a system. This means that if the pipe area A changes, then the flow velocity V must change accordingly.

Hydraulic energy is based on the pressure on fluid property. This is the energy that gets affected by friction and minor losses. No matter how long a pipe is, if the area does not change, the velocity will not change because of continuity (i.e., like intensity in electricity). Pressure, on the other hand, can be severely affected (i.e., like voltage when current goes through a resistance). This is why we need pumps to increase the hydraulic energy of the flow, so it reaches all the taps in a community. In fact, let us define how we estimate how much energy is lost because of friction losses.

6.1.2.2 Friction Losses Flows in closed conduits go through two main types of losses: minor losses (at pipe junctions and bends) and friction losses (from shear stresses exerted by the walls of the pipe). For large pipe systems, minor losses are negligible, and we often omit them. We will therefore focus solely on friction losses. To estimate the *head loss* H_L due to friction (i.e., energy loss due to friction), we commonly use the Darcy-Weisbach equation:

$$H_L = f \cdot \frac{L}{D} \cdot \frac{V^2}{2g} \tag{6}$$

where f is the friction factor that we will introduce next, L is the length of the pipe, and D is the diameter of the pipe. We can see that equation (6) is somehow related to kinetic energy. Basically, the faster the flow is, the bigger the losses, and the relationship is nonlinear—remember nonlinearities from the end of chapter 2? Moreover, the longer the pipe is, the larger the friction losses, and the larger the pipe is, the smaller the losses. This equation is overall fairly intuitive.

The friction factor f is a parameter that is determined based on two things. The first is the surface roughness of the pipe since a rougher surface will cause more friction. Pipes are given a surface roughness ε that is expressed in [mm] and that depends on the material. To make it dimensionless, we divide it by diameter D such that our relative roughness coefficient becomes:

$$\frac{\varepsilon}{D} \tag{7}$$

In some water distribution systems, older pipes are made of cast iron, but new pipes tend to be made of ductile iron, with an initial $\varepsilon = 0.06$ mm, which can go up by one order of magnitude with corrosion.

The second thing is the Reynolds number that is itself based on flow velocity, pipe diameter, and fluid property. The Reynolds number Re is typically used to define whether a flow is laminar or turbulent, and every engineer knows that a flow becomes turbulent if Re > 2,000. The Reynolds number is a dimensionless coefficient that takes the form:

$$\mathrm{Re} = \frac{\rho V D}{\mu} = \frac{V D}{\nu} \tag{8}$$

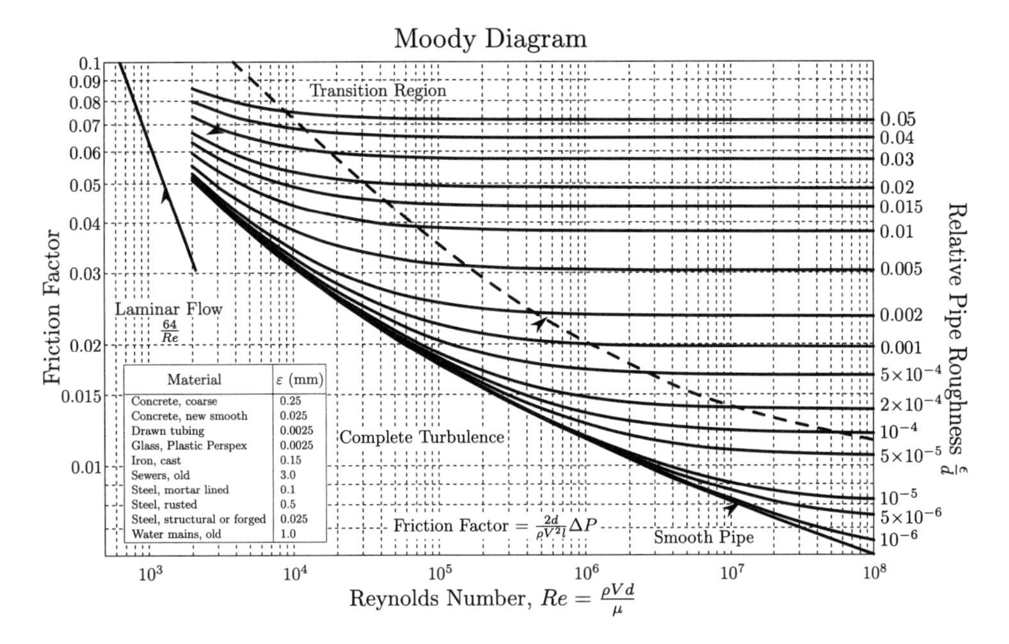

Figure 6.7
Moody diagram.
Source: Wikimedia Commons.

where μ is the viscosity of water, and $\nu = \mu/\rho$ is the kinematic viscosity of water. The latter is preferable to use, and numerically it is simply 10^{-6} m^2/s.

To determine the friction factor f from the relative roughness and the Reynolds number, we use another engineering favorite: the Moody diagram (figure 6.7). On the x-axis we select the Reynolds number, and on the second y-axis, located on the right-hand side, we select the relative roughness. The friction factor can then be read at the intersection of the two, on the y-axis located on the left-hand side of the x-axis. We just have to be careful and follow the solid lines since the friction factor increases for lower velocities because of the nonlinear impact of velocity. Once we have the friction factor, we can then calculate the friction losses and determine how big of a pump is needed.

6.1.2.3 Pumps Let us first recall the conservation of energy principle that states that no energy can be created or disappeared; it is always transferred,[8] and the energy at state 2 is equal to the energy at state 1 minus the losses. Or, in equation form:

$$H_1 = H_2 + H_L \tag{9}$$

where H_1 and H_2 are the total energies at states 1 and 2.

Using the energy equation (equation (5)) and the friction losses (equation (6)), we can now determine the total head of a flow somewhere down the pipe. Frequently, the flow cannot actually reach state 2 because the losses are large and/or state 2 is located above state 1, thus the flow at state 1 does not have enough energy to reach state 2. This is where the pump comes into play to add a head H_P to the flow, which gives us:

$$H_1 + H_P = H_2 + H_L \tag{10}$$

The power P_w that needs to be given to the water to add a head H_P to the flow is defined as:

$$P_w = \gamma \cdot Q \cdot H_P \tag{11}$$

This equation is fundamental since it tells us how much electrical power we need to add the necessary energy to the flow so that water reaches individual buildings at a sufficient pressure—and like in electricity, power is related to energy (H_P) times a flux (Q). To be more accurate, we also need to add an efficiency factor η since the total electrical power is not transferred to the water, and the "shaft" power P_s that we must generate is:

$$P_s = \frac{P_w}{\eta} = \frac{\gamma \cdot Q \cdot H_P}{\eta} \tag{12}$$

This efficiency varies depending on the flow, and every pump is designed to work optimally based on a certain flow rate and a certain power.

Accounting for power is important since powering the pumps that distribute water to their consumers may make up about 85% of all the electricity consumed by water distribution systems (Sedlak 2014, p. 175). At the same time, we saw that electricity production requires a great deal of water as well—that is, the energy-water nexus—and these two systems are highly interdependent.

6.1.2.4 Pipe Networks Naturally, water distribution systems have more than one pipe. We therefore need to discuss pipe networks. More specifically, we can look at two cases: pipes in series and pipes in parallel. Remembering continuity—that is, $Q = VA$—we can rewrite the Darcy-Weisbach equation slightly differently:

$$H_L = K \cdot Q^2 \tag{13}$$

with:

$$K = f \cdot \frac{L}{D} \cdot \frac{1}{2gA^2} \tag{14}$$

This version of Darcy-Weisbach is useful because the flow rate Q has to be constant in a pipe network; akin to electricity, the water has to go somewhere.

Example 6.2

Pumping Water in Civitas

Civitas gets its water from its main river, Vita. The average Civitian consumes 260 L of water per day. The treatment plant is located 250 m away and is at the same elevation as the river Vita. The pressure at the river is 0 kPa, and it needs to have a pressure 50 kPa when it gets to the plant. Estimate (1) the flow rate, (2) the dimension of the pipe to be used (take $V=1.5$ m/s and round up to nearest 25 mm[9]), (3) the friction losses in the pipe (the pipe is made of ductile iron) and the power required to bring the flow needed to the plant, (4) the total head needed, and (5) the shaft power for the pump (taking $\eta=0.7$). Considering the average Civitas household consumes, on average, 1 kW of power, (6) estimate how many households could be powered with the same power.

Solution

(1) There are 60,000 Civitians, and they consume, on average, 260 L/day, therefore:

$Q_{per\ person} = 260 \times 10^{-3} / (24 \times 60 \times 60) = 3.01 \times 10^{-6}$ m³/s per person.

Thus, $Q = 3.01 \times 10^{-6} \times 60{,}000 = 0.18$ m³/s.

(2) Continuity tells us that $Q = VA$, and $V = 1.5$ m/s. Therefore, $A = 0.18 / 1.5 = 0.12$ m². Since $A = \frac{1}{4}\pi D^2$, $D = (4A/\pi)^{1/2} = (4 \times 0.12/\pi)^{1/2} = 0.39$ m.

Rounding up to the nearest 25 mm, we set $D = 0.4$ m, which means that $V = (0.18 / \frac{1}{4}\pi 0.4^2) = 1.43$ m/s.

(3) The Darcy-Weisbach is: $H_L = f \cdot \frac{L}{D} \cdot \frac{V^2}{2g}$.

We first need the friction factor f. From the Moody diagram, the roughness of ductile iron is 0.06 mm, and therefore $\varepsilon / D = 0.06 / 400 = 0.00015$. The Reynolds number $Re = 1.43 \times 0.4 / 10^{-6} = 572{,}000$. From the Moody diagram, this gives us $f = 0.0145$. Therefore:

$$H_L = 0.0145 \times \frac{250}{0.4} \times \frac{1.43^2}{2 \times 9.81} = 0.944 \text{ m}$$

(4) The total head needed H_P is equal to the head loss H_L plus the hydraulic head required at the plant H_H (i.e., a pressure of 50 kPa), thus $H_H = P/\gamma = 50 / 9.79 = 5.10$ m. Therefore, $H_P = H_L + H_H = 0.944 + 5.1 = 6.04$ m.

(5) The shaft power required is $P_s = \frac{P_w}{\eta} = \frac{\gamma \cdot Q \cdot H_P}{\eta}$. Therefore,

$P_s = 9.79 \times 0.18 \times 6.04 / 0.7 = 15.21$ kW. Note that the power here is in kW because the specific weight is in kN/m³.

(6) Since the average power rating of a household is 1 kW, we have

Number of houses = 15.21 / 1 = 15.21 households

This means that the power required to displace the water 250 m to the treatment plant is equivalent to the average power consumed by 15.21 households. Obviously, much more power is needed to distribute the water across an entire city, especially if it has to deal with hills and slopes.

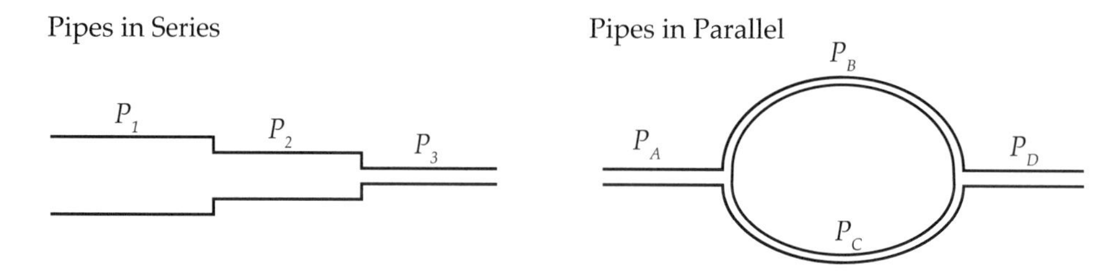

Figure 6.8
Examples of pipes in series and pipes in parallel.

Our two case studies take the form shown in figure 6.8. If we assume both systems have the same flow rate, then we need to have $Q_1 = Q_2 = Q_3 = Q_A = Q_{BC} = Q_D$. Note the term Q_{BC} since it is the sum of the flow rates in pipes *B* and *C* that amounts to the same flow rate—that is, $Q_A = Q_{BC} = Q_B + Q_C$.

What we like to do is to determine an equivalent *K*-factor from equation (14) since all the *Q*s are the same and then apply equation (13). How do we do that? It is actually not that hard. For pipes in series, because the flow rates have to be identical in the three pipes, K_{series} is simply the sum of the individual *K*s (this is why we isolated *Q* in equation (13) in the first place):

$$K_{series} = K_1 + K_2 + K_3 = \sum_i K_i \tag{15}$$

For pipes in parallel, it is a little different. Here, the *Q*s are not identical in all the pipes since the flow is split in the parallel section. What we must realize, however, is that the losses in pipes *B* and *C* are identical. Indeed, similar to energy, electricity, force, and so on, water takes the path of least resistance, and although the flow rates in pipes *B* and *C* are different, the losses must be equal. Here is a loose analogy: imagine you are with a friend at the airport, queuing at security. You are told to take one line and your friend is told to take another. Both of you then take ten minutes to go through security. How much time did the whole process take? It is ten minutes, not twenty minutes, and the same principle applies to losses in pipes.

Coming back to water and using continuity and equation (13), we get:

$$Q_{BC} = Q_B + Q_C$$
$$\sqrt{\frac{H_{L_{BC}}}{K_{BC}}} = \sqrt{\frac{H_{L_B}}{K_B}} + \sqrt{\frac{H_{L_C}}{K_C}}$$

We now also know that all losses have to be equal, thus $H_{L_{BC}} = H_{L_B} = H_{L_C}$, which gives us:

$$K_{BC} = \frac{1}{\left(\frac{1}{\sqrt{K_B}} + \frac{1}{\sqrt{K_C}}\right)^2} \tag{16}$$

Generalizing for $K_{parallel}$:

$$K_{parallel} = \frac{1}{\left(\sum_i \frac{1}{\sqrt{K_i}}\right)^2} \tag{17}$$

Moreover, similar to Kirchhoff's laws of current and voltage, at a node and in a loop, all the water that arrives must also leave, and the sum is 0. To solve complex pipe network problems, you can follow the Hardy Cross method manually, but software can more easily do the work for us. In particular, the U.S. Environmental Protection Agency (EPA) has a free and user-friendly software package called EPANet that can be downloaded from their website.[10]

Furthermore, similar to pipe networks, pumps can be installed in series or in parallel. Any single pump increases both the head and the flow rates of the flow. Pumps in series further increase the head of the flow, and pipes in parallel further increase the flow rate of the flow.

Armed with these equations, we can solve a fairly large array of problems. As importantly, we get a better understanding of how water distribution systems work and where energy is being consumed. We will now concentrate on flows that do not have pumps and that have one surface exposed to the air: open channels.

6.1.3 Flow in Open Channels

There are two big differences between flows in open channels and flows in closed conduits. The first one is that the water is not pressurized, and the hydraulic energy is simply the water depth. The second is that pumps are not used in open channels. Instead, the main driver is gravity, which the Romans had already mastered, as we saw at the beginning of this chapter. Open channels can therefore be just about anything from rivers and gutters to trapezoidal man-made channels and partially full pipes. Figure 6.9 shows a few examples.

In this section we will first cover the Manning equation, which will enable us to quantify flow rates in open channels, and we will then dive more into open channel fundamentals by discussing energy, critical flow, and the Froude number.

6.1.3.1 The Manning Equation The same basic laws of physics apply to flows in open channels as in flows in closed conduits. This means that continuity still applies, and if

Example 6.3

Pipe Network in Civitas

The water main leaving the Civitas water treatment plant located in zone 3 (see figure 1.5) has a diameter $D_m = 0.5$ m, a length $L_M = 100$ m, and an average flow rate $Q_M = 0.2361\ \text{m}^3/\text{s}$. The main is then split into two service zones, 2 and 4. The two pipes have the following characteristics: $D_2 = 0.2$ m, $L_2 = 300$ m, $D_4 = 0.25$ m, $L_4 = 500$ m. Assuming a friction factor for all the pipes of $f = 0.016$, determine (1) the losses in the water main, (2) an equivalent K-factor for the parallel pipe system, (3) the losses in the parallel pipe system, and (4) the flow rate in the two pipes going to zones 2 and 4.

Solution

(1) Using the Darcy-Weisbach equation (from equations (13) and (14)), the losses H_{L_M} in the water main are:

$$H_{L_M} = 0.016 \times \frac{100}{0.5} \times \frac{1}{2 \times 9.81 \times \left(\frac{\pi \times 0.5^2}{4}\right)} \times 0.2361^2 = 0.24\ \text{m}$$

(2) Using equation (14):

$$K_2 = 0.016 \times \frac{300}{0.2} \times \frac{1}{2 \times 9.81 \times \left(\frac{\pi \times 0.2^2}{4}\right)} = 1239$$

$$K_4 = 0.016 \times \frac{500}{0.25} \times \frac{1}{2 \times 9.81 \times \left(\frac{\pi \times 0.25^2}{4}\right)} = 677$$

Therefore:

$$K_{24} = \frac{1}{\left(\frac{1}{\sqrt{K_2}} + \frac{1}{\sqrt{K_4}}\right)^2} = \frac{1}{\left(\frac{1}{\sqrt{1239}} + \frac{1}{\sqrt{677}}\right)^2} = 224$$

(3) Using Darcy-Weisbach: $H_{L_{24}} = 224 \times 0.2361^2 = 12.49$ m.

(4) Since $H_{L_{24}} = H_{L_2} = H_{L_4}$, we have:

$$Q_2 = \sqrt{\frac{H_{L_{24}}}{K_2}} = \sqrt{\frac{12.49}{1239}} = 0.1\ \text{m}^3/\text{s}$$

$$Q_4 = \sqrt{\frac{H_{L_{24}}}{K_4}} = \sqrt{\frac{12.49}{677}} = 0.1358\ \text{m}^3/\text{s}$$

Checking: $Q_2 + Q_4 = 0.1 + 0.1358 = 0.2361 = Q_M$. The results are therefore correct.

Figure 6.9
Five examples of open channels.

the dimension of the channel changes or if the water depth increases or decreases, the velocity must change accordingly such that $Q=AV$.

The fundamental equation that we use to calculate the flow rate Q of an open channel is called the Manning equation, and it is defined as:

$$Q=\frac{C_m}{n}AR^{2/3}S_f^{1/2} \tag{18}$$

where n is the Manning channel roughness, which depends on the type of channel; a river will typically be rougher than a concrete channel, for instance, and a typical clay pipe used for sewers has $n=0.014$ compared to 0.012 for a concrete channel. C_m is a correction factor that is used to adjust n when the variables are not in a metric unit ($C_m=1$ for metric and 1.49 for nonmetric), A is the area of the channel, R is the hydraulic radius, and S_f is the slope of the water surface. These last two need some further details.

The hydraulic radius R is the ratio of the area A and the wetted perimeter P. The wetted perimeter is exactly what it sounds like. Since the water is exposed to air, it is basically the perimeter of the channel that is wet, where shear stresses act.

The slope of the water surface S_f is the gradient in the water surface—that is, the change in the water depth plus the change in the slope of the channel, S_0. It is expressed as a dimensionless number by dividing the change in elevation by the distance (i.e., for a 1% slope, $S=0.01$). In some cases, $S_f=S_0$, which happens when the flow is uniform, but in most cases it is not. The slope of the channel can in fact be controlled so that the flow velocity increases or decreases downstream. If the velocity changes, then the area A changes as well, which means that the water depth changes, therefore affecting the slope of the water surface. This slope is essential because it is the only variable that represents the impact of our main driver: gravity.

The Manning equation can be used in a number of ways. We may use the equation to determine the flow rate, but often we already know what we anticipate the flow rate to be. However, sometimes we need to determine the size of the channel, expressed by $AR^{2/3}$, that we need to build to accommodate a Q and S_f, and other times we need to determine the slope of the channel to accommodate a Q and maximum flow depth. These two examples are critical for us. In cities and using IDF curves, we can estimate a certain flow rate by multiplying our rain intensity with the area of the city. More accurately, we need to account for the area of impervious surfaces, like roofs, sidewalks, and roads, where rainfall inevitably leads to surface runoff that ends up in stormwater channels, like gutters, that are subject to the Manning equation. Moreover, some permeable surfaces also cannot accommodate high flows of water, and the remaining rainfall also leads to runoff. This is the case for grass, for instance, which cannot absorb much more than 3 cm (a little more than one inch) of rain.

6.1.3.2 Energy, Critical Flow, and the Froude Number We should take one more element into consideration, but we will not spend much time on it. Instead of talking about total energy, in open channels, we prefer to talk about the specific energy of a flow at a certain cross-section in the channel. Essentially, this means we are not considering the elevation z at that point, and since the hydraulic head is essentially reduced to the depth y (i.e., no built-in pressure), the specific energy of a flow becomes:

$$E_s = y + \frac{V^2}{2g} \tag{19}$$

The specific energy of a flow is therefore dictated by two components: the hydraulic energy (i.e., depth) and the kinetic energy (i.e., velocity). This can be seen from figure 6.10.

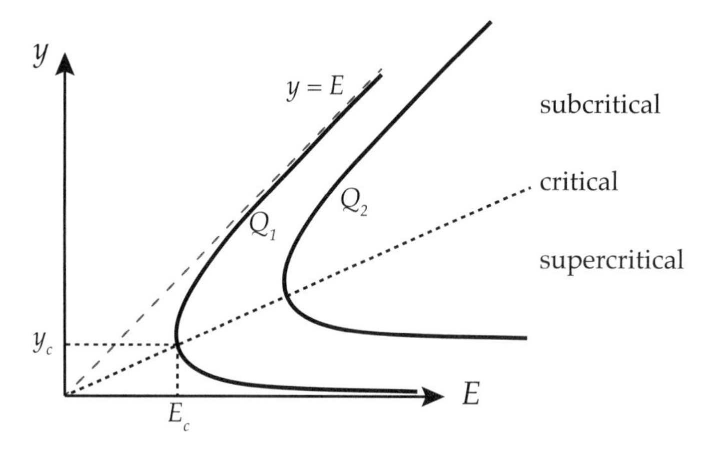

Figure 6.10
Energy in open channels.

For a given flow rate Q, if the flow depth y is high, then the velocity is low, and most of the energy is stored as hydraulic energy. In contrast, if the flow depth is low and the velocity is high, most of the energy is stored as kinetic energy. Between these two extremes, there is a point where both energies are low, which is where the energy of the flow is at the minimum. When this happens, the flow is said to be *critical*. When the velocity increases, it becomes *supercritical*—that is, the lower part in figure 6.10 since the flow depth y is smaller—and when the velocity decreases, it becomes *subcritical*.

In some cases we prefer to avoid supercritical flows, and this can therefore be another limiting factor. Supercritical flows exert a tremendous force on the bottom of the channel, and in natural channels this means the sediment is transported downstream, potentially affecting wildlife and other natural processes. For sanitary and stormwater sewers, this means the flow may arrive too rapidly at the treatment plant. On the contrary, supercritical flows are shallower by definition, which can be helpful to prevent a channel from overtopping (i.e., prevent flooding).

To determine whether a flow is critical or not, we use another famous dimensionless factor in fluid mechanics, the Froude number:

$$Fr = \frac{V}{\sqrt{gD}} \tag{20}$$

where D is the hydraulic depth of the flow, which is defined as the area A divided by the top width T (i.e., essentially the surface exposed to air, the opposite of the wetted perimeter). For a rectangular channel, the top width is equal to the bottom width of the channel, and the hydraulic depth is reduced to the water depth.

In the Froude number equation, $\sqrt{gD}$ represents the speed of gravity waves.[11] The Froude number is in fact nearly identical to the Mach number in aeronautics to determine the speed of an object relative to the speed of sound.[12] Therefore, if the Froude number is greater than 1, the flow is supercritical—that is, the flow is faster than gravity waves—if it is exactly 1, it is critical, and if the Froude number is lower than 1, the flow is subcritical.

Open channels are everywhere in real life. In cities, they are instrumental in carrying sanitary and stormwater sewers, using purely gravity. Rivers themselves are open channels that carry water to and from cities. These rivers are often the primary sources of drinking water for cities. They are also instrumental to generating electricity, as we saw in the previous chapter. In the absence of surface water from rivers or lakes, or when these cannot provide the capacity needed to supply the city, water resources engineers turn to groundwater, which is the topic of the next section.

Example 6.4

Designing a Stormwater Channel in Civitas

You are asked to design a stormwater channel for Civitas. The channel should accommodate a flow rate equivalent to a 24-h 100-year rain, and it will receive the runoff from an area of 20 km^2. The channel will be rectangular with a bottom width $B=7$ m, a slope of 0.001, and a Manning roughness of 0.012. Determine (1) the flow rate, (2) the depth of the water y (assuming the flow is uniform), and (3) whether the flow is subcritical, critical, or supercritical.

Solution

(1) From example 6.1, we determined that a 24-h 100-year rain represents an intensity of 7.36 mm/h. Moreover, since the channel will receive the runoff from an area of 20 km^2, this means the flow rate will be:[13]

$Q=7.36\times10^{-3}\times20\times10^{6}\ /\ (60\times60)=40.88\ m^3/s$

(2) Assuming a uniform flow, $S_f=S_0=0.001$, the area is $A=7y$, the wetted perimeter is $P=B+2y=7+2y$, $n=0.012$, and $C_m=1$, therefore:

$$40.88=\frac{1}{0.012}(7y)\left(\frac{7y}{7+2y}\right)^{2/3}0.001^{1/2}$$

Rearranging the unknowns:

$$\frac{(7y)^{5/3}}{(7+2y)^{2/3}}=\frac{0.012\times40.88}{0.001^{1/2}}=15.51$$

Iterating with different values of y gives us: $y=1.92$ m.

(3) Considering $V=Q/A$ and $y=1.92$, we have $A=7y=7\times1.92=13.44\ m^2$, and $V=40.88\ /\ 13.44=3.04$ m/s.

Using the Froude number on a rectangular channel: $Fr=3.04\ /\ (9.81\times1.92)^{1/2}=0.70$. The flow is therefore subcritical.

6.1.4 Groundwater Engineering

The topic of groundwater engineering can be extremely complicated because water is mixed with soil. What we will do here is first define our system by covering the principles of groundwater hydrology. We will then learn about the main equation that we should use to quantify groundwater flows—that is, an equivalent of the Manning equation but for groundwater. Finally, we will look at the physics of pumping water from the ground, which is particularly important for us since many cities directly depend on their groundwater.

6.1.4.1 Groundwater Hydrology Beneath our feet the ground is composed of solid matter, but this matter is most often porous. This means that the underground is full of little voids that are usually filled with either air or water (or oil and gas in oil and gas wells). Because water is heavier than air, the voids that are deepest in the ground get saturated with water only, and the voids at shallower depths are essentially filled with air. The demarcation between the two zones is called the *water table*, and its depth is simply governed by how much water is in the ground. These underground water reservoirs are called *aquifers*, and they are characterized as *unconfined* because the top surface is exposed to air, and no pressure is built up.

Sometimes, however, a layer of rock that is impervious to water (called *aquicludes*) forces the water to build some pressure, and these reservoirs are called *confined aquifers*. Thanks to this pressure, if we drill down, the pressure will force the water to spring out; these types of wells are more commonly known as *artesian wells*. These properties of groundwater fit within the field of groundwater hydrology, and they are illustrated in figure 6.11.

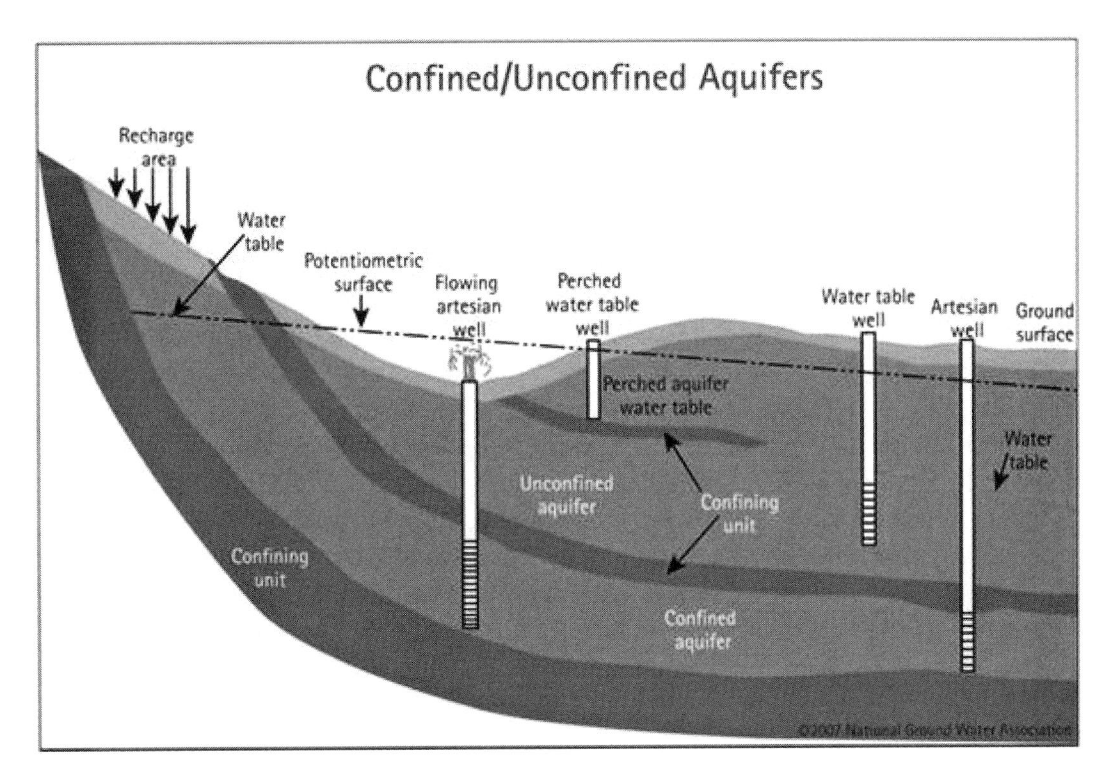

Figure 6.11
Groundwater hydrology. Reprinted from NGWA.org with permission of the National Ground Water Association (n.d.).

6.1.4.2 Darcy's Law The governing equation in groundwater engineering is known as Darcy's law (the same Darcy from the Darcy-Weisbach equation). Similar to the Manning equation, it finds a relationship between flow rate and system properties:

$$Q = KA\frac{\phi_1 - \phi_2}{L} \tag{21}$$

where A is the cross-section area—that is, perpendicular to figure 6.11—and L is the length of the section we are considering—that is, the distance between a point 1 and a point 2. The variables ϕ_1 and ϕ_2 are the pressures (in [m]) at points 1 and 2, respectively, which are the water depths for unconfined aquifers and the actual pressures for confined aquifers. The term $(\phi_1 - \phi_2)/L$ essentially represents the pressure gradient (a little like S_f for open channels). Finally, K is the hydraulic conductivity and is expressed in [m/day]. More precisely, K is defined as:

$$K = k\frac{\gamma}{\mu} \tag{22}$$

where k is the intrinsic permeability of the solid matrix—that is, how much fluid the solid matrix lets through—and is expressed in darcy,[14] and γ and μ are the specific weight and viscosity of the fluid traveling through the matrix (e.g., water).

Naturally, the water does not flow in the total area A but only in the voids, and we first define the specific discharge q as:

$$q = \frac{Q}{A} = K\frac{\phi_1 - \phi_2}{L} \tag{23}$$

which is expressed in [m/day], similar to velocity. The velocity v, however, is larger since only the area of the pores should be included. Defining n_e as the effective porosity of the ground—the ratio of the area of the pores to the total area—v becomes:

$$v = \frac{q}{n_e} \tag{24}$$

6.1.4.3 Pumps Groundwater is often a source of freshwater for communities. Some major cities in the world are entirely dependent on their groundwater, and the most famous case was Beijing in China—we will see what then happened in Beijing shortly. These aquifers possess a finite amount of water, however, and the pumping rate must be smaller than the rate needed for the aquifer to recharge. Within this section we are therefore interested in determining pumping rates. Figure 6.12 illustrates how water is pumped in unconfined and confined aquifers.

Physically, for unconfined aquifers,[15] the pumping rate Q_u is determined by:

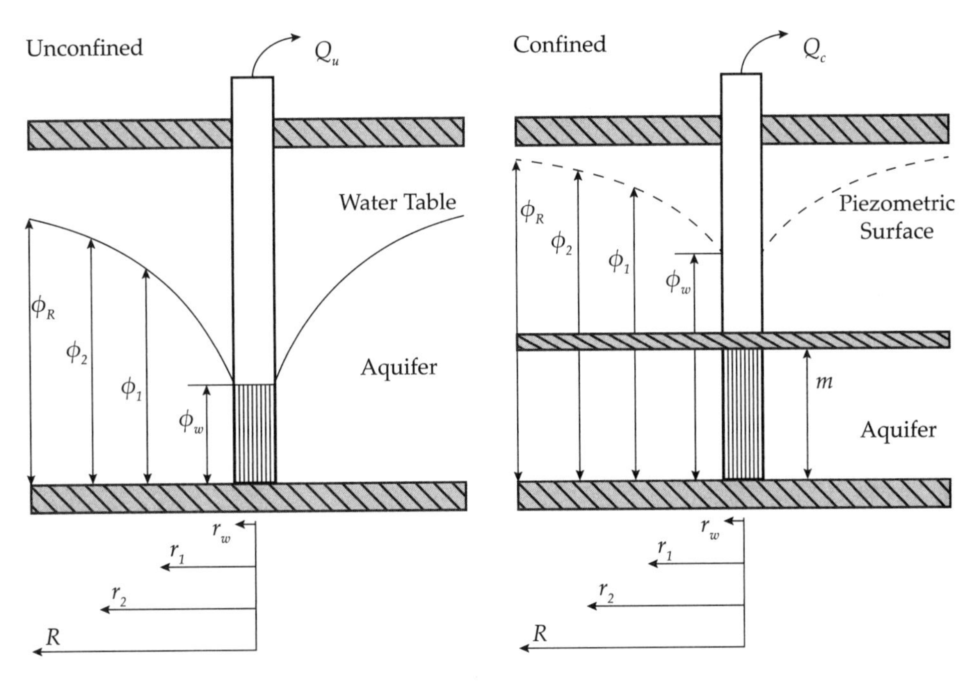

Figure 6.12
Wells in groundwater engineering.

$$Q_u = \frac{\pi K(\phi_2^2 - \phi_1^2)}{\ln\left(\frac{r_2}{r_1}\right)} \tag{25}$$

where r_1 and r_2 are the radii considered for the pressures ϕ_1 and ϕ_2. Most often r_1 is the radius of the well, r_w, and r_2 is the radius of influence R (where the water table is at the height it should be with no pumping). In the numerator the difference in pressure, $\phi_1 - \phi_2$, is often called the drawdown.

For a confined aquifer, the pumping rate Q_c is:

$$Q_c = \frac{2\pi Km(\phi_2 - \phi_1)}{\ln\left(\frac{r_2}{r_1}\right)} \tag{26}$$

where m is the saturated thickness—that is, the "height" of the aquifer. All other variables are the same as the unconfined case. Here, r_1 and r_2 also tend to be r_w and R. Moreover, the radius of influence is the radius at which the pressure is where it should be with no pumping.

Example 6.5

Groundwater in Civitas

Civitas is contemplating tapping into its groundwater to supply some of its population with freshwater. Assuming that 10,000 people who consume, on average, 300 liters of water per day were to be supplied water from this well, (1) determine the pumping rate needed. Two wells can potentially be tapped; one is located in a confined aquifer with a saturated thickness of 3 m, and the other one is located in an unconfined aquifer with a water table height of 15 m. For both aquifers, the hydraulic conductivity is 40 m/day and the radii of the well are 10 cm. Moreover, in both cases, the drawdown should not be more than 12 m or the pump will run inefficiently (and make the confined aquifer unconfined). The radii of influence are 300 m and 800 m for the confined and unconfined aquifers, respectively. Based on this information, calculate the maximum pumping rate for (2) the confined aquifer and (3) the unconfined aquifer and (4) determine which well should be tapped.

Solution

(1) $Q = 10{,}000 \times 300 \times 10^{-3} = 3{,}000\ \text{m}^3/\text{d}$.

(2) In the confined aquifer case, using equation (26) with $K = 40$ m/d, $m = 3$ m, $(\phi_2 - \phi_1) = 12$, $r = 0.1$ m, and $R = 300$ m:

$$Q_c = \frac{2\pi Km(\phi_2 - \phi_1)}{\ln\left(\frac{r_2}{r_1}\right)} = \frac{2\pi \times 40 \times 3 \times 12}{\ln\left(\frac{300}{0.1}\right)} = 1130\ \text{m}^3/\text{d}$$

(3) In the unconfined aquifer case, using equation (25) with $K = 40$ m/d, $\phi_2 = 15$ m, $\phi_1 = 15 - 12 = 3$ m, $r = 0.1$ m, and $R = 800$ m:

$$Q_u = \frac{\pi K(\phi_2^2 - \phi_1^2)}{\ln\left(\frac{r_2}{r_1}\right)} = \frac{\pi \times 40 \times (15^2 - 3^2)}{\ln\left(\frac{800}{0.1}\right)} = 3020\ \text{m}^3/\text{d}$$

(4) From the results, the unconfined aquifer should be tapped as $Q_u > Q > Q_c$.

If the pumping rate is too high, the aquifer may dry out (or a confined aquifer might become unconfined), and the rate of consumption becomes larger than the rate of production, which is not sustainable. For communities located close to the sea, a large drawdown may cause seawater to enter the well;[16] a similar process can occur next to contaminated areas. Monitoring the pumping rate is therefore critical.[17] For a large city such as Beijing, for instance, the availability of groundwater was not enough to supply the needed amount of water, and the country had to develop an extensive program, whose sustainability is questionable, to remediate this issue.[18]

We are now equipped with all the fundamentals we need to start designing water distribution systems.

6.2 Water Demand

In this section we will first look at some general water consumption trends. We will then see how water consumption typically varies by end use and how the number of people per household affects water consumption. We will also learn how the demand for water varies during the day. To learn more about the history of water supply and wastewater treatment, Sedlak (2014) offers a great general resource from an engineering perspective, and Melosi (2008) provides an excellent account of problems related to the United States specifically from a historian's perspective.[19]

6.2.1 Water Consumption Trends

Many different sectors require water for their activities. Figure 6.13 shows the proportional withdrawal of water per sector in 2015. The amount of water needed for thermal power plants accounted for 41% of the total withdrawal, which further reinforces this idea of the energy-water nexus. The second largest consumer of water was irrigation, with 37%. We will not discuss these two types of consumers, however, since the water is directly put back into the environment, to rivers for power generation and as groundwater for irrigation.[20]

For this chapter we are more interested in the 12% of public supply water withdrawal, which accounts for the water that is used to provide water to "at least 25 people or have a minimum of 15 connections. Public-supply water is delivered to users for domestic, commercial, and industrial purposes, and also is used for public services and system losses" (Dieter et al. 2018, p. 18). This is the water that gets treated and distributed to our taps.

Historically, until the 1800s people used to consume, on average, about 5 L/day (1 gal) (Guillerme 1988). With progress in water distribution systems, this number increased to approximately 20 L/day (5 gal) by 1850, and then to a significant 100 L/day (25 gal) by 1900 (Tarr 1988). Can you guess how much water we consumed in the early twenty-first century?

In 2015 about 150 billion liters of water were used every day in the United States for public supply only, with a served population of about 280 million people, representing 86% of the U.S. population (i.e., 325 million). The remaining were self-supplied with water. Focusing on the served population, these values amounted to

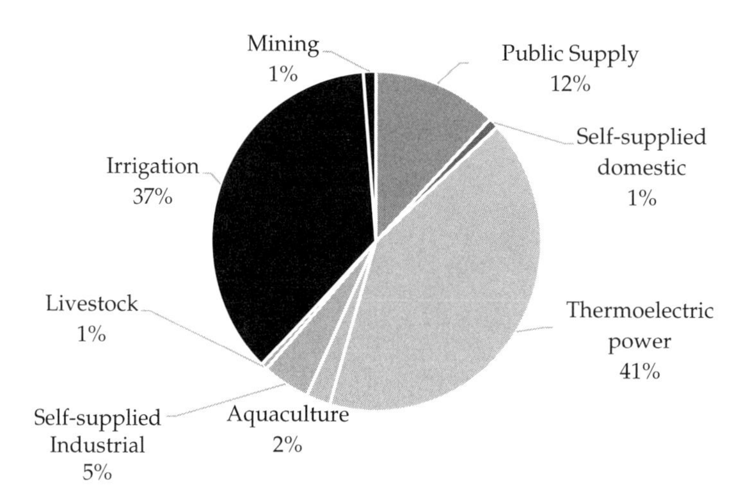

Figure 6.13
2015 water withdrawal in the United States.
Source: U.S. Geological Survey.

roughly 520 L per person (about 138 gal), but that includes all public water use. Out of these 520 L per person, 60% were used for residential/domestic purposes—as opposed to commercial, industrial, and other—amounting to about 311 L per person (about 82 gal). By the end of the twentieth century, we used to say that an average American consumed, on average, 100 gal per day (i.e., 380 L), and this number therefore decreased by close to 20% by 2015, directly contributing to our first sustainability principle.

The general trends since 1985 can be observed in figure 6.14, which again shows a steady decrease in water withdrawal from public use—although we are still far from the 100 L per day of 1900. The United States is also the largest per capita consumer of water in the world. There is no reason to believe that this decreasing trend cannot continue in future years, but it is unlikely to be substantial unless we drastically change how we consume water.

Naturally, not all states consume water evenly, and akin to power consumption, some states consume much less than others. Figure 6.15 shows water consumption by state. We can see that states on the East Coast tended to consume less water, with a minimum of 134 L/day in Connecticut. Typically, values hovered around the U.S. average, somewhere between 250 and 350 L/day. Mountain states consumed much more water, however, often above 400 L/day—the big exception is New Mexico, which was facing (and may still be facing) severe drought problems at the time.

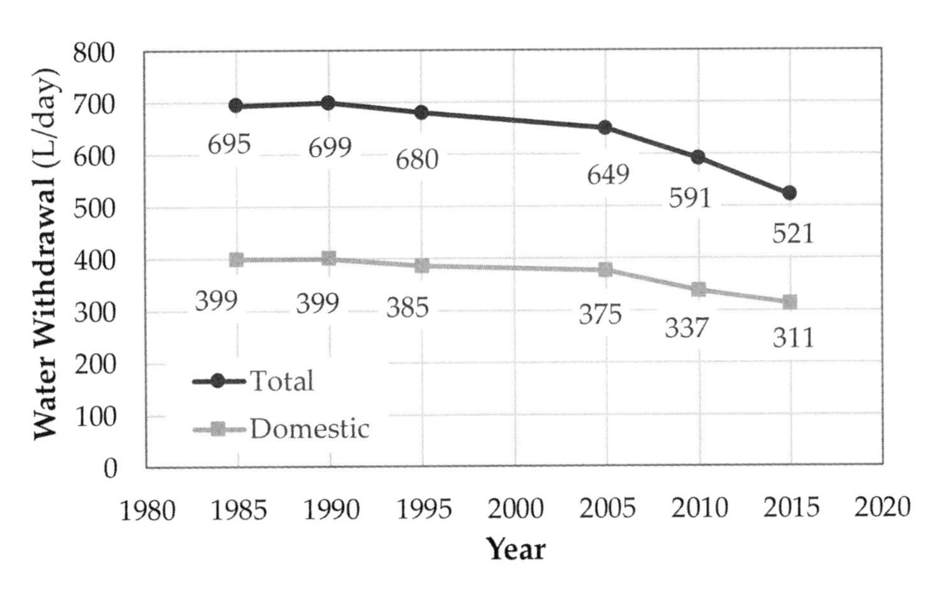

Figure 6.14
Evolution of public supply water consumption in the United States.
Source: U.S. Geological Survey.

Technically, the term *withdrawal* applies more accurately than *consumption*. Indeed, water distribution systems are large, and not all the water withdrawn from water bodies is consumed. In fact, leaks throughout water distribution systems can be significant. For the United States, Chin (2014) reports losses between 6% and 9%, and accounting for measurement and other errors, Thornton (2008) reports losses closer to 16%. This is fairly significant since a lot of energy goes into cleaning this water (as we will see later). Moreover, in cities with older systems, this number can increase substantially. In London, losses were estimated to be as high as 28% of the total withdrawal (Kennedy, Cuddihy, and Engel-Yan 2007), and in Rio de Janeiro and Sao Paulo, losses were estimated to account for more than 50% of the total withdrawal (Kennedy et al. 2015). Eliminating these losses requires a renewal of infrastructure—that is, changing the millions of kilometers of pipes beneath the streets.

6.2.2 Water Demand by End Use

The breakdown of the actual consumption by end use is given in table 6.1. Water consumption values are per day per household. Data were collected from more than 4,500 households around 2012. Details about the data can be found in DeOreo et al. (2016). We can note, however, that the data were collected from many different cities across

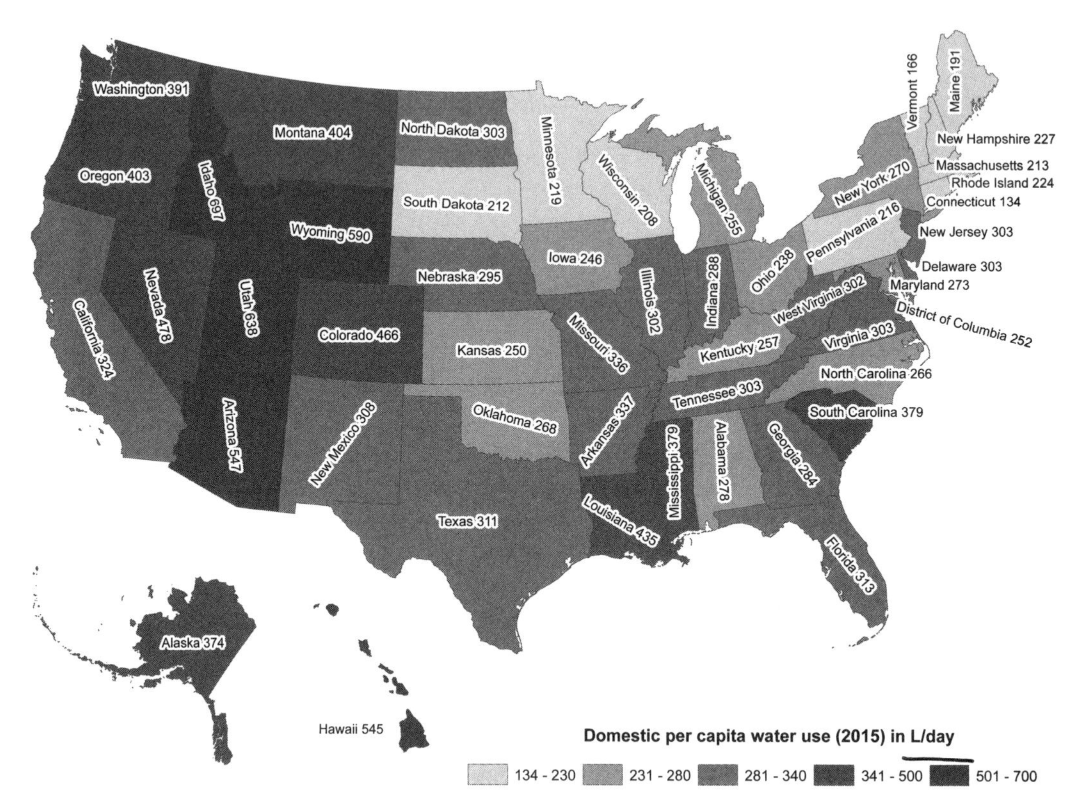

Figure 6.15
2015 domestic water withdrawal by state.
Source: U.S. Geological Survey.

the United States, and they show per household values. Considering households in the United States have, on average, 2.54 people, the per capita consumption is slightly more than 200 liters per day, which is therefore quite low.

From the table we can see that water consumption for toilets dominates at 24%, which can be reduced by using dual flush systems, for instance. Moreover, water from the toilet is referred to as *black water*, unlike water from the shower and faucet, which is referred to as *gray water*. One avenue to reduce our consumption is to use gray water to flush the toilet. We also note the strong presence of leaks (e.g., leaky faucets) that account for 12.3% of water consumption. There is therefore much potential to lower our current water demand for residential purposes, contributing toward our first sustainability principle.

Table 6.1
Water consumption by end use

	Water consumption (per household per day)		
Category	Gallons	Liters	Percentages
Toilet	33.1	125.3	24.0%
Shower	28.1	106.4	20.4%
Faucet	26.3	99.5	19.1%
Clothes washer	22.7	85.9	16.5%
Leaks	17.0	64.3	12.4%
Bath	3.6	13.6	2.6%
Dishwasher	1.6	6.1	1.2%
Other	5.3	20.1	3.8%
Total	137.7	521.2	100.0%

Source: DeOreo et al. (2016).

6.2.3 Water Demand by Household Size

From the list of end uses in table 6.1, we can see that some, like clothes washer and dishwasher, are shared across multiple people. Logically, the water consumption per person should decrease when more people live together, and this is actually what we observe. Figure 6.16 shows water consumption per person for household sizes ranging from one person to more than eight people. Akin to table 6.1, the data comes from the Water Research Foundation (see DeOreo et al. 2016 for more information).

The solid line in the middle is the median, and the gray area is bounded by the 25th percentile of the survey respondents at the bottom and the 75th percentile at the top. We can therefore see that in households with only one person, water consumption can vary quite a bit, but as more people live together, the difference in per person water consumption between the 75th and 25th percentiles gets narrower.

6.2.4 Water Demand by Hour

Water demand also varies during the day. Figure 6.17 shows daily water demand from the Jardine Water Purification Plant in the Chicago region for a typical summer day and a typical winter day.[21] Does the shape of the curves remind you of something? The demand is similar to that for power, which we saw in chapter 5 in figure 5.14. Demand is lowest at night, when people are sleeping, akin to the *base load* in electricity. It then increases in the morning, here around 6:00 a.m., which is essentially similar to the *morning ramp* in electricity. Demand then peaks in the afternoon or early

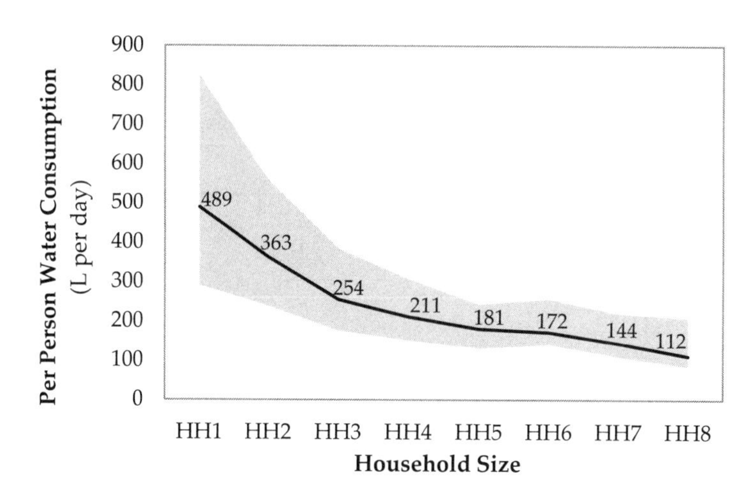

Figure 6.16
2016 per capita water consumption by household size.
Source: Water Research Foundation.

evening depending on the season and people's activities throughout the day. Finally, it decreases in the evening and goes back to the base load. We should note that, akin to electricity as well, demand profiles differ on weekends. The profiles shown in figure 6.17 correspond to a typical workday.

Moreover, we can see that demand is consistently higher in the summer—that is, about 20% higher in figure 6.17—which accounts for the use of water outdoors, such as for watering lawns.

In addition, focusing on the summer curve, the average demand was 38.80 m^3/s, and the peak demand was 43.16 m^3/s, which gives us a peak ratio of 1.11 for that day.[22] More generally in water distribution systems, Chin (2014) reports that the peak average demand factor is around 1.80—that is, the maximum water demand during a day in the year compared to the average daily demand—and that the peak maximum demand is around 3.00—that is, the maximum demand recorded at any point during a day compared to the average daily demand.

As a result, and again akin to electricity, the supply of water pressure can be adjusted during the day so that it stays around 300 kPa. This is especially important for larger water distribution systems since sudden drops in pressure can create water hammers that can puncture pipes. Smaller systems may not be impacted as much by large pressure differentials, and they therefore may not need to regulate

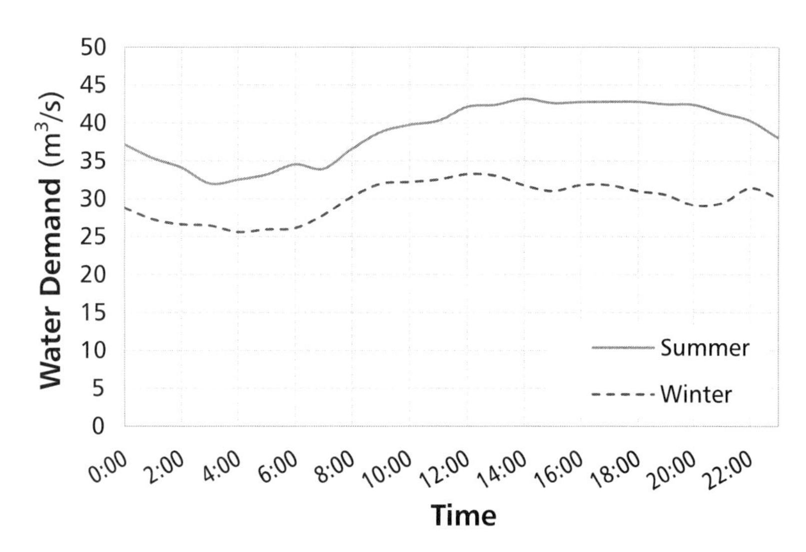

Figure 6.17
Typical daily water demand in the summer and the winter in the Chicago region.
Source: City of Chicago Department of Water Management.

their pressure, which can be acceptable as long as water pressures do not go below 140 kPa.[23]

Adjusting water pressure can be done in different ways. For fixed-rate pumps, a pressure regulator can be placed after the pumps to regulate the pressure. For variable-rate pumps, the pumping rate can be adjusted to provide the desired pressure (again, around 300 kPa).

Many water distribution systems also have reservoirs—such as water towers and standpipes—that can take the extra water when the pressure is high and supply water when the pressure is low. Moreover, if these reservoirs are elevated, pumping is not always required, ensuring a stable water supply that is not reliant on electricity or gas, which is more *resilient*. In fact, have you ever wondered why water towers have the height they have? Recalling our definition of hydraulic head h in meters for flows in closed conduits (i.e., $h = P/\gamma$), and since we need to ensure a pressure $P = 300$ kPa, we get $h = 300 / 9.79 = 30.6$ m, thus a height of about 30.6 m.

Back to our energy-water nexus, water distribution systems have a maximum pumping rate at peak demand in late afternoon/early evening, and this is therefore when they will also use the most energy for distribution. This is also when demand for electricity is high, if we remember figure 5.14.

6.3 Water and Wastewater Treatment

In this section we will briefly describe the process to make water potable so that we can safely drink it, and we will also briefly describe the process to polish wastewater before we put it back in natural waterways.

First, we must realize something pretty extraordinary. In the United States, everything related to drinking water is regulated by the Safe Drinking Water Act that became law in 1974. In contrast, everything related to wastewater is regulated by the Clean Water Act that became law in 1972. These are two completely different laws with different objectives and requirements. What separates the two laws? About 15 cm (6 in)! All the way until water reaches the end of the faucet, water is safe to drink, and it follows the Safe Drinking Water Act, and as soon as it hits the bottom of the sink and enters the drain, it becomes wastewater and follows the Clean Water Act. How extraordinary!

6.3.1 Water Treatment

Water treatment is a critical aspect of water supply and distribution systems, since without all the advances in water treatment, cities would never have been able to evolve. Water can be treated in many different ways. Figure 6.18 shows a relatively common way to make water potable.

First, the water is pumped in the plant. Second, the water is filtered through several screens to prevent debris and fish from entering the system. Third, an initial chemical

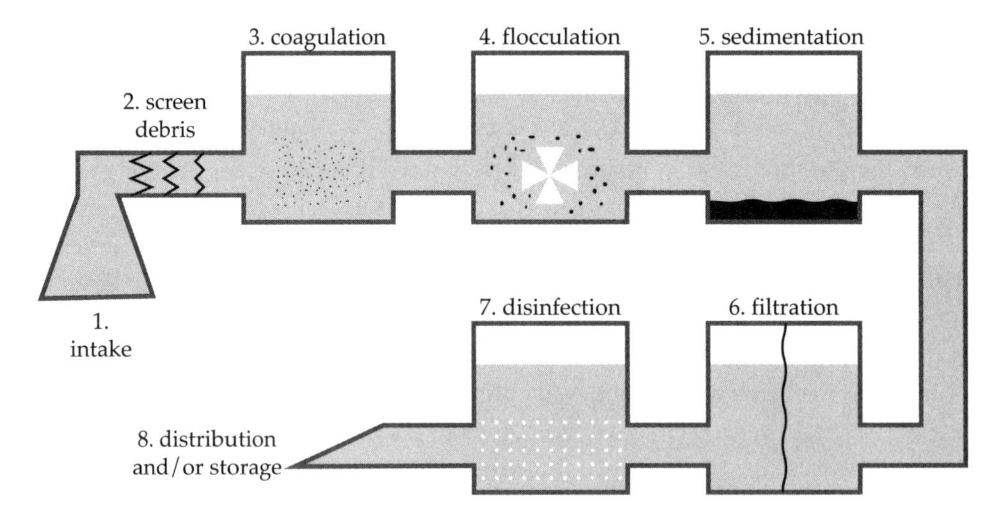

Figure 6.18
A common water treatment process.

treatment is applied to the water, and active carbon may be added to remove odor and taste. A coagulant is then added to prepare for the fourth step, flocculation, where clumps of suspended solids form that can be removed more easily. Fifth, these clumps settle at the bottom of a basin and are removed. Sixth, sand and/or coal is used to filter out the smaller particles. Seventh, chemicals, such as chlorine, are added to disinfect the water against any potential contimanants that might exist in water pipes.

Many other methods or variants of this method exist, such as using ozone and ultraviolet (UV) rays as disinfectants. These other methods will likely have a significant impact in the future. The current process to treat water takes anywhere between eight to twenty-four hours, and it is therefore quite slow. Increasing the speed of this process may help us become more responsive to changes in demand, which may enable us to better control the demand.

One potential future avenue is to stop this centralized process, which requires not only a long treatment time but skilled maintenance of a water distribution system to keep it in nearly perfect condition. This is because once contaminated water enters a treatment plant, it can jeopardize the entire system, requiring boiling-water orders. Future treatment plants may therefore be more decentralized, perhaps at the neighborhood scale, or perhaps even at the house scale. Every building would therefore have its own little treatment plant to purify water.

6.3.2 Wastewater Treatment

The treatment of drinking water is relatively energy intensive, and so is the treatment of sewage. Reducing the amount of wastewater that goes through a treatment plant is therefore highly desirable. There are also various ways to treat wastewater. Figure 6.19 shows a typical wastewater treatment process.

The wastewater is first screened to remove any large debris that might have found its way into the sewer (i.e., branches, bottles, small objects from stormwater runoff). Second, the wastewater is placed in a large, slow-moving basin where big and heavy solids/sediments can settle down in the tank and be removed while grease, oils, and light sludge, which float at the top, are also removed. Third, the sludge is taken away to be processed in digesters to remove odors and decrease the volume of the wastes, and it is then transported out of the treatment plant, either to be landfilled or used as fertilizer for agriculture depending on the quality of the sludge. Digesters also sometimes produce natural gas that can be burned to produce heat and/or electricity—we will learn about anaerobic digesters in chapter 9. The wastewater that has been rid of the sludge is then aerated (i.e., oxygen is added), which helps microorganisms degrade pollutants. Fourth, and similar to the second process—that is, sedimentation—the wastewater goes

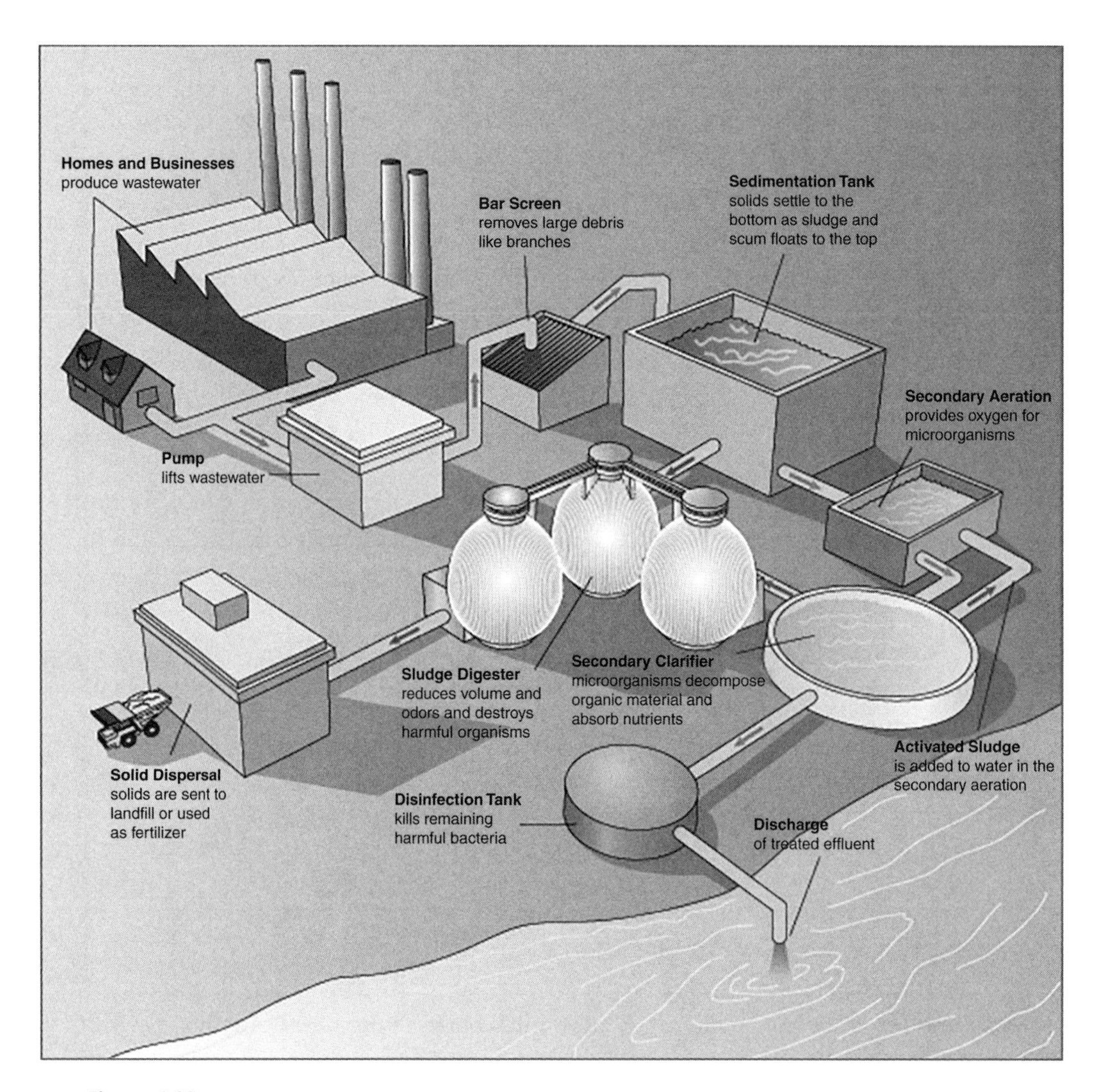

Figure 6.19
Typical wastewater treatment process. Credit: CK-12 Foundation.

through a process of *clarification*, in which smaller solids settle to the bottom of the tank (and are then removed akin to the second process). Fifth, the remaining water is disinfected (e.g., by using chlorine) and returned to a natural stream.

In the end, while the treated wastewater is not potable,[24] it can serve other purposes, such as watering plants and lawns. Some cities, like Paris and Saint Petersburg, Florida, have established a dual water distribution system, one with potable water and one with nonpotable water that can be used for lawns, for instance. This system is sometimes called *purple pipe* because of the color of the pipes generally used for it. Many golf courses are also located next to wastewater treatment plants, and they use treated wastewater for irrigation.

In communities that face severe shortages of water, as is the case for many cities in California, the treated water is put upstream of the water treatment plant. The water is further treated naturally in the stream before it enters the water treatment plant and is treated again. This process enables a significant reduction in total water withdrawal.

Energy needs to treat both water and wastewater are actually quite high. But before we talk about energy use in water, we must address one more important item, and it is stormwater management.

6.4 Stormwater Management

Stormwater management has become one of the most important issues that cities have to face. Put simply, poor management of stormwater is the leading cause of urban flooding, which is one of the most expensive and deadliest types of weather-related disaster in the United States (Kermanshah, Derrible, and Berkelhammer 2017). Indeed, the constant expansion and paving of cities (creating impervious surfaces), coupled with a higher frequency of extreme events, has directly led to more surface runoff. In this section we will learn about sewers and green infrastructure, and we will get introduced to two models for stormwater modeling.

6.4.1 Sewer Systems

There are two kinds of sewers. Sanitary sewers collect the wastewater from buildings, including black water and gray water from human use. Stormwater sewers collect runoff—that is, the excess rain that falls on roofs, sidewalks, and roads and that ends up in gutters and most often enters the sewer system through catch basins (e.g., located by street curbs). These two types of wastewater are very different. Sanitary sewer wastewater systematically needs to be treated before being put back in the environment. Stormwater, on the other hand, does not necessarily need to be treated.

Newer cities separate the two systems, but older cities often have combined sewer systems. In combined sewers all wastewater is treated the same way. Figure 6.20 shows a typical street cross-section with a combined sewer system. In the figure, we can see that both the sanitary sewer from the house and the stormwater sewer from the street end up in the same drain—note the small plugs labeled "flow restrictor" in the catch basins, which will be discussed later.

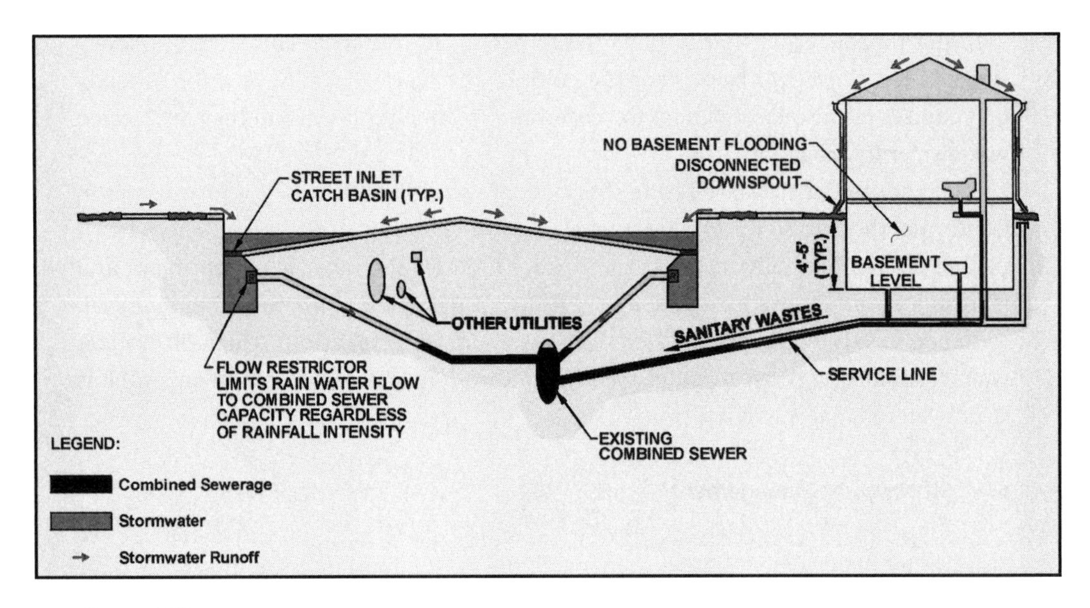

Figure 6.20
Typical street cross-section. Courtesy of Stantec.

Combined sewer systems therefore need to accommodate a significant influx of stormwater during heavy rain events, which can *overflow* easily; these events are called *combined sewer overflows* (CSOs). This overflow creates two major problems. First, local wastewater infrastructure is filled rapidly, causing local flooding throughout a city. Catch basins are full, and water starts to pond at curbs and in gutters. Even worse, some people have a basement sewer access that overflows a mix of sanitary water and stormwater. The second problem is that some of this untreated wastewater overflows directly into the natural streams.

CSOs are best illustrated in figure 6.21. For dry weather, the impact of combining the two sewer systems is low since the contribution of wastewater is minimal. For wet weather, however, the story is different. The systems easily reach the maximum possible capacity, which means that some of the wastewater is disposed of in a natural stream without being treated. As shown in the figure, a simple system is installed using a small bump/dam that retains the wastewater when the flow is low. During heavy rains, the flow easily passes over this bump. When both sewer systems are separated, the stormwater can go directly into the natural channel.

While CSOs would be okay if they were rare, they can actually happen frequently. In Chicago in 2017, for example, there were a total of 2,135 CSOs at different locations

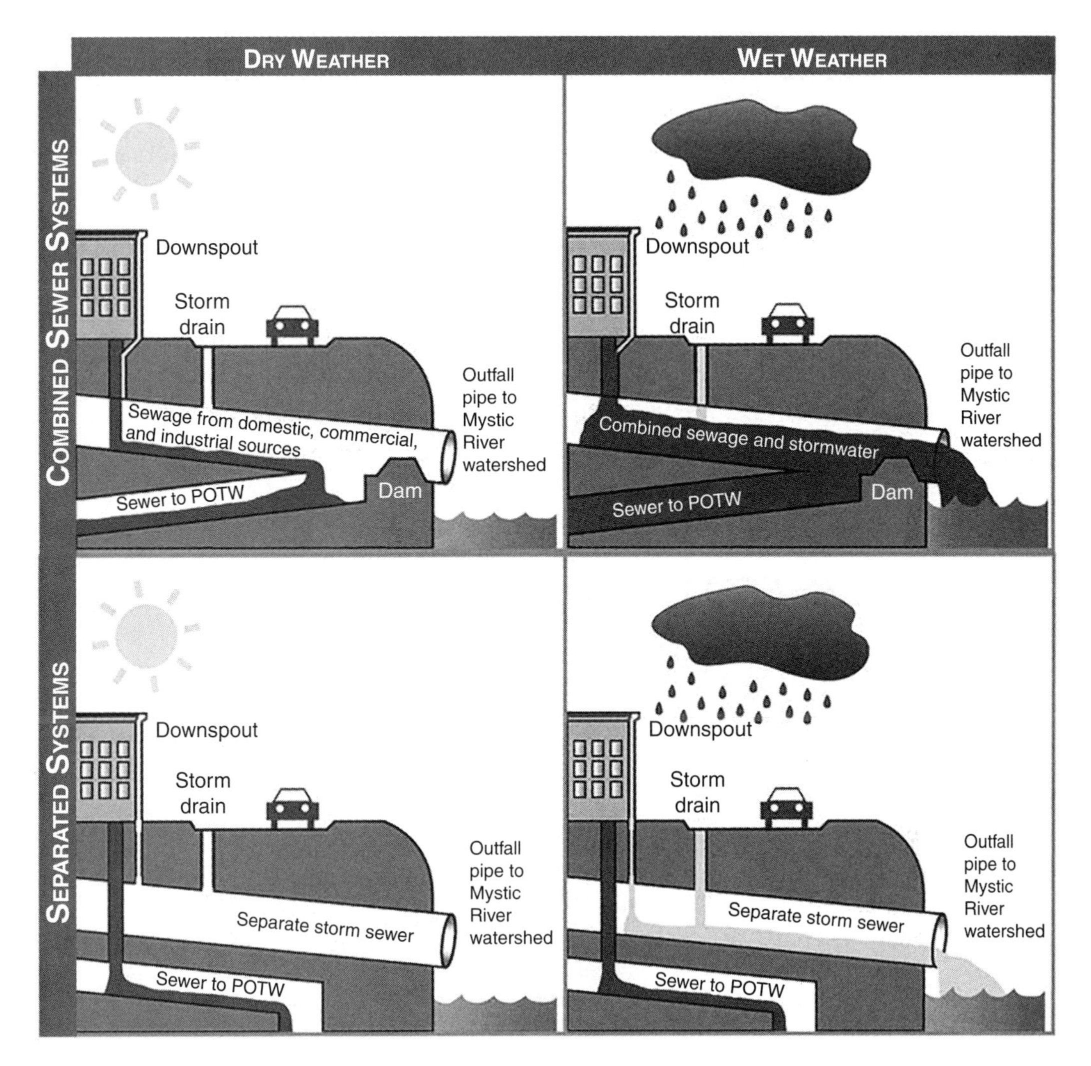

Figure 6.21
Combined and separated sewer systems. Credit: U.S. Environmental Protection Agency.

over sixty-eight rain events,[25] and unless something is done, the trend is likely to increase in the future.

Overall, handling stormwater runoff when heavy rains tend to be not only more frequent but also more intense is arguably the biggest water-related problem many cities have to face. Moreover, treating wastewater is incredibly energy-intensive. Separating both systems is therefore more desirable. This does mean that contaminated stormwater would not be treated, which has negative effects, but it remains preferable. Indeed, if CSOs were rare, dumping the wastewater to natural waterways might be acceptable, but they are not rare. These sewer systems are old, and they were built to accommodate conditions that are no longer valid. Cities have expanded significantly, therefore paving more surfaces that in turn create more runoff.

The options to solve this problem are limited. Ideally, combined systems can be retrofitted to separate the two. This is, however, generally not the preferred method since it requires that all roads in a city be opened and new pipes be added, which is also incredibly costly. Instead, some cities, like Chicago and Kansas City, have chosen to build large underground tunnels that store wastewater in excess of the capacity that can flow and be treated in a wastewater treatment plant. When the rain stops, the wastewater from these tunnels is then pumped to the treatment plant. These tunnels cost billions of dollars to build, however, and there is no guarantee they will prevent overflow in the longer term. An alternative is green infrastructure and low-impact development.

6.4.2 Green Infrastructure and Low-Impact Development

Green infrastructure (GI) and low-impact development (LID) refer to strategies that are generally small in scale to help manage stormwater runoff. Many cities have now chosen to implement GI and LID to address their flooding problems,[26] as opposed to undertaking massive infrastructure projects (and it can be much cheaper too[27]). This does not mean that GI cannot be implemented on a large scale—for example, one strategy can be implemented across a city—but individual projects are typically not massive in scale.

GI and LID can generally be defined as *retention* and *detention* basins. Retention basins retain the water and let it infiltrate into the ground. In contrast, detention basins detain and hold water to release it at a slower rate into the stormwater sewer system. Some GI projects can include elements of both retention and detention basins.

The simplest type of detention basin is created with a restrictor plate at the outlet of the basin. The restrictor plate is welded to the detention basin outlet pipe and has a small hole in it. This small hole restricts the release rate of water, allowing it to be detained in the pond while being released slowly. We already saw the effect of restrictor plates on runoff in figure 6.3. The plugs in figure 6.20 also represent restrictor plates.

Both retention and detention basins essentially help decrease the peak flow, but retention basins are preferable, when possible, since no (or less) water enters the stormwater system. In terms of size, these types of basins range from small boxes constructed on-site and often underground—for example, at the outlet of a parking lot—to open

Figure 6.22
Examples of permeable pavers (*left*) and freshly poured pervious concrete (*right*).

ditches next to roads, and they can even be designed as large open fields that can handle large volumes of water.

In general, GI and LID strategies try to make some of the current imperviousness of a city permeable again. As retention basins, they allow rainwater to infiltrate the ground and turn into groundwater. Impervious surfaces include roads, sidewalks, and roofs. As we said previously, even grass does not let water infiltrate easily, and any water in excess of about 3 cm (about 1 in) runs off to the nearest curb or simply ponds. To achieve this, permeable pavers and pervious concrete can be used, which essentially lets the water through; figure 6.22 shows one example of each.

Instead of grass, native plants absorb much higher quantities of water thanks to their deep and large roots. Rain gardens and bioswales can be installed on or adjacent to streets. These rain gardens essentially receive water from roofs and gutters. Moreover, they have a layer of gravel that can also store water before it percolates down to the water table (as a retention basin). If needed, they are also connected to the sewer for excess water (as a detention basin).

Figure 6.23 shows an example of a rain garden (*left*) versus the common practice (*right*). It is a little ironic that the grass patches we see in the common practice picture require an irrigation system while they literally have water flowing beneath them. Moreover, many designs of rain gardens exist. The one in figure 6.23 still shows the inlet of a catch basin right by the street curb, but to be more effective the inlet should

Figure 6.23
Example of rain garden (*left*) versus common practice (*right*).

be in the rain garden and raised slightly above the garden surface level so that it only accepts excess water. A better design is shown in figure 9.9 of chapter 9.

Naturally, many other types of GI and LID exist, from simply using sand as a retention basin to engineering stormwater wetlands that treat stormwater as well. In the same line of thought, in 2014, Copenhagen unveiled plans to build a "resilient" neighborhood, where the roads are actively designed as open channels to divert water to rain gardens during heavy rains. This kind of creative thinking remained the exception (as opposed to the norm) at the time of this writing, and it shows how infrastructure can and should be designed as integrated systems that solve multiple problems (e.g., traffic and stormwater runoff).

Generally, the cost of retrofitting cities with these types of LID infrastructure is also significant—although often less than conventional techniques[28]—but neighborhoods can be retrofitted in phases, bit by bit. Furthermore, many GI and LID strategies bring in a series of important additional benefits that can have significant impacts on communities. Rain gardens, for example, are known to help treat raw stormwater, to sequester carbon, to enhance the local biodiversity, and to improve general well-being in a community (Jato-Espino, Sañudo-Fontaneda, and Andrés-Valeri 2018).

Purely from a stormwater flow perspective, we should bear in mind that small-scale LID strategies such as rain gardens can only cope with as much stormwater as they can store, which is sufficient for day-to-day rains, such as most of the 2,135 CSOs that happened in Chicago in 2017, but it may not be sufficient for severe rainfalls, such as 100-year rains. As a result, it is desirable to accompany small-scale LID with other strategies—often integrating with multiple infrastructure systems, as is recommended in this book—that include emergency retention basins such as parks. A good example is the Cheonggyecheon River project in Seoul, as we saw in chapter 1, but other examples include recreational sports and other facilities that can be flooded without generating severe damages.[29] Sometimes, rerouting stormwater flows within sewer systems may also be sufficient. Kerkez et al. (2016) proposed to build smart stormwater systems by equipping sewer systems with more sensors and controllers to open and close gates to manage flows—and we will discuss the concept of Smart Cities in general a little in chapter 12. To design a great stormwater system, we first need to know how much stormwater to expect, which is our next goal.

6.4.3 Runoff Modeling

Considering how big of a problem stormwater runoff represents in urban areas, being able to estimate how much runoff a site has to handle is critical. Many modeling techniques can calculate runoff. Some methods are quite simple, and although they may not take into account the time dimension, they can be easily applied. We will be introduced to two methods in this section that are used extensively in the industry: the *rational method* (RM) and the *Natural Resources Conservation Service (NRCS) curve number model.*

Many other (and often better) methods are not necessarily more complex, but they are computational. Starting from fundamental principles of hydrologic modeling, and given enough details about the site being evaluated, computational models can simulate how a site responds to rainfall over time. These computational methods are outside the scope of this chapter, but they are not too difficult to use. One of the most popular software packages in the United States was developed by the U.S. Environmental Protection Agency (EPA). Simply called the Storm Water Management Model (pronounced "swim"), it is freely available on the EPA website.[30]

For now, we will focus on the simplest method to compute stormwater runoff: the RM.

6.4.3.1 Rational Method The RM dates back to 1889, and it was originally developed by Kuichling (1889). It offers a simple way to estimate the *peak discharge* Q from a certain rainfall—that is, the peak on a hydrograph. Because of the simplicity of the method, it can only be applied to relatively small areas to output any meaningful results—that is, roughly less than 1 km^2. The RM assumes that Q follows the following simple relationship:

$$Q = C \cdot i \cdot A \tag{27}$$

where C is the RM runoff coefficient that has no units, i is the rainfall intensity (e.g., in [mm/h]), and A is the watershed or draining area (e.g., in [m^2]). The units are particularly tricky here. The variables A and i are frequently given in units that do not match (e.g., A is in [km^2], and i is in [mm/h]). It is critical to convert them to the same standards so that Q has consistent units—for example, in [m^3/s].

The runoff coefficient C is essentially the proportion of the rainfall that becomes runoff. It is therefore bounded between 0 and 1. Moreover, it will be smaller for permeable surfaces and higher for impermeable surfaces. Table 6.2 shows values for typical urban areas. We can see that values range from 0.05 for flat lawns to 0.95 for downtown areas and concrete.

The RM is especially used to size sewer systems to handle peak discharge, and it is therefore quite important to select the right value for i. The peak discharge is essentially

Table 6.2
Runoff coefficients for urban areas

Description	Runoff coefficient C
Business	
Downtown areas	0.70–0.95
Neighborhood areas	0.50–0.70
Residential	
Single family	0.30–0.50
Multifamily detached	0.40–0.60
Multifamily attached	0.60–0.75
Residential suburban	0.25–0.40
Apartments	0.50–0.70
Parks, cemeteries	0.10–0.25
Playgrounds	0.20–0.35
Railroad yards	0.20–0.40
Unimproved areas	0.10–0.30
Drives and walks	0.75–0.85
Roofs	0.75–0.95
Asphalt	0.70–0.95
Concrete	0.80–0.95
Brick	0.70–0.85
Lawns; sandy soils	
Flat, 2% slope	0.05–0.10
Average, 2%–7% slope	0.10–0.15
Steep, 7% slope	0.15–0.20
Lawns; heavy soils	
Flat, 2% slope	0.13–0.17
Average, 2%–7% slope	0.18–0.22
Steep, 7% slope	0.25–0.35

Example 6.6

Runoff in Civitas Using the Rational Method

Civitas is considering turning an empty parking lot into a residential area. The area A covers 0.01 km^2. The RM coefficients of the parking lot and the residential area are 0.9 and 0.65, respectively. For a 20-year rain with a time of concentration T_C of 6 min, determine (1) the current flow of runoff $Q_{parking}$, (2) the projected flow of runoff $Q_{residential}$, and (3) the percentage of the flow of runoff that will not end up in the sewers.

Solution

(1) First, from example 6.1, we know that a 20-year rain with a duration of 6 min has an intensity i of about 200 mm/h, which is 5.56×10^{-5} m/s. Using an RM coefficient of 0.9 and an area of 10,000 m^2 (i.e., 0.01 km^2), we get:

$$Q_{parking} = 0.9\times5.56\times10^{-5}\times10{,}000 = 0.5\text{m}^3/\text{s}$$

With a coefficient of 0.9, most of the rainfall ends up as runoff (specifically 90%).

(2) Using the same technique as (1) with $C=0.65$, we get:

$$Q_{residential} = 0.65\times5.56\times10^{-5}\times10{,}000 = 0.36\text{m}^3/\text{s}$$

(3) The percentage of the flow of runoff saved is:

$$\frac{Q_{parking} - Q_{residential}}{Q_{parking}} = \frac{0.5-0.36}{0.5} = 0.28$$

Changing the land use from a parking lot to a residential area will reduce the flow of runoff by $0.5-0.36=0.14$ m^3/s, which is 28% less than the original flow of runoff. Naturally, this is completely captured by the change in the RM coefficient since $(0.9-0.65)$ / $0.9=0.28$. Note that technically, this is not entirely accurate because the RM looks at the flow at the peak discharge only.

the design criteria since it represents the greatest flow rate that the sewer system is designed to handle.[31] There are two steps to follow here. First, we need to know the type of rainfall that the system is designed to handle. The RM is often used to design minor sewer systems than can handle 10-year rains; we therefore have to use IDF curves as well. Second, we need to know the duration of the precipitation—that is, the value on the x-axis on the IDF curve or the t in equation (2). For this, we typically use the time of concentration T_C (as seen in section 6.1.1.1)—that is, time for the farthest point of the watershed to reach the outlet. Using T_C allows us to estimate the maximum discharge to expect because this is when all of the water from a watershed has reached the outlet.

Time is therefore important here. Although we will not cover it, many design studies first perform something called a *critical duration analysis* to determine the actual peak discharge.

Let us move on to the NRCS model now.

6.4.3.2 Natural Resources Conservation Service Curve Number Model The NRCS curve number model, also called the curve number (CN) method, is another simple method to apply, and it is incredibly popular in practice. It was originally developed in the 1980s by the U.S. Department of Agriculture (USDA), thus later than the RM.

The main goal of the model is to determine the rainfall excess Q (i.e., runoff), considering an initial abstraction I_a and a retention F. The initial abstraction I_a is the proportion of water that is "abstracted" automatically, essentially constituting any rain that can be captured before it turns into runoff—for example, by remaining on tree leaves. The retention F is the proportion of runoff that does not end up in the sewer—for example, thanks to infiltration and ponding. For a precipitation P, by applying continuity we get:

$$P = Q + I_a + F \tag{28}$$

All the variables in equation (28) are expressed as a depth, like precipitation—for example, in [mm]. This is a difference from the RM, which expresses Q as a flow rate in [m^3/s]. In fact, with the NRCS model, the rainfall excess Q is often multiplied by the area A of the watershed, giving us a total volume of water in [m^3] that has to be handled. This is important, and it highlights the main reason to use the RM or the NRCS model. We will say more about that after being properly introduced to the NRCS model.

The main assumption made by the NRCS curve number model is the following:

$$\frac{F}{S} = \frac{Q}{P - I_a} \tag{29}$$

where S is the *potential maximum retention*—that is, the largest F possible; S does not include I_a. In other words, we assume that the proportion of rain that is retained is equal to the proportion of rain minus the initial abstraction that ends up as runoff. For example, if 20% of all the water that could be retained is retained, equation (29) stipulates that 20% of the precipitation (less abstractions) will turn into runoff. Similarly, if 100% of all the water that could be retained is retained, then 100% of the precipitation (less abstractions) will turn into runoff. The logic here is that if the storage capacity S is large, then the runoff Q will be small (since a large proportion of the precipitation will be stored). In contrast, if the storage capacity S is low, then most of the precipitation will end up as runoff Q.

By substituting F from equation (28) into equation (29), we get:

$$\frac{P-I_a-Q}{S}=\frac{Q}{P-I_a}$$

$$P-I_a-Q=\frac{S\cdot Q}{P-I_a}$$

$$Q+\frac{S\cdot Q}{P-I_a}=P-I_a$$

$$Q\left(1+\frac{S}{P-I_a}\right)=P-I_a$$

$$Q=\frac{(P-I_a)^2}{P-I_a+S} \tag{30}$$

Some empirical evidence suggests that I_a tends to be about 20% of the precipitation maximum potential retention S,[32] therefore:

$$Q=\frac{(P-0.2S)^2}{P+0.8S},\ P>0.2S \tag{31}$$

Equation (31) will be the fundamental equation to use to estimate Q. Naturally, we also need to determine S, and this is where the CN comes in. First, we need to establish the type of soil with which we are dealing. The model categorizes soils into four hydrologic soil groups (HSG) with the properties shown in table 6.3. A full categorization of all U.S. soils can be found in *Urban Hydrology for Small Watersheds*, USDA Technical Report 55 (TR55), available in U.S. Department of Agriculture (1986).

From the HSG, the method then estimates the maximum potential retention S based on the curve CN of the land use type of the area. CN for urban areas are given in table 6.4. Equation (32) can then be used to find S:

$$S=25.04\left(\frac{1000}{CN}-10\right) \tag{32}$$

Table 6.3
Hydrologic soil groups

		Minimum infiltration rates	
Hydrologic soil group	Description	mm/h	in/h
A	High infiltration rates	>7.6	>0.30
B	Moderate infiltration rates	3.8–7.6	0.15–0.30
C	Low infiltration rates	1.3–3.8	0.05–0.15
D	Ultralow infiltration rates	0.0–1.3	0.00–0.05

Table 6.4
Runoff curve numbers for urban areas

Cover type and hydrologic condition	Percent impervious area	Curve number by HSG A	B	C	D
Fully developed urban areas (vegetation established)					
Open space (lawns, parks, golf courses, cemeteries, etc.)					
Poor condition (grass cover < 50%)		68	79	86	89
Fair condition (grass cover 50%–75%)		49	69	79	84
Good condition (grass cover > 75%)		39	61	74	80
Impervious areas					
Paved parking lots, roofs, driveways, etc. (excluding right-of-way)		98	98	98	98
Streets and roads					
Paved; curbs and storm sewers (excluding right-of-way)		98	98	98	98
Paved; open ditches (including right-of-way)		83	89	92	93
Gravel (including right-of-way)		76	85	89	91
Dirt (including right-of-way)		72	82	87	89
Western desert urban areas					
Natural desert landscaping (pervious areas only)		63	77	85	88
Artificial desert landscaping (impervious weed barrier, desert shrub with 1- to 2-inch sand or gravel mulch and basin borders)		96	96	96	96
Urban districts					
Commercial and business (downtown areas)	85	89	92	94	95
Industrial	72	81	88	91	93
Residential districts by average lot size					
1/8 acre or less (multifamily attached)	65	77	85	90	92
1/4 acre (multifamily detached)	38	61	75	83	87
1/3 acre (single-family, residential suburban)	30	57	72	81	86
1/2 acre	25	54	70	80	85
1 acre	20	51	68	79	84
2 acres	12	46	65	77	82
Developing urban areas					
Newly graded areas (pervious areas only, no vegetation)		77	86	91	94

Source: U.S. Department of Agriculture (1986).

The coefficient "25.04" is added since the *CN* outputs values in inches, and one inch is 25.04 mm. More *CN* values can be found in U.S. Department of Agriculture (1986).

To get a practical sense for the role of *CN*, figure 6.24 shows runoff versus precipitation for *CN* values ranging from 40 to 100, substituting equation (32) in equation (31). From this figure, for a precipitation of 50 mm (about 2 in), we can see that an area with a *CN* of 50 generates 8 mm of runoff versus 34 mm for an area with a CN of 90, which is significant.

To reiterate what was said earlier, the NRCS model provides a total amount of rainfall excess that can be converted to a total volume of water when multiplied by an area. The NRCS model is therefore useful to design detention and retention basins (since we are looking for a volume). In contrast, the RM is used to determine a peak discharge,[33] and it is therefore more useful to size sewer systems and weirs.

To make it even easier, the EPA has developed a user-friendly software package, the National Stormwater Calculator, to estimate the amount of stormwater from any area in the United States. It also estimates the amount of reduced runoff thanks to various LID strategies. It uses historical rainfall records along with local soil maps and land cover data (and it also uses the RM and the NRCS model). It is highly recommended and is available on the EPA website.[34] Finally, to know more about modeling, Beven (2012) wrote a whole book about it, aptly titled *Rainfall-Runoff Modelling: The Primer.*

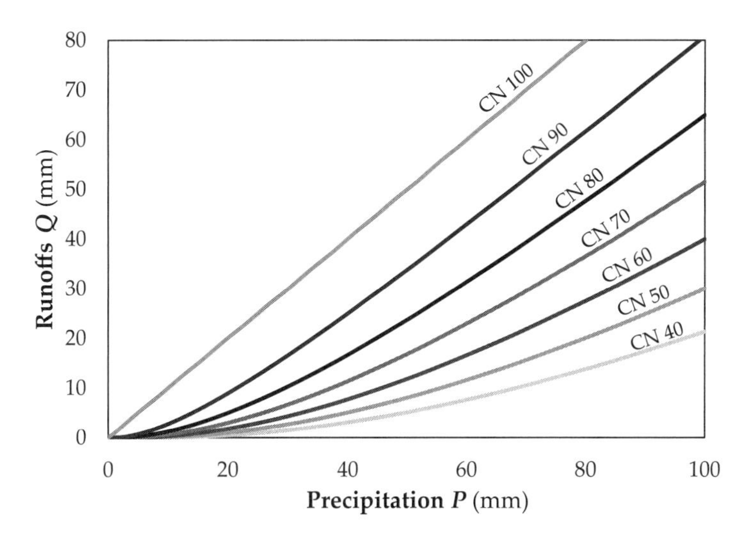

Figure 6.24
Impact of CN on runoff.

Example 6.7

Runoff in Civitas Using the NRCS Model

The goal of this example is the same as the one used in example 6.6. The residential area consists of multifamily attached houses (with 65% impervious area similar to example 6.6). Moreover, the HSG of the soil was qualified as C. For a 6 min 20-year rain, determine (1) the current amount of runoff $Q_{parking}$, (2) the projected amount of runoff $Q_{residential}$, and (3) the percentage of the amount of runoff that will not end up in the sewers. (4) Compare the results with example 6.6.

Solution

(1) Similar to example 6.6, from example 6.1, we see that a 20-year rain with a duration of 6 min has an intensity i of about 200 mm/h. The total precipitation is therefore $P=200\times(6/60)=20$ mm.

Furthermore, from table 6.4, a parking lot with HSG C has a curve number $CN=98$. Using equation (32), the maximum potential retention S is:

$$S=25.04\left(\frac{1000}{98}-10\right)=5.11$$

Since $P>0.2S=0.2\times5.11=1.02$ mm, we can use the NRCS method. Using equation (31), the runoff $Q_{parking}$ is:

$$Q_{parking}=\frac{(20-0.2\times5.11)^2}{20+0.8\times5.11}=15\,\text{mm}$$

Again, we see that most of the rainfall ends up as runoff. However, example 6.6 estimated that 88% of the rainfall ended up as runoff compared to three quarters (i.e., 75%) here. This is a sizable difference related to the fact that we account for abstraction with the NRCS model. Note that we cannot compare the two Qs yet since the units are different (Q in example 6.6 was in m^3/s).

(2) Using the same technique as (1) from table 6.4, we get $CN=90$, which gives us $S=27.82$ (and $0.2\times27.82=5.56$ mm$<P$) and therefore $Q_{residential}=4.93$ mm.

(3) The percentage of the amount of runoff saved is:

$$\frac{Q_{parking}-Q_{residential}}{Q_{parking}}=\frac{15-4.93}{15}=0.67$$

Changing the land use from a parking lot to a residential area will reduce the amount of runoff by $15-4.93=10.07$ mm, which is 67% lower than the original amount of runoff.

(4) Before comparing the savings alone, let us calculate the flows of runoff from example 6.6 in [mm]. In example 6.6, we had $Q_{parking}=0.5$ m^3/s for an area of 0.01 km^2. Therefore, for a duration of 6 min, the amount of runoff is $0.5\times6\times60\,/\,(0.01\times10^6)=18$ mm,

Example 6.7 (continued)

whereas we found 15 mm in example 6.7. Similarly, for $Q_{residential}$, we found 0.36 m^3/s in example 6.6, thus $0.36 \times 6 \times 60 \ / \ (0.01 \times 10^6) = 13$ mm compared to 4.93 mm in example 6.7.

We note two things. First, the RM estimated more runoff in all cases. Second, the percentage of runoff saved with the NRCS model is significantly higher. Again, this is partly due to the fact that the NRCS model accounts for initial abstraction. Moreover, it also accounts for soil type. In contrast, the RM is only applicable to estimate peak discharge, which is why the runoff results are higher in both cases. In practice, we simply need to be clear about what we want to do. If we want to design a detention or a retention basin (that requires a volume), then we use the NRCS model. In contrast, if we want to design a sewer, culvert, weir, or any type of infrastructure that requires a flow rate, then we use the RM.

6.5 Energy Use in Water

From the time water is first imported in a treatment plant to the time it is put back in the environment, a substantial amount of energy is used. In the year 2000, the electricity used to treat and distribute drinking water and to treat wastewater represented 4% of total electricity in the United States (Goldstein and Smith 2002). While the treatment of water has revolutionized the way water is being consumed in cities, the process has no room for error, and thousands of kilometers of pipes must work nearly perfectly to make sure water does not get contaminated. Reducing the amount of energy consumed will be one of the major contributions of Water 4.0, along with a better management of heavy rainfalls.

Table 6.5 shows some numbers that were collected in water and wastewater treatment plants and water distribution systems in California (California Energy Commission 2005). The energy values are given both in the common [kWh/MG] (i.e., megagallons) and in [Wh/L], and they are divided into five categories.

Supply and conveyance relates to the need to pump water in the treatment plant, which can be fairly substantial—for example, in Chicago, water is pumped close to 7 m (22 ft). Treatment relates to the actual treatment, which can require relatively little energy, depending on the technologies used, to an extremely high amount of energy if the water is pumped from contaminated areas or from the sea (as is the case for some treatment plants in California). As we have already seen, distribution generally requires a lot of energy to ensure a minimum pressure in the pipe system, and it can account for 85% of all energy used by water distribution systems (not accounting for wastewater

Table 6.5
Energy requirement for water distribution systems

	kWh/MG		Wh/L		Percentage	
Water cycle segments	Low	High	Low	High	Low	High
Supply and conveyance	0	4,000	0	1.06	0%	15.0%
Treatment	100	16,000	0.03	4.23	5%	61.0%
Distribution	700	1,200	0.18	0.32	37%	4.5%
Wastewater collection and treatment	1,100	4,600	0.29	1.22	58%	18.0%
Wastewater discharge	0	400	0	0.11	0%	1.5%
Total	1,900	26,200	0.50	6.92	100%	100.0%
Recycled water treatment and distribution for nonpotable uses	400	1,200	0.11	0.32	21%	5%

Source: California Energy Commission (2005).

Table 6.6
Energy of water treatment and water reclamation in greater Chicago

	kWh/MG	Wh/L	Percentage sub	Percentage total
Water (total)	2,700	0.71	100%	64%
Treatment	500	0.13	19%	12%
Distribution	2,200	0.58	81%	52%
Wastewater (total)	1,500	0.40	100%	36%
Collection	200	0.05	13%	5%
Treatment	1,300	0.34	87%	31%
Total	4,200	1.11		100%

Source: Mulvaney and Kozak (2010).

treatment). Finally, wastewater treatment can be the biggest consumer of energy—up to about 60% of the total energy used—but it naturally depends on whether the sanitary and stormwater sewers are combined.

Mulvaney and Kozak (2010) looked at energy use for most water and wastewater processes in the Greater Chicago area between 2005 and 2009. First they found some fairly significant fluctuations in energy use, showing how weather and seasons affect energy use (graphs not shown here). Table 6.6 reports overall trends.

Compared to table 6.5, we see that the Greater Chicago area is in the lower end of energy use for water and wastewater processes. Notably, this is because Chicago collects water from Lake Michigan and from groundwater wells, which is relatively clean. In contrast, the energy use for distribution is high, which could be because the distribution system in Chicago uses relatively few pumps, and water is distributed to a large area. Moreover, because of the flat topography, gravity is neither helping nor hindering the distribution process. We also see that looking specifically at water treatment and distribution, distribution accounts for about 80% of all energy use, which is typical. In terms of wastewater collection and treatment, we see here as well that energy use values are at the low end compared to the values in table 6.5.

Overall, the energy use in water and wastewater processes can fluctuate significantly based on local topography and the quality of the raw water collected. In 2018, Chini and Stillwell (2018) reported the energy intensities of over 160 utilities in the United States. Figure 6.25a shows energy intensities for water, and figure 6.25b shows energy intensities for wastewater processes. The values are shown in [Wh/L], and the full data is openly accessible from the authors.[35] We can see that values range from 0 to 1.25 Wh/L for water and from 0 to 1.70 Wh/L for wastewater. The mean and median values from the data set are shown in table 6.7.

In chapter 2 we adopted an energy consumed per liter of 1 Wh. We can see that this number roughly represents the mean values given in table 6.7. Sometimes it may be desirable to divide this energy intensity further since combined sewer systems also treat surface runoff. From table 6.7, as a general approximation, we can say that water collection, treatment, and distribution requires about 0.45 Wh/L and that wastewater treatment requires about 0.55 Wh/L. These values depend highly on local conditions, and they will be wrong most of the time (as it is for the case for Chicago), but they can help us get a general idea of the impact of the water sector on our total carbon footprint.

Therefore, if we assume a personal consumption of 311 L and an energy intensity of 1 Wh/L, we get a total daily use of 311 Wh = 0.311 kWh, thus an annual use of 114 kWh = 0.114 MWh. If we include all public supply water with an estimated 520 L of daily water consumption, the yearly energy use increases to 0.190 MWh. These values are therefore much lower than electricity consumption. In the United States, an average household consumes 10.8 MWh per year, and assuming an average of 2.54 people per household, this gives us a per person electricity yearly consumption of 4.25 MWh. The energy consumption from water is therefore roughly 4.5% of the energy consumption from electricity. Note that these numbers are averages, however, and they can be quite different from person to person. Moreover, the actual carbon footprint actually depends on the regional power grid emission factor.

Here again, akin to electricity, reducing the energy footprint from water consumption and disposal can involve little change in consumption behavior, although we can reduce our general water demand—see figure 6.13. Getting rid of the inefficiencies and

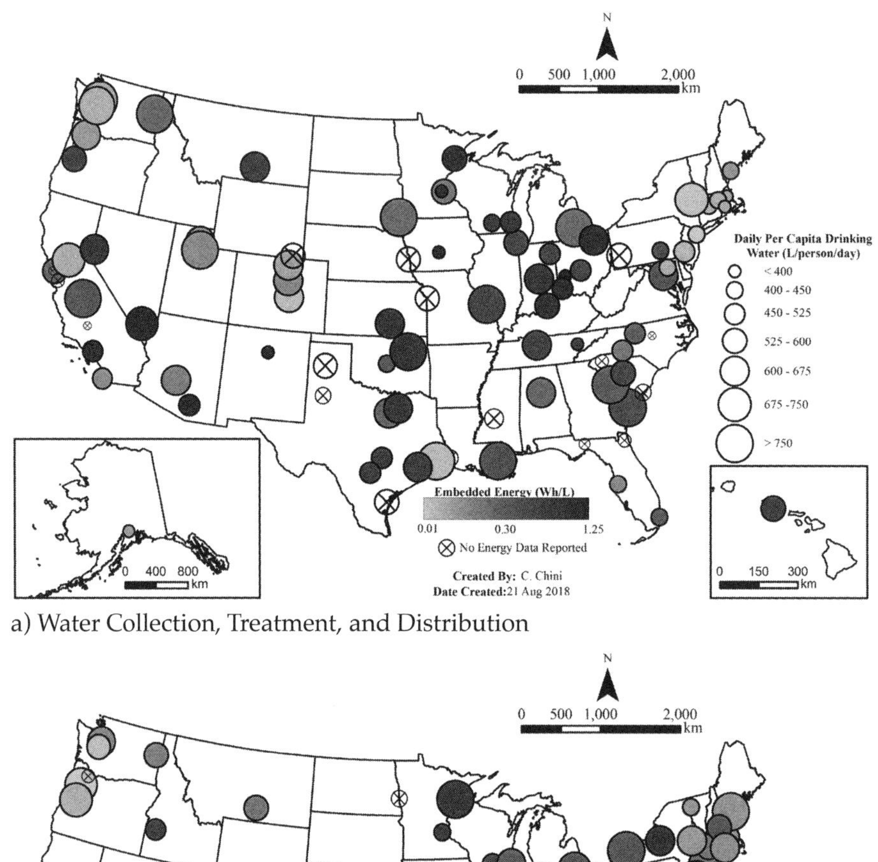

a) Water Collection, Treatment, and Distribution

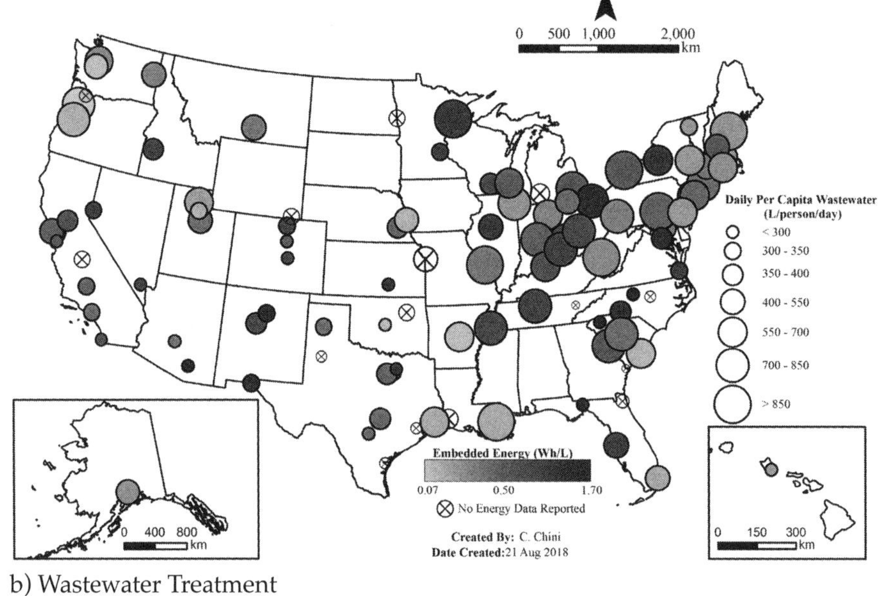

b) Wastewater Treatment

Figure 6.25

Water distribution and wastewater systems energy intensities of over 160 U.S. cities. Courtesy of Christopher Chini.

Table 6.7
U.S. average and median energy intensities of water distribution and wastewater systems

	Water (Wh/L)	Wastewater (Wh/L)
Median	0.34	0.46
Mean	0.43	0.53

Source: Chini and Stillwell (2018).

losses in the systems and implementing GI and LID strategies to reduce the amount of treated stormwater can quickly reduce the energy needed substantially.

In the longer term, if we manage to treat water more locally, then the energy footprint will decrease again. Along with a low-emission electricity grid, imagining urban water systems that have a low carbon footprint is therefore not far-fetched.

6.6 Conclusion

This concludes our chapter on water. In this chapter we initially recalled four major fundamentals of water resource engineering that govern a lot of the constraints with which we have to deal. First of all, we recalled several concepts of surface water hydrology, from defining what a watershed was and learning about hyetographs and hydrographs to estimating IDF curves. We then recalled the fundamentals of flow in both closed conduits (such as pipes) and open channels (such as sewers). Distributing water can be an energy-intensive process to overcome the losses induced by shear stresses. Open channels use gravity to drive the flow, but if the velocity gets too high, the flow may become supercritical. Finally, we recalled the fundamentals of groundwater engineering, notably learning about wells and pumping rate.

The second section of this chapter focused on water distribution systems, and we first looked at water consumption trends. We saw that by 2015 domestic consumption had notably decreased by close to 20% from the historical average of 380 L (i.e., 100 gal), putting daily consumption at about 311 L per person (i.e., 82 gal). Water consumption, however, varies heavily by state, with eastern states consuming less water. We also broke down water consumption by end use—briefly discussing the potential of reusing gray water. Moreover, we learned that per person water consumption tends to decrease as more people live together. We also saw that daily water demand varies quite significantly, akin to electricity consumption.

The third section focused on water and wastewater treatment. We have come a long way in a relatively short period of time, but treatment processes are likely to evolve quite significantly in the future. In particular, the two processes are quite energy-intensive, especially wastewater treatment.

The fourth section focused on stormwater management. We learned that there are two types of sewers, sanitary or stormwater sewers, that are sometimes combined and sometimes separated. The latter is much more desirable because treating wastewater requires a lot of energy, and stormwater does not necessarily have to be treated. Nonetheless, the amount of runoff that reaches sewer systems can also be significantly reduced by implementing GI and LID strategies that help rainwater to infiltrate the ground. We therefore do not have to spend excessive amounts of money retrofitting sewer systems and instead can retrofit neighborhoods and sidewalks by adding rain gardens and bioswales. We also learned two simple methods to model runoff: the rational method and the Natural Resources Conservation Service curve number model.

Finally, we discussed the energy breakdown by process from the initial intake of water to disposal. While numbers vary fairly substantially—based primarily on location and weather—water distribution and wastewater treatment tend to be the largest consumers of energy. Reducing consumption and fixing leaks can therefore improve the former, and implementing LID strategies along with reusing gray water can improve the latter. For illustrative purposes, in this book, we use an approximate energy value of 1 Wh/L.

After learning about electricity and water—two fields that require few changes in behavior—we move to transport, which requires the exact opposite. If we are to have a large impact in the realm of transport, we will definitely have to change the way people move in cities, and as we will see, the number of ways to achieve this goal is limited.

Problem Set

6.1 In line with quantifying your personal carbon footprint, the goal of this question is to quantify your personal footprint from your consumption of water. (1) From an online water footprint calculator, such as the one offered by http://www.watercalculator.org/, applied to indoor and outdoor water use only (not virtual), estimate your average daily and yearly water consumption in liters (using the total water footprint provided). (2) Assuming an energy rating of 1 Wh/L of water, calculate your yearly energy footprint from domestic water use. (3) Assuming all energy is provided from your regional power grid, compute your estimated GHG emissions footprint (use the power grid emission factor used in chapter 5). (4) Determine three ways you could reduce your personal water consumption and calculate the potential energy and emissions savings.

6.2 In your own words, describe the urban water cycle.

6.3 In your own words, describe what a watershed is.

6.4 In your own words, describe what a hyetograph and a hydrograph are and draw typical examples.

6.5 List three physical characteristics that affect storm flow and sketch their impacts on a hydrograph.

6.6 In your own words, describe what an IDF curve is and illustrate a typical example.

6.7 From the NOAA website, download precipitation data for the Chicago O'Hare station (search on the web "NOAA precipitation data"); note the option to download the data as a .csv file at the bottom of the page. (1) Select the return periods 1, 5, 10, 50, 100, and 200 and interpolate for a return period of twenty years. (2) Convert the values from inches to millimeters. (3) Convert the values to intensities in [mm/h]. (4) Plot the IDF curves and reproduce figure 6.6.

6.8 From the NOAA or another relevant website, download precipitation data for any location in the world. (1) Select the return periods 1, 5, 10, 50, and 100 and interpolate for a return period of thirty years (if needed, convert the values from inches to millimeters). (2) Convert the values to intensities in [mm/h]. (3) Plot the IDF curves and produce a figure similar to figure 6.6.

6.9 Recall the three types of energies of any pipe flow and express them in meters.

6.10 Discuss how the flow characteristics and the physical characteristics of a closed conduit affect friction losses.

6.11 A straight pipe has a diameter of 25 mm, a roughness of 0.01 mm, a length of 100 m, and is inclined upward at an angle of 17°. If the pressure at the end of the pipe is to be maintained at 500 kPa, (1) develop an expression for the pressure at the beginning of the pipe P_1 as a function of the flow rate Q and the friction factor f. Determine P_1 for flows of (2) 5 L/min and (3) 20 L/min.

6.12 By searching the web or any other source, discuss how pumps in series and pumps in parallel affect flow rate and pressure.

6.13 A community of 30,000 people consumes, on average, 300 L of water per person per day. The community is planning to use a nearby river for its water needs. The river is located 1 km away and is 10 m below the community level. (1) Calculate the average flow rate required from the river in [m^3/s]. (2) Assuming a maximum velocity of 1.5 m/s, determine the pipe diameter that will be needed, round up to the nearest 25 mm, and recalculate the velocity. (3) Assuming a friction factor of 0.018, calculate the head loss in the pipe due to friction losses. (4) If the water is to arrive at the community with a pressure of 50 kPa, determine the head required from the pump at the river. (5) With an efficiency of 0.8, compute the shaft power required from the pump.

6.14 A community of 200,000 people who consume, on average, 500 L of water per person per day is planning to use a nearby river as its drinking water source. The river is located 7 km away from the community and is 7 m below the community level. (1) Determine the average flow rate required from the river in [m^3/s]. (2) Assuming a maximum velocity of 1.5 m/s, determine the pipe diameter that will be needed, round up to the nearest 25 mm, and recalculate the velocity. (3) The pipe is made of ductile iron (ε=0.08 mm). Determine the friction factor to be used. (4) Based on this information, estimate the friction losses in the pipe. (5) If the water is to arrive at the community with a pressure of 50 kPa, determine the head required from the pump at the river. (6) With an efficiency of 0.85, determine the shaft power required from the pump and estimate the number of equivalent houses.

6.15 The horizontal system below consists of four pipes, two of which are in parallel. At point A, the discharge Q is 0.5 m^3/s, and the pressure P_A is 50 kPa. The characteristics of the system are shown in the figure. (1) Calculate the head loss h_{L1} due to friction losses in pipe 1. (2) Two different materials are used for the parallel pipe system. Calculate an equivalent loss coefficient K_{BC}. (3) Calculate the discharges Q_2 and Q_3 in each parallel section. (4) Compute a total loss coefficient K_{AC} that takes into account pipes 1, 2, and 3. (5) Compute the pressure P_C of the water at C. (6) Determine the length of pipe 4, L_4, so that all of the energy head is exhausted.

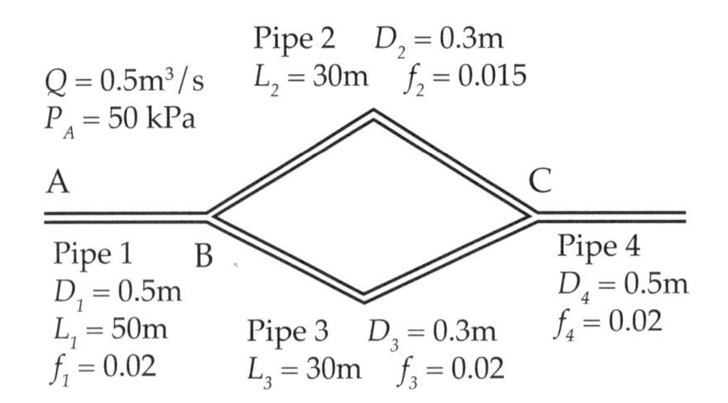

6.16 The pressure P_A in the horizontal parallel pipe system shown below is 100 kPa, and the flow rate is 0.06 m^3/s. Neglect minor losses and assume a friction factor of 0.02 for all pipes. Based on this information, determine (1) the friction loss coefficient K_{234} of the parallel ABC pipes, (2) the pressure P_C, (3) the fraction of the flow Q_2, Q_3, and Q_4 in each pipe, and (4) how far the water will flow before all of its pressure is exhausted.

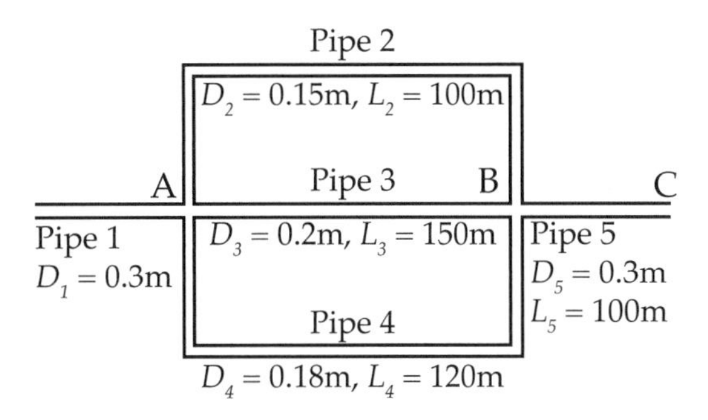

6.17 In your own words, describe the main characteristics of an open channel.

6.18 In your own words and by doing your own research, describe the differences between (1) a uniform versus nonuniform flow, (2) a steady versus unsteady flow, and (3) a prismatic versus nonprismatic flow.

6.19 In your own words, describe and illustrate how energy varies with flow depth for a given flow rate.

6.20 A natural rectangular channel (n=0.03) has a bottom width of 1 m and a discharge rate of 20 m^3/s. Using Microsoft Excel or another commercial software, (1) print a table with depth, area, wetted perimeter, hydraulic radius, velocity, energy, and slope for depths ranging from 1 m to 15 m in 0.5 m intervals, and (2) plot the specific energy diagram (y vs. specific E—also show the 45° line). From the diagram, determine (3) the critical depth, (4) the minimum specific energy, (5) the specific energy for a water depth of 10 m, and (6) the water depths when the specific energy is 5.27 m. Moreover, identify (7) the channel slope that results in critical flow.

6.21 You are tasked with estimating the size of a stormwater channel to accommodate a 24-h 100-year rain in Chicago, which the IDF curve suggests to be 7.36 mm/h. The city of Chicago is 606 km^2. The channel will be rectangular with a bottom width of 20 m and a slope of 0.005. Assuming a Manning roughness of 0.012, determine (1) the flow rate, (2) the flow depth, and (3) whether the flow is subcritical, critical, or supercritical and comment on whether or not this type of flow is desirable.

6.22 A rectangular open channel ($n=0.015$) has a bottom width $B=10$ m, a discharge of 100 m^3/s, and a slope of 0.001. Calculate (1) the normal depth in the channel. Considering the actual depth of the flow is 2 m, calculate (2) the Froude number of the flow and determine whether the flow is subcritical, critical, or supercritical. Based on these results, (3) determine the slope of the water surface S_f.

6.23 By performing your own research, derive the pumping rate equations for (1) confined and (2) unconfined aquifers.

6.24 A community of 30,000 people consumes, on average, 300 L of water per person per day. The community is planning to tap into its groundwater to provide the city with all its water needs. Both a confined and an unconfined aquifer are available with hydraulic conductivities of 30 m/day each, and one single 20 cm diameter well should be drilled. (1) Determine the average flow rate required from the well in m^3/day. (2) The confined aquifer has a saturated thickness of 20 m and a radius of influence of 200 m. Determine the drawdown from the well. (3) The unconfined aquifer has a water table height of 30 m and a radius of influence of 2 km. Determine the drawdown of the well. (4) Based on this information, select the preferred type of well if the goal is to purely minimize the drawdown.

6.25 A new development with 10,000 people who are expected to consume, on average, 400 L of water per day will need to supply 80% of its water needs with groundwater. Determine (1) the daily flow rate required from groundwater Q_{req} in m^3/day. Two wells are to be installed, both in soils that have hydraulic conductivities of 34 m/day. The first 10-cm-diameter well will tap a confined aquifer with a saturated thickness of 9 m and a radius of influence of 20 m. If the drawdown is not to exceed 3 m, calculate (2) the maximum pumping rate allowed in m^3/day. The second 10-cm-diameter well will tap an unconfined aquifer with a water table height of 20 m. With a radius of influence of 45 m, (3) determine the drawdown that will be caused by pumping the rest of the water needed. In your mind, (4) are these plans sustainable?

6.26 A truck crashed 500 m away from farmland, and it has contaminated an unconfined aquifer. A previous study found that this aquifer has a water table height of 30 m, a hydraulic conductivity of 170 m/day, an effective porosity of 0.3, and a hydraulic gradient of 0.005. Calculate (1) the discharge of water in the aquifer per kilometer of width, (2) the specific discharge of the flow, and (3) the seepage velocity of the flow. Based on this information, determine (4) how much time it takes for the water to reach the farmland. (5) In order to prevent the contamination of the farmland, 0.3-m-diameter wells with radii of influence of 20 m are installed. If the contaminant has spread along 74 m of width, estimate the number of wells needed and how far apart they should be spaced. (6) If the pumps have a maximum pumping rate of 0.5 m^3/s, determine whether this will be sufficient to prevent contamination (i.e., determine whether the drawdown can equate the water table height).

6.27 In your own words, describe and illustrate the treatment process to convert freshwater into drinking water.

6.28 In your own words, describe and illustrate the process to treat wastewater.

6.29 In your own words, describe the difference between combined and separated sewer systems.

6.30 In your own words, (1) describe what GI and LID strategies are and (2) give four examples.

6.31 A 5,000 m^2 empty lot in a downtown area will be turned into a condominium tower. The runoff coefficients for the current and future designs of the lot are 0.3 and 0.95, respectively. Considering a rainfall intensity of 75 mm/h and using the RM, calculate (1) the current and forecasted peak discharge of runoff due to the condominium tower and (2) the difference between the two that will need to be accommodated. Then, (3) suggest strategies to accommodate it.

6.32 A currently developed area of 20,000 m^2 will be retrofitted with GI in an effort to reduce stormwater runoff. For a 15-year storm, the area follows the rainfall intensity equation shown below. The properties of the current design are $C=0.75$ and $T_c=10$ min, and the properties of the proposed design are $C=0.40$ and $T_c=15$ min (1) Based on this information and using the RM, calculate the percentage reduction in peak discharge thanks to the new design. (2) In addition, if the municipal stormwater ordinance stipulates that any peak discharge below 0.2 m^3/s does not require any additional detention basin, determine whether the new design should also incorporate a detention basin.

$$i=\frac{7.14}{(1.2+t)^{0.63}},\ t \text{ is in min and } i \text{ is in mm/min}$$

6.33 A 5,000 m^2 empty lot in a downtown area will be turned into a condominium tower. The lot has a hydrologic soil group type C. The empty lot is essentially an open space in fair condition, and the condominium tower is essentially a commercial urban district. Considering a rainfall intensity of 75 mm/h for 30 min and using the NRCS curve number model, calculate (1) the current and forecasted rainfall excess due to the condominium tower, (2) calculate the difference between the two that will need to be accommodated, and (3) suggest strategies to accommodate it.

6.34 A community is being retrofitted as a residential district with an average lot size of one-fourth acre, and it will need to update its stormwater infrastructure to abide by the local stormwater ordinance (i.e., be able to handle a 10-year storm for 2 h). The community covers an area of 0.25 km^2, it has a hydrological soil group type B, and the equation for the 10-year rainfall intensity is shown below. Based on this information, (1) calculate the volume of water that the community will need to be able to handle and (2) suggest strategies to accommodate it.

$$i = \frac{4.97}{(6.8+t)^{0.54}}, \ t \text{ is in min and } i \text{ is in mm/min}$$

6.35 Using the EPA Stormwater Calculator, you are asked to estimate how much runoff is generated by your home (whether a house or an entire apartment building). Download and install the software on your computer. (1) Enter appropriate values that best qualify your household (that includes the building you live in plus the surrounding green areas that belong to the property owner; include garage ways but not the streets). For soil type, drainage, and topography, make educated guesses. For climate change, select "Median Change" and "Near Term." For land cover, represent your property as best you can. For LID Controls, do not enter anything. In Results, click on "Refresh Results," click on "Use as Baseline Scenario," and (1) report your results by clicking on "Print Results to PDF File," which should indicate all your assumptions. (2) In the LID tab, where LID stands for low-impact development, adjust the different LID strategies to possible scenarios. Provide three possible scenarios and report your results. (3) Assuming wastewater treatment requires an energy of 1 Wh/L and considering the savings in "Average Annual Runoff" and the "Site Area" of your property, estimate how much energy is saved by the three scenarios from (2) and how many GHGs have been saved (taking your regional power grid emissions factor).

Notes

1. Unless it is pressurized in closed conduits.

2. With a slope generally between 1% and 2%, as some may remember from their geometric design class.

3. The first tipping bucket was invented by Christopher Wren in 1662. The idea is therefore old, although modern tipping buckets have become "smart" and use electronics.

4. In stream hydrology, the term *stage* is used for "depth" and the term *discharge* is used for "flow rate." A typical stage-discharge graph (or rating curve) plots the stage of a stream (e.g., in [m]) versus the discharge of the same stream (e.g., in [m^3/s]).

5. Technically, there is actually a 36.60% chance that it does not happen, a 36.98% chance that it happens once, and an 18.48% chance that it happens twice in 100 years. To calculate this, we use the binomial distribution defined as probability $P(N,n)=\frac{N!}{n!(N-n)!}p^n(1-p)^{N-n}$, where N is the total number of years (i.e., 100), n is the number of occurrences we are investigating (i.e., 0, 1, and 2), and p is the probability that it happens (i.e., one over the return period, 1/100 = 0.01).

6. Examples other than the Sherman equation shown include the Bernard equation $\left(i=\frac{a}{t^c}\right)$, the Kimijami equation $\left(i=\frac{a}{b+t^c}\right)$, and the Talbot equation $\left(i=\frac{a}{b+t}\right)$.

7. The minimum pressure is around 140 kPa to ensure that water leaks out of the pipes and never in from potentially contaminated groundwater. Moreover, adequate pressure must be kept for fire protection, and a pressure of about 300 kPa can reach a height of $h=P/\gamma=300/9.79=30.6$ m.

8. And I always like to say it in its original French version from the famous French scientist Antoine-Laurent de Lavoisier, who contributed to the original *Encyclopedia*: "Rien ne se perd, rien ne se crée, tout se transforme."

9. As friction losses are directly related to the square of flow velocity, we generally prefer to have a lower velocity, and Chin (2014) recommends a maximum flow velocity of 1.5 m/s. Moreover, pipes with only certain diameters are available (e.g., for every 25 mm), which is why we calculate the right diameter and round up (not down) since the velocity will be even lower in a larger pipe.

10. Available at https://www.epa.gov/water-research/epanet and that many municipalities use (accessed August 16, 2018).

11. Not gravitational waves, which are completely different. In liquids, a simple way to see gravity waves is to think about the speed of the ripples when you throw a rock in a pond, for example.

12. To some extent, every water resources engineer is a rocket scientist.

13. This is actually an oversimplification of the problem. We could use the RM to estimate the rate of water inflow, and we will learn about the RM in section 6.4.3.1.

14. The darcy is a unit here, and 1 darcy = 0.987×10^{-12} m^2.

15. The two pumping equations can easily be derived from Darcy's law. Look it up.

16. In fact, this is a problem in many cities in California.

17. As a general (and free) guide to learn more about dewatering, groundwater control, and groundwater wells, see U.S. Army, U.S. Navy, and U.S. Air Force (1983).

18. Read more about China's gigantic plan that includes 4,350 km of open channels, roughly the distance between the West and East Coasts of the United States (Kuo 2014).

19. The book also offers a great account of the history of solid waste management in the United States as well, which we will cover in chapter 9.

20. Note, however, that the use of fertilizer in irrigation easily contaminates the water table, which can then flow and contaminate rivers and lakes.

21. The Jardine Water Purification Plant is one of the largest treatment plants in the world, treating and distributing nearly 1 billion gallons of water per day.

22. These numbers may not look impressive but they are. An average demand of 38.80 m^3/s amounts to more than 3.35 million m^3 of water that were treated and distributed on that day, which is close to the size of 1,350 Olympic swimming pools.

23. Otherwise, contaminated water can enter the system. See note 7.

24. Which is not totally true. There are some systems in the United States where the wastewater is perfectly treated and reused directly in the water distribution system.

25. For Chicago, you can check for yourself in real time or download past data easily by visiting http://istheresewageinthechicagoriver.com/ (accessed August 16, 2018).

26. Chicago has several great examples, like the Pilsen Sustainable Street (Cermak Road—see figure 6.22, *left*) and Argyle Street. Moreover, Philadelphia has also become a leader in green infrastructure in the United States.

27. The City of Granby in Quebec chose to install LID infrastructure, as opposed to traditional massive drains, to address its sewer backup problem. Not only did the LID do the job, but the capital costs were halved (Carbonneau 2016).

28. See note 29.

29. In Arlington Heights near Chicago, overflows of stormwater sewers are directed toward a baseball diamond during heavy rains to avoid sewer backups. This is possible because Arlington Heights does not have a combined sewer system.

30. Available at https://www.epa.gov/water-research/storm-water-management-model-swmm (accessed March 2, 2019).

31. Depending on the sewer system, higher discharges may provoke a CSO.

32. New empirical evidence finds a number close to 10%, but we will keep 20%, as is often adopted.

33. Although we could use peak discharge to calculate a volume, we would end up with too large a volume since we would assume the peak discharge applies at all times.

34. Available at http://www2.epa.gov/water-research/national-stormwater-calculator (accessed March 2, 2019).

35. The authors were extremely generous to produce the maps shown in figure 6.25. The data can be found at https://www.hydroshare.org/resource/df04c29d0ff64de0ace2d29145dd7680/ (accessed August 22, 2018).

References

Beven, K. J. 2012. *Rainfall-Runoff Modelling: The Primer.* Chichester, UK: Wiley.

California Energy Commission. 2005. *Integrated Energy Policy Report.* CEC-100-2005-007CMF. Sacramento: California Energy Commission.

Carbonneau, B. 2016. "Stormwater Management—Saint-André East—City of Granby." Federation of Canadian Municipalities (V/Ref: 13023). https://data.fcm.ca/home/programs/green-municipal-fund/funded-initiatives.htm?project=180cbdb5-7db6-e311-9ea6-005056bc2614.

Chin, David A. 2014. *Water-Resources Engineering.* Upper Saddle River, NJ: Pearson Education.

Chini, Christopher M., and Ashlynn S. Stillwell. 2018. "The State of U.S. Urban Water: Data and the Energy-Water Nexus." *Water Resources Research* 54(3): 1796–1811.

DeOreo, William B., Peter Mayer, Benedykt Dziegielewski, and Jack Kiefer. 2016. *Residential End Uses of Water Version 2.* Denver: Water Research Foundation.

Dieter, C. A., M. A. Maupin, R. R. Caldwell, M. A. Harris, T. I. Ivahnenko, J. K. Lovelace, N. L. Barber, and K. S. Linsey. 2018. "Estimated Use of Water in the United States in 2015." *U.S. Geological Survey Circular* 1441: 65.

Goldstein, R., and W. Smith. 2002. *Water & Sustainability*, vol. 4: *US Electricity Consumption for Water Supply & Treatment—the Next Half Century.* Palo Alto, CA: Electric Power Research Institute.

Guillerme, A. 1988. "The Genesis of Water Supply, Distribution, and Sewerage Systems in France, 1800–1850." In *Technology and the Rise of the Networked City in Europe and America*, edited by Joel A. Tarr and Gabriel Dupuy, 91–115. Philadelphia: Temple University Press.

Jato-Espino, Daniel, Luis A. Sañudo-Fontaneda, and Valerio C. Andrés-Valeri. 2018. "Green Infrastructure: Cost-Effective Nature-Based Solutions for Safeguarding the Environment and Protecting Human Health and Well-Being." In *Handbook of Environmental Materials Management*, edited by Chaudhery Mustansar Hussain, 1–27. Cham, Switzerland: Springer International.

Kennedy, C., I. Stewart, A. Facchini, I. Cersosimo, R. Mele, B. Chen, and M. Uda et al. 2015. "Energy and Material Flows of Megacities." *Proceedings of the National Academy of Sciences of the United States of America* 112(19): 5985–5990.

Kennedy, Christopher, John Cuddihy, and Joshua Engel-Yan. 2007. "The Changing Metabolism of Cities." *Journal of Industrial Ecology* 11(2): 43–59.

Kerkez, Branko, Cyndee Gruden, Matthew Lewis, Luis Montestruque, Marcus Quigley, Brandon Wong, and Alex Bedig et al. 2016. "Smarter Stormwater Systems." *Environmental Science & Technology* 50(14): 7267–7273.

Kermanshah, A., S. Derrible, and M. Berkelhammer. 2017. "Using Climate Models to Estimate Urban Vulnerability to Flash Floods." *Journal of Applied Meteorology and Climatology* 56(9): 2637–2650.

Kuichling, Emil. 1889. "The Relation between the Rainfall and the Discharge of Sewers in Populous Districts." *Transactions of the American Society of Civil Engineers* 20(1): 1–56.

Kuo, Lily. 2014. "China Is Moving More than a River Thames of Water across the Country to Deal with Water Scarcity." *Quartz*, March 6. http://qz.com/158815/chinas-so-bad-at-water-conservation-that-it-had-to-launch-the-most-impressive-water-pipeline-project-ever-built/.

Melosi, M. V. 2008. *The Sanitary City: Environmental Services in Urban America from Colonial Times to the Present*. Pittsburgh: University of Pittsburgh Press.

Mulvaney, Peter, and Joseph Kozak. 2010. "The Carbon and Energy Footprint of Water Treatment, Distribution and Reclamation and Waterway Management in Greater Chicago." Paper presented at the 2010 AIChE Midwest Regional Conference, Chicago.

National Ground Water Association. n.d. "Confined or Artesian Groundwater." https://www.ngwa.org/what-is-groundwater/About-groundwater/confined-or-artesian-groundwater (accessed March 2, 2019).

Sedlak, David L. 2014. *Water 4.0: The Past, Present, and Future of the World's Most Vital Resource*. New Haven, CT: Yale University Press.

Tarr, Joel A. 1988. "Sewerage and the Development of the Networked City in the United States, 1850–1930." In *Technology and the Rise of the Networked City in Europe and America*, edited by Joel A. Tarr and Gabriel Dupuy, 159–185. Philadelphia: Temple University Press.

Thornton, J. 2008. *Water Loss Control Manual*. 2nd ed. New York: McGraw-Hill.

U.S. Army, U.S. Navy, and U.S. Air Force. 1983. *Dewatering and Groundwater Control*. TM 5-818-5/AFM 88-5, chapter 6/NAVFAC P-418. Washington, DC: Departments of the Army, the Navy, and the Air Force. https://www.wbdg.org/FFC/ARMYCOE/COETM/ARCHIVES/tm_5_818_5.pdf.

U.S. Department of Agriculture. 1986. *Urban Hydrology for Small Watersheds*. Technical report 55. Washington, DC: U.S. Department of Agriculture. http://www.nrcs.usda.gov/Internet/FSE_DOCUMENTS/stelprdb1044171.pdf.

U.S. Department of Commerce. 1961. "Rainfall Frequency Atlas of the United States for Durations from 30 Minutes to 24 Hours and Return Periods from 1 to 100 Years." Technical paper 40. Washington, DC: U.S. Department of Commerce. https://www.nws.noaa.gov/oh/hdsc/PF_documents/TechnicalPaper_No40.pdf.

U.S. Geological Survey. 2016. "What Is a Watershed?" Reston, VA: U.S. Geological Survey. https://water.usgs.gov/edu/watershed.html.

Ward, A. D., S. W. Trimble, S. R. Burckhard, and J. G. Lyon. 2015. *Environmental Hydrology*. 3rd ed. Boca Raton, FL: CRC Press.

7 Transport

All models are wrong, but some are useful.

—Box 1987, p. 424

Transport is often compared to the human cardiovascular system (i.e., the bloodstream). While blood may not be an organ that performs an important function (e.g., like the kidneys), it enables the transfer of critical components (e.g., oxygen, nutrients, etc.) between and to organs, thus making it as critical, if not more critical, than an organ. In fact, we can live without certain organs, but we cannot live without blood. Transport does the same thing by enabling the movement of people and goods within and between cities, thus allowing us to go to work, buy food, engage in multiple activities, and so on. In the context of this book, transport has one major specific property that makes it quite unique. Unlike the other systems, a major reduction in energy use and emissions from transport will inevitably require a major change in lifestyle.

As a small semantic argument, the terms *transport* and *transportation* mean the same thing. The term *transportation* tends to be used in the United States, and *transport* is used everywhere else in the world. In this book I will prefer *transport*, not because it is more international but because it turns out that adding *ation* to a word actually consumes a lot of energy in terms of computer memory and paper for books.[1] Therefore, we will try to be a bit more sustainable by using *transport*.

Even if every nonelectric vehicle was substituted with an electric vehicle (EV) and even if the electricity used came from completely clean sources, the problem of congestion remains a problem for sustainability. Remember that sustainability occurs at the nexus of people, planet, and profit, and even "clean" congestion would still greatly affect people and profit. From a different perspective, traffic can be seen as consumption, and the capacity of roads can be seen as production, and if we are not increasing our rate of production—that is, if we are not building more roads (and we should not, as we will see in section 7.3)—then the rate of consumption—that is, traffic—should not increase either. This does not mean that we should not "travel" more, but it means that the use of automobiles that take a lot of space should decrease, especially when occupied by only one person over long distances. As a result, and similar to electricity and water, the transport realm will have to change and adapt to new conditions to embark on Scenario B. While the twentieth century saw the rise of the private automobile, the twenty-first century will likely see a return to basics, with more walking and cycling for short distances and more public and shared transport for longer distances. This is not to say that the private car will cease to exist, but as cities get bigger (both in surface area and in population), they simply cannot accommodate more cars. Naturally, the big players here, if we try to think about the future of transport, are autonomous vehicles, but there was too much uncertainty at the time of this writing to be able to predict / imagine their impact in a realistic manner.[2]

In general, transport accounts for a significant portion of greenhouse gas (GHG) emissions. From chapter 5, figure 5.1 showed that transport accounted for about 28.5% of GHG emissions in the United States in 2016. If we look at individual cities in the world, this proportion can change quite significantly. Figure 7.1 shows the per capita GHG emissions by sector for twenty cities in the world. The cities are ranked from left to right based on absolute GHG emissions from transport, but the labels show the

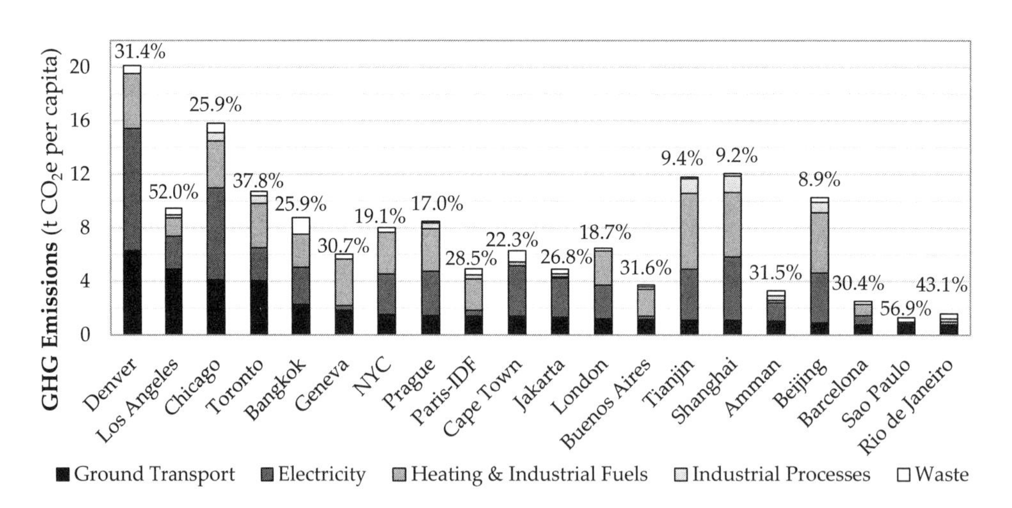

Figure 7.1

GHG emissions by sector in twenty cities. Where not shown, industrial processes are unknown and not zero.

Source: Kennedy, Ibrahim, and Hoornweg (2014).

relative proportion of transport GHG emissions attributed to transport. In absolute terms, transport emissions per capita range from 6.31 t CO_2e to 0.69 t CO_2e, but in relative terms they range from 52% to 8.9%, hence showing quite a lot of variability that partially depends on local climate conditions (colder places will require more heating), industrial activity, and electricity emission factors.

Beyond that, we can see that cities that rely heavily on car use, such as Denver, Los Angeles, Chicago, and Toronto, have higher GHG emissions attributed to transport in absolute values but also often in relative terms. Other cities with strong transit systems like New York City, Paris, and London manage to have relatively low emissions. These are the types of cities that we should aim for, in which families tend to own one (or no) car and pursue many of their activities by walking and cycling, and by riding transit.

Again, we will start by listing several fundamentals of transport that essentially set the limits of what can and cannot be achieved. First, we will focus on the fundamentals of traffic flow, pedestrian flow, and transit planning. Second, we will learn what demand means in transport and how it is quantified. Third, we will discuss the importance of land use in transport, which are metaphorically two sides of the same coin. Land use is in fact the number one system that we should consider to solve a significant part of the problems related to transport. Fourth, we will discuss how to model demand for transport—generally called *travel demand forecasting*—by learning about the traditional four-step urban transport–planning model, which will help us estimate our carbon footprint from transport. The quote shown at the beginning of this chapter is famous in transport, and every transport modeler knows it. Ultimately, any modeling exercise that we take part in outputs wrong results, but these results can still help us determine effective strategies for the future, which is why it is important for us to get introduced to elements of transport demand forecasting.

But for now, we will focus on some fundamentals.

7.1 Fundamentals of Transport

Like all engineering fields, transport is subject to fundamental laws, and again like all engineering fields, these laws actually tend to be derived from laws of physics. In fact, the first transport engineers were physicists. In transport, the most basic element is the *person* or the *vehicle*, and people/vehicles can be assimilated to particles that move along a road, much like the flow of water or the flow of electrons. In addition, and again similar to other engineering fields, some phenomena can be difficult to capture theoretically, but they can be observed and measured, resulting in empirical results (like the Darcy-Weisbach equation to calculate friction losses in a closed conduit).

In this section we will specifically focus on traffic flow, pedestrian flow, and transit planning. We start with traffic because the science of pedestrian flow was still young at the time of this writing and it had been greatly inspired by traffic flow. Transit service, on the other hand, tends to be more centrally coordinated, and it is therefore not subject to the same laws; we will focus on the impact of acceleration and deceleration rates and stop/station spacing to service a demand.

The main urban passenger mode we are missing is the bicycle. Although cycling is absolutely common (and environmentally friendly), we have few established practices, and they tend not to be universally accepted. That being said, the National Association of City Transportation Officials is working hard on the topic, and you can visit their website to learn more.[3] Moreover, we will discuss cycling in the two other sections. For now, we will concentrate on traffic flow.

7.1.1 Traffic Flow Theory

Traffic flow theory contains three fundamental metrics: (1) flow, (2) speed, and (3) density. At its most basic level, traffic flow q expressed in vehicles per unit of time is defined as:

$$q = \frac{n}{t} \tag{1}$$

where n is the number of vehicles traversing a road point in a time interval, and t is the duration of the time interval. Traffic flow q is in fact similar to the concept of current in electricity and flow rate in water resources engineering.[4] In some cases, q is referred to as a *volume* when the time is not instantaneous but is expressed over a period of time like an hour, resulting in number of vehicles per hour.[5] Naturally, the time interval t chosen can have a large impact. The flow is preferably measured during the peak period to account for particularly heavy traffic.

Speed is our second important metric. There are essentially two ways to measure speed. One way is to measure the average speed of vehicles passing at a road point, and the other way is to measure the average speed of one or multiple vehicles over a length of road. The first method is called *time-mean speed* $\bar{v}_t$ and is defined as:

$$\bar{v}_t = \frac{\sum v_i}{n} \tag{2}$$

where v_i is the speed of individual vehicles. This speed is easy to measure, but often we are more interested in knowing how much time it takes to traverse a certain distance. The second (and preferred) speed is *space-mean speed* $\bar{v}_s$. It is defined as:

$$\bar{v}_s = \frac{l}{\bar{t}} \tag{3}$$

where l is the distance traveled, and $\bar{t}$ is the average time it takes to travel this distance. Specifically, it is defined as:

$$\bar{t} = \frac{1}{n}\sum t_i \tag{4}$$

where n is now the number of vehicles that traveled the distance l of the roadway—our standard definition of n from now on.

If we substitute equation (4) in equation (3), we get:

$$\bar{v}_s = \frac{l}{\frac{1}{n}\sum t_i} = \frac{1}{\frac{1}{n}\sum\left[\frac{1}{(l/t_i)}\right]} \tag{5}$$

And since $v_i = l/t_i$, equation (5) shows us that space-mean speed is simply the harmonic mean of speed as opposed to the arithmetic mean captured in time-mean speed.

The third important metric is density k, expressed in vehicles per unit length and simply defined as:

$$k = \frac{n}{l} \tag{6}$$

It does not take too much effort to then relate equations (1), (3), and (6) on flow, space-mean speed, and density such that:

$$q = vk \tag{7}$$

Thinking in terms of units, q is often expressed in [veh/h], v is expressed in [km/h] (note that we dropped the bar and the subscript s for notational convenience since we will use it extensively), and k is expressed in [veh/km].

We can now go one step further and think about the relationship between v and k. Here, if many vehicles are present, then the average speed must be low, and in contrast, if there are few vehicles, then the average speed can be high. Put differently, if k is high then v is low, and if k is low then v is high.[6] One of the assumptions we can make is that the relationship between v and k is linear, taking the form:

$$v = v_f\left(1 - \frac{k}{k_j}\right) \tag{8}$$

where v_f is the free flow speed and k_j is the jam density. Both values are actually constant and properties of the system that depend on the speed limit and the number of

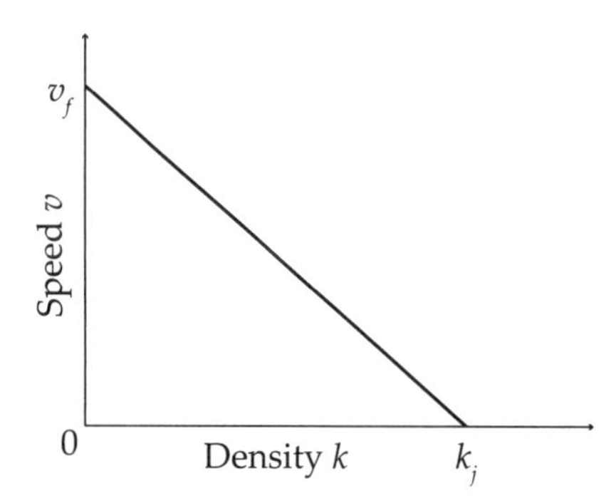

Figure 7.2
The speed-density relationship.

lanes. This linear assumption is the starting point of the Greenshields model developed by Greenshields et al. (1935). It is one of the first traffic flow models ever developed and it is still in use.

Diving in a bit more, equation (8) states that velocity v is equal to the free flow velocity v_f minus the proportion k/k_j of the free flow velocity. When the density is small, k/k_j is small, v is close to v_f, and the opposite is true when k get closer to k_j. The relationship is illustrated in figure 7.2.

Plugging equation (8) in equation (7), we get a quadratic flow-density relationship of the form:

$$q = v_f\left(k - \frac{k^2}{k_j}\right) \tag{9}$$

Similarly, we get a quadratic flow-speed relationship of the form:

$$q = k_j\left(v - \frac{v^2}{v_f}\right) \tag{10}$$

Both equations (9) and (10) are parabolas. For small and large values of k and v, q will be small, and in the middle, q will be maximum. In other words, there is one point in figure 7.2 for a specific k_{cap} and v_{cap}, where the number of vehicles that can travel across a certain distance is maximum: q_{cap}. To find out, we simply need to differentiate both equations. Doing so for equation (9), we get:

$$\frac{dq}{dk} = v_f\left(1 - \frac{2k}{k_j}\right) = 0 \tag{11}$$

Since $v_f > 0$, this means the second component needs to be 0, therefore:

$$1 - \frac{2k_{cap}}{k_j} = 0$$

$$\frac{2k_{cap}}{k_j} = 1$$

$$k_{cap} = \frac{k_j}{2} \tag{12}$$

The maximum (or capacity) flow q_{cap} therefore occurs when the density k is half the jam density k_j. In reality, empirical observations suggest that the maximum flow occurs closer to 20–25% of the jam density. The discrepancy between equation (12) and the reality is due to the fact that we assumed a linear relationship between speed and density.

Adopting a similar procedure but differentiating over v in equation (10), we then obtain the maximum flow q_{cap} when the speed v is half the free flow speed v_f as:

$$v_{cap} = \frac{v_f}{2} \tag{13}$$

Finally, using equations (7), (12), and (13), the flow capacity is calculated as:

$$q_{cap} = v_{vap} \cdot k_{cap} = \frac{v_f}{2} \cdot \frac{k_j}{2}$$

$$q_{cap} = \frac{v_f k_j}{4} \tag{14}$$

We can in fact link flow, density, and speed in one figure, as illustrated in figure 7.3. This is typically referred to as the *fundamental flow diagram*. The one shown is called the Greenshields model, again because it was developed by Greenshields et al. (1935), and it assumes a linear relationship between speed and density. Empirical data do not capture a quadratic relationship between flow, speed, and density, and other forms of relationships are commonly used. In fact, large traffic flow models with many vehicles do not use the quadratic form of the relationship in equations (9) and (10). Many models exist, including the Greenberg model and the California model. Another popular model is the triangle model, which assumes an initial linear increase up to about 20% to 25% of the jam density (following empirical observations), followed by a linear decrease (almost exactly like a hydrograph[7]). Practically, the triangle model assumes free flow speeds until the peak is reached, at which point the capacity flow (i.e., maximum) is achieved, as opposed to half the free flow speed with the parabolic

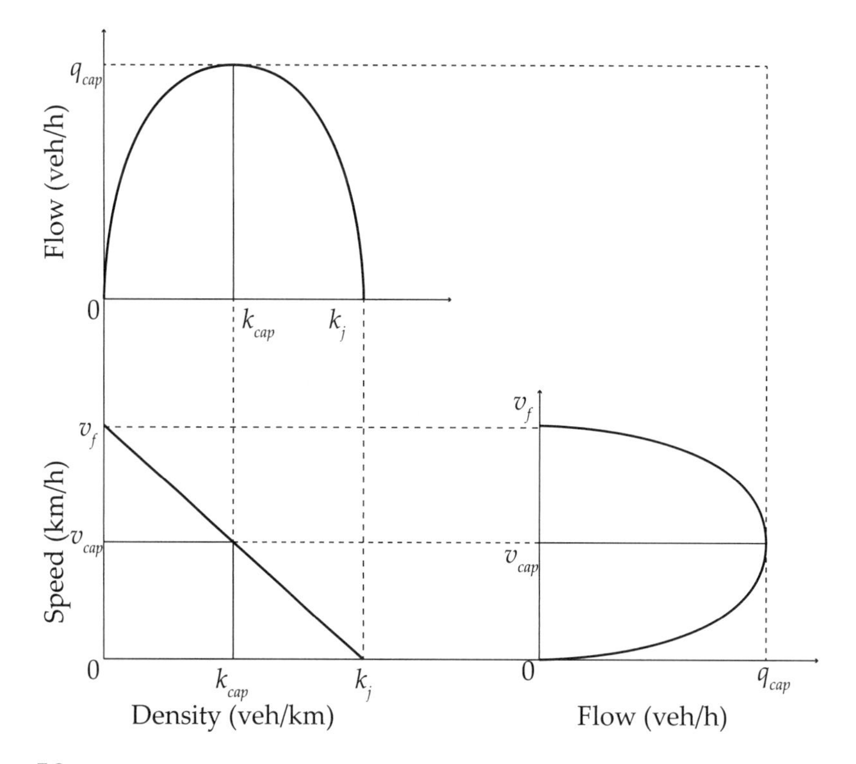

Figure 7.3
Flow-density, speed-density, and speed-flow relationships.

relationship.[8] Determining which model is more "realistic" is a large debate in the traffic flow community.

Beyond the model used, we see that for a given flow we have two possible speeds and two possible densities. This is why multiple variables need to be collected in practice, either flow and speed or flow and density. This is typically done with loop detectors embedded in the asphalt that sense when a car is passing over a specific area.[9] Flow is calculated by counting the number of passing cars in a given time. Speed and density can be measured in a variety of ways from *occupancy*, which measures the percentage of time a loop detector is activated in a given time (Hazelton 2004). New sources of data (e.g., from smartphones) can also be used to measure traffic flow (Herrera et al. 2010).

Furthermore, and beyond focusing on individual corridors, some evidence reveals that a similar fundamental diagram may also exist over entire areas, suggesting the existence of a network fundamental diagram (Geroliminis and Daganzo 2008). This is exciting and could be used in a future Science of Cities, although we do not discuss it in chapter 11.

In practice, traffic flow theory can also be used to rate corridors based on their congestion levels. This is called level of service (LOS). Specifically, LOS captures the performance of a street given certain traffic conditions. In the United States, transport LOS is divided into six standard categories:

- LOS A: free-flow conditions. Drivers can easily reach the posted limit and switch lanes.
- LOS B: near free flow. Drivers can reach the posted limit, but the presence of others can be felt, and drivers should be a little bit careful when switching lanes.
- LOS C: stable flow. Drivers can still reach the posted limit, but it becomes slightly harder to switch between lanes. Incidents can happen and queues can start to form.
- LOS D: stable but moderate flow. Drivers cannot reach the posted limit and have to be careful when switching lanes. The level of comfort goes down, and queues can easily form.
- LOS E: unstable flow. The roadway capacity is reached or is close to being reached. Drivers drive slower than the posted limit. The addition of new vehicles (e.g., from a ramp) easily creates delays. Incidents happen and queues exist.
- LOS F: breakdown flow. The roadway is over capacity, queues build up, and switching lanes becomes very difficult. The road is experiencing a constant traffic jam.

Figure 7.4 relates LOS with flow and speed for a basic freeway segment, and it is simply called the LOS diagram or *q-v* diagram (for flow and speed). Other types of roads and flow conditions have different LOS diagrams.

The units in figure 7.4 are in kilometers per hour [km/h] and in vehicles per hour per lane [veh/h/ln]. The density is also shown with the dotted lines in vehicles per kilometer per lane [veh/km/ln]. A version of the diagram with imperial units along with the table of values used to make the diagram is available in the appendix. The original data can be found in the Transportation Research Board's (2010) *Highway Capacity Manual 2010* and in Mannering (2013); technically, the diagram should show "passenger cars" as opposed to "vehicles"—that is, these conditions actually do not apply to large vehicles.

In practice, most engineers in the United States tend to use the Highway Capacity Software (HCS) program that automatically determines the LOS of a road based on input traffic conditions. The HCS program can be downloaded at the website of the McTrans Center at the University of Florida.[10] As a last note, we can see that the figure looks a little like the Moody diagram that we saw in the last chapter.

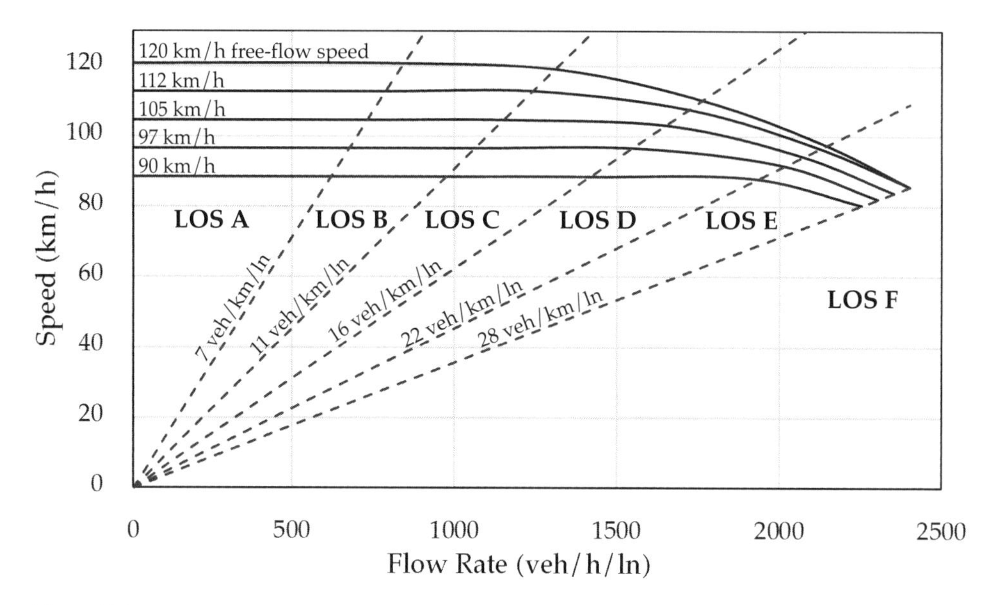

Figure 7.4
LOS based on traffic flow and speed.

Although there is much more to know about traffic flow, this will conclude our review of the fundamentals of traffic flow theory. For a more detailed introduction, the reader is invited to see Transportation Research Board (2010) and Mannering (2013).

7.1.2 Pedestrian Flow

Ultimately, we are all pedestrians, whether we walk to work or walk to our preferred mode of transport. It would therefore be logical for us to have an intricate understanding of pedestrian flow, and yet, we actually know very little. In fact, the 2010 version of the main transport manual for any transport engineer in the United States—the famous *Highway Capacity Manual* by the Transportation Research Board (2010)—still used a 1975 study by Pushkarev and Zupan (1975) and a 1971 study by Fruin (1971) to characterize pedestrian flow.

Unsurprisingly, the fundamental approach and the fundamental concepts are identical to traffic flow theory. In this section we will recall some of these fundamentals.

The first piece of information that we may need is the average space taken by a pedestrian. Forming an ellipse of an approximate 65 cm ($\approx$ 2 ft) for the shoulder breadth and an approximate 50 cm ($\approx$ 1.5 ft) of body depth, a pedestrian takes about one quarter of a squared meter of space (i.e., 0.25 m^2 or about 3 ft^2).

Example 7.1

Traffic in Civitas

A section of a small two-lane highway close to Civitas has a free-flow speed $v_f = 100$ km/h and a flow capacity (i.e., maximum flow) $q_{cap} = 3{,}000$ veh/hour. Using the Greenshields model to answer all questions, (1) determine the jam density k_j. (2) During a typical day, 2,000 vehicles were counted in one hour at one point on the highway. Estimate the two possible average speeds of the vehicles (i.e., under free-flow and congested conditions). (3) Based on this information, determine the LOS of the highway.

Solution

(1) Using equation (14), $q_{cap} = \frac{v_f k_j}{4}$. Therefore: $k_j = 4 \times 3000 / 100 = 120$ veh/km.

(2) Using and rearranging equation (8), we get:

$$\frac{k_j}{v_f} v^2 - k_j v + q = 0$$

Substituting the values:

$$\frac{120}{100} v^2 - 120v + 2000 = 1.2v^2 - 120v + 2000 = 0$$

Solving for v results in $v_1 = 21.13$ km/h or $v_2 = 78.87$ km/h.

(3) Since the highway has two lanes and a flow rate of 2,000 veh/h, the flow per lane is 1,000 veh/h/ln. Question (2) gives us two different speeds. Using figure 7.4 to determine the LOS:

- $v_1 = 21.13$ km/h; we get a LOS of F.
- $v_2 = 78.87$ km/h; we get a LOS of C.

To make it easier to read the LOS diagram, we can also calculate the two densities: $k_1 = 1{,}000/21.13 = 47.32$ veh/km/ln and $k_2 = 1{,}000/78.87 = 12.67$ veh/km/ln.

In the end, a speed of 78.87 km/h gives us an acceptable LOS, while a speed of 21.13 km/h does not. It is therefore quite important to take the actual average speed of the vehicle into account.

The second piece of information is the average walking speed of a pedestrian. Figure 7.5 shows the cumulative walking speed for young and older pedestrians in the United States. The average sits at about 5 ft/s, which is about 1.5 m/s or 5.5 km/h.

In a nearly identical manner to traffic flow theory, the flow and speed of pedestrians have a direct relationship. Adding the subscript p for pedestrians to equation (7), the relationship is captured by:

$$q_p = v_p k_p \tag{15}$$

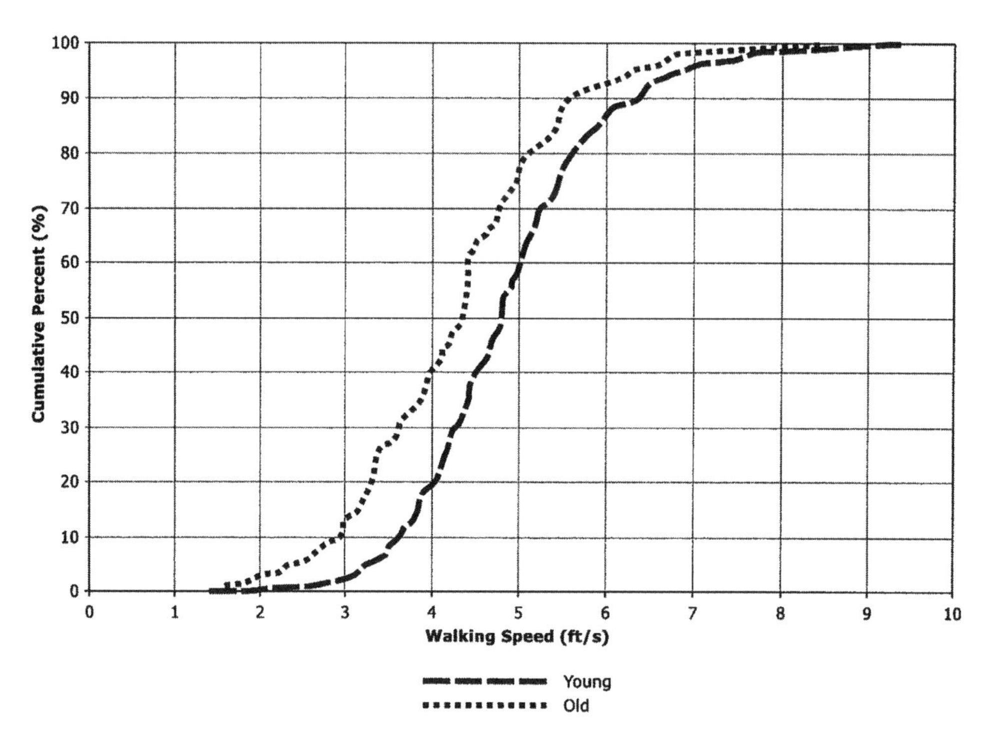

Figure 7.5

Cumulative walking speed. Reprinted with permission from Highway Capacity Manual (exhibit 4-12 in Transportation Research Board [2010]) by the National Academy of Sciences. Courtesy of the National Academies Press, Washington, DC.

where q_p is the unit flow rate in people per second per meter [pers/s/m].[11] Note the addition of *per meter*, while in traffic flow q tends to be per lane or per direction. The variable v_p is the pedestrian speed in [m/s], and k_p is the pedestrian density in people per square meter [pers/m^2].

Another way to look at equation (15) is to substitute the pedestrian density by the amount of space, s_p, taken by a pedestrian in [m^2/pers]—that is, not the size of a pedestrian, but the number of pedestrians divided by the surface area—resulting in:

$$q_p = \frac{v_p}{s_p} \tag{16}$$

Figure 7.6 illustrates the relationship between flow and space in a similar way to the fundamental diagram of traffic flow. Flow is therefore maximized when people take between 5 to 9 ft^2/pers or between 0.45 and 0.84 m^2/pers in the metric system.

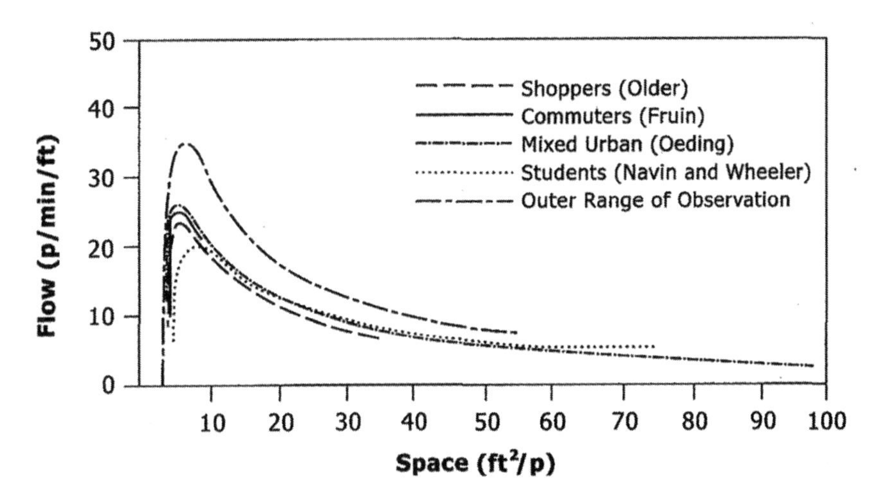

Figure 7.6
Flow-space relationship in pedestrian flow. Reprinted with permission from Highway Capacity Manual (exhibit 4-14 in Transportation Research Board [2010]) by the National Academy of Sciences. Courtesy of the National Academies Press, Washington, DC.

Naturally, by having more pedestrians, we also increase the number of potential crossflows that disrupt the speed-flow relationship, and we increase the potential for "conflict." Figure 7.7 shows empirical evidence for the probability of conflict versus space. Above 3.25 m^2/pers (i.e., 35 ft^2/pers), no conflicts happen. Below 3.25 m^2/pers, the probability of conflicts quickly rises to 50% at about 3 m^2/pers; it then stays fairly stable and increases again rapidly at 1.6 m^2/pers. When we reach 1.15 m^2/pers (about 12 ft^2/pers), then the probability of conflict is 100%. Note that 1.15 m^2/pers is higher than the maximum flow from figure 7.6, which means that under maximum flow conditions, the probability of conflict is 100%.

Although quite dated,[12] these results can help us design sidewalks, roads, and public places if we can estimate average pedestrian flows. This can be useful to promote walking and avoid constant pedestrian congestion. Avoiding the discomfort produced by congestion is certainly important, and we will now learn the fundamentals for public transport.

7.1.3 Public Transit Planning

All the major metropolitan areas of high-income countries in the world have public transport systems that millions of people use every day. After the private automobile, public transport tends to be the most heavily used transport mode in the high-income world,[13] and the popularity of transit—either as we know it or in a different form—will

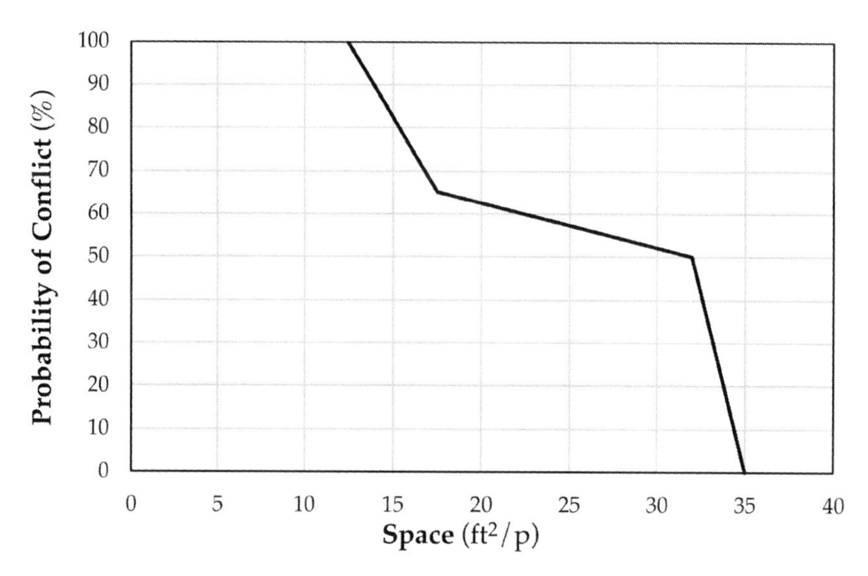

Figure 7.7
Pedestrian conflict and space. Adapted from figure 3.5 in Fruin (1971).

Example 7.2
Boardwalk in Civitas

The Civitian government is considering modernizing the riverfront of Vita by adding a 100-m-long and 3-m-wide boardwalk. The objective is to keep the probability of conflict to 99% under maximum occupancy. Considering an average walking speed of 4 km/h, estimate (1) the maximum flow of people per unit of width, (2) the maximum flow of people in one hour, and (3) the number of pedestrians allowed under maximum occupancy.

Solution

(1) Probability of conflict of 99% consists of an approximate space $s_p = 1.1$ m²/pers.
The pedestrian speed is $v_p = 4$ km/h $= 1.1$ m/s. Therefore, the maximum flow per unit width is $q_p = 1.1 / 1.1 = 1$ pers/s/m.

(2) The boardwalk is 3 m wide, thus the flow of people across the boardwalk is $q = 1 \times 3 = 3$ pers/s. Moreover, since $q = n/t$, where n is the number people and t is the time, the maximum flow of people in one hour is:

$n = q \cdot t = 3 \times (60 \times 60) = 10{,}800$ people.

(3) Considering the boardwalk is 3 m wide and 100 m long, the surface area of the boardwalk is $A = 3 \times 100 = 300$ m². Moreover, since the minimum space allowed is $s_p = 1.1$ m²/pers, the maximum occupancy $= 300 / 1.1 = 273$ people.

likely increase significantly in the future (we will learn partially why in section 7.3). Transit systems must therefore be well planned to ensure existing users are satisfied with the service, and potential new users are encouraged to use transit as their preferred mode of travel.

So far, we have considered transit as if only one mode existed, but this is not true. The American Public Transportation Association (APTA) identifies eighteen modes. Typically, we differentiate between rail and bus modes (both of which have their fans), but even within these two families, multiple modes exist. The most common modes include the following:

- Bus
- Trolleybus
- Bus rapid transit (BRT)
- Streetcar
- Tram/tramway
- Light rail transit (LRT)
- Metro/subway
- Regional/commuter rail
- Paratransit
- Ferry boat

Defining a mode is surprisingly hard. Trams, for instance, are light like streetcars, but in most places they have their own right-of-way (ROW) except at grade crossings. In essence, trams are operated like LRT systems, except that they enjoy a technology that is closer to streetcars (i.e., usually smaller and with lower capacity). Vuchic (2005) argues that instead of selecting and defining modes based on their technology, we should first think about the ROW used. He identifies three types:

- ROW A: fully separated (e.g., metro, subway, regional rail)
- ROW B: partially separated (e.g., light rail, tramway, BRT)
- ROW C: shared with vehicular traffic (e.g., bus, streetcars)

Based on the potential demand for transit (that we will see in section 7.2), we can first select which ROW applies and then, based on this ROW, select the type of service that we would like to offer. Some essential characteristics of transit modes are listed in table 7.1, including line capacity, which we can use to select the best transit service considering an anticipated demand, and operating speed, which we will use to determine station/stop spacing.

Table 7.1
Transit mode characteristics

Mode	ROW	Mode	Wagons per transit unit	Line capacity (sps/h)	Operating speed V_o (km/h)
Street transit	C	Bus	1	3,000–6,000	8–12
	C	Streetcar	1–3	10,000–20,000	8–14
Semirapid transit	B	BRT	1	6,000–24,000	16–20
	B	Tram/LRT	1–4	10,000–24,000	18–30
	A	AGT[a]	1–6 (10)	6,000–16,000	20–36
Rapid transit	A	LRRT[b]	1–4	10,000–28,000	22–36
	A	Metro	4–10	40,000–70,000	24–40
	A	Regional rail	1–10 (14)	25,000–40,000	30–55

Adapted from Vuchic (2005).
Notes: [a]Automated guided transit (AGT); [b]Light rail rapid transit (LRRT).

The values given in table 7.1 are only indicative. Sometimes we may also prefer to offer two parallel bus services (i.e., serving parallel corridors) rather than building one single tramway line. Rail modes tend to be preferred, however, because they have been shown to give a sense of "permanence" that attracts more riders and real estate development in the longer run (more on this later as well).

After selecting the mode, we need to determine the frequency of service and the station spacing. The total cycle time T (i.e., two-way) for a transit mode is defined as:

$$T = T_{o,1} + T_{o,2} + t_{t,1} + t_{t,2} = 2(T_o + t_t) \quad (17)$$

where $T_{o,1}$ and $T_{o,2}$ are the one-way operating times, and $t_{t,1}$ and $t_{t,2}$ are the terminal times.

Operating times are often identical in each way, and we simply use $2T_o$ and $2t_t$, but they do not have to be when operating on one-way streets. We also need to add the terminal times that are negotiated in labor contracts, and they are typically 15% of the operating time. Therefore, T can generally be reduced to:

$$T = 2(T_o + 0.15 \cdot T_o) = 2.3 \cdot T_o \quad (18)$$

Here, we can relate the operating time with the operating speed V_o, using the one-way route length L as:

$$T_o = \frac{L}{V_o} \quad (19)$$

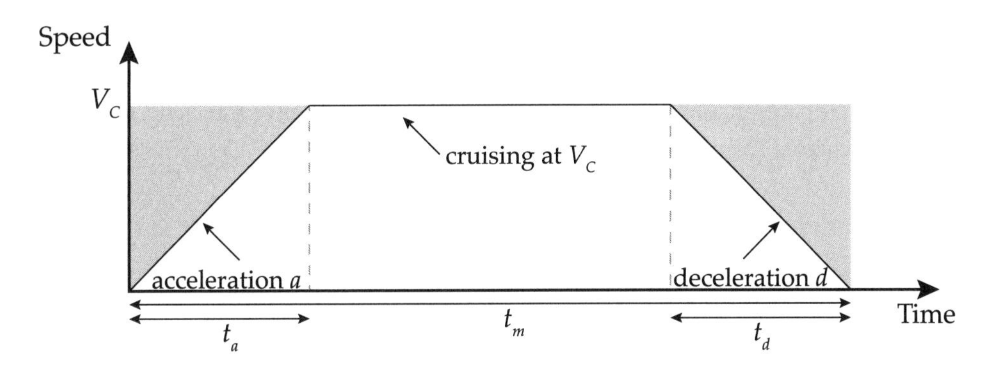

Figure 7.8
Interstation operating time.

The route length L is determined in the planning phase. The operating speed can be estimated with table 7.1, but it is usually computed from the operating travel time T_o that is calculated as:

$$T_o = t_m + t_s + t_i \tag{20}$$

where t_m is the time when the vehicle is in motion, t_s is the standing time (a.k.a., dwell time—that is, time for passengers to board or alight), and t_i is the time spent idling at intersections waiting for the green light. The in-motion time t_m includes the time to accelerate and decelerate and the cruising time when the vehicle has reached its cruising speed V_c in meters per second [m/s]. The process between two stations is shown in figure 7.8.

The times to accelerate t_a and decelerate t_d are defined similarly based on the acceleration rate a and deceleration rate d in [m/s^2]. Because a and d are the slopes in figure 7.8, we have $V_c = a \cdot t_a$, and $V_c = d \cdot t_d$ and therefore:

$$t_a = \frac{V_c}{a} \tag{21}$$

$$t_d = \frac{V_c}{d} \tag{22}$$

We also need to realize the area on the graph represents length (since length = speed × time). The total area under the curve plus the areas in the two shaded triangles—which is $V_c \cdot t_m$—represent the length that would be traveled if the cruising speed was kept at all times. Instead, we need to remove the two shaded triangles. The shaded area A_{shaded} for acceleration is defined as:

$$A_{shaded} = \frac{1}{2} V_c \cdot t_a = \frac{1}{2} V_c \cdot \frac{V_c}{a} = \frac{V_c^2}{2a} \tag{23}$$

The same equation applies for deceleration. The actual distance traveled L can therefore be calculated as:

$$L = V_c \cdot t_m - \frac{V_c^2}{2a} - \frac{V_c^2}{2d} \tag{24}$$

By manipulating equation (24), we can get the in-motion time t_m as:

$$\frac{L}{V_c} = t_m - \frac{V_c}{2a} - \frac{V_c}{2d}$$

$$t_m = \frac{L}{V_c} + \frac{V_c}{2}\left(\frac{1}{a} + \frac{1}{d}\right) \tag{25}$$

We now need to expand equation (25) to the entire line. To make it simple, we define n_s as the number of stations/stops. This means that we need to apply equation (25) to $n_s - 1$, because transit vehicles stop at the last terminal. This is illustrated in figure 7.9, where we have $n_s = 5$ stations in total but only four interstation spacings.

If we define the entire line route length as L and add the $n_s - 1$ stations/stops, the in-motion time becomes:

$$t_m = \frac{L}{V_c} + (n_s - 1)\frac{V_c}{2}\left(\frac{1}{a} + \frac{1}{d}\right) \tag{26}$$

This equation assumes that the cruising speed is reached. But this is not always the case. Although we will not go through the derivation, the in-motion time when the cruising speed is not reached can be calculated from:

$$t_m = \sqrt{\frac{2(a+d)l}{a \cdot d}} \tag{27}$$

where l is the space traveled. Equation (27) needs to be repeated as often as the cruising speed is not reached on the line, and this space traveled must be subtracted from the length L used in equation (26).

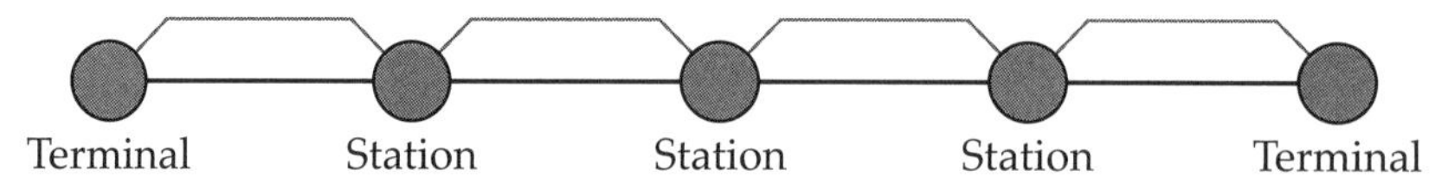

Figure 7.9
Number of transit stations.

Standing or dwell time t_s depends on the number of passengers boarding and alighting. The average standing time t_{avg} is typically taken as twenty seconds for initial modeling purposes. The total standing time is therefore:

$$t_s = n_s \cdot t_{avg} \approx 20 n_s \tag{28}$$

Note that here we need to use n_s as opposed to $n_s - 1$ since we account for the time it takes to board at the first station and the time to alight at the last station.

Finally, the idling time at intersections t_i is the time spent idling at red lights—t_i does not account for signal clearance time but purely applies to intersections with road traffic. If we assume that the red light lasts half the traffic signal cycle length $t_{traffic}$ and a uniform distribution of arrival rate at a red light, the average idling time is one-fourth of the cycle length, and the total idling time is:

$$t_i = n_c \cdot \frac{t_{traffic}}{4} \tag{29}$$

where n_c is the number of intersections/crossroads. Equation (29) does not account for transit signal priority systems that are relatively popular and that either make a green light longer or a red light shorter when a transit vehicle is arriving.

Using equations (21)–(29), we can now estimate the operating time T_o and hence estimate the operating speed V_o. Using equation (17) or (18), we can also calculate a cycle time T.

The next step is to determine the number of transit vehicles n_{tr} needed to operate the line. For this, we first need to set a frequency f (number of vehicles per hour), or more commonly, we set a headway h, which is the inverse of frequency and which tells us the time interval between two vehicles—for example, a vehicle every five minutes. Considering a vehicle is used for a time T, the total number of vehicles n_{tr} needed is then calculated as:

$$n_{tr} = w \cdot T \cdot f = w \cdot \frac{T}{h} \tag{30}$$

where f, T, and h must have the same units (e.g., in [min] or [h]), and w is the number of wagons/transit unit per service.

The last essential piece of information that we need to take into account is the time spent by transit riders to access and egress a station/stop and the time spent waiting for the bus/train to arrive.

The times to access and egress a station are t_{access} and t_{egress}, respectively. These are the times that transit users spend walking to and from a transit stop/station, and a trade-off exists here. For a generic walk time t_{wk} to and from a stop/station, considering a walk over a distance d_{wk} with a speed v_{wk},[14] t_{wk} is defined as:

$$t_{wk} = \frac{d_{wk}}{v_{wk}} \tag{31}$$

With many stations (i.e., large n_s), the access and egress times are small (i.e., t_{wk} is small), but the total in-motion time t_m is large, unnecessarily making the time spent in transit too long. Conversely, with few stations, the operating time T_o is short, but the walk time t_{wk} becomes too long (because d_{wk} is large). Therefore, there exists an "optimal" number of stations that make the entire travel time as short as possible. Figure 7.10 illustrates this trade-off and the presence of an optimal point.

The interstation spacing S is defined as:

$$S = \frac{L}{n_s - 1} \tag{32}$$

where the "– 1" reflects the presence of the last terminal; see figure 7.9.

Naturally, knowing exactly where potential line users live in the planning process is impossible, but this trade-off is an important limiting factor for our purpose.

The other time incurred by transit riders is that spent waiting for the bus or the train to arrive. This is the wait time t_w. Generally, for headways lower than ten minutes, riders do not pay attention to schedules and arrive at any time, following a normal distribution. The average wait time in this case is half the headway. For headways larger than ten minutes, however, riders tend to pay more attention to schedules—although with

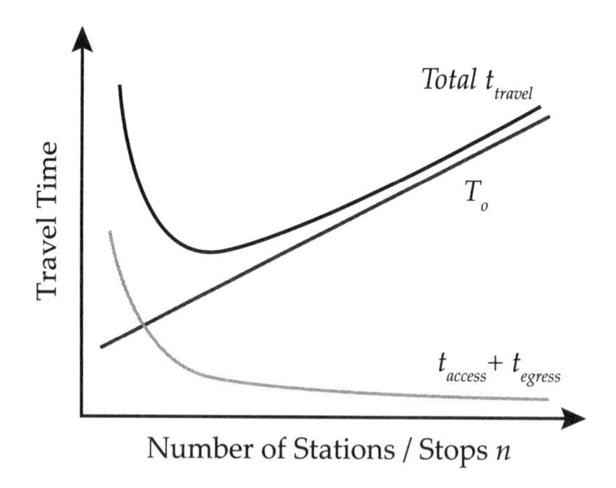

Figure 7.10
Time-spacing trade-off.

Example 7.3
Tramway line planning in Civitas

A new tramway line is being planned in Civitas that will run from west to east. The entire line will be $L = 10$ km; the cruising speed will be $V_c = 80$ km/h with acceleration and deceleration rates of 1.2 m/s^2. The standing time at each station is expected to be $t_s = 15$ s. Intersections are present at about every 200 m, with traffic signal cycle lengths of $t_{traffic} = 1$ min. (1) Determine the one-way operating time T_o as a function of the number of stations (assume the tramway always reaches cruising speed) and calculate T_o using $n_s = 10$. (2) Assuming a walking speed of 5.5 km/h, and assuming the average user accesses from and egresses to a distance of one-fourth of the interstation spacing S, express access and egress time as a function of the number of stations. (3) Assuming the trip of an average user is 10 km (i.e., L), plot the user travel time as a function of n_s and calculate the optimal number of stations. (4) Using the value of n_s found in (3), recalculate the operating time and calculate the operating speed V_o. (5) Assuming terminal times of 15% of the operating time, calculate the cycle time T and the number of vehicles needed n_{tr} considering a headway of 10 min and $w = 1$.

Solution

(1) Plugging equations (21)–(29) into equation (20) and converting all variables to meters and seconds, we obtain:

$$T_o = \frac{10 \times 1000}{80/3.6} + \frac{80}{2 \times 3.6}(n_s - 1)\left(\frac{1}{1.2} + \frac{1}{1.2}\right) + 15n_s + \frac{1 \times 60}{4} \times \frac{10 \times 1000}{200}$$

$$T_o = 450 + 18.52n_s - 18.52 + 15n_s + 750$$

$$T_o = 33.52n_s + 1181.48$$

Using $n_s = 10$: $T_o = 33.52 \times 10 + 1181.48 = 1516.68$ s $= 25.28$ min.

(2) In this case, $t_{access} = t_{egress}$, and $d_{wk} = S/4 = L\,/\,[4(n_s - 1)]$, therefore, combining equations (31) and (32), we get:

$$t_{access} = t_{egress} = \frac{S/4}{5.5/3.6} = \frac{L}{4(n_s - 1)} \times \frac{3.6}{5.5} = \frac{10 \times 1000 \times 3.6}{4(n_s - 1) \times 5.5} = \frac{1636.36}{n_s - 1}$$

(3) Using the expression from (1) and adding access and egress times, the user travel time t_{travel} is found as:

$$t_{travel} = T_o + t_{access} + t_{egress}$$

$$t_{travel} = 33.52n_s + 1181.48 + 2 \times \frac{1636.36}{n_s - 1}$$

See plot below of t_{travel} as a function of n_s.

(continued)

Example 7.3 (continued)

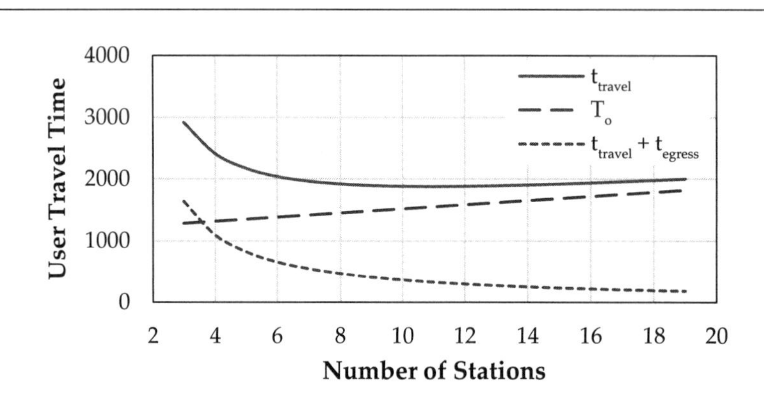

The travel time is minimized when the number of stations is 11. This translates into a user travel time $t_{travel}=1877.47$ s $=31.30$ min and an interstation spacing $S=10/(11-1)=1$ km.

(4) Using $n_s=11$, the operating time T_o becomes $T_o=33.52\times 11+1181.48=1550.2$ s $=25.84$ min. The operating speed is

$$V_o=\frac{L}{T_o}=\frac{10}{25.84/60}=23.22$$

$V_o=23.22$ km/h, which is within the range given in table 7.1.

(5) The total cycle time T is $T=2.3\cdot T_o=2.3\times 25.84=59.43$ min. Considering a headway $h=10$ min, the number of vehicles needed is $n_{tr}=1\times 59.43\ /\ 10=5.94$.

The number of needed vehicles is 6.

smartphone applications, it becomes easier to arrive at a station/stop shortly before the transit vehicle arrives. Some studies suggest taking the square root of the headway in this case as the average wait time. In equation form, the wait time t_w is:

$$t_w=\begin{cases}\frac{h}{2}, & \text{when } h\le 10 \text{ min}\\ \sqrt{h}, & \text{when } h>10 \text{ min}\end{cases} \tag{33}$$

We must further note that not all times are "felt" the same way. In fact, in general a minute of waiting feels longer than a minute of access/egress time, which in turn feels longer than a minute in the transit vehicle that is in motion.

This concludes our section on the fundamentals of transport. Now, we will learn what demand means in the context of transport so that we can apply our first sustainability principle.

7.2 Travel Demand

Just like electricity and water, demand for transport can be estimated and/or measured. In this section we will see how we can quantify the demand for transport by first defining and studying the concept of a *trip*. We will then give a length to these trips by looking at distance traveled, and subsequently, we will divide these trips by mode. Afterward, we will tag emission values to these trips based on the mode used and the distance traveled, and finally, we will see how we can group all of this information for an entire city, learning about OD matrices. But first, let us learn about the basic unit of travel demand: the trip.

7.2.1 Trips

The unit of analysis in transport has traditionally been the *trip*. A trip is the simple process of traveling from one location to another. The most famous type of trip is the work trip, to go from one's home to work. Short trips are sometimes combined together as *trip chains*. Moreover, when the last destination is the same as the first origin, then a series of trips is called a *tour*.[15] To collect data on trips and travel behavior in general, the National Household Travel Survey (NHTS) is a great resource in the United States. At the time of this writing, the latest survey available was for 2017 (Federal Highway Administration 2017).

From the 2017 NHTS survey, we find that the average number of daily trips per person was 3.37 in 2017, which is actually quite low. Table 7.2 shows historical averages, and we can see that after a peak in 1995, the average number of daily trips has been declining. Figure 7.11 shows how the number of trips was distributed in 2017, from zero to fifteen trips per day,[16] and we can see that about 17% of the population actually made no trip—that is, they did not leave their homes—while 26% made two trips—for example, from home to work and back.

Because many trips originate from home, trips can also be categorized as home-based and nonhome-based. Moreover, these trips have an associated *purpose* since few

Table 7.2
Evolution of average number of daily trips

Year	1977	1983	1990	1995	2001	2009	2017
Average number of daily trips	2.92	2.89	3.76	4.30	4.09	3.79	3.37

Source: McGuckin and Fucci (2018).

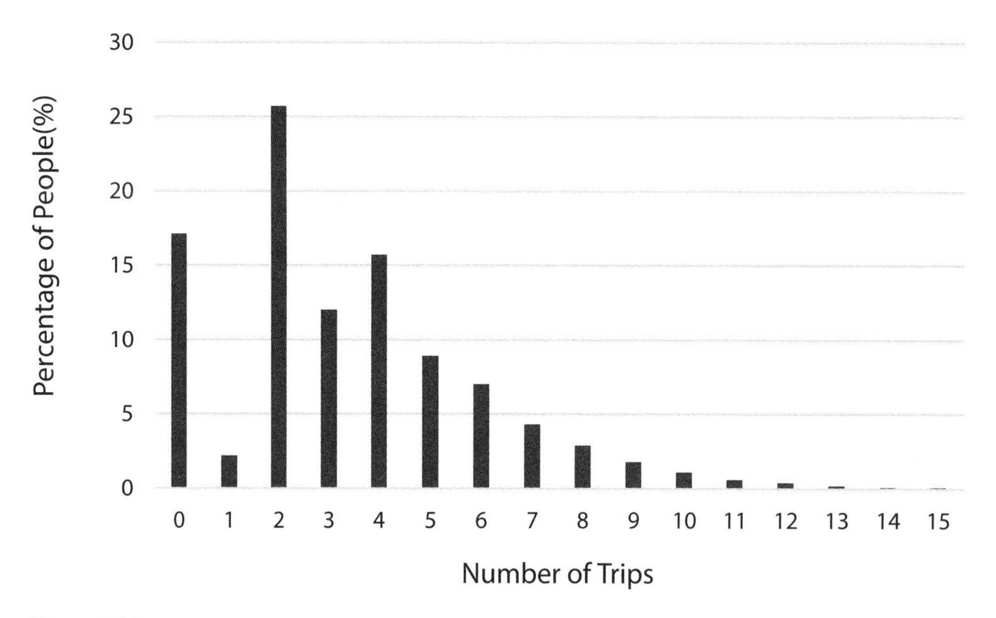

Figure 7.11
Distribution of the number of trips in 2017 in the United States.
Source: Federal Highway Administration.

people travel for the pleasure of traveling. Here are the ten most common reasons for a trip, as reported in the NHTS and their percentages:

- Home (35.2%)
- Work (12.5%)
- School/day care/religious activity (6.1%)
- Medical/dental services (1.5%)
- Shopping/errands (17.8%)
- Social/recreational (10.7%)
- Transport someone (6.8%)
- Meals (7.3%)
- Something else (2.1%)

In addition, these trips also have an inherent temporal feature—that is, a start time and an end time. Figure 7.12 shows a possible series of five trips for a day, starting at 8:00 a.m. and ending at 6:45 p.m.

More generally, figure 7.13 shows the number of annual trips by start time and purpose in 2017. Figure 7.13a shows the total number of trips, and figure 7.13b shows the

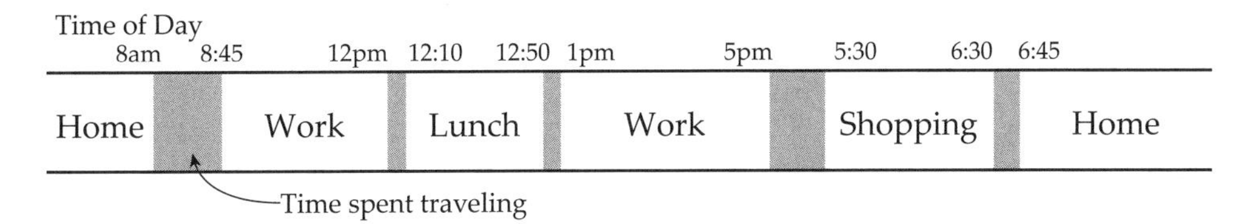

Figure 7.12
Possible series of daily trips.

number of trips by trip purpose. The figure clearly shows that few trips are made at night. Figure 7.13a shows a typical bimodal curve, which represents the morning and afternoon rush-hour periods. This bimodal feature is even more distinguishable in figure 7.13b when looking at the curve for the work trip. A similar feature is present for the category "Other" as well. Expectedly, most nonhome trips occur in between the two rush-hour periods, especially around 12:00 p.m., capturing the trips made to go to lunch, for example.

A caveat has to be mentioned here. If we applied our first sustainability principles directly, then our aim would be to reduce the number of trips. This is actually not correct. In fact, a higher number of trips tend to produce higher economic activity. We generally want more trips! However, the number of trips is only one side of the demand for transport. Making three trips by walking a total of one kilometer or driving a total of three kilometers is preferable than making one trip driving thirty kilometers (from a triple bottom line of sustainability point of view). Distance is therefore paramount here, as we are about to see.

7.2.2 Distance Traveled

Distance is indeed key since it has a direct impact on energy use and GHG emissions. Distance is defined in two ways in transport. One measure of distance accounts for vehicles, and it is called VKT for vehicle kilometers traveled (or VMT for vehicle miles traveled), and the other measure accounts for people, and it is called PKT for passenger kilometers traveled. The two indicators have a direct relationship based on the occupancy of the vehicle. Put simply, if N_o is the number of occupants in a vehicle, then:

$$PKT = N_o \cdot VKT \tag{34}$$

In 2017, across all trips and purposes, the average vehicle occupancy in the United States was 1.67.

Between VKT and PKT, the measure that we prefer depends on our analysis. VKTs are important since in the end it is the vehicles that are emitting GHGs, not the occupants.

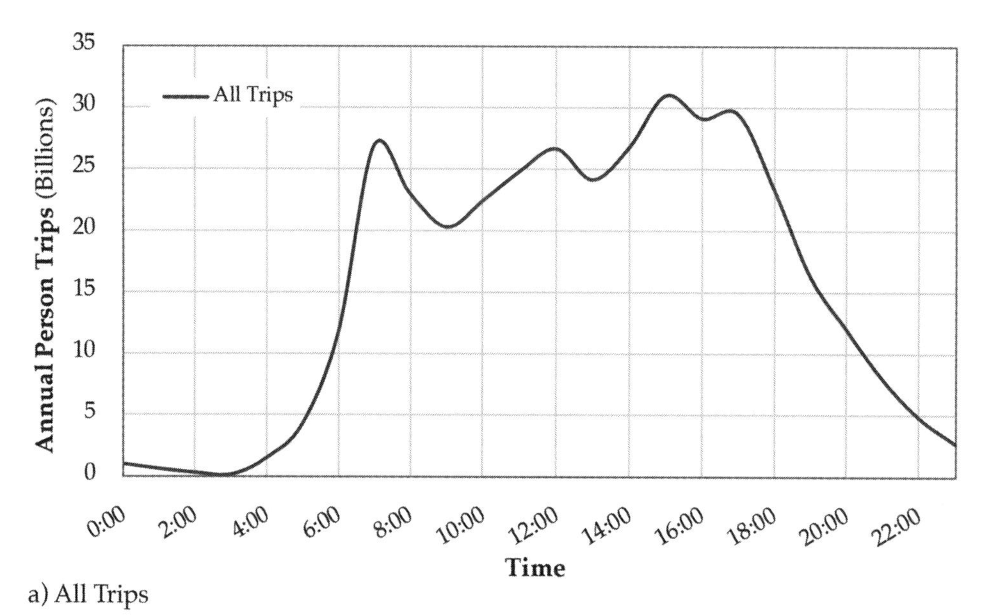

a) All Trips

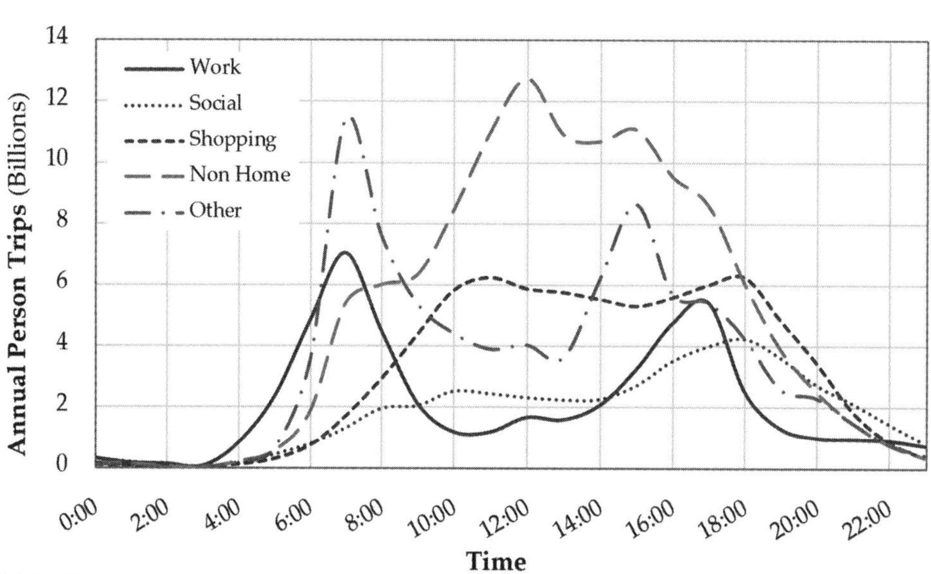

b) By Trip Purpose

Figure 7.13
Number of trips by start time and purpose in 2017 in the United States.
Source: Federal Highway Administration.

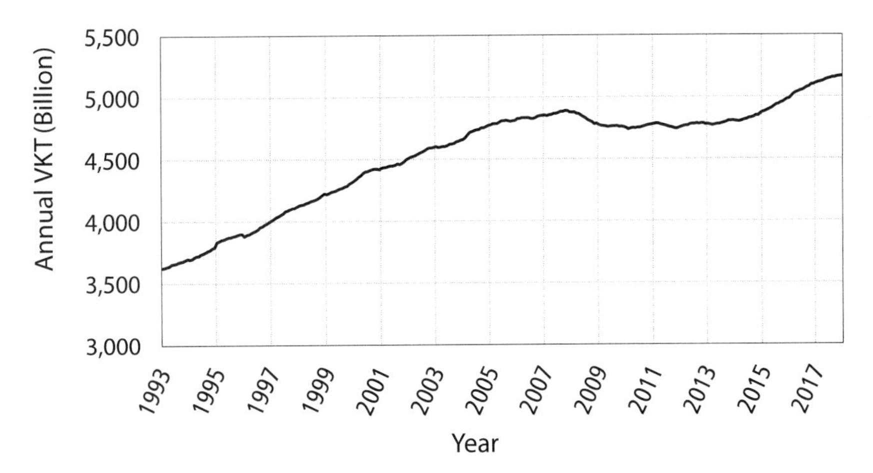

Figure 7.14
Annual U.S. VKT from January 1993 to December 2017.
Source: Federal Highway Administration.

That being said, we like to tag emissions to a usage, and if there are two occupants in a vehicle, then the emissions are split in two so that half is attributed to each person. This is, notably, why despite the fact that an individual bus emits much more GHG emissions than an individual car, if the bus contains fifty passengers and the car only one, then the per passenger emission will be often lower for the bus—we will talk about emissions in section 7.2.4.

Figure 7.14 shows how total annual VKT has evolved in the United States from January 1993 to December 2017. After a fairly significant increase from 1993 to 2008, annual VKT from 2008 to 2013 remained fairly constant, but it started to increase again in 2014. While it is a sign that the economy is doing better (after the 2008 crisis),[17] additional VKTs imply additional GHGs. This is also partly why transport has become the leading cause of GHGs in the United States, above electricity generation, as we saw in chapter 5.

Reducing the increasing trend while keeping a sound economy is therefore a top priority. If we link this distance feature to our trips, we can calculate the average distance per trip. In 2017 in the United States, the average PKT per trip was quite high at 17.21 km (10.7 mi), and the average VKT per trip for private vehicles was 15.37 km (9.55 mi). The fact that the average PKT and private VKT are close suggests that most trips are made by single-occupant private vehicles—that is, vehicles with the driver only, no passengers. As we saw above, the average vehicle occupancy in 2017 was 1.67, which supports that point, but this number also suggests that most trips were made with private vehicles as opposed to transit, which we will discuss in the next section.

Table 7.3
Evolution of PKT for work trip by mode

Average commute distance (km)	1977	1983	1990	1995	2001	2009	2017
All modes	14.58	13.74	17.14	18.72	19.49	18.97	19.66
Private vehicle	15.47	14.26	17.73	19.05	19.47	19.46	20.45
Public transit	12.04	14.48	20.52	20.73	18.88	16.38	19.46
Walk	—	—	1.34	1.19	1.46	1.58	1.91

Source: McGuckin and Fucci (2018).

Table 7.4
Average distance per trip and per person (km) in the Chicago region in 2008

Geographic location	Average distance per Trip	Average distance per Person (km)	Number of trips
Central Chicago	4.97	18.23	3.67
North Chicago	6.45	24.35	3.77
South Chicago	6.65	23.13	3.48
North Cook County	7.19	29.11	4.05
West Cook County	5.71	22.58	3.95
South Cook County	8.37	33.02	3.95
Lake County	9.03	34.91	3.87
DuPage County	8.59	35.94	4.18
McHenry, Kendall, western Kane Counties	10.80	41.65	3.86
Eastern Kane County	8.21	31.64	3.85
Will County and Grundy County	10.56	41.49	3.93
Region	7.64	29.61	3.87

Source: Parry (2010).

Table 7.3 shows that PKT by private vehicle for the work trip solely (i.e., commute trip) was the largest in 2017, with more than 20 km. The table also shows that people who walk to work travel, on average, 1.91 km, which is much lower. It suggests that while people do not mind spending long periods of time in a vehicle, they do not like to walk extensively to get to work (partly because of uncertain weather conditions). Naturally, these values vary greatly by geographic location as well.

Table 7.4 shows the average distance traveled by trip and by person in the Chicago region. Clearly, the people living in Central Chicago have much shorter trips since most destinations are simply closer, as opposed to the residents of the peripheral counties, who have to drive fairly extensively to reach their destinations. In fact, the average distance by trip and person is roughly half as long in Central Chicago as in a peripheral county. Even compared to Cook County—the county where the city of Chicago is located—average distances in Chicago are roughly two-thirds smaller. Table 7.4 also shows the average number of daily trips. Most regions in Chicago are close to the 2009 national average of 3.79—the Chicago data was collected around 2008—although two areas are above four trips per day.

One feature we have not discussed explicitly about travel demand so far is the choice of transport mode, although it is quite relevant for us since transit and active transport require a lot less of energy. So before getting into energy and emissions, let us discuss the notion of mode share.

7.2.3 Mode Share

As the name suggests, mode share accounts for the use of different transport modes. The terms *mode split* or *modal share* are often used and mean the same thing. The term *mode choice* is used mainly in modeling, and we will briefly discuss it in section 7.4. Modes are categorized into three main families:

- Private vehicles that include cars, sport utility vehicles (i.e., SUVs), pickup trucks, vans, and sometimes motorcycles
- Public transport that includes all modes discussed in section 7.1.3, and sometimes even taxi service
- Active transport that mainly includes walking and cycling

The three families are then divided based on the type of survey that is being carried out. Another important source of transport data in the Unites States is the American Community Survey, which does not differentiate between cars, light trucks, and vans. It also groups motorcycles with bicycles, and it only tracks the usual mode of travel to work.

Mode share can be computed in two ways, which, unsurprisingly, depend on whether we look at trips or distance traveled. Most commonly, mode share values are calculated based on the number of trips made by mode. Table 7.5 shows the evolution of nationwide mode share values in the United States from 1969 to 2017. We can see the share of private vehicles has largely dominated since 1969.

Here again, location matters and denser urban places are much more likely to have a much higher share of transit and active transport. Table 7.6 shows the mode share values for the various counties of the Chicago region in 2008. Unsurprisingly, transit has a

Table 7.5
Evolution of commute mode in the United States.

Usual commute mode	1969	1977	1983	1990	1995	2001	2009	2017
Auto, truck, van, or SUV	90.8	87	88.6	87.8	91	90.8	89.4	87.5
Public transit	8.4	6	5.3	5.3	5.1	5.1	5.1	6.9
Walk	N/A	4.1	4.3	4	2.6	2.8	2.8	2.8
Other	0.8	2.9	1.8	2.9	1.3	1.3	2.7	2.7

Source: McGuckin and Fucci (2018).

substantial share in Central Chicago, but we can see that about one quarter of all trips are made by walking. This is significant and highly desirable. The share of walk trips rapidly decreases when we move away from the center of the region. Transit remains high in all of Chicago, which is supported by an extensive transit system (the Chicago Transit Authority). Bike mode share is higher in Chicago but remains relatively low on average. The share of private vehicles increases from about half of all trips in Central Chicago to more than 90% in peripheral counties. These observations are fairly common for large metropolitan areas. Nonprivate vehicle trips are more numerous in city centers, but the car rapidly dominates in the suburbs.

Differences can also be made by looking at world cities. Figure 7.15 shows mode share values for multiple cities, ranked from lowest private vehicle mode share, *left*, to highest, *right*. The cities on the left tend to have high walk and transit mode shares. These cities are mainly European and Asian. Unsurprisingly, the largest private vehicle users are found in the United States, Canada, and Australia. This is indeed not surprising considering their low population densities, as we will see in section 7.3.

A caveat must be noted about these values collected from Wikipedia.[18] The city boundaries were not defined. Tokyo, for instance, is reported to have only a 12% private vehicle mode share, and this may be true for central Tokyo (i.e., 23 wards of the Tokyo Metropolitan Government) but not for the entire region. Similarly, Chicago is reported to have a 13% transit mode share, which likely corresponds to the city of Chicago, not the entire Chicago region, as shown in table 7.6. As urban engineers, we must be careful and make sure we know what the data that we have actually measures.

So why are walking, cycling, and transit better than private vehicles? For starters, they take much less space and therefore do not generate as much congestion. In addition, and very importantly for us, their per passenger emissions are also much lower, as we will learn in the next section.

Table 7.6
2008 mode share in the Chicago region

Geographic location	Private vehicle	Transit	Walk	Bike	Other
Central Chicago	53.3%	16.5%	26.4%	2.0%	1.6%
North Chicago	68.6%	13.2%	15.3%	1.5%	1.3%
South Chicago	68.6%	16.6%	13.2%	0.1%	1.4%
North Cook County	84.6%	3.6%	8.3%	1.4%	2.1%
West Cook County	80.0%	6.0%	11.0%	0.9%	2.0%
South Cook County	85.8%	4.6%	6.7%	0.5%	2.6%
Lake County	89.9%	1.8%	4.2%	0.3%	3.8%
DuPage County	87.5%	2.7%	6.6%	1.5%	1.8%
McHenry, Kendall, and western Kane Counties	92.1%	1.3%	2.8%	0.3%	3.4%
Eastern Kane County	86.2%	2.1%	9.4%	0.1%	2.1%
Will County and Grundy County	90.6%	1.7%	3.8%	0.6%	3.4%
Region	79.5%	6.9%	10.4%	1.0%	2.3%

Source: Parry (2010).

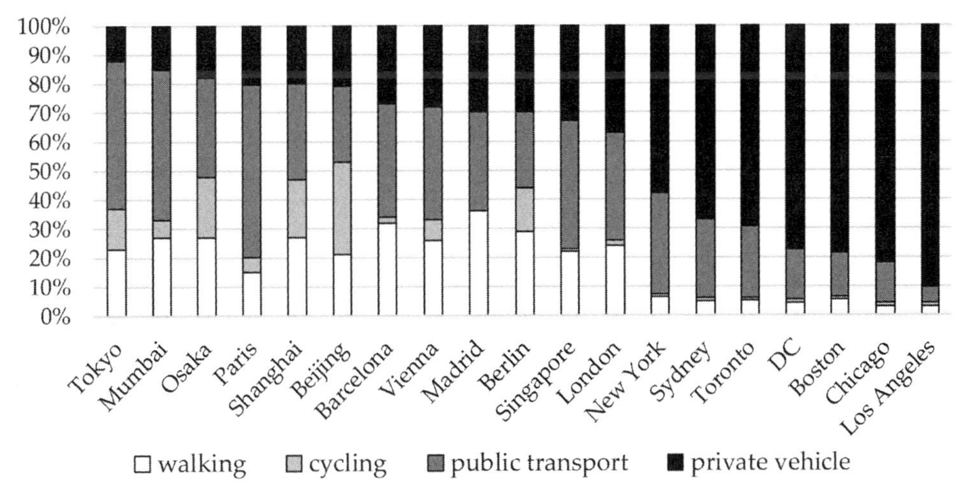

Figure 7.15
Mode share in various cities around the world. Years of data collection vary by city. Note that no formal definition of city boundaries was provided.
Source: Wikipedia.

7.2.4 Greenhouse Gas Emission Factors

Sustainability applied to transport has two main components. The first one is congestion, which in itself is not sustainable and must be managed. The second one is the environmental component related to GHG emissions, and it is the focus of this section.

Put simply, motorized transport requires energy,[19] and it comes mainly in two forms: fossil fuel and electricity. Fossil fuels have direct emissions and electricity has upstream emissions, as we discussed at length in chapter 5. At the time of this writing, fossil fuels were used in most private vehicles, including cars and SUVs, but also in conventional buses and diesel-powered regional trains. In the United States, transport accounts for more than two-thirds of all oil consumption (Greene 2004). Electricity is used mainly in trolleybuses, streetcars, tramways, metros, and in a growing share of private vehicles.

The main metrics that we use to quantify these emissions are grams of CO_2 equivalent per kilometer traveled (i.e., [g CO_2e/km]) for fossil fuel-powered modes and kilowatt-hours of energy consumed per kilometer traveled for EVs (i.e., [kWh/km]). For both measures, distance traveled is the common denominator. Alternatively, we can also look at the distance that can be traveled per unit of energy—that is, PKT or VKT per [kWh] or [g CO_2e].

Beyond looking at distance, other possible measures include time. After all, when stuck in traffic, we are still emitting GHGs, even when we are idling. In fact, emissions from congestion are significant, and the EPA has developed the software package MOtor Vehicle Emission Simulator (MOVES) to quantify emissions from distance travel while accounting for congestion. It is available for download on their website.[20] We will overlook this aspect here for simplicity.

Thinking purely in terms of energy—that is, in [kWh] as opposed to GHG—table 7.7 shows standard values to compare private vehicle (i.e., *car* in the table) with transit vehicles. The last column displays energy efficiency values in [PKT/kWh] (i.e., how far can we travel with 1 kWh), and we can see that transit systems perform much better when they are well attended, easily consuming ten times less energy than private vehicles.

The next step is to turn these values into actual emissions. Table 7.8 shows the emissions used by the EPA for their nationwide GHG inventory—the values were collected in 2018, but the EPA periodically updates its emission factors.[21] In this case, all values are cited in [g CO_2e/km], although the unit of measure changes from [veh-km] (i.e., VKT) to [prs-km] (i.e., PKT) for transit modes. The definition of the modes here also differs slightly from the NHTS data used previously—see the definitions below the table.

Assuming private vehicles have only one occupant, buses have fewer GHG emissions than private vehicles. We do have to be a bit careful about these types of claims,

Table 7.7
Transport energy consumption by mode

Mode	Vehicle capacity [sps/veh]	Vehicle occupancy [prs/veh]	Energy consumption	Energy efficiency	
			[veh-km/L]	[VKT/ kWh]	[PKT/ kWh]
Car (standard)	5	1.2–2.8	3.8–60	1.04–1.64	1.2–4.6
Car (compact)	4	1.2–2.8	7.2–8.1	1.96–2.20	2.4–6.2
Carpool	6	2.0–6.0	3.8–6.0	1.04–1.64	2.1–9.8
Bus	45–70	10–70	1.3–2.2	0.35–0.60	3.5–42
			[kWh/veh-km]		
Trolleybus	45–70	10–70	2.2–4.1	0.24–0.46	4.6–32
Streetcar/Tram	80–200	15–200	1.6–5.1	0.20–0.62	2.9–125
Metro	150–200	25–200	3.5–5.1	0.20–0.29	4.9–57

Source: Vuchic (2007).

Table 7.8
EPA transport emission factors

Vehicle type	g CO_2e/km	Unit
Passenger car[a]	215	veh-km
Light-duty truck[b]	297	veh-km
Medium- and heavy-duty truck	914	veh-km
Motorcycle	120	veh-km
Bus	35	prs-km
Transit rail (i.e., subway, tram)	74	prs-km
Regional rail	101	prs-km
Intercity rail (i.e., Amtrak)	88	prs-km

Source: U.S. Environmental Protection Agency.
Notes: [a]Includes passenger cars, vans, SUVs, and small pickup trucks; [b]Includes full-size pickup trucks, full-size vans, and extended-length SUVs.

however. Emission factors vary greatly with ridership. Moreover, emissions vary based on many factors, from occupancy and driver behavior to whether the air conditioning or heat is turned on. The dividing line about which mode is more environmentally friendly is not as clear-cut as it seems. However, buses are generally considered to be more sustainable, and we will see why in section 7.3.

Naturally, electric transit modes depend on the nature of the power grid, and a tramway in Vermont will emit much less than a tramway in Texas (remember the power grid emission factors from chapter 5). As a general approximation, however, table 7.8 can help calculate initial estimates.

When performing an inventory, we need to get all the modes and all the emissions together. For this, we need to come up with the energy consumed and the GHG emitted for each transport mode and sum them. The IPAT equation and the Kaya identity that we learned in chapter 2 can help us. If we recall the definition of the Kaya identity:

$$E = P \cdot A \cdot T \cdot \varepsilon \tag{35}$$

where E represents the total GHG emissions in grams or metric tons [CO_2e], P is the population, A is the activity [prs-km], T is the impact of technology [kWh/prs-km], and ε is the conversion factor from energy to GHG emissions [g CO_2e/kWh]. For electric modes, equation (35) works well. For traditional vehicles that have combustion engines, we can simply merge T and ε together to get units of [g CO_2e/prs-km].

We then have two ways of calculating equation (35). If P is the total population, then A needs to be the average PKT over the entire population as opposed to the average trip distance, which is not what we learned. Alternatively, if A is the average trip distance, then P needs to account only for the population that uses the specific mode.

One way to keep population P is to add a parameter with mode share. If we define S_m as the share of mode m, then equation (35) becomes:

$$E = \sum_m S_m \cdot P \cdot A \cdot T \cdot \varepsilon \tag{36}$$

Mode also affects the activity A and the technology-emission $T \cdot \varepsilon$. As we saw, the different modes have different distance traveled and emission factors. We can therefore add the subscript m for mode here as well.

Finally, depending on the data that we have and the quantity that we are looking for, we need to account for differences between VKT and PKT. For this, we simply need to account for the relationship between VKT and PKT, as we saw in equation (34). In other words, if the activity A is in PKT, then the $T \cdot \varepsilon$ needs to be in PKT as well, and the same holds for VKT. Equation (36) therefore takes the following form to account for transport emissions:

$$E = \sum_m S_m \cdot P \cdot A \cdot (T \cdot \varepsilon)_m \tag{37}$$

The combination of T and ε together seems a bit cumbersome, but it will be easy to apply in practice. Again, we just need to ensure that we get [g CO_2e] or [t CO_2e] in the end. Equation (37) is sometimes expressed as the ASIF equation as originally defined by Schipper and Marie-Lilliu (1999).

We will now learn about one more concept so that we can estimate emissions from entire regions.

7.2.5 Origin-Destination Matrix

The final concept that we will cover in this section is the origin-destination (OD) matrix. Except for national data that use odometer readings, we generally do not have exact information on the number of kilometers driven by each and every car. We therefore need to get an estimate of the total travel demand for a region. For this, we first divide an area in zones—in urban studies, these zones are often called TAZ for traffic analysis zones—and count/estimate how many trips go from one zone to another. In these OD matrices, trip origins are represented by the rows, and trip destinations are represented by the columns.

If a trip T_{ij} originates from zone i and terminates at zone j, then the total number of trips O_i that originate from zone i are:

$$O_i = \sum_j T_{ij} \quad (38)$$

And the total number of trips D_j that end at zone j are:

$$D_j = \sum_j T_{ij} \quad (39)$$

Finally, the total number of trips T made in the city is defined as:

$$T = \sum_i \sum_j T_{ij} = \sum_i O_i = \sum_j D_j \quad (40)$$

For instance, table 7.9 shows the OD matrix for Civitas for the work trip only. Civitas has five zones, as we saw in chapter 1. The rows in the first column show zones where trips originate, and the columns in the first row show the zones where trips end. Civitas

Table 7.9
Work trip OD matrix for Civitas

O\D	1	2	3	4	5	Total
1	7,354	3,476	1,934	2,456	1,017	16,237
2	4,267	1,567	1,143	476	938	8,391
3	2,102	478	1,112	554	280	4,526
4	2,432	235	673	995	508	4,843
5	2,678	823	231	778	1,493	6,003
Total	18,833	6,579	5,093	5,259	4,236	40,000

Table 7.10
Average trip distance between zones in Civitas

Zones	1	2	3	4	5
1	5.1	7.8	8.2	7.2	9.7
2	7.8	7.2	7.4	14.6	10.1
3	8.2	7.4	6.3	8.8	13.9
4	7.2	14.6	8.8	5.9	10.4
5	9.7	10.1	13.9	10.4	6.8

has 40,000 workers, and the OD matrix shows $T=40{,}000$ trips. Note that many of these trips originate and end in the same zone.

From the matrix, we can see that zone 1 produces the largest number of trips with 16,237 trips. At the same time, it attracts the most number of workers since it has 18,833 trip ends. Based on equations (38) and (39), $O_1=16{,}237$ and $D_1=18{,}833$. Zone 1 is therefore highly residential and commercial.

The second largest producer of trips is zone 2, with $O_2=8{,}391$, and the second largest attractor of trips is zone 2 as well, with $D_2=6{,}579$. Both zones 1 and 2 are therefore highly residential and commercial, and they are therefore mixed use, which is desirable to provide eyes on the street, as we saw in chapter 4.

We can have a similar matrix that shows the average trip distance d_{ij} between each zone. This is most often done by measuring the distance between the centroid of one zone to the centroid of another zone. Table 7.10 shows the average trip distance between zones in Civitas; the values shown are in kilometers. This means that the average trip distance within zone 1 is 5.1 km, and the average trip distance between zone 1 and 3 is 8.2 km.

It is then easy to compute total distance traveled by the entire workforce by multiplying the number of trips between zones by the appropriate distance. We then need to consider mode share and emission factors, and we can estimate the total number of GHGs that are emitted every day for the work trip of 40,000 workers in Civitas.

The analysis might seem a little crude with only five zones since most trips may be much shorter. However, cities typically have many more zones. The Chicago Metropolitan Agency for Planning has defined close to 1,700 such zones in the form of TAZs for the region, which therefore represent $1{,}700\times1{,}699=2{,}888{,}300$ possible origin-destination trips. At this scale, looking at distances between centroids (i.e., zone centers) outputs reasonable estimates.

Example 7.4

Transport Emissions in Civitas from the Work Trip

The aim is to estimate the GHG emissions of the 40,000 Civitian workers in their daily commute (one way). Use the travel demand values given in tables 7.9 and 7.10, and the emission factors given in table 7.8. The work trip mode share is as follows:

- Private vehicle: 56%, of which 70% are passenger cars, and 30% are light-duty vehicles. Assume an average occupancy of 1.2.
- Transit: 34%, of which 50% are bus riders, 30% are tram riders, and 20% are regional rail riders.
- Active transport: 10%, of which 70% walk and 30% bike.

Compute (1) the total distance traveled between each zone in Civitas and (2) the average work trip distance. (3) Assuming the mode share values apply for every single trip and that distances calculated apply to all modes, calculate the total emission of every single mode as well as the total emissions from Civitians. (4) Assuming 250 workdays in a year, compute the average annual per capita emission from Civitian workers for their morning commutes.

Solution

(1) The total distance traveled between each zone is simply obtained by multiplying the cell values in the OD matrix from table 7.9 with the distance values between and within each zone from table 7.10. The answer is:

O\D	1	2	3	4	5	Total
1	37,505	27,113	15,859	17,683	9,868	108,028
2	33,283	11,282	8,458	6,950	9,470	69,443
3	17,236	3,537	7,006	4,875	3,892	36,546
4	17,510	3,431	5,922	5,871	5,280	38,014
5	25,977	8,312	3,211	8,091	10,155	55,746
Total	131,511	53,675	40,456	43,470	38,665	307,777

The matrix shows us that a total of 307,777 km are traveled every morning by Civitians to go to work.

(2) Considering there are 40,000 trips and the total distance traveled is 307,777 km, the average work trip distance = 307,777 / 40,000 = 7.69 km.

(3) We need to apply the modified IPAT equation—that is, equation (37)—to each mode. Moreover, because we are given the share of each technology within each mode, we need to split S_m into two components. The emissions of each mode are:

(continued)

Example 7.4 (continued)

$$E_{passenger\ car} = (0.56 \times 0.7) \times 40000 \times 7.69 \times 215 = 25.92 \times 10^6 \text{ g } CO_2e$$
$$E_{light-duty} = (0.56 \times 0.3) \times 40000 \times 7.69 \times 297 = 15.35 \times 10^6 \text{ g } CO_2e$$
$$E_{bus} = (0.34 \times 0.5) \times 40000 \times 7.69 \times 35 = 1.83 \times 10^6 \text{ g } CO_2e$$
$$E_{tram} = (0.34 \times 0.3) \times 40000 \times 7.69 \times 75 = 2.32 \times 10^6 \text{ g } CO_2e$$
$$E_{regional\ rail} = (0.34 \times 0.2) \times 40000 \times 7.69 \times 101 = 2.11 \times 10^6 \text{ g } CO_2e$$
$$E_{walk} = 0 \text{ g } CO_2e$$
$$E_{bike} = 0 \text{ g } CO_2e$$

Based on these calculations, an average occupancy of 1.2, and using equation (34), the total daily emissions in Civitas for the work trip is:

$$E = (25.92/1.2 + 15.35/1.2 + 1.83 + 2.32 + 2.11) \times 10^6 = 40.65 \times 10^6 \text{ g } CO_2e$$
$$= 40.65 \text{ t } CO_2e$$

(4) Assuming 250 workdays. $E_{annual} = 40.65 \times 250 = 10{,}163$ t CO_2e. Per Civitian worker, this is:

$$E_{annual} = 10{,}163 \ / \ 40{,}000 = 0.25 \text{ t } CO_2e \text{ per worker}$$

Expanding these results to a two-way trip, we find that the average annual emissions for the commute round trip is 0.5 t CO_2e. This likely represents roughly half to one-fourth of the total transport emissions, as people tend to make more nonwork trips than commute trips—although we should account for distance traveled.

7.3 Transport and Land Use

Behind all the travel demand that we have been learning about, one of the main drivers is land use. The way that we occupy and use the land has a substantial impact on the way we travel. For instance, if we do not have any grocery stores within a distance we are willing to walk, we will have to drive, take transit, or bike to a grocery store. The way the land is being used has direct impact on how much transport we need to "consume." If everything is close, we will not need to consume much, but if everything is far, we will have to consume a lot.

There is an obvious relationship here with the concept of *accessibility*. As discussed in chapter 4, the focus in transport planning has shifted from *mobility* to *accessibility* to a great extent. Accessibility can be defined in various ways. Hansen (1959) was one of the first to define accessibility mathematically, and he defined it as follows:

$$A_{im} = \sum_j E_j \cdot f(c_{ijm}) \qquad (41)$$

where A_{im} is the accessibility of zone i with mode m, E_j are the opportunities at point j, and $f(c_{ijm})$ represents a general "cost" c for someone in i to reach j with mode m. First, we see that accessibility here is defined for a zone (as opposed to a point). Second, opportunities can include anything that is relevant to what is being studied—sometimes opportunities only include employment (i.e., access to jobs), and sometimes they include hospitals, schools, shops, restaurants, and so on.

Finally, the cost function $f(c_{ijm})$ can take any representation and include any costs. Most costs c include travel time by mode as well as some monetary costs, such as fuel costs, tolls, and transit fare. The simplest form of cost may be distance (although it is not sensitive to travel mode). Costs, however, are rarely assumed to have a linear impact on accessibility. In other words, a destination that takes double the time to reach does not have half the accessibility. Common options include the use of one over the cost c to some power b—that is, c_{ij}^{-b}—or the inverse of an exponential with power b—that is, $\exp(-bc_{ij})$.

Equation (41) is also called a gravity-based measure, and we will understand why in the next section when we discuss trip distribution. For now, this is all we will see on accessibility, but I recommend the work of Professor David Levinson, who is a leading scholar on accessibility. Several of his books discuss accessibility directly, including *Planning for Place and Plexus: Metropolitan Land Use and Transport* (Levinson and Krizek 2018).

There is also an obvious paradox in transport and land use that we quickly mentioned in chapter 2 but that is worth repeating. When we complain about congestion, we often argue that we need more roads and parking spots. But roads and parking spots take space, and if too many roads are built, there will not be enough space left for the places we want to visit in the first place. Although this may sound ridiculous, it actually happened. Prior to the 1960s and 1970s, many, if not most, activities were located in city centers. With suburbanization came an increase in car use, and severe congestion quickly followed. To remediate that, many roads and highways were built and parking lots constructed, which left very little space for stores and other activities, and many downtowns "died" all around the United States.[22]

So what is the solution? What if the same space could transport more people? This is where the space that a car takes matters compared to transit or active transport. Figure 7.16 illustrates this point by showing the space taken by people on one street segment if they all ride a bus, ride bicycles, or use single-occupied cars. Many variations of this figure can be found around the world.

The other solution for people to take less space is for them to live closer to activity centers. What if we could all walk to our destination? For this, the population density needs to be high, and for the population density to be high, we need to live in smaller houses or taller buildings. This type of land use reflects much more the traditional European and Asian cities and lifestyles. The other benefit to an increase in population density is the fact that distances traveled are shorter, and thus transport energy use is

Example 7.5

Employment Accessibility in Civitas

Using the zonal employment numbers in Civitas from table 7.9, (1) calculate the employment accessibility of each zone in Civitas based on Hansen's definition of accessibility. Although travel time by mode is preferred, simply use the distance values from table 7.10, ignore travel mode, and take one over distance squared as the cost function. (2) Discuss the results.

Solution

(1) Using employment numbers from table 7.9, distances from table 7.10, and one over the travel distance squared as the cost function, the employment accessibility for zone 1 is

$$A_1 = \frac{18833}{5.1^2} + \frac{6579}{7.8^2} + \frac{5093}{8.2^2} + \frac{5259}{7.2^2} + \frac{4236}{9.7^2} = 1054$$

Following the same procedure for all zones, we get:

Zone	Employment	Accessibility
1	18,833	1,054
2	6,579	596
3	5,093	618
4	5,259	650
5	4,236	431

(2) From the results shown in (1), we can see that zone 1 possesses the highest employment accessibility, which is not surprising considering it is in the Civitas Central Business District. Zones 2, 3, and 4 have relatively similar employment accessibility values, and zone 5 is the least accessible zone in Civitas in terms of employment.

much smaller. In a seminal work from 1989, two Australian researchers, Peter Newman and Jeffrey Kenworthy, found that cities with higher population densities consumed a lot less energy from transport (Newman and Kenworthy 1989). Figure 7.17 shows per capita transport energy use versus population density in 1995, and we can clearly see that the relationship follows a power law. Their initial findings had a tremendous impact in the transport world—notably, pushing for a better understanding of the relationship between transport and land use. At the time of this writing, their latest work was available in Newman and Kenworthy (2015). The full data is also available in the Millennium Cities Database (Kenworthy and Laube 2001).

Figure 7.16
Road space by mode. Courtesy of We Ride Australia (www.weride.org.au).

Much more work is needed, however, to better capture the impact of land use on transport, and future strategies to control demand for transport (our first sustainable principle) must include a strong land use component.

This is all that we will discuss on the topic of transport and land use. Although this section is short, the conceptual relationship between transport and land use is exceptionally strong and must be remembered and used.

Now, we will move on to the final section for this chapter. We will tie everything together in transport modeling.

7.4 Transport Modeling and the Four-Step Model

When trying to model and understand cities, transport modelers were ahead of the curve at the time of this writing. As we mentioned at the beginning of the chapter, transport is the one sector that has a direct impact on the behavior of people, and modeling the impact of transport on people is therefore critical. In fact, the most famous quote in transport (cited at the beginning of the chapter) has actually nothing to do with transport itself but focuses on modeling instead.

Transport models have gotten very sophisticated by trying to simulate the daily activities of everyone and by trying to predict the impact of land use on travel demand.

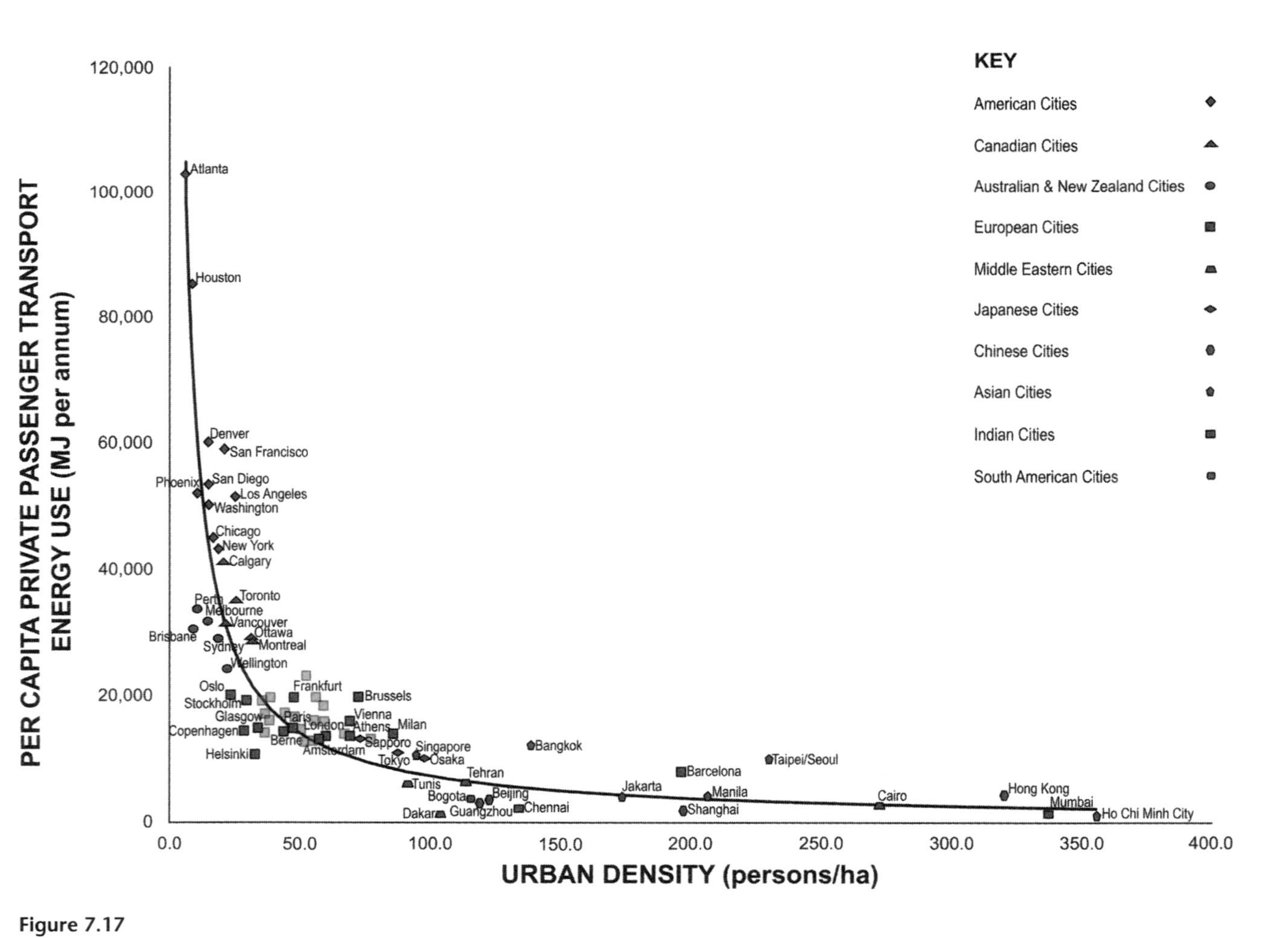

Figure 7.17

Population density and transport energy consumption. Courtesy of Peter Newman and Jeffrey Kenworthy.

For our purpose, we will content ourselves with a model first formulated in the 1950s and that is still in use. It is called the *urban transportation modeling system (UTMS)*. It is composed of four steps, and most people actually call it the Four-Step model, which we will also do. The four steps are as follows:

1. Trip generation
2. Trip distribution
3. Mode split
4. Assignment

The role of each step is illustrated in figure 7.18. Trip generation consists of determining how many trips are made in a city. Put simply, it is to determine the total number of trips originating and ending in each zone—that is, the row and column sums of the OD matrix. Based on these trip-end values, trip distribution consists of determining the flow of trips between each zone—that is, the value of each cell in the OD matrix. Mode split consists of estimating the mode that will be used for the trip, and it requires a fair amount of modeling to take into account travel time and cost. Finally, assignment consists of estimating the route that will be taken during the trip. Notably, the assignment step determines the level of congestion experienced. This congestion has an impact on travel time, and thus on mode split, and a loop feeds the information back to the mode choice model from step three until an equilibrium is found.

In this section, we will briefly introduce each step by outlining the main hypothesis and models used. For a thorough, yet simple, description of the four-step model, Meyer and Miller (2001) is recommended.

7.4.1 Trip Generation

The first component of the four-step model is to determine the number of trips that are produced by each zone and that end in each zone. We do not know yet where people go—that is, we do not know the value of each cell in the OD matrix—but we find the totals.

The easiest way to understand trip generation is to focus on the work trip. Trip generation consists of finding or estimating the residential population in each zone and the number of jobs in each zone. Table 7.11 shows the trip generation table for Civitas.

This type of information is most often gathered from the census for residential population. For employment, a survey must be made. In the United States, the Longitudinal Employer-Household Dynamics (LEHD) survey made by the U.S. Census Bureau is commonly used.[23] Naturally, how the distribution of population, employment, and other land use characteristics evolves depends on transport as well. When long-term

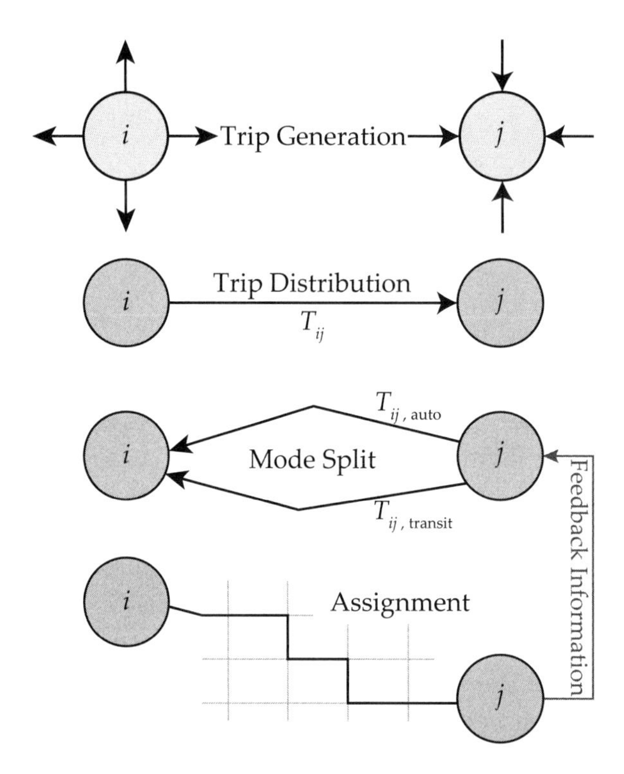

Figure 7.18

Schematic of the four-step model. Adapted from Meyer and Miller (2001).

Table 7.11

Trip generation for Civitas

Zones	Origins	Destinations
1	16,237	18,833
2	8,391	6,579
3	4,526	5,093
4	4,843	5,259
5	6,003	4,236
Total	40,000	40,000

travel forecasts are made (e.g., in twenty years), land use changes need to be forecasted as well. Like mode choice and trip assignment, this procedure is iterative.

For nonwork trips, regression models are conventionally used to estimate the trip rates per zones. Such variables as household income, auto ownership rate, and population density are used to predict the number of trips produced by zone, and variables such as zonal floor space and accessibility are used to predict the number of trips attracted to a zone. The Institute of Transportation Engineers (ITE) has an entire table filled with coefficients that can be used to estimate trip rates.

The next step is to fill in the blank cells in the OD matrix, and this is what we do in the trip distribution step.

7.4.2 Trip Distribution

In the trip distribution step, the main goal is to estimate where people actually go during their trip. Relating it back to the OD matrix (section 7.2.5), it is to estimate all the possible T_{ij}. For work trips, we may know or have an estimate of the actual workplace of every individual from the LEHD survey or from tax authorities, but for nonwork trips, we do not. We therefore need to find a clever way to estimate the trip flows between each zone.

Although various methods exist, the most popular by far is called the *gravity model*. It is called that because it greatly resembles Newton's law of universal gravitation. In its simplest form, it is defined as:

$$T_{ij} = K O_i D_j f_{ij} \tag{42}$$

where K is a normalizing factor that we will discuss below, O_i is the sum of all trips originating from zone i, D_j is the sum of all trips ending in zone j, and f_{ij} is a "friction" factor that encourages or discourages trip flows between two zones.

The friction factor is similar to the cost function from Hansen's accessibility equation (equation [41])—this is also why Hansen's accessibility is a gravity-based measure. As discussed with equation (41), many different types of functions and measures can be used for the cost function and for f_{ij} here. The travel time matrix is often used—whether we assume a negative power law or an exponential decay—and is fed in a generalized cost function that also includes monetary cost. For simplicity, here we will use one over the distance squared—that is, d_{ij}^{-2}—similar to example 7.5.

To find the value of K, we have several methods. First, we can simply ensure that the total number of trips T remains correct, but we let the sums of origins O_i and the sums of destinations D_j by zone that we found in step one fluctuate. In this case, we put equation (42) in (40), and we get a unique value of K for the entire matrix:

Table 7.12
Application of unconstrained gravity model in Civitas

O\D	1	2	3	4	5	Total
1	15,341	2,291	1,605	2,149	954	22,340
2	3,389	1,390	1,018	270	455	6,522
3	1,654	710	758	401	129	3,652
4	2,296	195	416	955	247	4,109
5	1,568	505	206	381	718	3,378
Total	24,248	5,091	4,003	4,156	2,503	40,001

$$K = \frac{T}{\sum_{ij} O_i D_j f_{ij}} \tag{43}$$

Based on the O_i and D_j from the trip generation matrix, we first calculate K and use it in equation (42). By applying equation (42) to Civitas and using d_{ij}^{-2} from table 7.10 for f_{ij}, we get $K=0.0013$, and by applying equation (42), we get the results shown in table 7.12. From the table, we can see that the results are not totally off. As modeled, the total number of trips T is still 40,000, but the O_i and D_j, however, have changed.

To remediate this problem, we need to set additional constraints. One way is to set constraints only on O_i or D_j. For instance, we can ensure that all the sums of origins remain intact by putting equation (42) in equation (38), which gives us a K_i for each zone of origin:

$$K_i = \frac{1}{\sum_j D_j f_{ij}} \tag{44}$$

Similarly, we can do the sum for all the destinations and get unique values for K_j:

$$K_j = \frac{1}{\sum_i O_j f_{ij}} \tag{45}$$

Finally, we can set multiple constraints so that both the origins and destinations remain correct, as well as the total number of trips. This requires a slightly modified version of the gravity model for which we now have two Ks:

$$T_{ij} = K_i K_j O_i D_j f_{ij} \tag{46}$$

If we use equations (38) and (39) again, then we get two equations with the unknown K_i in the denominator for K_j, and vice versa. We therefore cannot solve the Ks analytically in this case, and we will have to use an iterative method. Several methods exist, and the most famous uses the concept of statistical entropy initially developed in statistical mechanics and then famously used in information theory by Shannon (1948). Although we will not go over the theory—except a little bit in chapter 11—the general formulation is:

$$H = -\sum_i \sum_j p_{ij} \log p_{ij} \text{ with } \sum_i \sum_j p_{ij} = 1 \tag{47}$$

where p_{ij} are probabilities. In our case, p_{ij} are simply the proportions of trips compared to the total T.

7.4.3 Mode Split

The third step of the model is mode split. The goal is to come up with a way to determine which mode will be used on a trip. The difference here is that we do not look at zones anymore but at individual trips—that is, individual people. We therefore need to formulate a mode choice model that outputs the probability of using a mode over another. For instance, what is the probability of using a private vehicle over transit, walking, or biking?

For this step we introduce a famous econometric theory called *expected utility theory*. As it reads, expected utility theory measures the utility that one has for a given mode, and the goal is to select the mode that maximizes utility—hence the name *utility maximization theory*. Based on certain variables that we have collected, we define the utility U of a person i for mode m as

$$U_{im} = \beta_1 X_{1m} + \beta_2 X_{2m} + \cdots + \beta_n X_{nm} = \sum_i \beta_i X_{im} \tag{48}$$

where the β are parameters, and the X are independent variables. These variables can have attributes of the zone, such as population density, but they will tend to have mostly attributes of the household or individual, such as household income, size of household, number of cars, age, gender, distance to city center, and so on.

The parameters β will express the influence of each variable on the final utility, essentially acting as weights. To find their value, we need to "calibrate" the model. When we deal with a simple linear equation and when we can linearize a relationship,[24] we can simply use the ordinary least squares (OLS) method. Even if you have not heard the name, you have used OLS many times already, including in chapter 2. Here, however, we are dealing with a more complicated situation. Instead of the OLS method, we tend to use a method called maximum likelihood that we will not cover here. If you want to know more, Train (2003) explains it in a simple way.

Once calibrated, we can calculate the utility of all the different modes and calculate the probability P_{im} that an individual i using mode m has:

$$P_{im} = \frac{U_{im}}{\sum_m U_{im}} \tag{49}$$

The problem with this formulation is that we assume we know exactly the utility of an individual and that we have all the right variables. Put differently, we assume we know all the preferences and behaviors of everyone perfectly. This is obviously not true, and we therefore need to add an error term in equation (48) to account for unobserved variables, giving us:

$$U_{im} = \sum_i \beta_i X_{im} + \varepsilon_{im} = V_{im} + \varepsilon_{im} \tag{50}$$

where ε is some error. We have also symbolized all the observed variables as V_{im}. If we compare the probability of two modes 1 and 2, the probability of using mode 1 over mode 2 is:

$$P_{i1} = P\ (U_{i1} \geq U_{i2}) \tag{51}$$

If we substitute equation (50) in (51), we get:

$$P_{i1} = P(V_{i1} + \varepsilon_{i1} \geq V_{i2} + \varepsilon_{i2})$$

$$P_{i1} = P\ (V_{i1} - V_{i2} \geq \varepsilon_{i2} - \varepsilon_{i1}) \tag{52}$$

We can see that the probability of using mode 1 over mode 2 depends greatly on the error terms. By making some assumptions about the form of these error terms, we can derive an equation. Although we are tempted to assume that the errors are normally distributed—that is, we assume they follow a normal distribution—the resulting model is a little cumbersome to manipulate. Instead, if we assume a Gumbel type I distribution, we end up with the simple and elegant *logit* model (McFadden 1974):

$$P_{im} = \frac{e^{V_{im}}}{\sum_m e^{V_{im}}} \tag{53}$$

Basically, by simply taking the exponential of the V_{im} as opposed to the V_{im} themselves, we end up with a more sophisticated model that accounts for unobserved variables—this process of adding error terms is called *stochastic modeling*. Needless to say, the logit model has become extremely popular in economics and travel demand modeling. In fact, the person who first came up with the logit model, Daniel McFadden, received a Nobel Prize in Economics for his work.

Moreover, in large-scale modeling, for each individual a random number between 0 and 1 is drawn, and depending on the value, a mode is assigned to the individual. For example, say the probabilities are split as private vehicle, 30%; transit, 50%; and walk, 20%. If the random number generated is between 0 and 0.3, the private vehicle will be selected, if it is between 0.3 and 0.8, the transit will be selected, and if it is between 0.8 and 1, the walk mode will be selected. This process ensures that the mode with the highest utility has a higher chance of being selected, but it is not necessarily selected.

Furthermore, when calibrating the parameters β, the sign of the parameter—whether it is positive or negative—is important and informs us about whether the variable has a positive impact on the utility or a negative impact. Travel time, for example, has a negative impact, and therefore it has a negative sign, which means that a higher travel time translates into a lower utility.

Finally, the natural logarithm of the sum of the individual utilities in the denominator has been adopted as an indicator of accessibility in transport modeling. The larger the denominator is, the greater the accessibility of an individual. For instance, someone may choose to walk as opposed to drive, but if her utility for driving is high as well, she may have a higher "accessibility" than someone who has a small utility for both driving and walking. The accessibility A of individual i is defined as:

$$A_i = \ln\left(\sum_m e^{V_{im}}\right) \tag{54}$$

Equation (54) is so popular as an indicator of accessibility that we often simply call it the *logsum*.

We will not dive any further into the mode split step, but again, Train (2003) is a great resource to learn more about discrete choice modeling.

7.4.4 Assignment

The final step is assignment. Now that we know where people go and which mode they use, we need to determine which route they will actually select. Here, we essentially apply shortest-path and shortest-time algorithms.

To calculate a shortest-path, Dijkstra's and Floyd-Warshall's algorithms are often used. Although they are simple and they work well, they do not account for traffic congestion—that is, they assume unlimited capacity and therefore constant travel time on individual roads.

To be more accurate, when calculating traffic flow and using a simulation software package, other types of algorithms that account for congestion are preferred. There are many methods that are still relevant, including a fairly old method, initially developed

Example 7.6

Mode Split in Civitas

The observed utility equations below were calibrated for private vehicle (veh), transit (tr), and walking (wk) in Civitas, where t stands for travel time and c for cost (noting that the cost of walking is zero dollars).

- $V_{i,veh} = -0.1 \cdot t_{veh} - 0.2 \cdot c_{veh} + 0.5$
- $V_{i,tr} = -0.05 \cdot t_{tr} - 0.1 \cdot c_{tr}$
- $V_{i,wk} = -0.03 \cdot t_{wk}$

(1) Based on the values for each individual in the table below and using a logit model, determine the probability of selecting each mode for each individual. (2) Calculate the utility-based accessibility for all individuals and comment on the results.

Individual	t_{veh}	c_{veh}	t_{tr}	c_{tr}	t_{wk}
1	15	2	25	1	50
2	10	2	12	1	20
3	8	3	20	1	10
4	25	5	31	1	55
5	48	7	44	1	80
6	35	6	35	1	65

Solution

(1) We first calculate the utility for each mode using the calibrated utility equations given, and we then apply the logit model defined in equation (53). The results are shown in the table below. For instance, we can see that individual 1 has a fairly equal probability of choosing any mode, although private vehicle has a higher probability. Individuals 2 and 3 are more likely to walk considering the low travel time and the zero cost. Individual 4 is as likely to take transit as to walk. Despite the high walking time, the model tells us that individual 5 is more likely to walk than to drive because of the high cost of driving. Finally, individual 6 is more likely to take transit.

(2) For each individual, we calculate $\ln[\exp(V_{veh}) + \exp(V_{tr}) + \exp(V_{wk})]$. The results are shown in the table below. We can see that individual 3 has the highest accessibility, which is understandable since travel times and costs remain low for the three alternatives. Individual 2 follows with similar conditions. Individuals 5 and 6, however, have significantly low accessibilities since travel times and costs of all alternatives are high.

Individual	V_{veh}	P_{veh}	V_{tr}	P_{tr}	V_{wk}	P_{wk}	Accessibility
1	−1.4	0.34	−1.35	0.36	−1.5	0.31	−0.316
2	−0.9	0.28	−0.7	0.34	−0.6	0.38	0.373
3	−0.9	0.27	−1.1	0.22	−0.3	0.50	0.392
4	−3	0.11	−1.65	0.44	−1.65	0.44	−0.835
5	−5.7	0.02	−2.3	0.52	−2.4	0.47	−1.638
6	−4.2	0.05	−1.85	0.50	−1.95	0.45	−1.157

in the 1960s by the Bureau of Public Roads (1964),[25] that is often called the BPR method. Mathematically, the method defines the travel time t_i on a specific link (i.e., road segment) as:

$$t_i = t_{i,f}\left[1 + a\left(\frac{d_i}{c_i}\right)^b\right] \quad (55)$$

where $t_{i,f}$ is the free-flow travel time in [h], d_i is the demand for link i (i.e., traffic volume) in [veh/h], c_i is the capacity of link i in [veh/h], and a and b are constants. Essentially, equation (55) consists of the free-flow travel time plus an extra travel time related to the demand (i.e., volume) for the link. Taken from the *Highway Capacity Manual* (Transportation Research Board 2010), table 7.13 shows recommended values of a and b for a freeway and a multilane highway.

The goal then is to assign the trips so that the travel times are not unreasonably high. If we try to assign the trips manually, initially, we have no idea about link volumes; all we know is the free-flow travel time. We would therefore assign many trips to the same links, giving us an estimate of link volume, which then allows us to calculate a link travel time. We would then balance these trips and distribute them over the network to lower travel times in general and reiterate the process several times. Commonly, we try to achieve two principles developed in the 1950s, called Wardrop's principles (because they were developed by John Glen Wardrop, an English mathematician). The first principle is user equilibrium, and it dictates that at equilibrium no one can reduce his or her travel time by taking a different route. The second principle is system optimum, and it states that the total network travel time is minimized. The two principles therefore look at two different aspects of the system (individual versus

Table 7.13
Recommended BPR parameters for freeways and multilane highways

Road type	Free-flow speed (km/h)	Speed at capacity (km/h)	a	b
Freeway	120	86	0.39	6.3
	112	85	0.32	7
	104	83	0.25	9
	96	82	0.18	8.5
	88	80	0.1	10
Multilane highway	96	88	0.09	6
	88	82	0.08	6
	80	75	0.07	6
	72	67	0.07	6

the system as a whole). Several optimization algorithms can be used to assign trips in optimal ways. A typical method is the Frank–Wolf algorithm, but again, many methods exist. For simple and illustrative examples of user equilibrium methods, see Mannering (2013, chap. 8).

To improve on the BPR method and to more accurately simulate traffic, many software packages use dynamic traffic assignment (DTA) techniques. DTA tends to model individual cars that follow primarily three rules: car following, lane changing, and gap acceptance. DTA also considers traffic to be dynamic, and therefore a road may be congested only between 8:00 a.m. and 9:00 a.m., unlike static models such as BPR.

Once we have achieved everything and assigned all trips in an optimal way, we typically plug the values back in step three on mode split and recalculate the mode probabilities since some people might prefer to switch to a different mode now. We then go through this feedback process between steps three and four several times until the values do not change significantly.

Importantly for us, when carrying out a carbon inventory of an entire city, it is often the results of this last step that we use as distance traveled, to which we can then tag emission factors. Naturally, we have different types of vehicles with different emission factors, and these features need to be taken into account.

To get a general flavor for transport assignment, we quickly recall the procedure used in Dijkstra's algorithm in example 7.7. We can note that even this simple example takes some time, and we are not accounting for traffic volumes, which is why we need powerful computers to be able to assign all trips in an optimal way for a city.

As hinted at the beginning of the section, the transport modeling realm is vast, and the models developed have become quite sophisticated. At the time of this writing, the latest generations of models were called *integrated transport and land use models*. Not only did they attempt to capture land use effects—that is, by predicting when people move houses and where new land development occurs—they also tracked all of our activities during the day.

The next step for transport models is to integrate the consumption of nontransport activities like electricity, water, and gas, leading toward the simulation of entire cities. These models are unlikely to use a four-step model structure, however. In fact, they are even unlikely to use the simple logit model. Instead, the next generation transport models is bound to integrate elements of artificial intelligence and Machine Learning, carrying names like gradient-boosting machine, support vector machine, and neural networks. The calibration process may be faster or slower, but the accuracy should improve. These techniques are extremely powerful, and they will likely help to redefine

Example 7.7

Transport Assignment

The road system below possesses six intersections and eight roads. The numbers next to the roads are road lengths, but these could be any kind of costs in terms of distance, time, or money, for example. (1) Based on the information given, determine the shortest path from point A to point F.

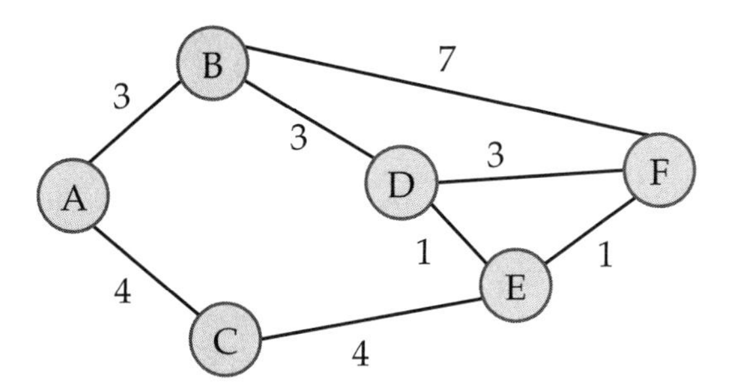

Solution

(1) Our goal is to find the quickest path to go from point A to F. Starting with point A, we look at the two adjacent nodes, B and C, and note their length: A–B = 3 km and A–C = 4 km. Since A–B is shorter, we select B and move to its adjacent nodes.

Now, we have A–B–F = 10 km and A–B–D = 6 km, both of which are longer than the original A–C. We therefore go back to C and look at its adjacent nodes, and we have A–C–E = 8 km.

Out of the three paths we have seen, A–B–D is shortest, and we therefore select D and move on to its adjacent nodes.

We now have A–B–D–F = 9 km and A–B–D–E = 7 km. So far, A–B–D–E is still the shortest, so we carry on to F and find that A–B–D–E–F = 8 km.

We had one more path with 8 km before, the path A–C–E. We check whether it is longer once we get to F, and A–C–E–F = 9 km. Thus, it is longer.

Although it goes through more intersections, the shortest path from A to F is therefore A–B–D–E–F. The procedure is illustrated in in the figure given.

Although this example is simple, it gives us an idea of the task required to solve large problems, especially when we take congestion into account, and we need to solve many simultaneous equations.

(continued)

Example 7.7 (continued)

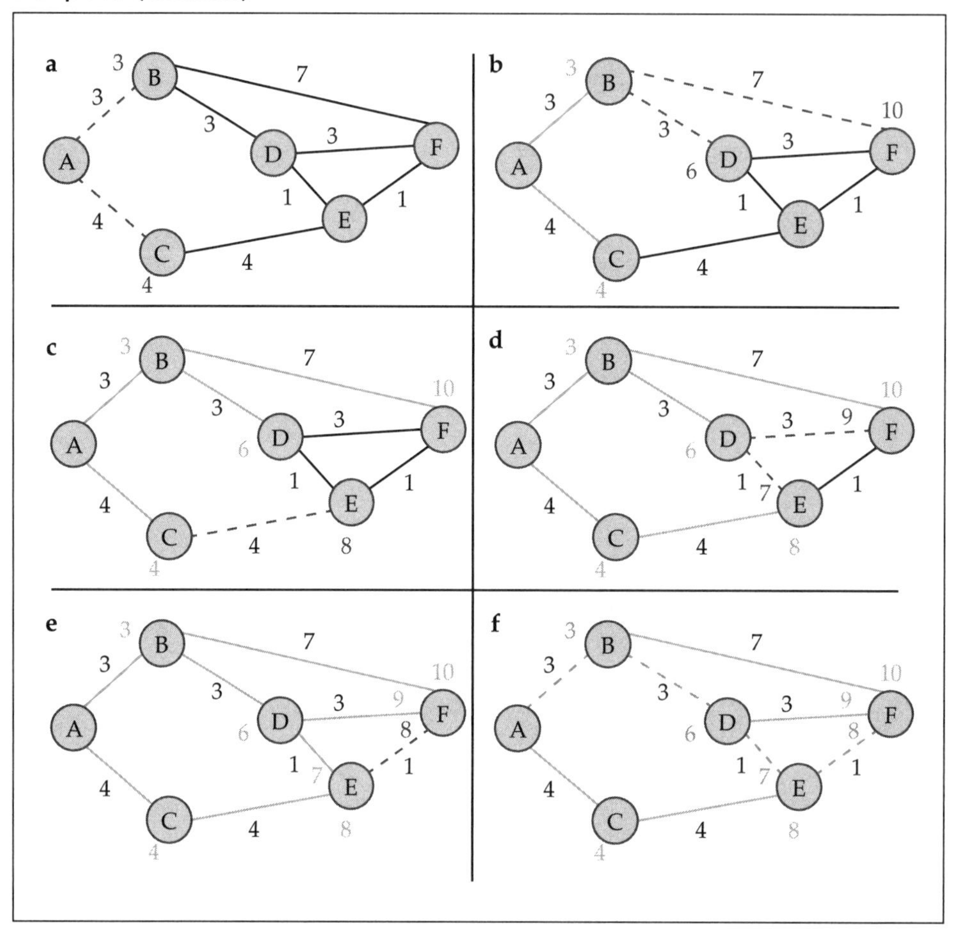

much of the modeling involved in urban engineering. We will actually get exposed to some of these techniques in chapter 11.

7.5 Conclusion

This concludes our chapter on transport. After setting the context for the chapter and looking at the importance of transport emissions in several cities in the world, we first learned several important fundamentals in transport. Starting with traffic flow theory, we learned the relationship between flow, density, and speed. Notably, we established the fundamental flow diagram and saw a way to determine the level of service of every road depending on traffic conditions. Using a similar methodology, we looked at pedestrian flow (or, more accurately, flux), and we studied how to design streets and sidewalks to avoid a high probability of conflict. Subsequently, we dwelled on public transport planning, learning about the various transit modes, their right-of-ways, and their varied characteristics. We learned about the trade-off that exists between interstation spacing and access time to a station.

Second, we defined what demand means in transport. We first looked at the concept of a trip and analyzed travel demand trends in the United States as a whole and throughout the Chicago region, specifically. We also looked at the temporal distribution of trips during a typical day. To these trips we then added distance, looking at PKT and VKT. While the number of trips in itself is not something we want to control (in fact, we encourage more trips), distance is crucial since congestion and emissions from transport get higher when trips are longer. We then specifically looked at mode split/share and the various alternatives that we can take when making a trip. In line with our goal, we defined typical emission factors for the various modes, and we saw that private vehicles consume more energy and are responsible for significantly more GHG emissions. Finally, we learned about OD matrices, which inform us about travel demand at a large scale.

Third, we briefly looked at the relationship between transport and land use, and we were introduced to one mathematical definition of accessibility—that is, Hansen's accessibility. We learned that transport and land use are intrinsically linked, that they can support each other and that they can harm each other. In particular, we discussed the paradox with our desire to have more roads, which can then simply remove our need to travel to a location. There is a vibrant research community around the topics of transport and land use, and although the section was a little short, it offers a good introduction.

Finally, we discussed the general field of transport modeling, and we focused on the conventional four-step model. Although it has been used since the 1950s, the four-step model is relatively complicated, and it is computationally intensive. In the four steps, trip generation informs us on the number of trips being produced by a zone and ending in a zone. Trip distribution links these trip ends together, and we modeled these flows using the gravity model. Mode split consists of determining the mode selected for a trip or, more accurately, the probability that a mode is selected, and we learned about the logit model. Finally, the assignment informed us on the individual routes taken by an individual during a trip, and we went through a simple example of Dijkstra's algorithm. Although this section offered only a quick introduction to the field of transport modeling, it gave us the necessary basics to help determine how transport functions in a city and how it affects our energy use and GHG emission levels.

Because it has a direct impact on lifestyle, transport is arguably the most challenging infrastructure system to tackle. As opposed to electricity and water, where changes do not necessarily affect the end consumer, transport systematically induces or encourages a change in behavior—for example, by switching modes.

As we learned, one solution to decrease our carbon footprint from transport is to simply live closer to one another, in smaller houses. In fact, this has many benefits, including less energy used for heating, cooling, and lighting, as we are about to learn in chapter 8.

Problem Set

7.1 In line with quantifying your personal carbon footprint, the goal of this question is to quantify your personal footprint from transport. (1) Using any online mapping platform, estimate your distance traveled for each mode during a typical day during the week and during the weekend. (2) Using the emission factors from table 7.8, calculate your GHG emissions from transport for a typical day during the week and during the weekend. (3) Multiplying your weekday value by 260 days and weekend day value by 105 days to count for a full year, estimate your annual carbon footprint from transport.

7.2 In your own words and based on the lessons of this chapter, discuss what the quote at the beginning of the chapter means: "All models are wrong, but some are useful."

7.3 In your personal opinion, describe what impact you think autonomous vehicles may have on the transport realm.

7.4 Discuss the assumptions made by the Greenshields model and illustrate the relationship between flow, speed, and density.

7.5 Select any traffic flow model other than the Greenshields model and discuss and illustrate it.

7.6 By searching the web, select a city and a street that has traffic-flow data available (in real time or past data), download the data, and determine its level of service at different times during the day.

7.7 While standing by a road, you notice that one car goes by every 3.8 s and that the cars are separated by about 70 m. Based on this information, determine (1) the flow, (2) the average speed, and (3) the density of the traffic on this road.

7.8 A highway has a measured capacity of 2,800 veh/h, the jam density is 110 veh/km, and the relationship between speed and density is expressed by the equation below.

Based on this information, determine (1) the average speed of the traffic at capacity (i.e., space-mean speed) and (2) the free-flow speed.

$$v = v_f\left[1-\left(\frac{k}{k_j}\right)^{2.8}\right]$$

7.9 From the q-v diagram, determine the LOS of the roads shown in the table below.

Case	Lanes	Speed (km/h)	Flow (veh/h)
1	1	48	196
2	1	92	852
3	2	73	1,430
4	2	102	812
5	3	86	3,222

7.10 A four-lane highway has the flow-density relationship below. Determine (1) the capacity of the highway and (2) the speed at capacity. Based on this information, determine (3) the LOS of the highway when it is at capacity.

$$q = 70k - \frac{k^2}{2}$$

7.11 A four-lane highway has a free-flow speed of 100 km/h and the flow-density relationship shown below. (1) Plot the relationship between flow and density. From this equation, determine (2) the flow capacity of the highway. Calculate (3) the jam density and (4) the speed at capacity and determine (5) the level of service when the highway is at capacity.

$$q = 100k - 0.5k^2$$

7.12 A two-lane road possesses the flow-density relationship shown below. Based on this information, determine (1) the jam density, (2) the capacity of the road, (3) the free-flow speed, (4) the speed at capacity, (5) the possible speeds when the flow is 3,000 veh/h, and (6) the LOS for the two possible speeds from (5).

$$q = 120k - 0.6k^2$$

7.13 In your own words, describe the difference between the flow-space relationship in pedestrian flow as shown in figure 7.6 with the flow-density relationship in traffic flow as shown in figure 7.3.

7.14 Organizers of an event are trying to determine how much time it would take to evacuate a room of 2,000 people with two 4-m-wide exits. Assuming an average walking speed of 5.5 km/h and a potential of conflict of 99%, determine (1) the flow of people per exit and (2) how much time it would take to evacuate the entire room.

7.15 To be certified to fly, airplanes have to be able to be evacuated within 90 s. The Airbus A380 has a capacity of 873 people. Assume that out of sixteen doors only eight can be opened and that only one person can go through the door at a time. Also, assuming an average walking speed of 5.5 km/h and a potential of conflict of 99%, calculate (1) the flow of people per exit and (2) how much time it would take to evacuate the entire airplane. During a real-life simulation,[26] it took 78 s for the entire plane to be evacuated. (3) Discuss why your model performed well or poorly.

7.16 A 40-m-long and 5-m-wide pedestrian bridge will be built. The objective is to keep the probability of conflict below 50% under maximum occupancy. Assuming an average walking speed of 4 km/h, determine (1) the time it takes to cross the bridge, (2) the maximum flow of people per unit width, (3) the maximum flow of people in one hour, and (4) the number of people allowed under maximum occupancy.

7.17 Search and report the official definition used by the APTA for the following transit modes: bus, trolleybus, LRT, heavy rail (i.e., metro), commuter rail, paratransit, and ferry boat.

7.18 Select a transit mode and discuss how it has evolved over time and in the world by conducting your own research.

7.19 From first principles, derive equation (27), which calculates the transit in-motion time when the cruising speed is not reached.

7.20 In your own words, describe the trade-off between access/egress time and in-motion time.

7.21 You are asked to determine the number of shuttles needed to carry 1,000 people from an event to their hotel located 5 km away. The operating speed of the shuttle is 30 km/h (accounting for intersections and all other types of delays), the loading and unloading times are estimated to take 5 min each, and each shuttle has a capacity of sixty people. Based on this information, determine (1) the two-way operating time of the shuttle and (2) the number of shuttles that will be needed if the entire operation is not supposed to take more than one and a half hours.

7.22 A new BRT line is being planned that will run for 8 km at a cruising speed of $V_c = 50$ km/h, with acceleration and deceleration rates of 1 m/s^2. The standing time is expected to be 20 s, and there are intersections at about every 200 m with traffic signal cycle lengths of $t_{traffic} = 1$ min. (1) Determine the one-way operating time T_o as a function of the number of stops (assume the BRT always reaches cruising speed) and calculate T_o using $n_s = 10$. (2) Assuming a walking speed of 5.5 km/h and assuming the average user accesses from and egresses to a distance of one-fourth of the interstation spacing *S*, express access and egress time as a function of the number of stops. (3) Based on the equation and keeping $L = 8$ km, find the optimal number of stations (*hint*: calculate the total travel time for $n_s = 8$, 9, 10, 11, and 12 and select the smallest). (4) Using the value of n_s found in (3), calculate the operating time and calculate the operating speed V_o. (5) Assuming terminal times of 15% of the operating time, calculate the cycle time *T* and the number of vehicles needed n_{tr} considering a headway *h* of 5 min and $w = 1$.

7.23 Assume an 18-km-long LRT line is currently being planned, and you are asked to find the optimal interstation spacing, *S*. The cruising speed of the LRT is 50 km/h, with acceleration and deceleration rates of 1 m/s^2. Expect standing time at each station to be 20 s. (1) Determine the one-way operating time T_o as a function of the number of stops (assume the tramway always reaches cruising speed) and compute the number of stations n_s assuming we are aiming for $T_o = 30$ min. (2) By looking at the route plans, you realize there are intersections at every 200 m with a traffic signal cycle length of 1 min. Come up with a new equation for T_o as a function and n_s and compute the new T_o considering the number of stations found in (1). From now on, consider the trip of an average user is 8 km (i.e., reset $L = 8$ km). (3) Assuming a walking speed of 5.5 km/h and assuming the average user accesses from and egresses to a distance of one-fourth of the interstation spacing *S*, express access and egress time as a function of the number of stops. (4) Assuming the trip of an average user is 8 km, plot the user travel time as a function of n_s, calculate the optimal number of stations (round up), and determine the interstation spacing *S* for this 8 km trip. (5) Using the value of *S* found in (4), determine the total number of stations and recalculate the operating time (i.e., reset $L = 18$ km) and the operating speed. (6) Assuming terminal times of 15% of the operating time, calculate the cycle time *T* and the number of vehicles needed n_{tr} considering a headway *h* of 10 min and $w = 2$.

7.24 In your own words, define the concept of *trip*, *trip chain*, and *tour*. Give five examples of *trip purpose*.

7.25 Break down one of your typical days in terms of trip and trip purpose and record the time each trip started and each trip ended.

7.26 By visiting the FHWA website dedicated to the NHTS available at http://nhts.ornl.gov/,[27] download the data and compile figures similar to those shown in figure 7.13.

7.27 By visiting the FHWA website dedicated to travel monitoring available at https://www.fhwa.dot.gov/policyinformation/travel_monitoring/tvt.cfm,[28] (1) update figure 7.14 in the chapter and (2) comment on the latest trend.

7.28 In the chapter, table 7.8 reports the emission factors for each travel mode commonly used in the United States. By searching the web, find similar factors for any country or city other than the United States.

7.29 The tables below show the OD matrix and the zone-to-zone distance data of a city. Based on this information, calculate (1) the total distance traveled, (2) the average trip length, and (3) the total GHG emissions, assuming an average emission factor of 127 g CO_2e.

Origin and destination matrix

O/D	1	2	3	Total
1	557	374	132	1,063
2	452	346	84	882
3	387	103	402	892
Total	1,396	823	618	2,837

Distances

Zones	1	2	3
1	1.23	3.42	4.01
2	3.42	1.87	4.49
3	4.01	4.49	2.12

7.30 The daily distance traveled per mode for a person are as follows: 5 km by car, 6 km by transit, and 2 km by walking. The GHG emission factors for the car and transit are 215 and 51 g CO_2e/km, respectively. Based on this information, calculate the GHG emissions (1) per day and (2) per year (assuming the same trips are repeated daily for a year). (3) Based on your knowledge of yearly GHG emissions in the United States, discuss the results.

7.31 You were asked to estimate the carbon footprint of a city of 50,000 people for a typical weekday. The city was divided into six zones; the OD matrix and zone-to-zone distance data are shown in the two tables below. For emission factors, use the values in table 7.8. The mode share of the city is as follows:

- Private vehicles: 84%, of which 60% are passenger cars, and 40% are light-duty vehicles. Assume an average occupancy of 1.1.
- Transit: 14%, of which 80% are bus riders, and 20% are tram riders.
- Active transport: 2%, of which 50% walk and 50% bike.

Compute (1) the total distance traveled between each zone and (2) the average trip distance. (3) Assuming the mode share values apply for every single trip and that distances calculated apply to all modes, calculate the total emission of every single mode as well as the total emissions from the residents. (4) Assuming the values apply for 365 days, compute the annual emissions from transport from the city residents and per city resident.

Origin and destination matrix

O/D	1	2	3	4	5	6	Total
1	4,808	10,625	2,110	2,463	6,234	4,897	31,137
2	10,625	1,897	5,375	1,574	2,178	5,492	27,141
3	2,110	5,375	7,176	7,439	9,762	2,185	34,047
4	2,463	1,574	7,439	3,034	10,556	10,674	35,740
5	6,234	2,178	9,762	10,556	9,877	9,670	48,277
6	4,897	5,492	2,185	10,674	9,670	2,314	35,232
Total	31,137	27,141	34,047	35,740	48,277	35,232	211,574

Distances (km)

Zones	1	2	3	4	5	6
1	1.50	6.62	3.27	1.72	9.05	9.41
2	6.62	2.25	8.13	3.53	5.80	7.64
3	3.27	8.13	1.01	7.16	1.80	6.29
4	1.72	3.53	7.16	0.99	8.69	4.82
5	9.05	5.80	1.80	8.69	1.57	10.51
6	9.41	7.64	6.29	4.82	10.51	1.75

7.32 In your own words, (1) explain the concept of *accessibility*, and (2) explain how Hansen's accessibility is defined and how it is able to capture accessibility.

7.33 For the city described in problem 7.31, calculate (1) the Hansen's accessibility of each zone using distance as a cost and the power function with exponent −2 as the cost function (i.e., t^{-2}).

7.34 The table below shows the travel time t data for a given mode for the same city as problem 7.29. Based on this information and the table given in problem 7.29, calculate Hansen's accessibility for each of the three zones assuming (1) a power cost function with coefficient −2 (i.e., t^{-2}) and (2) an exponential cost function with coefficient −0.3 (i.e., $\exp(-0.3t)$). (3) Compare the results.

Travel times (min)

Zones	1	2	3
1	12.10	32.70	39.20
2	32.70	18.60	45.30
3	39.20	45.30	22.40

7.35 The tables below show the travel time t data by mode for the same city as problems 7.29 and 7.34. (1) Based on this information and the table given in problem 7.29, calculate Hansen's accessibility for each mode and for each of the three zones and the total accessibility by summing the three accessibilities for each zone, assuming a power cost function with coefficient −2 (i.e., t^{-2}). (2) Discuss the results and argue which zone is the most accessible.

Private vehicle travel time (min)

Zones	1	2	3
1	17.50	22.70	29.40
2	22.70	16.50	31.40
3	29.40	31.40	18.90

Transit travel time (min)

Zones	1	2	3
1	13.80	23.40	29.40
2	23.40	22.70	46.50
3	29.40	46.50	27.40

Cycling travel time (min)

Zones	1	2	3
1	18.15	49.00	58.80
2	49.00	27.90	67.90
3	58.80	67.90	33.60

Walking travel time (min)

Zones	1	2	3
1	19.20	62.20	75.40
2	62.20	35.80	90.30
3	75.40	90.30	45.80

7.36 The observed utility equations below were calibrated for private vehicle (veh), transit (tr), and walk (wk) for an individual. Assuming $t_{veh}=15$, $c_{veh}=4$, $t_{tr}=25$, $c_{tr}=2$, and $t_{wk}=35$, calculate (1) the probability of using each mode and the (2) utility-based accessibility of the individual.

$$V_{i,veh}=-0.11\cdot t_{veh}-0.22\cdot c_{veh}+1.2$$

$$V_{i,tr}=-0.05\cdot t_{tr}-0.05\cdot c_{tr}$$

$$V_{i,wk}=-0.04\cdot t_{wk}$$

7.37 The observed utility equations below were calibrated for private vehicle (veh), transit (tr), and walk (wk) for a community.

$$V_{i,veh}=-0.09\cdot t_{veh}-0.14\cdot c_{veh}+0.66$$

$$V_{i,tr}=-0.08\cdot t_{tr}-0.09\cdot c_{tr}$$

$$V_{i,wk}=-0.019\cdot t_{wk}$$

(1) Based on the values for each individual in the table below and using a logit model, determine the probability of selecting each mode for each individual. (2) Calculate the utility-based accessibility for all individuals and comment.

Individual	t_{veh}	c_{veh}	t_{tr}	c_{tr}	t_{wk}
1	5	3	15	1	20
2	15	3	15	1	25
3	30	4	25	1	45
4	42	5	28	1	70
5	63	7	54	1	180
6	29	3	31	1	50

7.38 Using Dijkstra's algorithm, (1) find the shortest path(s) to go from A to F for the network shown below. (2) If multiple solutions exist, state to which one(s) you would assign the traffic in a modeling context.

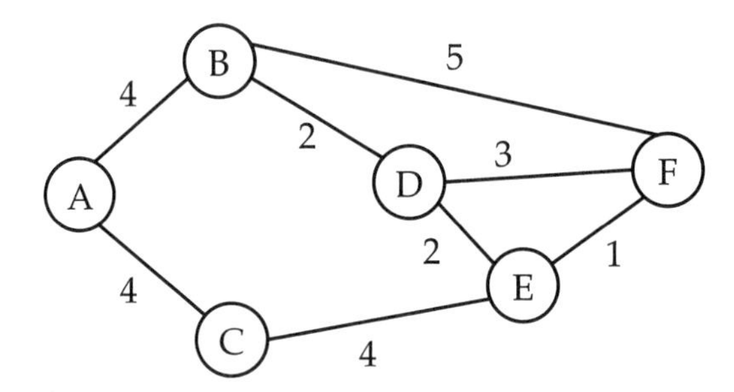

7.39 Using Dijkstra's algorithm, (1) find the shortest path(s) to go from A to L for the network shown below.

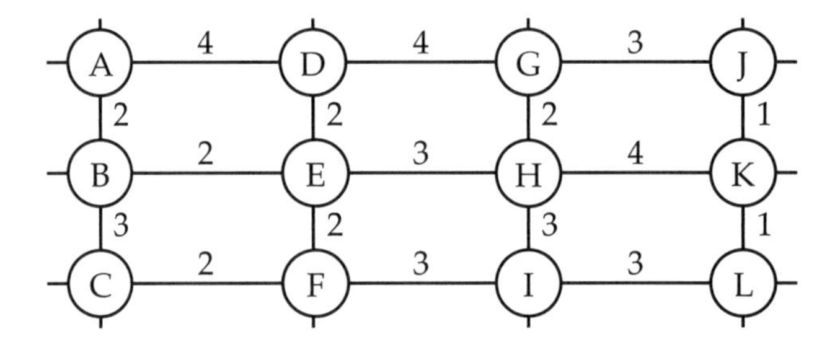

Notes

1. Professor David Levinson (whom we will see later) wrote a blog article about it in which he makes a pretty convincing case for the use of "transport." See https://transportist.org/2011/04/01/its_transportational/ (accessed August 16, 2018).

2. Many argue that congestion will get worse, and others argue that it will get better. While it is impossible to predict accurately what will happen, there is no doubt that the emergence of autonomous vehicles will have a disruptive impact on transport.

3. Available at https://nacto.org/ (accessed March 2, 2019).

4. In fact, did you notice that we use the symbol q for traffic flow, similar to Q for flow rate in water? Many of these concepts and laws have the same origins.

5. Here, we can make an obvious parallel between flow and the IDF curves from chapter 6. It could be interesting to determine whether IDF curves exist in traffic-flow theory—what is a 10-year traffic versus a 100-year traffic?

6. This is a little like area and velocity in closed conduits. When the diameter of a pipe decreases, the velocity of the flow increases, and vice versa because of continuity.

7. It is simply fascinating to see the technical overlap across all fields of engineering. Why did we separate the fields in the first place?

8. Another method assumes a truncated triangle (i.e., trapezoid) instead, which implies that maximum flows can be achieved at free flow or lower speeds.

9. On many roads you can see thick black lines that form somewhat of a rectangle; these are loop detectors that were installed after the road was built. When placed at intersections, they also help traffic-signal operation.

10. Available at https://mctrans.ce.ufl.edu/mct/ (accessed March 2, 2019).

11. Which technically is a *flux* rather than a *flow*, but we will not differentiate here.

12. But then, as a colleague of mine told me, the "machinery" of human walking has not evolved during this period, either.

13. In the low- and medium-income world, public transport often tends to be used more heavily than the private automobile.

14. To be precise, throughout the chapter, we are talking about "network" distance, following the streets, as opposed to the straight-line distance on a map (i.e., "as the crow flies").

15. Quite surprisingly, there are no formal definitions of trip chains and tours (McGuckin and Nakamoto 2004).

16. The raw data actually shows up to fifty trips per day, but the numbers after fifteen were negligible.

17. Noting that fuel prices also decreased.

18. The values were collected mostly in the 2010s. They were retrieved from https://en.wikipedia.org/wiki/Modal_share (accessed March 4, 2019), and they are continuously updated by Wikipedia contributors. The values currently shown are therefore likely different.

19. Nonmotorized, "human-powered" transport modes naturally also require energy, from food (in terms of calories or joules), but this is not the focus of this chapter.

20. Available at http://www.epa.gov/otaq/models/moves (accessed March 4, 2019).

21. Moreover, the EPA actually gives CO_2, CH_4, and N_2O emissions. Here the three GHGs were combined based on their global warming potentials given by the EPA in the same document. The values were retrieved from https://www.epa.gov/sites/production/files/2018-03/documents/emission-factors_mar_2018_0.pdf (accessed March 4, 2019).

22. Funnily, some of the activities that remain in city centers are banks, real estate, and insurance companies that themselves financed the flight to the suburbs.

23. Available at http://lehd.ces.census.gov/ (accessed August 17, 2018).

24. For instance, the power law can easily be linearized by taking the logarithm on both sides, as we saw briefly in chapter 3 and as we will see in chapter 11.

25. I have actually used this in my own research. In Wisetjindawat Wisinee et al. (2017), we modeled a massive earthquake and a typhoon to measure the vulnerability of the road system in the Nagoya region in Japan to these events.

26. The video of the real-life simulation is available at https://www.youtube.com/watch?v=XIaovi1JWyY (accessed March 4, 2019).

27. Accessed March 4, 2019.

28. Accessed March 4, 2019.

References

Box, George E. P. 1987. *Empirical Model-Building and Response Surfaces*. New York: Wiley.

Bureau of Public Roads. 1964. *Traffic Assignment Manual for Application with a Large, High Speed Computer*. Washington, DC: U.S. Dept. of Commerce, Bureau of Public Roads, Office of Planning, Urban Planning Division.

Federal Highway Administration. 2017. *2017 National Household Travel Survey*. Washington, DC: Federal Highway Administration, U.S. Department of Transportation. http://nhts.ornl.gov.

Fruin, J. J. 1971. *Pedestrian Planning and Design*. New York: Metropolitan Association of Urban Designers and Environmental Planners.

Geroliminis, Nikolas, and Carlos F. Daganzo. 2008. "Existence of Urban-Scale Macroscopic Fundamental Diagrams: Some Experimental Findings." *Transportation Research Part B: Methodological* 42(9): 759–770.

Greene, David L. 2004. "Transportation and Energy." In *The Geography of Urban Transportation*, 3rd ed., edited by Susan Hanson and Genevieve Giuliano, 274–293. New York: Guilford Press.

Greenshields, B. D., J. R. Bibbins, W. S. Channing, and H. H. Miller. 1935. "A Study of Traffic Capacity." *Highway Research Board Proceedings* 14: 448–477.

Hansen, Walter G. 1959. "How Accessibility Shapes Land Use." *Journal of the American Institute of Planners* 25(2): 73–76.

Hazelton, Martin L. 2004. "Estimating Vehicle Speed from Traffic Count and Occupancy Data." *Journal of Data Science* 2(3): 231–244.

Herrera, Juan C., Daniel B. Work, Ryan Herring, Xuegang Jeff Ban, Quinn Jacobson, and Alexandre M Bayen. 2010. "Evaluation of Traffic Data Obtained via GPS-Enabled Mobile Phones: The Mobile Century Field Experiment." *Transportation Research Part C: Emerging Technologies* 18(4): 568–583.

Kennedy, C., Ibrahim, N., and D. Hoornweg. 2014. "Low-Carbon Infrastructure Strategies for Cities." *Nature Climate Change* 4(5): 343–346.

Kenworthy, J. R., and F. Laube. 2001. *The Millennium Cities Database for Sustainable Transport.* Brussels: International Union of Public Transport. Murdoch, Australia: Institute for Sustainability and Technology Policy. CD-ROM.

Levinson, David M., and Kevin J. Krizek. 2018. *Planning for Place and Plexus: Metropolitan Land Use and Transport.* 2nd ed. New York: Routledge.

Mannering, Fred L. 2013. *Principles of Highway Engineering and Traffic Analysis.* 5th ed. Hoboken, NJ: Wiley.

McFadden, Daniel. 1974. "Conditional Logit Analysis of Qualitative Choice Behavior." In *Frontiers in Econometrics*, edited by P. Zarembka, 105–142. New York: Academic Press.

McGuckin, N., and A. Fucci. 2018. *Summary of Travel Trends: 2017 National Household Travel Survey.* Report Number: FHWA-PL-18-019. Washington, DC: Federal Highway Administration, U.S. Department of Transportation. https://nhts.ornl.gov/assets/2017_nhts_summary_travel_trends.pdf.

McGuckin, N., and Y. Nakamoto. 2004. "Trips, Chains, and Tours—Using an Operational Definition." Paper submitted for the National Houehold Travel Survey Conference, Novmber 1–2. http://onlinepubs.trb.org/onlinepubs/archive/conferences/nhts/McGuckin.pdf.

Meyer, Michael D., and Eric J. Miller. 2001. *Urban Transportation Planning: A Decision-Oriented Approach.* 2nd ed. Boston: McGraw-Hill.

Newman, Peter, and Jeffrey R. Kenworthy. 1989. *Cities and Automobile Dependence: A Sourcebook.* Brookfield, VT: Gower Technical.

———. 2015. *The End of Automobile Dependence: How Cities Are Moving beyond Car-Based Planning.* Washington, DC: Island Press.

Parry, Frank. 2010. *Chicago Regional Household Travel Inventory: Mode Choice and Trip Purpose for the 2008 and 1990 Surveys.* Chicago: Chicago Metropolitan Agency for Planning.

Pushkarev, B., and J. Zupan. 1975. *Urban Space for Pedestrians.* Cambridge, MA: MIT Press.

Schipper, Lee, and Céline Marie-Lilliu. 1999. "Transportation and CO_2 Emissions: Flexing the Link—a Path for the World Bank. World Bank Environmentally and Socially Sustainable Development." Washington, DC: World Bank. http://documents.worldbank.org/curated/en/826921468766156728/pdf/multi-page.pdf.

Shannon, C. E. 1948. "A Mathematical Theory of Communication." *Bell System Technical Journal* 27: 379–423, 623–656.

Train, Kenneth. 2003. *Discrete Choice Methods with Simulation.* New York: Cambridge University Press.

Transportation Research Board. 2010. *Highway Capacity Manual 2010*. Washington, DC: National Research Council. http://www.hcm2010.org/.

Vuchic, Vukan R. 2005. *Urban Transit: Operations, Planning, and Economics*. Hoboken, NJ: John Wiley & Sons.

———. 2007. *Urban Transit Systems and Technology*. Hoboken, NJ: John Wiley & Sons.

Wisetjindawat, Wisinee, Kermanshah, Amirhassan, Derrible, Sybil, and Motohiro Fujita. 2017. "Stochastic Modeling of Road System Performance during Multihazard Events: Flash Floods and Earthquakes." *Journal of Infrastructure Systems* 23(4): 04017031. https://doi.org/10.1061/(ASCE)IS.1943-555X.0000391.

8 Buildings

> Obviously one cannot afford to replace the building stock very often. ... This implies an awesome responsibility for city planners, architects, and engineers: *Do it well or the mistakes will haunt us for a long time.*
>
> —Kreider, Curtiss, and Rabl 2009, p. 2

To this point we have not directly focused on buildings, yet most of the energy consumption and greenhouse gas (GHG) emissions that we have discussed occurs in buildings. After all, people live in buildings, and this is where we consume electricity, natural gas, water, and so on, and this is where we drive/ride/walk/cycle to every day. Moreover, buildings are the places where we work, shop, and carry out many of our activities. In fact, on average, we spend almost 90% of our time inside buildings (Klepeis et al. 2001). All of these things combined actually sum to a significant amount of energy. Figure 8.1 shows the total energy use by sector in the United States in 2017. The data comes from the U.S. Energy Information Administration (EIA), which is a great resource for virtually any type of data on energy.

We can see that 29% of the energy we consume is in transport; 20% and 19% is used by the residential and commercial sectors, respectively (most of which occurs in buildings); and finally, 32% is used by the industrial sector (a certain portion of which is also used in buildings). Note that we are focusing on energy use here as opposed to GHG emissions. As we saw in figure 5.1, transport was responsible for 29% of the U.S. GHG emissions, and the similarity between the energy use and GHG emissions results is interesting and telling about the way energy is consumed—after all, motor engines, natural gas power plants, and coal-fired power plants have relatively the same efficiencies (between 30% and 40%).

Back to the buildings, figure 8.1 tells us that not accounting for industrial activities, about 40% of the energy use in the Unites States is consumed in buildings. At the

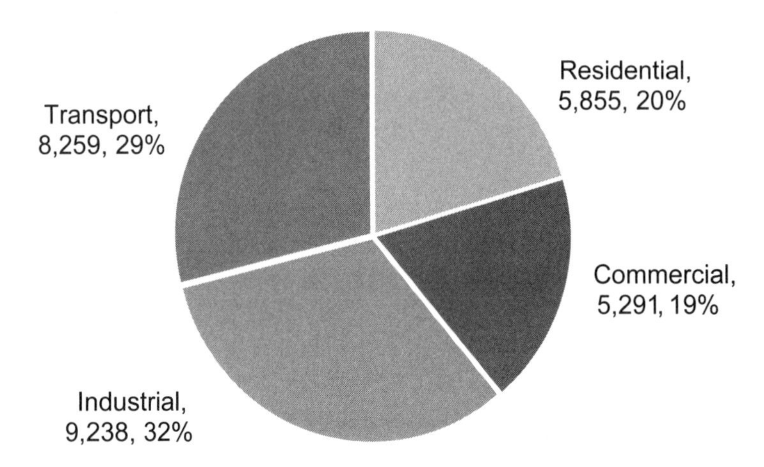

Figure 8.1
2017 U.S. total energy use in [TWh] by sector.
Source: U.S. Energy Information Administration.

international scale, despite all energy consumed in transport, agriculture, industrial processes, and all other sectors, buildings tend to account for 32% of the total energy used (Lucon et al. 2014).

In addition, although we have not directly discussed it, we have already looked at some of the energy consumption from buildings—namely, when we learned about electricity and water. Naturally, buildings consume more energy than simply in electricity and water. Figures 8.2a and 8.2b better inform us on how energy is consumed in residential and commercial buildings, respectively. The energy in residential buildings is divided into five types of uses, and the energy in commercial buildings is divided into ten types of uses. The percentage at the top of the bars shows the proportion of the total energy that is being used for space conditioning (i.e., space heating and cooling). This data comes from the EIA as well, specifically from the 2015 Residential Energy Consumption Survey (RECS) for residential buildings and from the 2012 Commercial Buildings Energy Consumption Survey (CBECS) for commercial buildings. Both RECS and CBECS contain tons of valuable and insightful information.

To be clear, in figure 8.2 we are seeing *site* energy use values as opposed to *source* energy use values. Site energy use is the energy used in the house, and it does not account for the energy used to generate electricity, for example. In contrast, source energy use includes the total energy used in the entire process. Because source energy use relies heavily on nonbuilding characteristics, we will focus specifically here on site energy use (that we will simply call energy use).

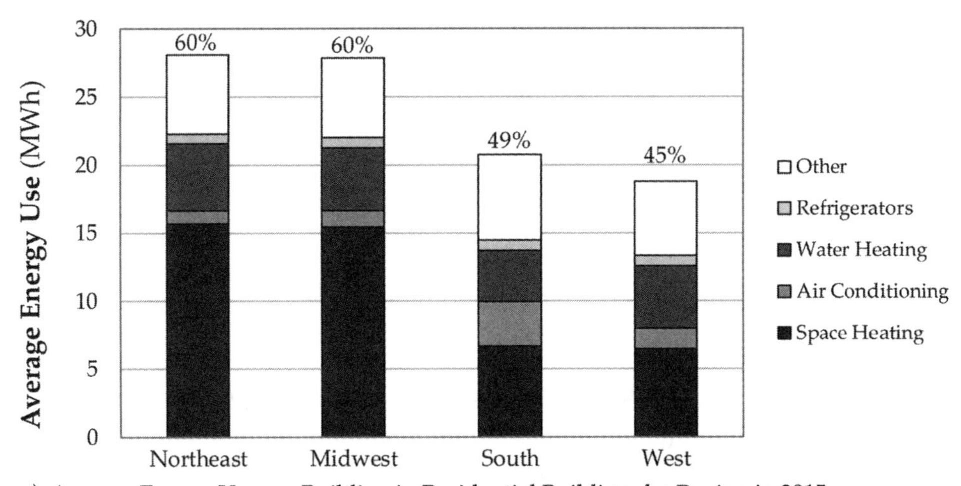

a) Average Energy Use per Building in Residential Buildings by Region in 2015

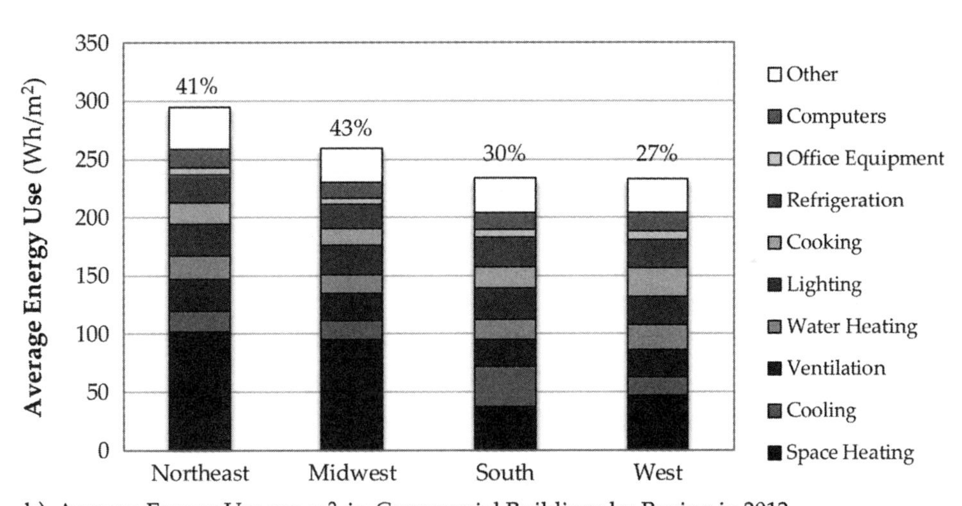

b) Average Energy Use per m^2 in Commercial Buildings by Region in 2012

Figure 8.2

Average energy use in residential and commercial buildings. The percentages at the top indicate the proportion of energy being used for space conditioning (i.e., space heating and cooling). *Source*: U.S. Energy Information Administration.

One of the big takeaways from figure 8.2 is that most of the energy used in buildings is for space conditioning. In particular, buildings in the colder Northeast and Midwest climates require a lot of energy for space heating. Buildings in the warmer South and West climates require energy both for heating and cooling, but they generally consume less energy than buildings in the Northeast and Midwest. As we will see, this is partly because "generating" heat is much more energy-intensive than "moving" heat, which is what air conditioners do by displacing the heat outside—akin to refrigerators, which displace heat to the back of the refrigerator. Moreover, the temperature difference to heat buildings in colder climates tends to be higher than the temperature difference in warmer climates to cool buildings.[1]

In this chapter we will therefore focus entirely on heating and cooling. The other types of uses are important, but we have already discussed the energy consumption of appliances in chapters 5 and 6. Additionally, although water heating is a significant consumer of energy and it could be reduced—for example, by recuperating the heat from hot water going down the drain—we will not discuss it here.

As with the other chapters, we will first review some fundamentals that set the limits on what we can and cannot achieve, and we will restrict this section to the fundamentals of heat transfer and thermal comfort. To better understand demand in the context of buildings, we will then inspect building energy use trends and learn about some metrics that we can use and general guidelines. Finally, we will establish some design and technology recommendations to lower energy needs for space conditioning, whether it is by designing buildings differently or by providing hot and cold air from less energy-intensive sources.

Whenever possible, we will use principles and metrics from the main authority on heating and cooling in the United States: the American Society of Heating, Refrigerating, and Air-Conditioning Engineers (ASHRAE).

8.1 Fundamentals of Thermal Comfort and Heat Transfer

Before looking more into how we can make our buildings more efficient, we need to understand what drives the use of energy in buildings. After all, why do we condition temperature in buildings? It is because we, as human beings, are in the buildings, and we want to control the temperature of our environment. We will therefore need to review some principles of thermal comfort. Thermal comfort implies that we seek to heat or cool a room. We will then review the essentials of heat transfer so that we will be able to understand how the material properties of walls and roofs—the famous *R* and *U* values—define the efficiencies of buildings. We will also look particularly at windows and air exchange in buildings since they are responsible for some of the largest heat losses. Finally, efficiencies for heating and cooling do not function the same way, and we will discuss the difference.

8.1.1 Principles of Thermal Comfort

Just as an engine, the human body generates heat. It has to do so because it constantly loses heat, which simply dissipates in the cooler environment. This is what the second law of thermodynamics tells us. This process actually has to function well since the body temperature cannot fluctuate much.

We control our body temperature by burning calories and generating heat from food, a process called *metabolism*. For our purpose, metabolism is best understood from the viewpoint of the metabolic rate M. The metabolic rate can be measured in units of *met*, but we will keep it in the same units as heat transfer in buildings: [W/m^2], and 1 met = 58.2 W/m^2. The metabolic rate is therefore the rate of heat loss—that is, the amount of energy per unit of time and therefore a "power"—that our body dissipates. M can be calculated as:

$$M = E + C + R \tag{1}$$

where E is the rate of heat exchanged by evaporation, C is the rate of heat exchanged by convection, and R is the rate of heat exchanged by radiation. In general, convection and radiation each account for about 40% of the heat exchanged, and evaporation accounts for 20% (Pohl 2011). We are also missing a term here related to respiration, but it is not as significant as these three, and we will ignore it.

Naturally, this metabolic rate depends on the activity that we are doing. Table 8.1 shows different values of M for various activities. As expected, the more active we are, the more power we produce, and therefore the more heat we lose. When we sleep we only dissipate 40 W/m^2, in contrast to when we play sports, during which our metabolic rate increases to 250–400 W/m^2—this difference represents an entire order of magnitude and is therefore significant. Note that the metabolic rate is per unit area as opposed to per unit volume, since the heat dissipates through the skin. This is actually very important for buildings as well—in terms of volume-to-surface ratio—as we will see later.

If the environment is too hot and the body cannot lose heat at these rates depending on the activity, then we are not comfortable. Similarly, if the environment is too cold and the body loses heat faster than these rates for an activity, we are not comfortable. The temperature of the environment in which most of us are "comfortable" at rest is contained within a small range that spans from approximately 20°C (68°F) to 24°C (75°F); outside this range and depending on the activity we are doing, we are typically not comfortable, although the range obviously varies for many conditions (including

Table 8.1
Metabolic rate of various activities

Activity	Metabolic rate: Met units	Metabolic rate: W/m²
Resting		
Sleeping	0.7	40
Reclining	0.8	45
Seated, quiet	1.0	60
Standing, relaxed	1.2	70
Walking		
0.9 m/s, 3.2 km/h, 2.0 mph	2.0	115
1.2 m/s, 4.3 km/h, 2.7 mph	2.6	150
1.8 m/s, 6.8 km/h, 4.2 mph	3.8	220
Office activities		
Reading, seated	1.0	55
Writing	1.0	60
Typing	1.1	65
Filing, seated	1.2	70
Filing, standing	1.4	80
Walking about	1.7	100
Lifting/packing	2.1	120

Activity	Metabolic rate: Met units	Metabolic rate: W/m²
Driving		
Automobile	1.0–2.0	60–115
Heavy vehicle	3.2	185
Miscellaneous occupational activities		
Cooking	1.6–2.0	95–115
Housecleaning	2.0–3.4	115–200
Seated, heavy limb movement	2.2	130
Machine work		
Sawing (table saw)	1.8	105
Light machine work (electrical industry)	2.0–2.4	115–140
Heavy machine work	4.0	235
Handling 50 kg (100 lb) bags	4.0	235
Pick-and-shovel work	4.0–4.8	235–280
Miscellaneous leisure activities		
Dancing, social	2.4–4.4	140–255
Calisthenics/exercise	3.0–4.0	175–235
Tennis, single	3.6–4.0	210–270
Basketball	5.0–7.6	290–440
Wrestling, competitive	7.0–8.7	410–505

Source: American Society of Heating, Refrigerating, and Air-Conditioning Engineers (2010a).

health, gender, diet, and so on). The only way we can be comfortable and even survive in areas with high temperature fluctuations is to shelter ourselves, hence the need for temperature-controlled buildings.[2]

At this point we have hinted toward concepts of heat transfer, briefly mentioning convection and radiation, but how does heat get transfered? What are the principal mechanisms? Let us move on to the fundamentals of heat transfer.

8.1.2 Fundamentals of Heat Transfer

The transfer of heat from one location to another can be done in only three ways: (1) conduction, (2) convection, and (3) radiation. All three are relevant for buildings. Conduction exchanges occur in buildings when heat is transferred through solid materials such as walls, roofs, and floors. Convection exchanges occur at interfaces between surfaces and fluids such as air and water—for example, between exterior walls and the outdoor air and between each layer of an air cavity within walls. Radiation exchanges occur between surfaces at different temperatures; between walls, ceilings, and floors indoors; and on outside walls and roofs. Here, we will review these three processes, starting with conduction.

For heat, we will use the conventional Q symbol.[3] The rate of change of heat over a distance dx (i.e., the heat exchange) is therefore dQ/dx, and in physics, we typically represent this rate with a dot •, therefore giving us $Q^{\bullet}$, which we will use in this chapter.

8.1.2.1 Conduction Conduction is the process of transferring heat from one molecule to another. Conduction especially occurs in solid and liquid media, where molecules are packed together, as well as in still gases. The interpretation of heat conduction in science is attributed to the famous French scientist Joseph Fourier, who was very active in the early nineteenth century. He defined heat transfer from conduction as:

$$Q^{\bullet} = -kA\frac{dT}{dx} \tag{2}$$

where k is the thermal conductivity of our medium in watts per meter times kelvin, [W/(m·K)], A is the area that we are considering, and T is the temperature.

What Fourier's law (equation (2)) tells us is that conduction is directly proportional to the change in temperature and inversely proportional to the distance that heat has to travel. We have actually seen this equation before in a different context. It is nearly identical to Darcy's law of groundwater flow expressed in equation (14) in chapter 6. Although we have not defined it like that, $Q^{\bullet}$ is also nearly identical to current in electricity, expressed by Ohm's law. In fact, Fourier's law is inspired by these relationships, and it is actually not a fundamental law of physics (like conservation of mass) but a simple property of matter. Fourier's law is therefore not always applicable. In particular, we can only apply it in steady-state conditions—that is, when the flow of heat transfer is constant, and there is no heat storage to account for.

Moreover, the presence of the negative sign (–) in front of Fourier's law is there since heat must flow from a higher temperature to a lower temperature (back to the second law of thermodynamics).

Table 8.2 shows typical k values for various materials. We can see that to minimize heat loss, we prefer small values of k. Of relevance, we see that (still) air has a low thermal conductivity, which explains why double- and triple-pane windows are much better insulators than single-pane windows (but we will talk about windows later). We also see that air has a lower k value than insulation material. But again, this only applies to

Table 8.2
Density and conductivity of various materials

Description	Density [kg/m³]	Conductivity—*k* values [W/(m·K)]
Air (still, convection exchanges occur for moving gases)		
Air at −50°C (−58°F)	1.534	0.0204
Air at 0°C (32°F)	1.293	0.0243
Air at 20°C (68°F)	1.205	0.0257
Building board		
Gypsum or plasterboard, 0.5 in	800	0.16
Plywood (Douglas fir)	545	0.115
Hardboard, medium-density	800	0.105
Insulation		
Cellular glass	136	0.05
Glass fiber, organic bonded	64–144	0.036
Expanded perlite, organic bonded	16	0.052
Expanded polystyrene extruded, smooth skin surface	29–56	0.029
Cellular polyurethane (R11 exp.) (unfaced)	24	0.023
Mineral fiber with resin binder	240	0.042
Mineral fiberboard, wet felted, core or roof insulation	256–272	0.049
Mineral fiberboard, wet felted, acoustical tile	288	0.05
Loose fill		
Cellulosic insulation (milled paper or wood pulp)	37–51	0.039–0.046
Sawdust or shavings	128–240	0.065
Wood fiber, softwoods	32–56	0.043
Perlite, expanded	80–128	0.053
Vermiculite, exfoliated	112–131	0.068
Masonry materials/concrete		
Cement mortar	1,860	0.72
Lightweight aggregates including expanded shale, clay or slate; expanded slags; cinders; pumice; vermiculite; also cellular concretes	1,920	0.75
Sand and gravel or stone aggregate (not dried)	2,240	1.73
Stucco	1,860	0.72
Masonry units		
Brick, common	1,920	0.72
Brick, face	2,080	1.3
Woods		
Maple, oak, and similar hardwoods	720	0.16
Fir, pine, etc.	510	0.12

Adapted from Kreider, Curtiss, and Rabl (2009).

still air. In fact, insulation materials are full of air and voids, essentially simply decreasing the impact of convection processes (fiberglass is 99% air, thus simply keeping air relatively still). Although not shown, other gases have even lower thermal conductivities, again only when they are still.

From table 8.2, we see that masonry materials/units and concrete have high k values (compared to wood, for example). This is actually quite important since these materials create a *thermal bridge* between the indoors and the outdoors, but we will talk more about that later.

For homogeneous media, and as illustrated in figure 8.3a, we can integrate equation (2) between the boundaries of the medium, giving us:

$$Q^{\bullet} = -kA\frac{\Delta T}{\Delta x} = kA\frac{T_1 - T_2}{\Delta x} \tag{3}$$

We can drop the negative sign by placing T_1 first, since T_1 is larger than T_2. Moreover, to compare various buildings with one another, we calculate the heat transfer per unit area $q^{\bullet}$, giving us:

$$q^{\bullet} = k\frac{T_1 - T_2}{\Delta x} = \frac{T_1 - T_2}{\Delta x / k} \tag{4}$$

The denominator works essentially like a resistance in electricity, and we can define it exactly as a resistance R:

$$R = \frac{\Delta x}{k} \tag{5}$$

This is the famous R value that is used in building science to qualify different materials. It is expressed in $[(m^2K)/W]$, and materials with larger R values are better insulators. Note that the R value accounts for thickness Δx, and therefore similar materials with different thicknesses have different R values. We can simply calculate R values for a material by dividing the thickness by the appropriate k value.

Back to our analogy with electricity, if we rearrange equation (4), we can also see that we get $\Delta T = Rq^{\bullet}$, which is essentially Ohm's law ($V = RI$), where the difference in temperature is the difference in electric potential—that is, V, the voltage drop—and the heat flow is the current I.[4]

Buildings have also combined construction materials with insulation materials, creating multiple layers of materials that cover buildings in what are called *building envelopes*. As illustrated in figure 8.3b (for layers A, B_1, and C), to account for these multiple layers of materials—and just like resistors in series in electricity—the R values are simply summed to calculate heat transfer per unit area:

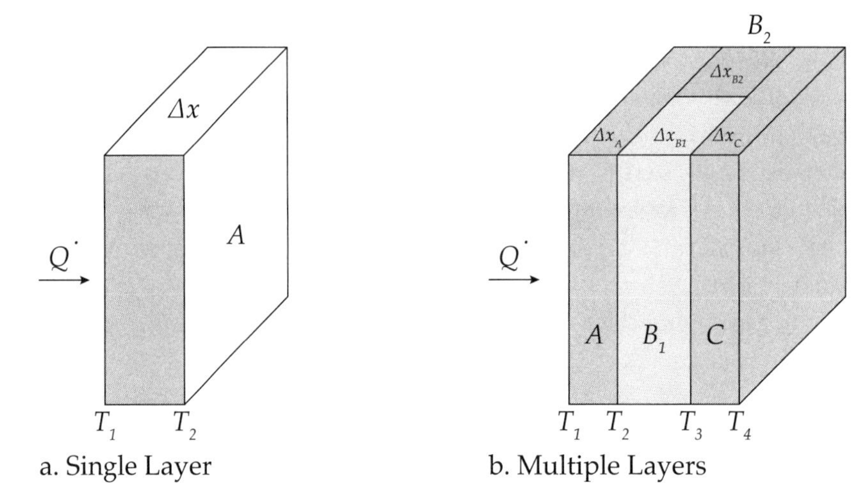

Figure 8.3
Illustration of heat conduction.

$$q^{\bullet} = \frac{\Delta T}{R_1 + R_2 + \cdots + R_n} \tag{6}$$

For n successive layers of materials, we can therefore define an effective resistance R_{eff} as:

$$R_{eff} = \sum_{i=1}^{n} R_i \tag{7}$$

We note that R is in the denominator of equation (6), and instead of [m^2K/W], we can more easily understand a property expressed in [W/m^2K]. We therefore define another famous metric in building science. The conductance U that gives us the U value of a material is simply defined as:

$$U = \frac{1}{R} = \frac{k}{\Delta x} \tag{8}$$

Like thermal conductivity, k, we prefer low values of U for better insulation. Defining an effective conductance $U_{eff} = 1/R_{eff}$, the per unit area heat loss is therefore defined as:

$$q^{\bullet} = U_{eff} \cdot \Delta T \tag{9}$$

And the total heat loss is defined as:

$$Q^{\bullet} = A \cdot U_{eff} \cdot \Delta T \tag{10}$$

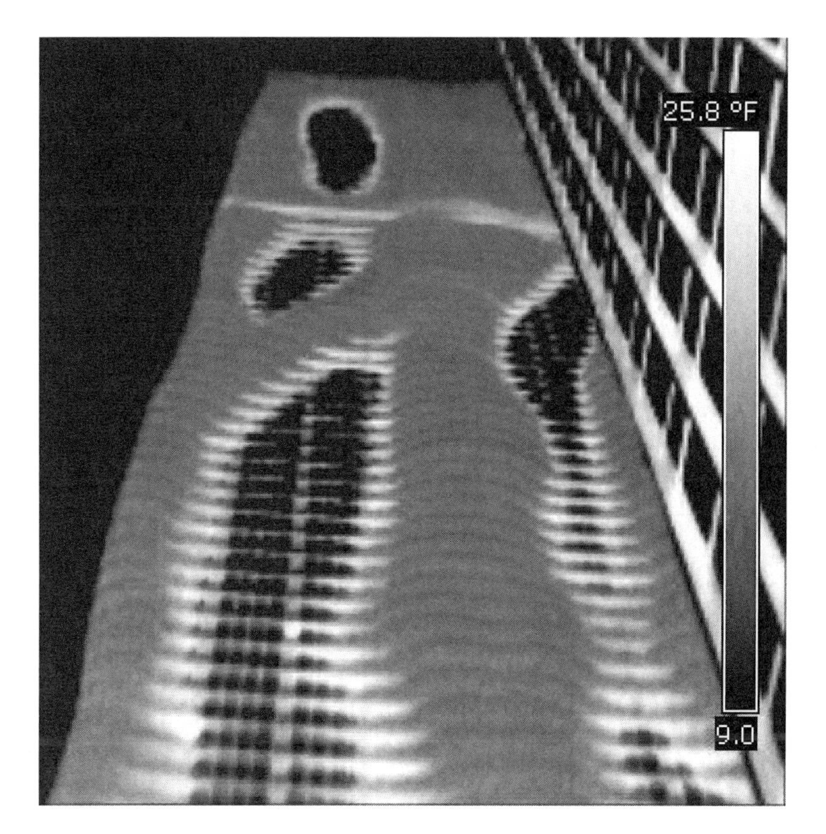

Figure 8.4
Example of a thermal bridge, Aqua building in Chicago.

This equation works well for media that have homogeneous layers of materials, but most walls have structural elements that make them heterogeneous, as illustrated in figure 8.3b (for layers B_1 and B_2). A typical wall will have studs, for instance, that need to be accounted for. Studs, concrete slabs, and other features of walls that are not insulated are also called *thermal bridges*—as briefly discussed above when talking about k values—and have a significant impact on how much heat dissipates.

Figure 8.4 shows a thermal image of the beautiful and complex Aqua building in Chicago on a winter day. The lighter surfaces are the balconies that act as thermal bridges. Despite this problem, the Aqua building performs relatively well. Look at the building on the right in the picture, it has even worse thermal bridges through its steel frame. Thermal bridges represent a significant problem, and this is why older buildings are often retrofitted by adding a thin insulating layer on the outside walls and why newer buildings have requirements for continuous exterior insulation.

The total heat flow $Q^{\bullet}_{total}$ of a wall composed of m different sets of layers is defined as:

$$Q^{\bullet}_{total} = \sum_{j=1}^{m} Q^{\bullet}_{m} \tag{11}$$

Instead of calculating all the various heat flows and summing them, we can first calculate an equivalent conductance U_{eq}. Here, we essentially weight the conductance of the various sets of layers of materials based on their proportional areas as:

$$U_{eq} = \frac{1}{A}\sum_{i=1}^{m} A_i U_i = \frac{1}{A}\sum_{i=1}^{m} \frac{A_i}{R_i} \tag{12}$$

where A_i is the area of the layer i and A is the total area of the wall.

Equation (11) is not complete, however. We are missing the mechanisms of heat transfer due to convection and radiation at the boundaries. Although we can neglect the heat transfers due to convection and radiation between solid layers (but not air cavities), we cannot neglect convection and radiation at boundaries (i.e., at the surface of the walls). Let us therefore move on to our second heat transfer mechanism: convection.

8.1.2.2 Convection Convection is the process of heat transfer by the movement of fluids. Although convection occurs at multiple scales in buildings, we will focus purely on the larger scale, studying the heat transfer between our solid medium (i.e., walls, floors, roofs) and the air. The physical process of convection is fairly complicated since the velocity of the air has a large impact on the rate of heat loss—this is why food gets colder more quickly when you blow on it—but we will approximate all convective heat losses using a simplified version of Newton's law of cooling:

$$Q^{\bullet} = h_c \cdot A \cdot \Delta T \tag{13}$$

where h_c is the convection coefficient in [W/m^{2}K], A is the area, and ΔT is the difference in temperature between the surface temperature and the fluid temperature away from the surface. For instance, ΔT is the difference between the temperature at the wall and the temperature in the middle of the room.

The heat transfer per unit area $q^{\bullet}$ is simply defined as:

$$q^{\bullet} = h_c \cdot \Delta T \tag{14}$$

Equations (13) and (14) greatly resemble Fourier's law that we saw for conduction, except that we do not have any thickness anymore. In fact, we can similarly calculate a convective resistance R value:

$$R = \frac{1}{h_c} \tag{15}$$

Example 8.1

Average Thermal Properties of a House in Civitas

The wall of a house in Civitas measures 10 m×6 m. The exterior has a 10 cm thick layer of common brick, and the interior has a 2.5 cm thick layer of plasterboard. In between, the wall is insulated with a 12 cm thick layer of polyurethane. Wood studs (hardwood) are placed every meter, with dimensions 12 cm×5 cm. (1) Calculate the equivalent conductance U_{eq} for the wall. (2) The average temperature during the year in Civitas is 10°C (50°F), while the temperature indoors is kept at 20°C (68°F). Determine the per unit area and average heat loss by the wall. (3) Assuming the four walls, roof, and floor of the house have the same area and heat losses, determine the total energy demand in the year.

Solution

(1) The various k values for brick, plasterboard, polyurethane, and hardwood are collected from table 8.2. Respectively, they are $k_{brick}=0.72$, $k_{pb}=0.16$, $k_{pu}=0.023$, and $k_{wood}=0.16$ W/(m·K).

We have two different series of layers for the wall. We therefore need to calculate two different U values. Let us start first with the R values. R_{stud} for the area with the studs is:

$$R_{stud} = R_{brick} + R_{wood} + R_{pb} = \frac{0.10}{0.72} + \frac{0.12}{0.16} + \frac{0.025}{0.16} = 1.045$$

R_{ins} for the area with insulation is:

$$R_{ins} = R_{brick} + R_{pu} + R_{pb} = \frac{0.10}{0.72} + \frac{0.12}{0.023} + \frac{0.025}{0.16} = 5.513$$

To combine them we first calculate the respective areas. The total area is $A=10\times6=60$ m^2. There are a total of 11 studs (one every meter), each of which is 5 cm wide and 6 m tall, therefore: $A_{stud}=11\times0.05\times6=3.3$ m^2. The area with the insulation A_{ins} therefore is $A_{ins}=60-3.3=56.7$ m^2. We can then calculate the equivalent conductance U_{eq} as:

$$U_{eq} = \frac{3.3}{60} \times \frac{1}{1.045} + \frac{56.7}{60} \times \frac{1}{5.513} = 0.224 \text{ W/m}^2\text{K}$$

(2) Based on the information calculated in (1), the per unit area heat loss is: $q^{\bullet} = U_{eq} \cdot \Delta T = 0.224 \times (20-10) = 2.24 \text{ W/m}^2$, and the average heat loss by the wall is $Q^{\bullet} = q^{\bullet}A = 2.24 \times 60 = 134.4$ W.

(3) The average heat loss for the house is:

$$Q^{\bullet}_{house} = 6 \times 134.4 = 806.4 \text{ W}$$

With 365 days per year and 24 hours per day, the yearly energy demand is $E_{year}=806.4\times365\times24=7.064\times10^6$ Wh$=7.064$ MWh.

Although we will not use R directly here, it is quite useful to calculate an effective R when combined with the R values for conduction, and we will see an example later.

Despite the fact that the relationship is simple, calculating the value of h_c is not so simple. We have to consider three factors:

1. Because hot air rises, we need to know whether the surface is tilted or not and if not, whether it is facing upward or downward.
2. We need to know whether the flow is "free" or "forced." Free flow happens indoors or when there is no wind. Forced flow happens on exterior surfaces due to the wind or on interior surfaces due to forced air movement from a fan or a heating, ventilation, and air-conditioning (HVAC) system.
3. Finally, we need to know whether the flow is laminar or turbulent.

Let us examine all cases. First, a free flow is said to be laminar when:

$$L^3 \Delta T < 63 \tag{16}$$

where L is the average length of the sides for horizontal surfaces or the length of the surface in the direction of buoyancy-driven flow for tilted surfaces—that is, the length of the side that changes in elevation (thus the height of a wall, not the width).

In the case of a forced flow, a flow is laminar when:

$$vL < 1.4 \tag{17}$$

where v is the wind velocity in [m/s].

Table 8.3 contains some empirical equations for h_c based on the three factors. Note that h_c here is dimensional, and it is expressed in [W/m²K]. All variables used in an

Table 8.3

Equations for convection coefficient h_c

Type	Laminar	Turbulent
Free flow		
Warm horizontal facing up/cold horizontal facing down.	$h_c = 1.32\left(\frac{\Delta T}{L}\right)^{1/4}$ (18)	$h_c = 1.52\ (\Delta T)^{1/3}$ (19)
Warm horizontal facing down/ cold horizontal facing up.	$h_c = 0.59\left(\frac{\Delta T}{L}\right)^{1/4}$ (20)	
Tilted surface. Applies for angles between 30° and 90°.	$h_c = 1.42\left(\frac{\Delta T \sin\beta}{L}\right)^{1/4}$ (21)	$h_c = 1.31\ (\Delta T \sin\beta)^{1/3}$ (22)
Forced Flow		
Applies to all surfaces. L is the length of the surface in the direction of the flow.	$h_c = 2.0\left(\frac{v}{L}\right)^{1/2}$ (23)	$h_c = 6.2\left(\frac{v^4}{L}\right)^{1/5}$ (24)

Example 8.2

Average Convective Heat Loss of a Roof in Civitas

The roof of the house in example 8.1 has the properties illustrated below, where L_h is the length of the roof perpendicular to the drawing. For all questions, assume a surface temperature of 30°C (86°F) and an air temperature of 25°C (77°F)—that is, $\Delta T = 5K$. (1) Assuming that free-flow conditions apply, calculate the convection coefficient of the roof h_{free}. (2) Assuming a wind speed of 5 m/s and a wind direction along L_h, calculate the convection coefficient of the roof h_{forced}. (3) Under both conditions, calculate the convective heat loss per unit area and total.

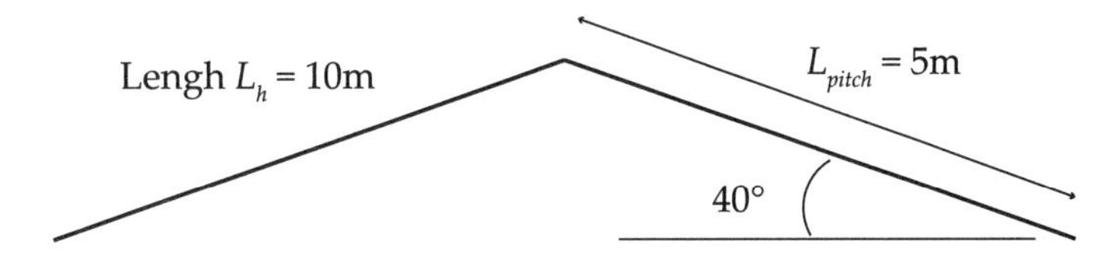

Solution

(1) First, we need to check whether the flow is laminar or turbulent. For this, we apply equation (16), noting that the length of the buoyancy-driven flow is $L_{pitch} = 5$ m. We have $L^3 \Delta T = 5^3 \times 5 = 625 > 63$. The flow is therefore turbulent, and we need to use equation (22), with $\beta = 40°$:

$$h_{free} = 1.31(5 \sin 40)^{1/3} = 1.93 \text{ W/m}^2\text{K}$$

(2) Here again, we need to check whether the flow is laminar or turbulent by applying equation (17): $vL_h = 5 \times 10 = 50 > 1.4$. The flow is also turbulent, and we need to use equation (24):

$$h_{forced} = 6.2\left(\frac{5^4}{10}\right)^{1/5} = 14.18 \text{ W/m}^2\text{K}$$

(3) Applying equation (14) in both cases, free flow: $\dot{q}_{free} = 1.93 \times 5 = 9.65$ W/m² and forced flow: $\dot{q}_{forced} = 14.18 \times 5 = 70.9$ W/m². The total area A of the roof is $A = 2 \times 5 \times 10 = 100$ m², therefore, free flow: $\dot{Q}_{free} = 100 \times 9.65 = 965$ W and forced flow: $\dot{Q}_{forced} = 100 \times 70.9 = 7{,}090$ W.

We can see that these values are much higher than those we saw in example 8.1. Convection can indeed have a significant impact on heat loss. We should note, however, that this example happens on a hot summer day on the roof because of radiation (see next section). On interior walls or exterior walls not warmed by the sun, the surface temperature is nearly identical to the air temperature. In fact, convection has a much lower impact on interior walls, as we will see.

equation must therefore use the metric system and be in meters [m] and kelvins [K]. These equations were collected from Kreider, Curtiss, and Rabl (2009).

An example of a warm horizontal surface facing up is a horizontal roof under the sun in the summer, and a warm horizontal surface facing down is a horizontal ceiling in the winter.

Typical convection coefficients of air range from 0.5 to 5 W/m^2K in free flow and from 5 to 50 W/m^2K in forced air. This represents a difference of two orders of magnitude and is therefore significant—again, this is why you blow on hot food to cool it down.

Although some approximations had to be done, these equations will suffice for our purpose. We will now learn about our third type of heat transfer mechanism: radiation, which will enable us to determine, for example, why roofs get so warm in the summer.

8.1.2.3 Radiation The final heat transfer process is radiation that transfers heat through electromagnetic waves. There are actually two types of electromagnetic waves: shortwaves (visible light, ultraviolet, and near infrared) and long waves (infrared light). The color (for shortwaves) and material property (for long waves) of the surface have significant impacts. The radiation energy E_b of a purely emitting body (referred to as a *black body*) is defined as:

$$E_b = \sigma T^4 \tag{25}$$

where σ is the Stefan-Boltzmann constant that is equal to 5.67×10^{-8} W/m^2K^4. Note that the temperature is raised to the power 4, and radiation is therefore highly nonlinear.

Black bodies, however, do not exist in real life. That being said, we can measure how close to a black body a normal *gray* body is. We simply define the emissivity ε as the ratio of the emitting power of a gray-to-black body as:

$$\varepsilon = \frac{E}{E_b} \tag{26}$$

where E is the emitting power of the gray body. Table 8.4 contains typical emissivity values ε of various materials. Emissivity is especially important for long waves. It is also directly related to another property of materials: absorptivity α, and in fact they have the same value for gray bodies. Unlike emissivity, absorptivity is important for shortwaves. Materials that can absorb radiation can also emit radiation, and under steady-state conditions (i.e., during a sunny day), the amount of radiation absorbed is the same as the amount emitted. But again, this is only true for gray bodies, and many surfaces are not gray. Table 8.4 includes absorptivity values for nongray surfaces, and

Table 8.4
Typical emissivity and absorptivity values

Surface	Emissivity	Absorptivity
Asphalt	0.93	
Brick, red & rough	0.93	0.55–0.68
Clay	0.91	
Concrete	0.85	0.60
Concrete, rough	0.94	0.60
Glass, smooth	0.92–0.94	
Ice, smooth	0.97	
Iron, polished	0.14–0.38	
Iron, plate rusted red	0.61	
Iron, dark gray surface	0.31	
Marble, white	0.95	0.44
Paint, various oils	0.92–0.96	
Paper	0.93	
Plaster, rough	0.91	
Plastics	0.90–0.97	
Sandstone	0.59	0.62–0.73
Soil	0.90–0.95	
Snow	0.80–0.99	
Water	0.90–0.99	
Wood, beech, planed	0.935	
Wood, oak, planed	0.885	
Wood, pine	0.95	

Adapted from The Engineering Toolbox (2017a, 2017c).

we can see, for example, that concrete and brick material have emissivity values around 0.9, but their absorptivity is around 0.60.

Another valuable property is reflectivity ρ, which represents the amount of rays that are directly reflected back without being absorbed. Reflectivity applies to shortwaves. Light surfaces in particular have high shortwave reflectivity, thus they help prevent high cooling needs in the summer.[5]

For nonopaque surfaces like glass, transmissivity τ is important as well. It is essentially the amount of radiation that is transmitted through the material. This is why we do not tan when we are inside a building, even if the sun is shining, since not all radiation is transmitted through windows.

Absorptivity α (or emissivity for gray bodies), reflectivity ρ, and transmissivity τ of a surface are actually linked such that:

$$\alpha + \tau + \rho = 1 \tag{27}$$

So what is better, high absorptivity/emissivity or high reflectivity? It is actually not that simple a question since both have advantages and disadvantages. In addition, we need to think about material properties for conduction and convection processes. As a general guideline, we will prefer light surfaces, regardless of material property, to reflect heat, unless we are in a cold climate and need the additional absorbed solar radiation, which we will discuss in section 8.3.

To go further, let us first consider the simple case in which two surfaces are parallel, close to each other, and they have the same area A. In this case the radiation $Q^{\bullet}_{12}$ from surface 1 to surface 2 is defined as:

$$Q^{\bullet}_{12} = \frac{A\sigma}{1/\varepsilon_1 + 1/\varepsilon_2 - 1}(T_1^4 - T_2^4) \tag{28}$$

where T_1 and T_2 are the temperatures of surfaces 1 and 2, respectively, and ε_1 and ε_2 are the emissivity values of surfaces 1 and 2. The per unit area of heat transfer between surface 1 and 2 easily becomes:

$$q^{\bullet}_{12} = \frac{\sigma}{1/\varepsilon_1 + 1/\varepsilon_2 - 1}(T_1^4 - T_2^4) \tag{29}$$

In order to facilitate the calculation of heat transfer in buildings due to conduction, convection, and radiation, we define a new radiation coefficient h_r, similar to the convection coefficient that we want to be:

$$h_r = \frac{q^{\bullet}_{12}}{T_1 - T_2} \tag{30}$$

In practice, we can simplify the calculation of h_r one more time, when T_1 and T_2 are not too far apart—that is, within about 10°C (20°F)—we get:

$$h_r = \frac{4\sigma T^3}{1/\varepsilon_1 + 1/\varepsilon_2 - 1} \tag{31}$$

where T is simply the average of T_1 and T_2. We can now use an equation identical to equation (13) for convection to estimate heat transfer due to radiation:

$$Q^{\bullet} = h_r \cdot A \cdot \Delta T \tag{32}$$

And the heat transfer per unit area is defined as:

$$q^{\bullet} = h_r \cdot \Delta T \tag{33}$$

Most surfaces are not parallel and close, and they do not have the same area, however. In these cases we need to calculate a *view factor F*, also called an *angle factor*, a

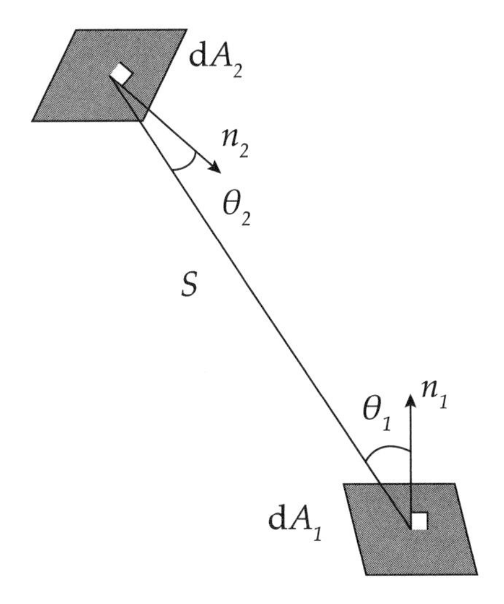

Figure 8.5
Angle factor in radiation heat transfer.

configuration factor, or a *shape factor*—for us, shape factor will mean something else later. This is essentially a factor between 0 and 1 that we can apply to equation (33), such that:

$$q^{\bullet} = F \cdot h_r \cdot \Delta T \quad (34)$$

From figure 8.5 the factor $F_{1,2}$ that needs to be used to calculate the proportion of the radiation from surface 1 to surface 2 is defined as:

$$dF_{1,2} = \frac{\cos\theta_1 \cdot \cos\theta_2}{\pi S_{1,2}^2} dA_2 \quad (35)$$

where θ_1 and θ_2 are the angles from surfaces 1 and 2 to each other, A_1 and A_2 are the areas of surfaces 1 and 2, and $S_{1,2}$ is the distance between surfaces 1 and 2.

Equation (35) makes a few important assumptions. In particular, it assumes that all surfaces are gray or black and that they are uniform (i.e., uniform radiation). Although these assumptions are not technically correct, in most cases the method outputs reasonable estimations.

One consequence of these assumptions is the reciprocity rule that gives us:

$$F_{1,2} A_1 = F_{2,1} A_2 \quad (36)$$

Figure 8.6 shows a graph of view factors when two surfaces are either parallel or perpendicular rectangles. The equations used to make the graphs on figure 8.6 or similar

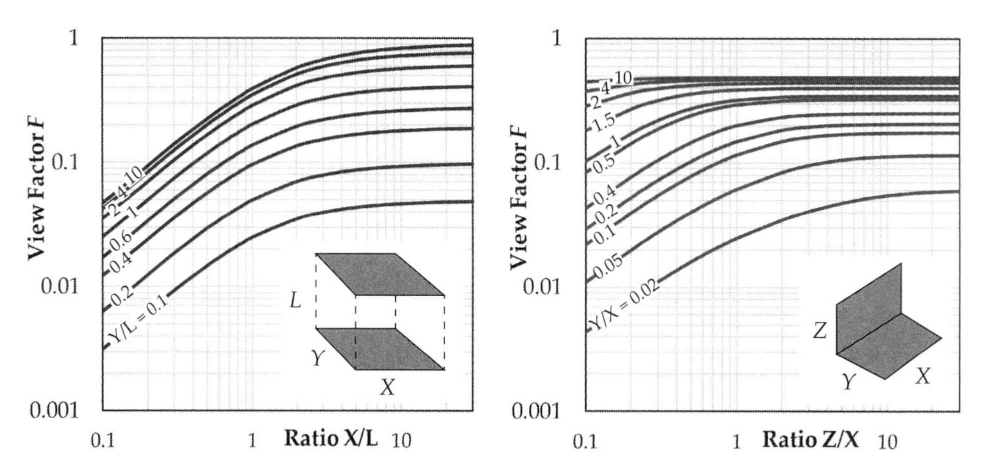

Figure 8.6
View factors for parallel and perpendicular rectangles.

graphs, as well as equations for many other geometries, can easily be found online, but the main book to refer to on the topic is *Thermal Radiation Heat Transfer* by Howell, Menguc, and Siegel (2016).

We have to be careful, however, since these equations cannot be used to estimate the heat transfer from the sun—can you imagine the radiation heat transfer if we used the sun's temperature as T_2? They can only be used to estimate the radiation heat transfer from other surfaces in the environment from long wave radiation.

To account for shortwave heat gains from solar radiation $q^{\bullet}_{solar}$, we use:

$$q^{\bullet}_{solar} = \alpha \cdot I_{solar} \tag{37}$$

where α is absorptivity, and I_{solar} is the total shortwave solar radiation flux incident on the surface, and it accounts not only for the direct solar radiation flux but also for the diffuse radiation flux from the sky and from the ground. Notably, the variable I_{solar} depends on the angle of the exposed surface in relation to everything else. We also see that I_{solar} has the same units as $q^{\bullet}_{solar}$—that is, in [W/m^2]. To learn more, see McClellan and Pedersen (1997).

To determine typical I_{solar} values for the United States, we can use the National Renewable Energy Laboratory (NREL) map for concentrating solar power that we saw in chapter 5—see figure 5.19. A rough average gives us about 400 W/m^2, but it can go up to 1,000 W/m^2 on a clear summer day and down to 50 W/m^2 on a cloudy winter day. Heat gains from solar radiation can therefore be significant, especially in the summer.

Example 8.3

Radiation in Civitas

The roof of a house ($\alpha_{roof}=\varepsilon_{roof}=0.95$) in Civitas has an area of 80 m^2. During a hot summer day, it receives an average solar radiation of 300 W/m^2 over 10 hours. (1) Calculate the total amount of energy, in terms of heat, received by the roof during the day due to solar radiation. By the end of the day, the roof temperature reaches 50°C. At night the roof radiates heat back into the night sky ($T_{sky}=3.9$°C and $\varepsilon_{sky}=0.74$). (2) Calculate the total amount of energy that radiates back into the sky.

Solution

(1) Using equation (37), we get a solar radiation heat transfer of:

$$q^{\bullet}_{solar} = \alpha_{solar} \cdot I_{solar} = 0.95 \times 300 = 285 \text{ W/m}^2$$

Since the roof measures 80 m^2, and the sun shone for 10 hours, the total energy E_{solar} received from the sun is:

$$E_{solar} = q^{\bullet}_{solar} \cdot A \cdot t = 285 \times 80 \times 10 = 228{,}000 \text{ Wh} = 228 \text{ kWh}$$

(2) Using equation (31) with $T=0.5(3.9+273.15+50+273.15)=300.1$ K, we get:

$$h_r = \frac{4 \times 5.67 \times 10^{-8} \times 300.1^3}{1/0.95 + 1/0.74 - 1} = 4.366 \text{ W/m}^2\text{K}$$

Therefore, the radiation heat transfer from the roof is:

$$Q_{roof} = h_r \cdot A \cdot \Delta T = 4.366 \times 80 \times (50 - 3.9) = 16{,}102 \text{ W}$$

And the total energy E_{roof} radiating back is:

$$E_{roof} = Q_{roof} \cdot t = 16{,}102 \times 14 = 225{,}428 \text{ Wh} = 225.428 \text{ kWh}$$

We can see that nearly as much heat is radiating back into the night sky as was gained during the day. This is actually normal since the temperature of the Earth is relatively constant, so all the "energy in" must equal the "energy out." This phenomenon is called night sky radiation, and it explains why some objects at night can be cooler than the ambient air; night sky radiation is particularly strong on cloudless and dry nights.[6]

This brief introduction is enough to enable us to get a general understanding of how radiation works. More importantly, in the next section we will see that we can combine h_r and h_c. For more information on radiation in general, see Kreider, Curtiss, and Rabl (2009).

8.1.2.4 Combining Heat Transfer Processes While the three heat transfer processes are physically different, we managed to find similar equations for all of them, which

Example 8.4

Revisiting the Average Thermal Properties of a House in Civitas

For the house in Civitas from example 8.1, we had calculated $U_{eq}=0.224$ W/m^{2}K. (1) Add the combined convection and radiation coefficients recommended by ASHRAE and determine the system conductance U_{sys}. (2) Calculate the new heat loss per unit area and the heat loss by the wall. (3) Calculate the impact of convection and radiation as a percentage of the total heat loss.

Solution

(1) Using equation (39) and the equivalent conductance calculated in example 8.1, we calculate U_{sys} as:

$$\frac{1}{U_{sys}}=\frac{1}{h_i}+\frac{1}{U_{eq}}+\frac{1}{h_o}=\frac{1}{17.6}+\frac{1}{0.224}+\frac{1}{34}=4.55$$

This means that $U_{sys}=1 / 4.55=0.22$.

Note that even though it does not seem to make sense, we expect U_{sys} to be lower since we are adding an additional "barrier" to prevent heat from dissipating. In reality, we should account for an increase in the temperature difference because of convection and radiation.

(2) Using U_{sys}, the per unit area heat loss $q^{\bullet}$ is:

$$q^{\bullet}=U_{sys}\cdot\Delta T=0.22\times(20-10)=2.2\ \text{W/m}^2$$

And the average heat loss by the wall is:

$$Q^{\bullet}=q^{\bullet}A=2.2\times 60=132\ \text{W}$$

(3) In example 8.1, we calculated $q^{\bullet}=2.24$ W/m^2. This means that convection and radiation are responsible for: $(2.24-2.2) / 2.24=0.0179=1.79\%$ of the heat loss.

In this case the losses due to convection and radiation amount to 1.79%, but these numbers vary greatly depending on how well insulated a building is.

allows us to combine the three into one equation. We can express the total heat transfer as:

$$Q^{\bullet}_{sys}=U_{sys}\cdot\Delta T=\frac{\Delta T}{R_{sys}} \tag{38}$$

where the subscript *sys* refers to the entire system under consideration. For a single layer of material, we can simply sum the R values to get R_{sys} and U_{sys}:

$$R_{sys}=\frac{1}{U_{sys}}=\frac{1}{h_i}+R_{layer}+\frac{1}{h_o} \tag{39}$$

Table 8.5
Typical values of conductance (h_i in W/m^{2}K) and resistance (R in m^{2}K/W)

Values for still air

		Nonreflective		Reflective			
		$\varepsilon=0.90$		$\varepsilon=0.20$		$\varepsilon=0.05$	
Position of surface	Direction of heat flow	h_i	R	h_i	R	h_i	R
Horizontal	Upward	9.3	0.11	5.17	0.19	4.32	0.23
	Downward	9.1	0.11	5	0.2	4.15	0.24
Sloping at 45°	Upward	8.3	0.12	4.2	0.24	3.35	0.3
	Downward	7.5	0.13	3.41	0.29	2.56	0.39
Vertical	Horizontal	6.1	0.16	2.1	0.48	1.25	0.8

Values for moving air for nonreflective surface with $\varepsilon=0.90$

Position of surface	Direction of heat flow	h_i	R
6.6 m/s wind (for winter)	Any	34.1	0.03
3.3 m/s wind (for summer)	Any	22.7	0.044

Source: Kreider, Curtiss, and Rabl (2009).

where h_i and h_o are the indoor and outdoor combined convection and radiation coefficients, and R_{layer} is the sum of the conduction R values for all materials. Note that only R values can be summed—do not make the rookie mistake of summing U values.

To make it simpler and as a general guideline, the American Society of Heating, Refrigerating, and Air-Conditioning Engineers (2010b) recommends using these values for most design conditions:

- $h_i=17.6$ W/m^{2}K
- $h_o=34$ W/m^{2}K at 2.6 m/s wind speed

Table 8.5 contains other possible values for h_i specifically that can vary significantly from the typical value by ASHRAE. Note that the American Society of Heating, Refrigerating, and Air-Conditioning Engineers (2010b) contains many tables from which h_i, h_o, and R values of various materials can be simply looked up.

It is now time to focus on windows and air exchange, as these two are often responsible for the largest heat losses in buildings.

8.1.3 Windows and Air Exchange

Heat transfer through walls, floors, and roofs accounts for about 50% of all heat transfer occurring in buildings. The other 50% is due to windows and air exchange (i.e., air leakage, forced ventilation, and opened areas), each accounting for about 25% (Huang, Hanford, and Yang 1999). It is therefore quite important to understand how these two processes function. We will start with windows.

8.1.3.1 Windows In windows, heat is conducted in the glass, convection occurs not only on both sides of the window but also between panes in multipane windows, and heat radiates both on and through windows since they are not opaque. Finally, windows have frames that inevitably let some air through as leakage. Overall, depending on how insulated and airtight a building is, windows tend to be significant causes of heat loss.

Because we cannot simply sum several R values in windows, we will have to use U values given by manufacturers (which actually makes things a lot easier for us). Table 8.6 shows typical U values for several types of windows—many more values can be found in American Society of Heating, Refrigerating, and Air-Conditioning Engineers (2010b). Like material U values, we prefer low U values for windows. The same equation for windows can be applied as the one seen earlier:

$$Q^{\bullet}_{w} = U_{w} \cdot A_{w} \cdot \Delta T \tag{40}$$

where $Q^{\bullet}_{w}$ is the heat transfer from the window, U_w and A_w are the U values and area of the window, respectively, and ΔT is the temperature difference between the two sides of the window.

From the table, we see that windows with multiple panes perform much better than single-pane windows. Double-, triple-, and quadruple-pane windows roughly have one-half, one-third, and one-quarter the U values, respectively, of single-pane windows. We also see that the type of frame can have a significant impact on the U value, and wood frames perform better than aluminum frames (partly because aluminum has a higher thermal conductivity than wood).

Moreover, we can see that some of the types of windows included have a low-emittance *e* coating, which lowers the emissivity of the window—in terms of radiation heat transfer—helping to keep the heat/cold inside and helping to prevent the heat/cold from the outside from coming in. Low-*e* coatings, notably, help decrease the *solar heat gain coefficient* that is calculated based on solar transmissivity and solar absorptivity, but we will not discuss it here.[7]

Finally, we can see that different gases between panes in multiple-pane windows have an impact. Air performs well but argon performs better, and krypton performs even better than argon.

This is all we will learn about windows; let us move on to air exchange.

Table 8.6
Typical *U* values for different window types

Glazing type (all windows are operable)	*U* factor (W/m²K) Aluminum without thermal break	Aluminum with thermal break	Wood or vinyl frame
Single-pane, 0.31 cm (1/8 in)	6.98	6.08	5.17
Double-pane, 1.25 cm (1/2 in) air space	4.32	3.29	2.84
Double-pane, e=0.40, 1.25 cm (1/2 in) air space	3.92	2.95	2.56
Double-pane, e=0.20, 1.25 cm (1/2 in) air space	3.69	2.73	2.33
Double-pane, e=0.20, 1.25 cm (1/2 in) argon space	3.46	2.56	2.16
Triple-pane, e=0.10 on two panes, 1.25 cm (1/2 in) argon spaces	2.67	1.70	1.48
Quadruple-pane, e=0.10 on two panes, 0.65 cm (1/4 in) krypton spaces	2.56	1.65	1.36

Source: American Society of Heating, Refrigerating, and Air-Conditioning Engineers (2010b).

8.1.3.2 Air Exchange Akin to windows, air exchange is a major cause of heat transfer—again, about 25% of the total heat transfer. No matter how well insulated a wall is, if the window is not airtight or if the door is open, significant heat losses will occur—although we should note that, in general, you actually want some air exchange, if only to bring in oxygen for people to breathe.

In buildings, air is exchanged with the exterior environment through three main mechanisms:

- Wind
- Stack effect
- Induced ventilation (e.g., using mechanical devices)

Wind essentially creates a pressure drop between the outside and the inside of a building, forcing air out of the building, whether through large open areas like open windows and doors or through leaks. Stack effect relates to the fact that hot air rises, thus inviting hotter air to leave the building through cracks/leaks around ceilings in the winter and inviting cooler air to enter a building, often from floors. Induced ventilation (e.g., fan, HVAC system) has become common in commercial and office buildings, and it is increasingly used in houses. Essentially, it artificially creates a pressure difference to move air.

Example 8.5

Effect of Adding Windows to Civitas House

The 10 m × 6 m wall of the Civitas house that we saw in example 8.1 with U=0.224 W/m²K actually has four windows of size 60 cm × 1.5 m. The four windows are double paned with e=0.20 and 1.25 cm argon space, and they have a wood frame. Based on this information, (1) calculate the new equivalent U_{eq} for the wall and (2) determine the impact of the windows of the previous U value.

Solution

(1) The four windows occupy a combined area of 4 × 0.6 × 1.5 = 3.6 m² out of the total wall area of 10 × 6 = 60 m². Moreover, from table 8.5, we see that the U value of the four windows is 2.16 W/m². Using equation (12), the equivalent U value of the wall, U_{wall}, is:

$$U_{wall} = \frac{3.6}{60} \times 2.16 + \frac{(60-3.6)}{60} \times 0.224 = 0.34 \text{ W/m}^2\text{K}$$

(2) The new equivalent U value is therefore 100 × (0.34 − 0.224) / 0.224 = 52% larger than the previous U value. Put differently, despite taking only 100 × 3.6 / 60 = 6% of the entire area, the windows account for 38% of the new U value since:

$$100 \times \frac{(3.6/60) \times 2.16}{0.34} = 38\%$$

Regardless of the mechanism, the volumetric flow rate of air $V^{\bullet}$ across an opening in [m³/s] can be calculated from:

$$V^{\bullet} = A \cdot c \cdot \Delta p^n \tag{41}$$

where A is the area of the open area in [m²] (e.g., summing all leakage areas), c is a flow coefficient in [m/(s·Pa^n)], Δp is the pressure difference in [Pa] between the outside pressure p_o and the inside pressure p_i, and n is an empirically determined exponent between 0 and 1 that is usually taken as 0.65.

Different methods, models, and graphs can be applied to determine the flow coefficient c and the pressure difference Δp, both of which are beyond the scope of this book. What we will do instead is to consider an air exchange rate a in [s^{-1}] (i.e., per second) that can be applied to the entire volume V of a room/building in [m³], such that:

$$V^{\bullet} = a \cdot V \tag{42}$$

Air exchange rates are standard in the industry, and they can even be regulated by various authorities. Table 8.7 shows typical values for various buildings/rooms in units of [h^{-1}] (i.e., per hour) that need to be converted to [s^{-1}] for equation (42). We can see that values range from 1 h^{-1} for residences up to 30 h^{-1} for bars. Understandably, as a

Table 8.7
Typical air exchange rates for different types of buildings/rooms

Building/room	*a* (1/h)
All spaces in general	4
Auditoriums	8–15
Banks	4–10
Bars	20–30
Cafeterias	12–15
Classrooms	6–20
Computer rooms	15–20
Department stores	6–10
Dining rooms (restaurants)	12
Homes, night cooling	10–18
Hospital rooms	4–6
Kitchens	15–60
Malls	6–10
Museums	12–15
Offices, public	3
Offices, private	4
Residences	1–2
Restaurants	8–12
Retail	6–10
Shopping centers	6–10
Theaters	8–15
Warehouses	2

Source: The Engineering Toolbox (2017b).

room gets crowded, the air exchange rate increases to replace the carbon dioxide that people breathe out with air for people to breathe in. Air exchange rates in new houses, however, are commonly around 0.35 h^{-1} or less—that is, changing the air roughly once every three hours thanks to mechanical ventilation systems—and the lower it is, the better in terms of energy use reduction.

Once we have estimated a volumetric flow rate, we have to estimate the heat transfer linked with air exchange $Q^{\bullet}_{air}$, and this is done by using the following equation:

$$Q^{\bullet}_{air} = \rho \cdot c_p \cdot V^{\bullet} \cdot \Delta T \quad (43)$$

where ρ is the density of air in [kg/m^3], c_p is the specific heat of air in [J/(kg·K)], and ΔT is the temperature difference between the outside and the inside in [K].

Example 8.6

Air Exchange in Civitas House

The house in Civitas from examples 8.1 to 8.5 has an exposed volume of 460 m^3, and the average air exchange rate of the total house is assumed to be 0.35 h^{-1}. (1) Determine the air exchange U value, U_{air}, of the house. Considering the total surface area of the house is approximately 400 m^2, (2) compare the heat loss coefficient from air exchange with other heat processes using the U_{wall} value of 0.34 W/m^2K from example 8.5 and (3) compute a new equivalent U value for the entire house and a new U_{eq} on a per surface area basis. Considering the floor area of the house is approximately 170 m^2, (4) calculate an equivalent U value per unit floor area U_{floor} (to be used later).

Solution

(1) Using equation (44) with $\rho \cdot c_p = 1{,}200$ J/(m^3K), we have:

$$U_{air} = 1200 \times 460 \times \frac{0.35}{60 \times 60} = 53.67 \text{ W/K}$$

(2) With a U_{wall} value of 0.34 W/m^2K and a total surface area of 400 m^2, we get a heat loss coefficient of $0.34 \times 400 = 136$ W/K.

The values suggest that air exchange is responsible for $100 \times 53.67 / (53.67 + 136) =$ 28.3% of the heat loss in the house. Although this example is greatly simplified, we can see that we are close to the 25% value given at the beginning of the section.

(3) Considering the two heat transfer processes are separate, we can simply sum the two U values, giving us a new U value of $53.67 + 136 = 189.67$ W/K.

On a per surface area basis, this gives us $U_{eq} = 189.67 / 400 = 0.47$ W/m^2K.

(4) From (3), we found that the total U value of the house was 189.67 W/K. Therefore, per unit floor area, we have

$U_{floor} = 189.67 / 170 = 1.12$ W/m^2K.

Under standard conditions (20°C), the component $\rho \cdot c_p$ is 1,200 J/(m^3K). Although it is often quoted as 1.2 kJ/m^3K, we prefer to use the value in joules since once we multiply it with $\rho \cdot c_p$ by $V^\bullet$ that is in [m^3/s], we will get [J/s·K], which is [W/K].[8]

As a last note, we can see that we can define a component U_{air} that can represent an air exchange U value similar to all heat transfer processes we have seen before, such that:

$$U_{air} = \rho \cdot c_p \cdot V^\bullet \qquad (44)$$

We have to be careful, however, because U_{air} depends on the volume of a building as opposed to the area of a surface. It is therefore expressed in [W/K] as opposed to in

[W/m^{2}K]. We can therefore sum U_{air} with U_{sys} but only after we have taken into account all surface areas within U_{sys}. An alternative is to divide U_{air} by the surface area of the building and then sum it with U_{sys}.

8.1.4 Heating and Cooling Efficiency

The final fundamental element to discuss before going to the energy demand section is to consider the efficiencies of heating and cooling. Two completely different mechanisms exist. Put simply:

- Heating: heat is generated
- Heating/cooling: heat is displaced

A furnace burns fuel, such as natural gas, to heat up air, oil, or water that in turn heats a room. Furnaces typically reach efficiencies in the range 0.80–0.98. This sounds high compared to the efficiencies we learned in chapter 5, but we have to remember that we are generating heat. In chapter 5, chemical energy was transformed into a mechanical movement, and losses occurred in terms of heat. In this case we only care about generating heat. The losses are therefore simply the heat that goes out the chimney and other small inefficiencies. The efficiency for furnaces is called annual fuel utilization efficiency (AFUE). If heat is generated using an electric radiator, then the efficiency is 1 since all the electrical energy is "lost" in heat, which is what we want.

The other common mechanism—used for both heating and cooling—is the vapor-compression cycle used in heat pumps such as those in refrigerators. It does not turn electricity into cold, but it uses electricity to remove heat from the inside of the refrigerator (and simply leaves it at the back). In this case, we talk about the coefficient of performance (COP). The COPs of air conditioners easily reach 2.8 to 4.3. For heat pumps that work in the opposite direction to provide heat (as opposed to removing heat), COPs can range even more, from 2 to 7, depending on whether heat is transferred from the air or from the ground. This may seem like the efficiency is greater than 1, but this is not the case. It means that with 1 kWh of electricity, we can displace 2–7 kWh of heat, but we are not converting energy to a different kind of energy (we are just displacing it).

In the United States, the instantaneous COP for air conditioners is also referred to as the energy efficiency ratio (EER), and it is subject to various climate conditions. With average climate conditions taken over the entire cooling season, the COP is instead referred to as the seasonal energy efficiency ratio (SEER). The EER and the SEER are typically calculated by dividing the instantaneous cooling output in [Btu/h] by the power rating in [W] of the air conditioner, thus in [Btu/h/W], which is the same as [Btu/Wh]. Using the conversion factor from chapter 1—that is, 1 Btu = 0.293071 Wh—we can convert the EER value into a unitless COP by multiplying it by 0.293071. Because the SEER is not instantaneous, it is a little more complicated, but the Pacific Gas and Electric Company recommends a conversion value of 0.9—that is, EER = 0.9 × SEER. The same company recommends air conditioners that have at least an EER or SEER value of 11 or 13 Btu/Wh, respectively, which represents a COP of 3.22 and 3.42, respectively. In this book, we will also cite EER and SEER values as dimensionless coefficients similar to the COP.

These efficiencies clearly indicate that cooling requires much less energy than heating. A small but insightful study showed that for the same living standards a house in Minneapolis, where it can get very cold, is 3.5 times more energy demanding than a house in Miami that has a hot and humid climate (Sivak 2013). We will therefore need to keep these mechanisms in mind when we look at energy demand.

The last important point here is to relate energy to GHG emissions. When electricity is used, we need to refer to the power grid emission factors that we discussed in chapter 5. For natural gas we can use a value of 1,923 g CO2e/m^3 or 172 g CO2e/kWh,[9] partly based on the emission factor given by the U.S. Environmental Protection Agency (2018).[10]

Accounting for the energy source is critical because it will dictate which technology is preferable when it comes to GHG emissions. Consequently, from this information, is it preferable to use natural gas or electricity to heat a home? Generally, an electric heat pump with a high COP is preferred, but given the choice between a gas furnace and electric radiators, the answer depends on the grid. Assuming an AFUE of 0.9, the emission factor from a natural gas furnace is 172 / 0.9 = 191 g CO_2e/kWh. Therefore, and since electric heaters have an efficiency of 1, if the power grid has an emission factor lower than 191 g CO_2e/kWh, it is preferable to use electricity for heating. On the contrary, if the power grid has an emission factor higher than 191 g CO_2e/kWh, it is preferable to have a natural gas furnace.

Back to our small example comparing energy demand in Miami and Minneapolis from chapter 5, we see that Florida and Minnesota have similar power grid emission factors. This suggests that not only is the energy demand 3.5 times higher in Minneapolis than in Miami, but the GHG emissions are also 3.5 times higher.

This completes our review of the fundamentals. We will now look at demand and review different metrics.

Example 8.7

Energy Required for Heating and Cooling

For the house in Civitas from example 8.6, we had calculated $U_{eq}=0.47$ W/m^{2}K. (1) Assuming we are heating a home with a furnace that has an efficiency AFUE=0.9, calculate the coefficient of energy required for heating. (2) Assuming we are heating a home from a ground-source heat pump (GSHP) with a coefficient of performance COP=5, calculate the coefficient of energy required for heating. (3) Assuming we are cooling a home from a room air conditioner with a coefficient of performance EER=3, calculate the coefficient of energy required for cooling. Note that we ignore solar radiation here for simplicity.

Solution

(1) For heating with a furnace, the energy coefficient $E_{h,f}$ needed for 0.47 W/m^{2}K is $E_{h,f}=0.47$ / 0.9=0.522 W/m^{2}K.

(2) For heating with a GSHP, the energy coefficient $E_{h,g}$ needed for 0.47 W/m^{2}K is $E_{h,g}=0.47$ / 5=0.094 W/m^{2}K.

(3) For cooling with a room air conditioner, the energy coefficient E_{ac} needed for 0.47 W/m^{2}K is $E_{ac}=0.47$ / 3=0.157 W/m^{2}K.

We can therefore see that per kelvin, heat pumps are much more energy efficient than furnaces.

8.2 Energy Demand in Buildings

From the fundamentals we noticed that heat transfer—and therefore building energy demand linked with space heating and cooling—is governed by three factors:

1. Material property, in particular for conduction and radiation
2. Area and geometry in general (length L for convection)
3. Climate, in the form of temperature differences between the indoors and the outdoors as well as the intensity and frequency of solar radiation

We are now going to evaluate the three factors, compile easy-to-use metrics (back to the IPAT equation and the Kaya identity), and look at current trends in the United States and the world. The first concept we will need to learn for that is the degree day.

8.2.1 Degree Days

The units of equivalent conductance U_{eq} are [W/m^{2}K]. The measure of temperature, here in kelvin, is in the denominator. We then multiply our formula by the temperature difference ΔT to get the rate of heat loss. What we get in the end is a rate—that is, a power, in [W] = [J/s]. What if we want to expand that to estimate the energy loss per day, per month, or per year? Multiplying our result by twenty-four hours for a day would mean that we are assuming the outdoor temperature to be constant, but this is obviously not true. What we need to do instead is to take the average temperature. This works well for a day, for a month, and even for a year, but we can compute one metric that can provide quite a lot of information about energy requirements. It is the degree day D that is defined as:

$$D = \sum_{days} (T_o - T_{ref}) \quad (45)$$

where T_o is the average daily temperature, and T_{ref} is a reference temperature that we choose and that represents the temperature above which cooling occurs and below which heating occurs in our building. Put simply, D is the sum of all the degrees that are above or below a certain temperature, and it has units of kelvin-days [Kd]. Because heating and cooling are not the same thing, we talk in terms of heating degree days (HDD) and cooling degree days (CDD).

For instance, if the average outdoor temperature in a day is 10°C (50°F) and our reference temperature is T_{ref} = 20°C (68°F), then the HDD for that day is 10. As an example over an entire year, in Chicago, setting T_{ref} = 20°C (68°F), the average yearly number of HDD between 2013 and 2017 was HDD = 3,939. Setting T_{ref} = 24°C (75°F), the average yearly number of CDD between 2013 and 2017 was CDD = 136. We can therefore see that in Chicago, much more energy is required for heating than for cooling. To collect these values, the practical website, DegreeDays.net was used.[11]

Table 8.8 shows the average HDD and CDD for various cities in the world. They are ranked based on their HDD. Colder places will require a substantial additional amount of energy compared to warmer places. Warm places, however, may require much energy due to heavy cooling loads. In general, temperate climates offer the best compromise for relatively low heating and cooling. In the table, Miami is the city with the least number of HDD and CDD.

The reference temperature can be set as the desired indoor temperature, but technically, it should be a little lower since a nonnegligible amount of heat is generated by the sun, the lights, and even the occupants. Based on the building and furnace properties, we can calculate a balance temperature, but we will not do it here.

We are now equipped with the right tools to estimate the annual energy requirements for a building. The degree day D incorporates both temperature and time, but it is in units of [Kd]. We can either leave it as is or multiply it by twenty-four hours to get [Kh], which will eventually give us [Wh]. We then only need to add our system conductance U_{sys} and the area A. The total energy needed E is calculated as:

Table 8.8
HDD and CDD for several world cities

City	HDD	CDD	Total
Calgary	5,424	30	5,454
Moscow	5,131	35	5,166
Chicago	3,939	136	4,075
Beijing	3,255	355	3,610
Seoul	3,134	252	3,386
London	3,097	19	3,116
Paris	3,071	57	3,128
New York City	3,011	177	3,188
Tokyo	1,989	273	2,262
Istanbul	1,978	212	2,190
San Francisco	1,961	20	1,981
Rome	1,930	224	2,154
Tehran	1,882	818	2,700
Mexico City	1,287	52	1,339
Sydney	926	153	1,079
Cairo	574	877	1,451
Miami	110	905	1,015
Singapore	—	1,447	1,447

Note: Reference temperatures were set at 20°C for heating and 24°C for cooling. The results are yearly averages for years 2013 to 2017, and they are ranked by heating degree day (HDD).

$$E = U_{sys} \cdot A \cdot D \tag{46}$$

Equation (46) is therefore expressed in watt-days [Wd]. If we multiply it by twenty-four, we get [Wh], and if we multiply it further by the number of seconds in an hour (3600), we get [J].

Note that sometimes we might use per unit floor area U value, U_{floor}, instead of U_{sys}, depending on the value of area that we have. Put simply, if A is the surface area, we use U_{sys}, and if A is the floor area, we use U_{floor}. This is mostly relevant when we compare different building types, as we will do later, since we are often given information on the floor area but rarely on the surface area of a building (i.e., walls and roof).

We can now expand the equation to an entire city or a country and use the IPAT equation. In the end we want I as the impact in [Wh] or [J]. P is still population in [pers]. The affluence in our case resides in two things. First, it is the number of degree days in [Kd] since it dictates how much heating and cooling we need. Second, we also need to take the area into account, which is part of the affluence variable, and we need

Example 8.8

Building Energy Consumption in Civitas

Civitas has a total population of 60,000 people, the average household size in Civitas is 2.6 people, and the average house floor area is 170 m^2. The average HDD is 2,902, and the average CDD is 104. In example 8.6, the per unit floor area *U* value U_{floor} was calculated as 1.12 W/m^2K. (1) Calculate the total number of households, *H*, and the average floor space per person, A_{avg}. (2) Calculate the per capita and per household energy required for heating and cooling for an entire year. (3) Assuming the furnace has an efficiency AFUE=0.9 and the air conditioners have an efficiency EER=3, calculate the input energy required for heating and cooling per household and per square meter.

Solution

(1) The total number of households is $H = 60{,}000 / 2.6 = 23{,}077$. The average residential building is 170 m^2 for 2.6 people. This means that the average floor space per person is $A_{avg} = 170 / 2.6 = 65.38\ m^2$.

(2) Using the IPAT equation, noting that we have two affluence metrics: A_{avg} and HDD and CDD. The per capita energy for heating and cooling are:

$$I_{h,pc} = A_{avg} \times \text{HDD} \times U_{floor} = 65.38 \times 2902 \times 1.12 = 212{,}501\ \text{Wd} = 5.10\ \text{MWh}$$

$$I_{c,pc} = A_{avg} \times \text{CDD} \times U_{floor} = 65.38 \times 104 \times 1.12 = 7{,}615\ \text{Wd} = 0.18\ \text{MWh}$$

The per household energy for heating and cooling are:

$$I_{h,ph} = P \times I_{h,pc} = 2.6 \times 5.10 = 13.26\ \text{MWh}$$

$$I_{c,ph} = P \times I_{c,pc} = 2.6 \times 0.18 = 0.47\ \text{MWh}$$

(3) The total energy required for space conditioning per household is $I_{ph} = I_{h,ph} / \text{AFUE} + I_{c,ph} / \text{EER} = 13.26 / 0.9 + 0.47 / 3 = 14.9$ MWh.

Considering an average household floor area of 170 m^2, the total energy required for space conditioning per m^2 is:

$$I_{space} = I_{ph} / 170 = 14.9 / 170 = 0.0876\ \text{MWh} = 87.6\ \text{kWh/m}^2$$

to include an average area per pers in [m^2/pers]. The technology *T* will include an average conductance and may be in units of [W/m^2K].

We should also note that very often, energy consumption per building already integrates the number of degree days as affluence *A* and technology *T* are combined in units of [kWh/year]. Although this makes using the IPAT equation easier, it prevents us from separating the impact of climate from the building properties.

Let us now look at one more measure that will help us understand what drives space conditioning in buildings.

8.2.2 Compactness and Shape Factor

We note that all the metrics we have used so far have considered the area of walls and roofs, as opposed to the volume of a building. This is in fact a major consideration since heat is lost through surfaces. A way to minimize heat loss is therefore to minimize the surface area for a given volume. We can define compactness C as:

$$C = \frac{V}{A} \tag{47}$$

where V is the volume of a building [m^3], and A is the total exposed area of the building [m^2]—that is, walls, floor, and roof. Although compactness is in units of length [m], it is really not related to any particular length. The best analogy is the hydraulic radius and hydraulic depth in open channels that we covered in chapter 6.

The optimal geometry for a high compactness is a sphere, which is impractical, but the next best geometry is a cube. Figure 8.7 shows two buildings with the same volume but two different values of compactness. The building on the left-hand side is a cube with a high compactness, while the building on the right-hand side is more rectangular, exposing more surface area to the environment, and thus having a lower compactness.

A relative compactness RC can also be calculated when the compactness of a building C is compared to the compactness C_{ref} of a reference building such that:

$$RC = \frac{C}{C_{ref}} \tag{48}$$

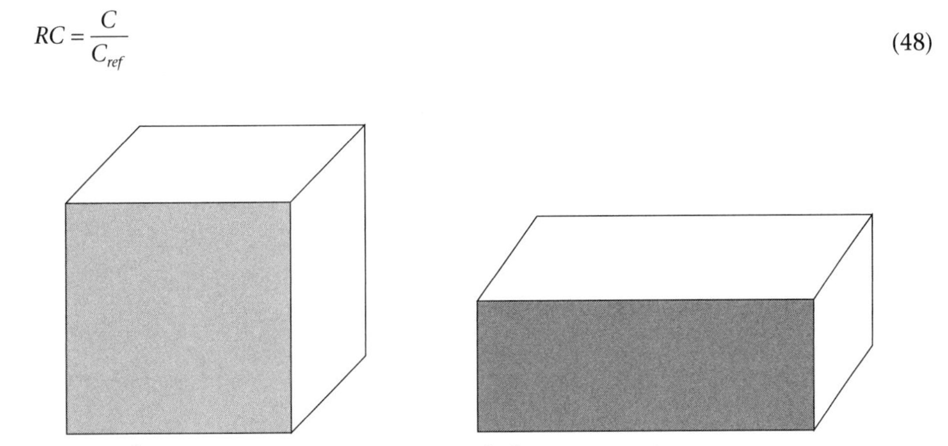

Figure 8.7
Examples of high and low building compactness.

With compactness in mind, another useful measure is the shape factor S that simply divides the length of the building L by the width W:

$$S = \frac{L}{W} \tag{49}$$

Although we will not discuss the geometry of the building further, we will keep in mind that reducing the exposed area for a given volume is desirable to reduce energy needs. We will now review some trends building energy demand.

8.2.3 Building Energy Demand Trends

Before comparing buildings with one another, we can compare how building energy demand varies by region. Table 8.9 shows an average residential building energy use for the four census regions in the United States in 2015, and it also shows an average value for the United States collected from RECS 2015. Figure 8.8 shows the same data. The data shown is also the same that was shown in figure 8.2a, except that we are not looking at energy use per building but per [m^2] instead.

Unsurprisingly, we see that houses in the Northeast and Midwest consume more energy per [m^2] than houses in the South and West purely due to space heating. Space heating in the Midwest, for instance, requires 73 kWh/m^2 compared to 37 kWh/m^2 in the South; this is a difference of 36 kWh/m^2. However, space cooling is higher in the South, with 18 kWh/m^2 compared to 6 kWh/m^2 in the Midwest; this is a difference of 12 kWh/m^2. Not only does space cooling require less energy, but when taking into account AFUE and COP, space heating becomes much more energy-intensive—and as we saw before, a house in Minneapolis requires 3.5 times more energy than a house in Miami.

From these values we can also see that the result we obtained from example 8.8—that is, 87.6 kWh/m^2 for space conditioning—is close to the value found in the Northeast.

Table 8.9

2015 regional residential building energy use in the United States

	Average size		Energy intensity (kWh/m^2)					
Region	ft^2	m^2	Total	Space heating	Air conditioning	Water heating	Refrigerators	Other
Average	2,008	187	121	55	11	23	4	32
Northeast	2,090	194	142	81	5	26	3	30
Midwest	2,278	212	131	73	6	22	4	28
South	1,939	180	112	37	18	21	4	35
West	1,791	166	106	39	9	27	5	33

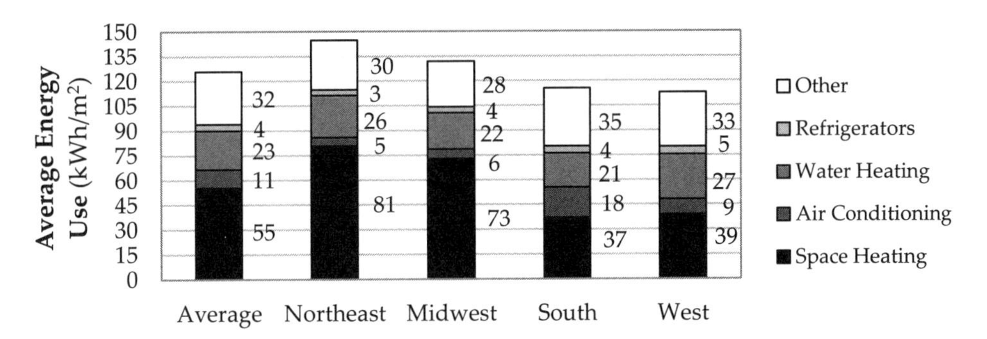

Figure 8.8
Average energy use per square meter in residential buildings by region in 2015.

This is partly because we assumed the same *U* value for one wall with windows applied to all surfaces, which is not accurate, as floors and roofs tend to have much lower *U* values, but we still calculated a reasonable value.

In reality, we can get much lower than the values shown in table 8.9 and even close to 0! In fact, the concepts of passive houses and net zero–energy homes are relatively popular.[12] For instance, two similar houses were built in Florida. One was a conventional construction, and the other had various features, some of which we will discuss in the next section, to make it less energy-intensive. Over one year the conventional house consumed 22,600 kWh, compared with 6,960 kWh for the new house. Furthermore, solar panels installed on the roof produced 5,180 kWh of electricity, therefore decreasing the energy consumption of the house to 1,780 kWh, which is 7.8% of the conventional house. This means that the new house saved 92% of its total energy consumption.

These savings do not only apply to warm places that require mainly space cooling. A similar project was undertaken in Washington, DC (using ground-source heat pumps for heating), and the house consumed a total 10,585 kWh while producing 7,510 kWh of electricity from solar panels, thus its net consumption was 3,075 kWh. Although no control house was built in this case and considering the house was 238 m^2, 3,075 kWh represents 12.9 kWh/m^2. Therefore, its energy consumption was reduced 88.5% compared to values given in table 8.9 for houses in the South region (where Washington, DC, is located as defined by the Census). Similar results were obtained in California and Colorado, as well as in other states in the United States (Parker 2009).

Commercial buildings typically consume more energy than residential buildings. Table 8.10 shows average total and space conditioning energy uses for the United States by building type, collected from CBECS 2012. The value for an average residential

Table 8.10

Average total and space conditioning energy use per square meter in commercial buildings by building type in 2012

	Total energy use		Energy for space conditioning	
Building type	kWh/m^2	Ratio	kWh/m^2	Ratio
Average residential	121	1.00	66	1.00
Religious worship	120	0.99	62	0.94
Retail (other than malls)	211	1.74	67	1.02
Education	217	1.79	101	1.52
Office	245	2.03	84	1.27
Average commercial	252	2.08	87	1.32
Hotels	305	2.52	57	0.87
Enclosed and strip malls	345	2.85	85	1.28
Hospitals	730	6.03	307	4.65
Restaurants	891	7.37	132	2.00

building was added to compare the two sectors. We can see that restaurants tend to consume much more total energy per [m^2], and this is mainly because of cooking—again, because heat needs to be generated. When looking at the energy use for space conditioning, it is relatively low, especially compared to hospitals that need to keep a strict temperature and high ventilation rates throughout the day.

The data presented in table 8.10 is the same as the data used for figure 8.2b. Space heating is here again the main consumer of energy. Although space cooling seems to have a minor role, it really does not. Space cooling plays a major role. In fact, it is much too common to walk in commercial buildings on a hot summer day and to feel too cold or even to wear a sweater. Air-conditioning units are typically oversized and poorly operated. In fact, we have estimated that more than 100 million MWh (more than 100 TWh) of energy were lost in 2012 due to overcooling commercial buildings in the United States. This represents 7.83% of the total electricity used by commercial buildings. Financially, this was a loss of more than $10 billion, and environmentally, this was a loss of more than 55 Mt CO_2e (Derrible and Reeder 2015). These results can only let us imagine how much energy is lost in heating because of leaks in buildings.

To estimate the energy demand from buildings, several software packages exist. In particular, BEopt™ (Building Energy Optimization) from the NREL, which uses the package EnergyPlus, and eQuest from the U.S. Department of Energy are free and easy to use while offering a formidable panoply of options.[13]

This completes our section on trends in building energy demand. Unfortunately, this section does not contain any international examples. Beyond general energy demand values, it is actually quite difficult to gather information on building energy simply because energy highly depends on building properties. Unlike electricity, water, and transport, for which we can calculate meaningful average demand metrics at the city scale (i.e., the average consumption of electricity and so on), this is not typically possible for buildings. In fact, the RECS and CBECS data sets provided by the EIA are quite unique in the world.

In the next section, we will list various strategies that we can take to reduce building energy use, therefore lowering demand for energy in buildings in line with our first sustainability principle.

8.3 Building Design and Technology Recommendations

The famous French writer and pilot Antoine de Saint-Exupéry (1939) has a great quote that is pertinent to the goal we are trying to achieve:[14] "Perfection has been attained not when there is nothing more to add, but when there is nothing more to take away" (p. 60).[15] Although it applies to many, if not most, designs in life, buildings might just be the exception. In buildings, we need to provide additional layers to avoid undesirable heat transfers.

In this section we will review some common strategies to decrease building energy use. There are actually two sets or families of strategies that we can adopt:

1. Passive design strategies to reduce energy requirements
2. Active technologies to reduce energy use

The first family relates to the way buildings are designed in the first place—for instance, by minimizing the outer surface for a certain volume. The second family relates to the technologies that we use—for instance, by using ground-source heat pumps (defined later) as opposed to conventional furnaces. In this section we will look at strategies in both families, starting with better designs.

Naturally, many more strategies exist beyond the scope of this book. *The New Net Zero: Leading-Edge Design and Construction of Homes and Buildings for a Renewable Energy Future* by Maclay (2014) is a great reference.

8.3.1 Better Designs

Many strategies exist to reduce energy use, thanks to good design. This is not a new field. In fact, we used to be really good at it before we had powerful heating and cooling technologies. Even the very first treatise on architecture, the famous *De Architectura* by Vitruvius (1999), written around 15 BCE and that we briefly discussed in chapter 4, examines building design extensively to take into account not only sun exposure but also wind that creates drafts. Here we review several practices, starting with a surprisingly simple one.

8.3.1.1 Size When thinking about designing a building, the first question that comes up is size. We want to make sure that the building is not too small to serve its purpose. At the same time, a building too large may require more energy for its daily operation while space is inappropriately used.

We can relate building size back to what we learned in transport. A building too large with few people will take unnecessary land that could be used for other purposes.

To illustrate this point, I remember a work of art that I saw in Seoul, South Korea, during a visit in summer 2014 (figure 8.9). Titled *Home within Home within Home within Home within Home* (2013) by Korean artist Do Ho Suh, the piece was exhibited in the Seoul Box gallery at the National Museum of Modern and Contemporary Art. The artist reproduced the real size of his Korean home before leaving for the United States and placed it inside the American home he had lived in during his studies in Boston. The difference in size is impressive. Clearly, the large building requires a lot more energy, but at the same time it likely houses more people. Thinking of it in terms of space per person [m^2/pers] and energy per person [Wh/pers] can help.

In line with the main message of this book on urban infrastructure integration, we have to think about the amount of energy required in terms of electricity and water, as well as the energy required for transport. Generally speaking, a large building full of people in a dense location is likely to consume less energy per person than a small empty building far from everything that demands very long trips.

8.3.1.2 Compactness The second recommendation is one that we have seen before. As discussed, heat dissipates through surfaces, and limiting the area of a building is the second strategy that we can think of. Naturally, we do not necessarily make all buildings into a cube like the one we saw in figure 8.7, and we enjoy longer buildings with windows, but it is a practice that we can keep in mind. Sometimes, however, we will expose more surface on purpose, as we will see in shading.

8.3.1.3 Orientation Orientation is a commonly used practice by architects. In the Northern Hemisphere, ensuring that one of the walls faces south, therefore receiving the greatest amount of radiation heat from the sun, is favorable in colder climates and undesirable in warmer climates.

In houses, areas such as livings rooms are often located on the southern side of the house to receive sun during daily activities, while bedrooms are on the northern side since we tend not to use them as much during the day.[16]

Figure 8.9
Home within Home within Home within Home within Home (2013) by Do Ho Suh, Seoul Box gallery at the National Museum of Modern and Contemporary Art, Seoul, South Korea.

In commercial buildings, it may not be as important to have certain rooms on the southern side. Nevertheless, equipped with solar panels to produce electricity or with vertical gardens to absorb heat in the summer, south facades can become important. But we will discuss technologies later.

8.3.1.4 Shading Shading is arguably the least used design strategy, although it used to be common before the extensive use of air conditioners. Radiation from the sun is

Figure 8.10
Shading: outside shades (*left*) and indoor shades with air gap (*right*).

desirable in the winter, but solar rays are weak, and the heat gains are not significant. In the summer and in warm places, radiation from the sun is not desirable. Shading can therefore help to prevent or reduce radiation.

For a house, the easiest shading technique is to have trees that can substantially help drop the surface temperature. Similarly, a large structure can be placed between the sun and the house. Roofs can also be extended to provide shading to living areas. Another strategy is to use materials, windows, or paints with a high reflectivity so that the radiation bounces back into the atmosphere instead of being absorbed—although the neighbors may not be happy since they will likely have to bear the radiation. We will discuss some of these technologies in the next section.

Another design for houses is to have window shades outside of the house, which is common in Europe (figure 8.10, *left*). Once the sun's rays enter a building, it is too late, and closing the interior shades simply lets the heat stay around the windows.

On the topic of indoor shades, mainly for commercial buildings, some buildings with floor-to-ceiling windows now leave a gap between the floors and the walls, as illustrated in figure 8.10, *right*. This way, when the shades are down and since warm air rises, the warm air simply evacuates at the top of the building. This technique can in fact be quite effective, short of having outside shades.

Staying with commercial buildings, self-shading used to be a common practice. In fact, most skyscrapers built until the 1950s incorporate many self-shading designs. One of the best examples is the famous Empire State Building in New York City (figure 8.11,

Figure 8.11
Self-shading of buildings: building in New York City (*left*) and Empire State Building (*right*).

right). The "groove" in the middle of the building is a self-shading design feature, so the building does not get too warm in the summer. The building on the left in figure 8.11 shows this feature more clearly, again with a small groove in the middle. Other methods include reflecting the radiation from the sun or having something to absorb it, such as plants (green roof) or water (blue roof), but we will look at those later.

This concludes our short review of design recommendations. Naturally, many more designs exist, such as playing with shading and opening windows to encourage air ventilation in the summer (e.g., opening a window downstairs and upstairs in a multistory house to create a draft). This chapter is too short to review all of them, but it will suffice.

8.3.2 Technologies

Good design is always preferable to the use of technologies—back to Tainter's diminishing marginal returns from chapter 1—but in addition to good design, technologies can help to further decrease building energy use. In this section we review several technological recommendations, starting from the most simple and mainstream to more sophisticated technologies.

In addition, toward the end of the section, we will analyze how some of these technologies (individually and combined) perform by comparing their internal rates of return and their potential to reduce GHG emissions. We will also briefly discuss the Leadership in Energy & Environmental Design (LEED) rating.

Again, we will overlook technologies linked specifically to appliances, such as efficient lighting and water heating, and will instead focus on technologies that have a direct impact on heating and cooling.

8.3.2.1 Turning Off and Down Equipment Although it sounds ridiculous, the first technology that we must cite is "the finger," or if we want to be a little more advanced, we can think of a smart thermostat. So much heat is generated every year to heat empty rooms and buildings, and likewise, air conditioners work hard to cool empty rooms and buildings. Similarly, rooms can be overheated and overcooled. This energy cost of overheating and overcooling is substantial.

The main textbook in the field of heat and cooling that we quoted earlier sums it up well: "This brings us to a fundamental recommendation: *Turn equipment down or off when there is no need for it*" (Kreider, Curtiss, and Rabl 2009, p. 718). We feel the lack of application of this rule very often when walking into an overheated or overcooled building.

8.3.2.2 Sealing Leaks Air leaks are not only common, they are ubiquitous. Fixing all or at least most air leaks is paramount. Investing in expensive insulation to have all the hot and cold air leak out because of small gaps is a frequent problem. Fixing leaks does not take much time, and it is often not costly.

8.3.2.3 Windows The third biggest recommendation is to carefully select windows. We learned that heat transfer greatly depends on the conductance U of a wall, and windows have significantly higher U values than well-insulated walls. The norm is to have at least double-pane windows, if not triple-pane windows. The air gap between the panes acts as a great insulator, thanks to the low k value of air.

8.3.2.4 Insulation Remaining on conductance, materials with better insulating properties will have a higher R value and therefore a desirable low U value. Providing that all or most of the leaks in a building have been sealed and windows have double or triple panes, investing in better insulation is then the next best thing. In fact, for a typical house, following these first four recommendations may very well be enough.

8.3.2.5 Reflecting Material/Paint As we saw previously, materials or paints that reflect the sun's rays can be useful in the summertime or in warm places to reduce heat gains from the sun. This is as easy as having a white/light exterior finish on the wall of a building or light/gray asphalt for the roof. Note, however, that a light surface will simply reflect radiation. More sophisticated techniques are presented next.

8.3.2.6 White-Blue-Green Roof In summertime, the roof is the main area of a building that receives radiation from the sun. Adding a reflecting material to the roof—for

Figure 8.12
Green roof on the Chicago City Hall. Courtesy of the City of Chicago.

example, by painting it white—will help decrease the amount of heat transfer in the building. As we saw earlier, however, light-colored surfaces simply reflect radiation—for example, to neighboring buildings. An alternative is to put something on the roof that absorbs these radiations without transferring the heat accumulated to the house. Two options are fairly popular.

The first is to put plants out on the roof that directly use the sun's energy to grow. These are called *green roofs*. They are more popular for commercial buildings, and they can offer attractive roof gardens as well. How effective are they? Figure 8.12 shows a building in downtown Chicago where half is used as the Chicago City Hall, which has a green roof, and the other half as the Cook County Clerk's Office, which does not. The picture on the right shows an infrared image of the building. We can see that temperatures on the Cook County Clerk's Office go as high as 66°C (151°F), while they remain at a cool 23.3°C (74°F) for the Chicago City Hall, thanks to its green roof. We should note, however, that it is not because of a hot roof that heat enters the building; it also depends on the adequacy of the building's insulation.

The second option is the blue roof. Blue roofs have aggregates—that is, gravel, typically of a light (reflective) color—that get filled with water during rain events. When the sun shines, most of the energy then goes toward evaporating the water located in the aggregates or simply in heating the aggregates as opposed to having heat enter the building. Although they may not be as effective as green roofs, blue roofs are easy to maintain and quite common, but they can only be installed on flat roofs. The aggregates on blue roofs also act as a weight to keep roof shingles in place during strong winds, and they also help extend the life of roof shingles since sun exposure can induce cracks and general wear. Figure 8.13 shows a typical blue roof in the foreground and

Figure 8.13
Examples of blue (*foreground*) and green (*background*) roofs in Madison, Wisconsin.

a green roof in the background (that also acts as a terrace). The blue roof is part of the Wisconsin State Capitol in Madison. The green roof is above two of the best restaurants in Madison, which may in fact use the green roof to grow vegetables and herbs.

8.3.2.7 Solar Water Heating Roofs can also be used to specifically warm things that we would like to warm, like water, for instance. In Southeast Asia, most roofs of large buildings have large tanks full of water. The water settles in the tank until it is used and simply warms up. Figure 8.14a shows an example of a solar heating water tank in Cambodia.

Naturally, this is not possible all year round in colder places like Chicago because the water in the tank would freeze in the winter. Nonetheless, tubes filled with water can also be used seasonally on roofs. The water is then prewarmed before being heated in a water heater. Many houses throughout the world use this method, including in Beijing, where the picture in figure 8.14b was taken.

a) Siem Reap, Cambodia

b) Beijing, China

Figure 8.14
Solar water heating.

8.3.2.8 Solar Photovoltaic Although we will not dwell long on the subject since we partially covered it in chapter 5, roofs offer a prime spot for solar photovoltaic (PV) panels. The solar panels use the radiation from the sun to generate electricity. The electricity generated can then be used to meet the electrical needs of the building—for example, to run a water heater or an air conditioner.[17] This means that solar panels prevent significant heat gains to the house, while providing electricity that can be used to further cool the house. In fact, air conditioners are most needed on very hot days when the radiation from the sun is the strongest—and thus when more electricity can be generated.

8.3.2.9 Vertical Gardens In line with roof gardens, vertical gardens are somewhat popular. Vertical gardens are essentially plants that can grow on vertical surfaces. These types of gardens are especially popular next to expressways since the CO_2 emitted from vehicles can be used directly by the plants, but they can be used on building walls as well to absorb the radiation from the sun, although their impact can be limited (Susorova, Azimi, and Stephens 2014).

8.3.2.10 Air-Source and Ground-Source Heat Pumps As discussed, displacing heat is much more energy-effective than generating heat. A method to warm up a building is therefore to displace heat from one place to another using a heat pump.

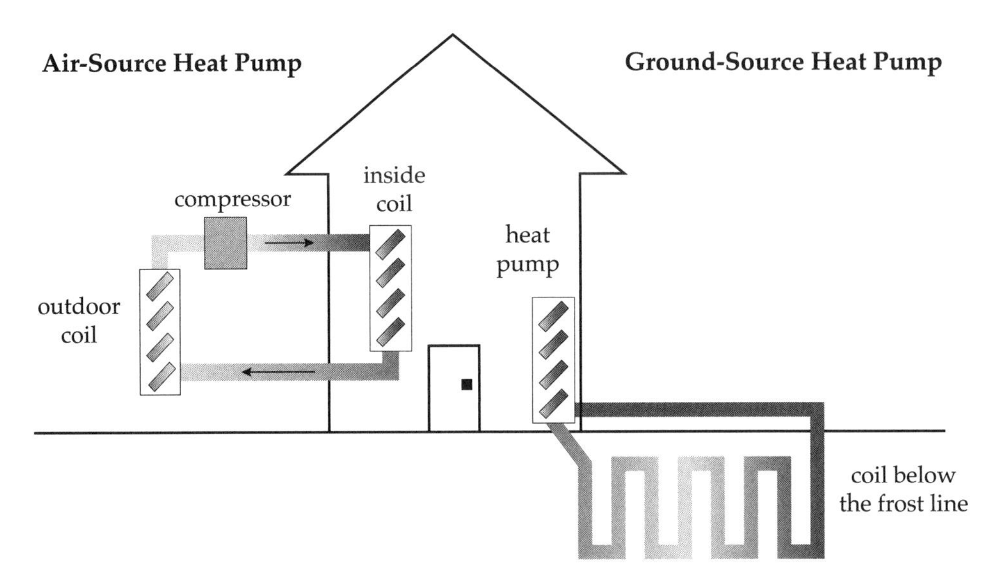

Figure 8.15
Air-source heat pump (*left*) and ground-source heat pump (*right*).

One option is simply to use outside air. Essentially, "heat" is extracted from the outside air and is displaced inside. This also works when the outside air is very cold. The process makes the outside air even colder and displaces the heat captured inside. These types of systems are air-source heat pumps (ASHPs). A sketch is shown on the left-hand side of figure 8.15. Although they work in all climates, their performance can be poor in cold climates, and an alternative is to use the ground.

Soil below the frost line—about 1.8–2 m (6–7 ft) underground—keeps a relatively constant temperature all year round, around 10°C (50°F) depending on geographical location. The temperature differential with the outside air can be useful. In the summer the soil underground can be used to cool a building, and in the winter it can be used to heat a building. The systems used here are ground-source heat pumps (GSHPs). The right-hand side of figure 8.15 shows a schematic of how a GSHP works.

For a GSHP, pipes are buried in the ground, either horizontally (e.g., in a soccer field) or vertically (e.g., in bore holes, and they can also be placed in foundation piles; Amatya et al. 2012). GSHPs have a SEER of about 4 for heating and 6 for cooling.

Depending on the nature of the underground, heat can even be stored and used later. There are two types of underground thermal storage systems: (1) the aquifer thermal energy system and (2) the bedrock thermal energy system. In both cases the medium is warmed up in the summer because the outside air is warmer. In the winter the trapped heat is then used to heat buildings. The medium is therefore cooled, and this cold is used the next summer to cool down buildings.

Overall, ASHPs and GSHPs can be very effective at providing thermal comfort with a minimum amount of energy. Coupled with a clever design, no leaks, efficient materials, and an effective operation, they can significantly help achieve passive house or net zero-energy homes, as discussed in section 8.2.3.

8.3.2.11 District Heating and Cooling The last technology that we will discuss is a little more advanced than the others, and it expands further than a simple building. District heating and cooling systems capitalize on economies of scale by using one or multiple large boilers or chillers to serve multiple buildings. These systems are common on campuses, complexes, and places with multiple large buildings like business districts. They do not resolve the problems we have seen earlier, and they may be deterrent to resilience, but they can be effective. Economies of scale do stop when losses in the systems become too large, however.

An example of a district heating system is a combined heat and power (CHP) system, as briefly discussed in chapter 5. CHPs generate electricity and the heat produced from the gas turbine can be used to heat entire districts in the form of steam or water at high temperatures.

For cooling, deepwater sources that tend to have a fairly uniform temperature at around 4°C (40°F) all year round can be used. In Toronto, Canada, most commercial buildings in the central business district are cooled using water from Lake Ontario, for example.

8.3.2.12 Technologies and Internal Rate of Return The previous technologies were listed in somewhat of a hierarchy, from simple to more complicated, but we can go one step further now. Usually, you will ask yourself two questions before selecting a technology. First, what is the GHG reduction potential of the technology? Second, when will I start saving money thanks to the new technology?

Bristow and Bristow (2017) wondered the same thing. They wanted to retrofit their house when they lived in Southern Ontario (Canada), and they were not sure which technology to select. They therefore decided to simulate the performance of many of them, both individually and by combining some, using BEopt. Quite naturally, the indicator for GHG reduction potential is the percentage of GHG reduction (compared to their existing energy consumption levels). In terms of economics, and like large infrastructure projects, they decided to look at the internal rate of return. Ideally, we would

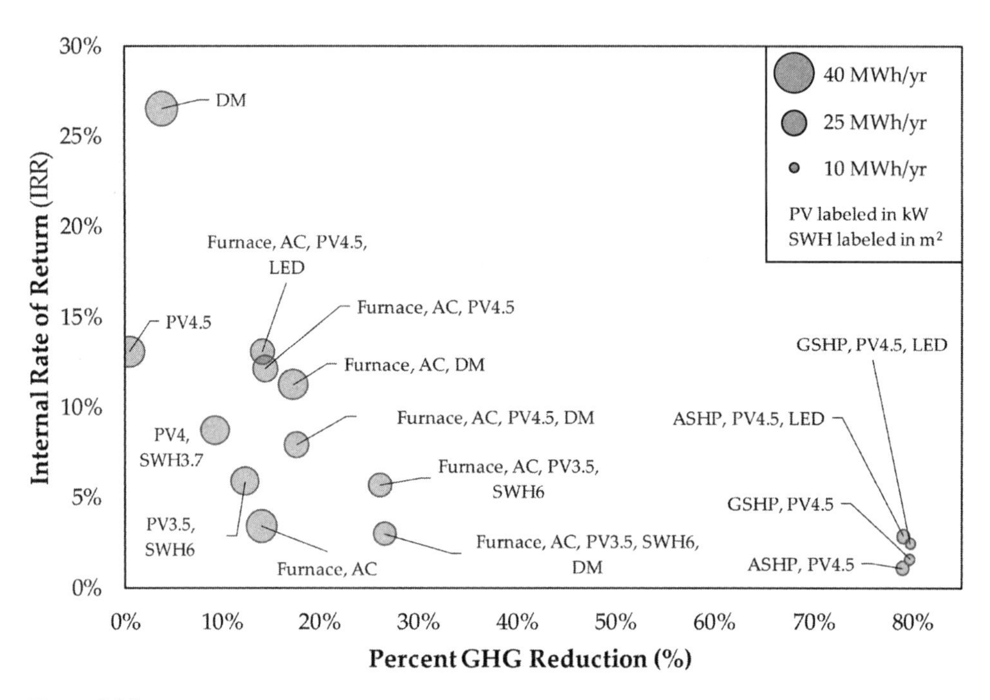

Figure 8.16
Comparing the IRR versus the GHG reduction potential of several technologies for a house in Southern Ontario. Adapted from Bristow and Bristow (2017).

prefer both GHG reduction potential and internal rate of return (IRR) to be high. The results are shown in figure 8.16. The full study is available at Bristow and Bristow (2017), and it includes a part for transport as well.[18]

In the figure, *LED* stands for light-emitting diodes (the only lighting technology we will see in this chapter); *DM* stands for demand management and includes light, insulation, and appliance upgrades; the GSHP has a SEER of 4.86;[19] the ASHP is given a SEER of 5; the furnace is given an AFUE of 0.98; the air conditioning (*AC*) is given a SEER of 6.15; PV4.5, PV4.0, and PV3.5 represent solar PV with power ratings of 4.5, 4.0, and 3.5 kW, respectively; and SWH6.0 and SWH3.7 represent solar water heating with a surface area of 6.0 and 3.7 m^2, respectively. The size of the bubbles represents the total energy used by the house over a year, and we can see that the size of the bubbles decreases from close to 40 MWh (*top left*; close to business as usual) to close to 10 MWh (*bottom right*; hence the nearly 80% GHG reductions).

First, from the figure we can see that there are no obvious "winners." Either the IRR is high or the GHG reduction potential is high, but we cannot have both at the same time.

Put differently, to get significant GHG reductions, we need to make serious investments. Implementing significant but basic demand management strategies—that is, switching to LED lights, installing better insulation, upgrading appliances—may be "easy" and reasonably cheap, but it is not nearly as effective as making substantial investments in a technology such as an ASHP or GSHP.

Second, we see that the most effective technologies deal specifically with space conditioning. Southern Ontario gets cold in the winter, and switching to ASHPs or GSHPs can help substantially. This is specifically the case in Ontario, which gets most of its electricity from nuclear and thus has a relatively low power grid emission factor of about 110 g CO_2/kWh at the time of this writing.

Third, we see that combining multiple technologies tends to improve the GHG reduction potential as well (naturally, at the expense of the IRR).

As an additional note, when it comes to solar, the figure also shows us that solar water heating is more effective than solar PV. Indeed, despite being in a cold climate, the PV3.5, SWH6 strategy achieves higher GHG reduction than the PV4.0, SWH3.7 strategy. Therefore, before thinking of installing PV panels on your roof, you might want to consider installing a solar water heater instead.

8.3.2.13 Leadership in Energy & Environmental Design Rating The Leadership in Energy & Environmental Design (LEED) rating system was developed by the U.S. Green Building Council (USGBC) in the 1990s to help architects design more efficient buildings. The USGBC is a private nonprofit organization. It established a rating system to assess the energy performance of buildings based on their design. The rating goes from certified to platinum depending on the number of points scored during the design phase, many of which may have come from predicted energy savings.

There has been a lot of controversy surrounding LEED, and we will not discuss the LEED system too much. One of the greatest concerns with the LEED rating was that the score was based purely on the initial design, so it did not account for problems in the construction and errors from contractors (e.g., the presence of significant leaks). This seemed to be changing at the time of this writing, however. After all, the LEED rating is relatively new and therefore not exhaustive, and it is frequently updated. On the positive side, the rating system has been quite successful in terms of adoption, and being LEED certified is well regarded, encouraging more architects to design ever more energy-efficient buildings.

The score sheet used by the USGBC to assess the performance of a building can easily be downloaded from the USGBC website, and it can serve as an effective checklist to determine whether a building is on the right track to energy efficiency.

8.4 Conclusion

This concludes our chapter on buildings. After looking at how energy is used in buildings, we decided to focus purely on space conditioning: the heating and cooling of buildings. After all, buildings are mainly used because they offer a temperature-controlled environment, and we therefore first reviewed the principles of thermal comfort.

From thermal comfort, we quickly moved on to review the fundamentals of heat transfer, detailing the three processes: conduction, convection, and radiation. For conduction, we used Fourier's law, which is nearly identical to Darcy's law in groundwater engineering but is applied to heat transfer. Notably, we learned the meaning of *R* values and *U* values, which capture the property of materials to conduct heat. We then used a simplified version of Newton's law of cooling to define convection. We saw that convection is largely governed by a convection coefficient, h_c, that depends on a number of properties of air in our case (including air velocity). Finally, we examined the principles of radiation, and through various approximations, we defined a similar radiation coefficient, h_r. We then combined the three heat transfer processes to assess the heat transfer properties of buildings. Subsequently, we paid particular attention to windows and air exchange in buildings, which are major sources of heat loss. In the first section, we also discussed thermal efficiency, learning that generating heat is much more energy-intensive than displacing heat.

In the second section, we focused on analyzing energy demand in buildings, and we first learned about the concept of degree day that we use to estimate how much a building will have to be heated/cooled to reference temperatures during a year. We then discussed the notion of compactness and shape factor, which often drive the demand for energy in buildings. We finally looked at some trends in the United States, listing successful examples of low-energy buildings and citing two software packages that can be used to estimate the energy demand of buildings.

Finally, the last section was devoted to design and technology recommendations to lower the energy use of buildings. The first set of recommendations focused on better designing buildings, discussing the impact of size, compactness, orientation, and shading. The second set of recommendations dealt exclusively with technologies, from the simple and yet poorly used "finger" to turn equipment off or down to better insulation, green roofs, solar panels, and GSHPs.

If we go back to the quote at the beginning of the chapter, we should not forget that the life of a building is long. This means that a poorly designed or constructed building will have a lasting negative impact on energy demand. Adopting desirable standards, techniques, and technologies right now is therefore essential for us to have an efficient building stock in fifty years. In fact, the building sector is the only sector where growth can translate to lower energy use, as long as the buildings are energy efficient. We saw that we could improve energy use by 90% fairly easily with off-the-shelf technologies and strategies.

We have now reviewed the big four types of infrastructure systems—electricity, water, transport, and buildings—but we are not done yet. There is one more aspect of urban engineering that is fundamental and that we must address: trash. We generate a significant amount of trash every year, and managing the collection, processing, and final disposal of trash is far from trivial. We will dedicate an entire chapter to it. It is time to discuss solid waste and solid waste management.

Problem Set

8.1 The main purpose of this problem is to estimate your personal carbon footprint from your home energy use. (1) Download and install the software packages eQuest or BEopt, enter your home properties to the best of your ability, run the simulation, and report your yearly household space heating and cooling loads, plus the total energy use. (2) Compare your estimated gas and electricity use from the software calculation with the values you calculated in problems 1 and 2 of chapter 5. Note that both your electricity and gas consumption from chapter 5 should be larger since they included consumption from electric appliances and from cooking. (3) Considering a natural gas emission factor of 0.172 t CO2e/MWh and using your regional grid emission factor, estimate your household emission from building energy use and your personal consumption. (4) Add three additional features to your house (e.g., solar panels), rerun the simulation, and report the savings.

8.2 Download the latest data available from the EIA's *Monthly Energy Review*.[20] Update figure 8.1 with the latest (1) monthly and (2) annual data.

8.3 Download the latest data available from the EIA's *Monthly Energy Review*.[21] (1) Plot pie charts similar to figure 8.1 but applied to monthly data for the past twelve months and (2) comment on the results.

8.4 Based on the metabolic rates given in table 8.1, (1) estimate all the activities you do during a typical day and determine for how long (in hours). (2) Convert these power values to energy values in [Wh/m^2]. Multiply these values by an estimated surface area of your body and (3) estimate how much energy in [Wh] you use per day. Considering that 1 Wh is about 0.86 calories, (4) calculate how many calories you burn in a typical day.

8.5 In your own words, define the three types of heat transfer exchanges: (1) conduction, (2) convection, and (3) radiation.

8.6 Two insulating materials are available for a new construction project. Material A has a thermal conductivity k value of 0.05 W/(m · K) and costs $2.99 for a sheet with a thickness of 10 cm. Material B has a k value of 0.036 W/(m · K) and costs $3.99 for a sheet with a thickness of 8 cm. Based on this information, determine the material that offers the best performance for the price.

8.7 The back wall of a store measures 30 m wide × 10 m high. The exterior has a 20 cm thick uninsulated concrete block (k = 0.45 W/(m · K)) that also supports the wall, and the interior has a 5 cm thick layer of gypsum. The insulation in between is composed of a 40 cm thick layer of cellular glass. (1) Calculate the equivalent conductance U_{eq} for the wall. Assuming that the average temperature difference in the year is 15°C, determine the (2) per unit area and (3) average heat loss by the wall.

8.8 A house has a wall that measures 8 m wide × 6 m high. The exterior has a 7-cm-thick layer of brick (common) and the interior has a 1.25-cm-thick layer of gypsum. The insulation in between consists of a 9-cm-thick layer of cellular glass. Wood studs (hardwood) are placed every 1 m and are 4.5 cm wide × 9 cm deep (similar to the cellular glass). Five windows measuring 90 cm × 1.5 m are located on the wall, with U_w values of 2.84 W/m^{2}K. (1) Based on this information, calculate the equivalent conductance U_{eq} for the wall. (2) The average temperature difference in the year is 8°C. Determine the per unit area and average heat loss by the wall. (3) Assuming the four walls, roof, and floor of the house have the same area and heat losses, determine the total energy consumed in the year in Wh.

8.9 The wall of a store measures 12 m wide × 8 m high. The exterior has a 7 cm thick layer of brick (face), and the interior has a 5 cm thick layer of stucco. The insulation in between is composed of a 30 cm thick layer of glass fiber (organic bonded). Wood studs (hardwood) are placed every 4 m and are 5 cm wide × 10 cm deep against the brick, and the remaining 20 cm are filled with the same insulating material as the rest of the wall. Finally, two large 3 m × 3 m windows are located on the wall, with U_w values of 2 W/m^{2}K. (1) Calculate the equivalent conductance U_{eq} for the wall. (2) The average temperature difference in the year is 13°C. Determine the per unit area and average heat loss by the wall. (3) Assuming the four walls, roof, and floor of the store have the same area and heat losses, determine the total energy consumed in the year in [Wh].

8.10 Based on the equations shown in table 8.3, (1) determine whether it is preferable to have a roof with a small or large angle in free-flow conditions. (2) In forced-flow conditions, determine how the convection coefficient scales with wind velocity.

8.11 Based on the equations in table 8.3 and assuming a horizontal surface in laminar-flow conditions, determine whether convective heat transfer processes are higher in the summer when the roof is warmer or in the winter when the ceiling of the room inside is warmer.

8.12 You would like to know how effective blowing on a hot horizontal surface can be to cool it down. The surface is square and measures 0.1 m × 0.1 m, and the temperature difference is 40°C. Calculate the convective heat transfer in (1) free-flow conditions and in (2) forced-flow conditions (assume $v = 10$ m/s). (3) Compute the ratio to determine how much more effective it is to blow on the plate.

8.13 A store has a 10 m × 10 m flat roof. On a cold winter day, after conduction heat losses, the air temperature is 20°C colder than the roof temperature. (1) Assuming free-flow conditions apply, calculate the convection coefficient of the roof h_{free}. (2) Assuming a wind speed of 10 m/s, calculate the convection coefficient of the roof h_{forced}. (3) Under both conditions, calculate the convective heat loss per unit area and total. (4) Assuming this temperature differential applies to the whole year, and taking the heat transfer rate for free flow, calculate the total energy lost in the year.

8.14 A house has a roof with a gable structure, a 32° angle, and a pitch length L_{pitch} of 9 m. The length of the roof is $L_h = 12$ m. On a warm spring day, the temperature on the roof gets 15°C warmer than the air temperature. (1) Assuming free-flow conditions apply, calculate the convection coefficient of the roof h_{free}. (2) Assuming a wind speed of 6 m/s and a wind direction along the length of the roof, calculate the convection coefficient of the roof h_{forced}. (3) Under both conditions, calculate the convective heat loss per unit area and in total. (4) Assuming this temperature differential applies to the whole year, and taking the heat transfer rate for free flow, calculate the total energy lost in the year.

8.15 In your own words and by searching the web, describe what short and long electromagnetic waves are.

8.16 By searching the web, report the emissivity of ten materials not listed in table 8.4.

8.17 A 10 cm × 10 cm dish ($\varepsilon_d = 0.95$ and $T_d = 20$ C) is put in a 50 cm × 50 cm oven ($\varepsilon_o = 0.6$ and $T_d = 200$ C). The oven is heated from the top and the bottom, and the dish is placed in between both surfaces at a distance of 20 cm (all surfaces are parallel).

Calculate (1) the view factor coefficient between the two surfaces and the dish and (2) the per unit area and total radiative heat transfer from the oven to the dish, considering a temperature difference of 100°C.

8.18 Using figure 5.20 from chapter 5 or any other source, select an area in the United States (or elsewhere) and (1) report the average annual solar radiation per unit area and per day. Assuming a roof is made of asphalt, (2) calculate the typical heat transfer solar radiation. Based on your knowledge, (3) discuss whether this result is significant or not.

8.19 A person ($\varepsilon_p = \alpha_p = 0.99$ and $T_p = 30°C$) is lying in the sun with a shortwave solar radiation flux of $I_{solar} = 800$ W/m^2. The person's surface area can be roughly approximated as a 1.7 m by 40 cm rectangle. (1) Calculate the per unit area and total radiation heat transfer that the person receives from the sun. Because it is too hot, the person moves to the shade and is now perpendicular to a 3 m high wall ($\varepsilon_w = 0.85$ and $T_w = 45°C$) that directly receives sunlight. Assuming the person measures 1.7 m (parallel to the wall) by 40 cm (perpendicular to the wall), calculate (2) the view factor coefficient between the person and the wall and (3) the per unit area and total heat transfer from the wall to the person. Based on these results, (4) determine the percentage reduction in radiative heat transfer.

8.20 Considering a building has an U_{eq} of 0.5 W/m^{2}K and that the average temperature difference in the year is 13°C, (1) add the combined convection and radiation coefficients recommended by ASHRAE and determine the system conductance U_{sys}. (2) Calculate the new heat loss per unit area and the heat loss by the wall.

8.21 A wall measuring 10 m×3 m has three 2 m×1 m single-pane windows with aluminum frames (no thermal break); the rest of the wall has a U value of 0.4 W/m^{2}K. To better insulate the building, the windows are going to be replaced with triple-pane windows with vinyl frames. Calculate (1) the current U value of the entire wall, (2) the future U value of the entire wall, and (3) the percentage improvement in insulation thanks to the new windows.

8.22 A new regulation is about to pass to change the required air exchange rate for a restaurant from 12 to 14 h^{-1} during operation (8 h per day). A particular restaurant has an exposed volume of 400 m^3, and the average temperature difference with the outside is 10°C. Calculate the (1) current and (2) future heat transfer linked with air exchange and compute (3) how much energy in [kWh] and money (assume 1 kWh is $0.12) will be lost in a year because of the regulation.

8.23 A project to retrofit a building is considering switching the gas furnace (AFUE = 0.9) with an ASHP (COP = 5). The building owners pay about $5,600 per year in gas, and 1 kWh of gas costs $0.14. (1) Calculate how much energy is used for heating per year. (2) Considering 1 kWh of electricity costs $0.12, calculate the expected costs of yearly electricity use. (3) The building is located in an area with a power grid emission factor of 356 g CO_2e/kWh. Based on your knowledge of emission factors for gas, calculate the GHG emission savings.

8.24 A house has a total exposed area of 422 m^2 (including walls, windows, roof, and floor) and a U value of 0.37 W/m^2K (not including air exchange). The house has a volume of 450 m^3 and an air exchange rate of 0.5 h^{-1}. (1) Calculate the system conductance U_{sys} that includes air exchange. The average temperature difference with the outside is 8°C. (2) Calculate the energy consumed for heating in a year. The house currently has a gas furnace with an AFUE of 0.92, but there are plans to install a GSHP with a COP of 6 (the power grid emission factor is 300 g CO_2e/kWh). Calculate the (3) current and (4) future energy consumption and GHG emissions related to heating the house and (5) compare the results and estimate the monetary savings (assume $0.12 per [kWh] for both gas and electricity).

8.25 The wall of a building is 9 m long × 6 m high. The exterior has a 5 cm thick layer of brick (face), and the interior has a 5 cm thick layer of gypsum. The insulation in between is composed of a 10 cm thick layer of glass fiber, and wood studs (hardwood) are placed every 3 m and are 10 cm deep by 5 cm against the brick. Finally, three 2 m × 2 m windows are located on the wall, with U_w values of 2.2 W/m^2K. (1) Calculate the equivalent conductance U_{eq} for the wall. (2) Add the combined convection and radiation coefficients recommended by ASHRAE (h_i = 17.6 and h_o = 34 W/m^2K) and calculate the new equivalent conductance U_{eq}. (3) The average temperature difference in the year is 15°C. Determine the per unit area and average heat loss by the wall. (4) Assuming the four walls, roof, and floor of the house have the same area and heat losses, determine the total energy consumed in the year. (5) Assuming this energy is provided by a heat pump (COP of 5) and the grid emission factor is 0.454 t CO_2e/MWh, determine the total electricity consumed and the associated GHG emissions.

8.26 The average temperature for a week in a city are shown in the table below. Considering a reference temperature of 20°C, calculate the (1) number of HDD and (2) number of CDD for this city.

Day	Temperature
Monday	17
Tuesday	15
Wednesday	19
Thursday	21
Friday	23
Saturday	25
Sunday	20

8.27 Select any city in the world as long as it is not present in table 8.6. (1) Collect and report the number of HDD (T_{ref}=20°C) and CDD (T_{ref}=24°C) for this city from any source (e.g., http://www.degreedays.net/), selecting a five-year average.

8.28 A community has a population of 25,000 people, the average number of people per household is 2.5, the average house floor area is 183 m^2, and the average per unit floor area conductance U_{floor} was estimated to be 0.78 W/m^{2}K. The community is located in an area with average HDD of 2,347, and an average CDD is 371. Based on this information, (1) calculate the total number of households, H, and the average floor space per person, A_{avg}. (2) Calculate the per capita and per household energy required for heating and cooling for an entire year. (3) Assuming the furnace has an efficiency AFUE=0.92 and the air conditioners have an efficiency EER=4, calculate the input energy required for heating and cooling per household and per square meter of floor area.

8.29 A house has a total exposed surface of 154 m^2 with an equivalent conductance U_{eq}=0.61 W/m^{2}K. The number of HDD=2,634, and the number of CDD=78. (1) Calculate the total amount of energy used for heating and cooling per year in [MWh]. (2) Considering an AFUE of 0.9 and a SEER of 3, calculate how much energy in [MWh] is used for space conditioning. (3) Calculate the energy intensity of the house in [kWh/m^2] and comment on whether this value is high or low based on your knowledge of energy intensities for the different regions of the United States. (4) Assuming an emissions factor of 0.172 t CO_2e/MWh for natural gas and 0.372 t CO_2e/MWh for electricity, determine the amount of GHGs emitted for space conditioning the house. (5) List three design criteria that have an important impact on building energy use. (6) List five technologies that have an important impact on building energy use.

8.30 Select any city in the world as long as it is not present in table 8.6. (1) Collect and report the number of HDD (T_{ref}=20°C) and CDD (T_{ref}=24°C) for this city from any source (e.g., http://www.degreedays.net/), selecting a five-year average. Report the total population of the city. (2) Collect or simulate in BEopt or eQuest and report average U values in [W/m^{2}K] for residential buildings in this city (*hint: if simulated, simply keep default values for a house*). (3) Look up or estimate potential AFUE and EER or SEER values to use for this city and report them. (4) Unless you find better values for your city, use an approximate 50 m^2 of floor space per people and calculate the energy use for space heating and cooling per person in the city for one year.

8.31 Select any two buildings (e.g., two houses in your neighborhood). Then, (1) estimate their compactness and (2) discuss how you think they compare in terms of building energy use.

8.32 Select any building (e.g., your house, your department, a store) and list five ways the building could be retrofitted to decrease its energy use for space conditioning.

8.33 By searching the web, find an example of a low-energy-use house (e.g., a passive house) and list the strategies undertaken to achieve such a low energy use.

8.34 Pick any technology or combination of technologies used in figure 8.16 and model it in BEopt or eQuest for your area to determine whether you get the same IRR and GHG reduction potential.

Notes

1. Assuming we like to have the inside temperature around 22°C (72°F). The difference with 0°C (32°F) in the winter for colder climates is 22°C, whereas the difference with 35°C (95°F) in the summer for warmer climates is only 13°C.

2. In the United Kingdom, before the spread of "cheap" energy, the average room temperature in the winter was 13°C (55°F), thus quite chilly (MacKay 2009).

3. Once again, we use Q like we did in the water and transport chapters. In electricity, we use I for intensity, but we could very well use Q there as well.

4. Is it not fascinating how scientific laws in one field can simply apply to other fields? It really makes me wonder why we divided engineering into various disciplines in the first place and why most students are asked in which field they prefer to specialize.

5. This is also why we wear sunglasses in the summer since the reflectivity of light surfaces, such as light-colored buildings or sand, is high—otherwise, we need to squint.

6. I personally remember feeling night sky radiation when I visited the Grand Canyon. As soon as the sun set, the temperature suddenly got a lot cooler.

7. But it is important. When you buy windows, aim for low U values and low solar heat gain coefficients.

8. Remember that 1 W is defined as 1 J/s.

9. In chapter 5, we learned that natural gas power plants have an emission factor of 500 g CO_2e/kWh and an efficiency of about 42%. This means that if we were to fully capture all the energy from burning the natural gas, the emission factor would be about $500 \times 0.42 = 210$ g CO_2e/kWh. This is higher than what we use here because we accounted for the full life-cycle emissions in chapter 5.

10. Similar to the comment made in the transport factor, the U.S. Environmental Protection Agency actually gives CO_2, CH_4, and N_2O emissions that need to be combined based on their global warming potentials provided in the same document. The information is given both in [m^3] of gas consumed and in [kWh] since gas meters typically provide a volume of gas consumed, but from using the equations in the chapter, we get [W] and [Wh] values. We should also note that the conversion from [m^3] to [kWh] is not straightforward since gas is highly volatile—at different temperatures and pressures, the energy content will be different—so some assumptions had to be made.

11. Available at http://www.degreedays.net/ (accessed March 7, 2019).

12. Technically, *passivehaus* in the original German.

13. BEopt (Building Energy Optimization) is available from https://beopt.nrel.gov/; EnergyPlus is available from https://energyplus.net/; and eQuest is available from http://www.doe2.com/equest /(accessed March 7, 2019).

14. Saint-Exupéry was already quoted in chapter 1 about how water is essential to life.

15. Original French: "Il semble que la perfection soit atteinte non quand il n'y a plus rien à ajouter, mais quand il n'y a plus rien à retrancher."

16. And we tend to prefer bedrooms that receive less light so as not to be woken up by the sun every morning.

17. Solar panels have a relatively low efficiency, and to warm water it is much more preferable to use a solar water heating technology discussed in section 3.2.7.

18. Anecdotally, they then moved to British Columbia and were unable to use the results they had generated.

19. The value given in the article is 16.6, but it is likely in [Btu/Wh], which we can convert to COP, as discussed in section 8.1.4. The other original values were 17 (for 5) and 21 (for 6.15).

20. Available at https://www.eia.gov/totalenergy/data/monthly/ (accessed September 30, 2018).

21. Available at https://www.eia.gov/totalenergy/data/monthly/ (accessed September 30, 2018).

References

Amatya, B. L., K. Soga, P. J. Bourne-Webb, T. Amis, and L. Laloui. 2012. "Thermo-mechanical Behaviour of Energy Piles." *Géotechnique* 62(6): 503–519.

American Society of Heating, Refrigerating, and Air-Conditioning Engineers. 2010a. *ANSI/ASHRAE Standard 55-2010: Thermal Environmental Conditions for Human Occupancy*. Atlanta, GA: American Society of Heating, Refrigerating, and Air-Conditioning Engineers. http://arco-hvac.ir/wp-content/uploads/2015/11/ASHRAE-55-2010.pdf.

———. 2010b. *Handbook of Fundamentals*. Atlanta: American Society of Heating, Refrigerating, and Air-Conditioning Engineers.

Bristow, David N., and Michele Bristow. 2017. "Retrofitting for Resiliency and Sustainability of Households." *Canadian Journal of Civil Engineering* 44(7): 530–538. https://doi.org/10.1139/cjce-2016-0440.

Derrible, Sybil, and Matthew Reeder. 2015. "The Cost of Over-Cooling Commercial Buildings in the United States." *Energy and Buildings* 108: 304–306. https://doi.org/10.1016/j.enbuild.2015.09.022.

de Saint-Exupéry, Antoine. 1939. *Terre Des Hommes*. Paris: Gallimard.

Engineering Toolbox, The. 2017a. "Absorbed Solar Radiation." http://www.engineeringtoolbox.com/solar-radiation-absorbed-materials-d_1568.html.

———. 2017b. "Air Change Rates in Typical Rooms and Buildings." http://www.engineeringtoolbox.com/air-change-rate-room-d_867.html.

———. 2017c. "Emissivity Coefficients of Some Common Materials." http://www.engineeringtoolbox.com/emissivity-coefficients-d_447.html.

Howell, John R., M. Pinar Menguc, and Robert Siegel. 2016. *Thermal Radiation Heat Transfer*. Boca Raton, FL: CRC Press.

Huang, Joe, James Hanford, and Fuqiang Yang. 1999. *Residential Heating and Cooling Loads Component Analysis*. Report LBNL-44636. Berkeley: Lawrence Berkeley National Laboratory.

Klepeis, Neil E., William C. Nelson, Wayne R. Ott, John P. Robinson, Andy M. Tsang, Paul Switzer, Joseph V. Behar, Stephen C. Hern, and William H. Engelmann. 2001. "The National Human Activity Pattern Survey (NHAPS): A Resource for Assessing Exposure to Environmental Pollutants." *Journal of Exposure Analysis and Environmental Epidemiology* 11(3): 231–252. https://doi.org/10.1038/sj.jea.7500165.

Kreider, J. F., P. S. Curtiss, and A. Rabl. 2009. *Heating and Cooling of Buildings: Design for Efficiency*. Rev. 2nd. ed. Boca Raton, FL: CRC Press.

Lucon, O., D. Ürge-Vorsatz, A. Zain Ahmed, H. Akbari, P. Bertoldi, L. F. Cabeza, N. Eyre, et al. 2014. "Buildings." In *Climate Change 2014: Mitigation of Climate Change*, edited by O. Edenhofer, R. Pichs-Madruga, Y. Sokona, E. Farahani, S. Kadner, K. Seyboth, A. Adler, et al., 671–738. Contribution of Working Group III to the Fifth Assessment Report of the Intergovernmental Panel on Climate Change. Cambridge: Cambridge University Press.

MacKay, D. J. C. 2009. *Sustainable Energy—without the Hot Air*. Without the Hot Air Series. Cambridge: UIT.

Maclay, B. 2014. *The New Net Zero: Leading-Edge Design and Construction of Homes and Buildings for a Renewable Energy Future*. White River Junction, VT: Chelsea Green Publishing.

McClellan, T. M., and C. O. Pedersen. 1997. "Investigation of Outside Heat Balance Models for Use in a Heat Balance Cooling Load Calculation Procedure." In *ASHRAE Transactions*, 469–484. Atlanta: American Society of Heating, Refrigerating and Air-Conditioning Engineers.

Parker, Danny S. 2009. "Very Low Energy Homes in the United States: Perspectives on Performance from Measured Data." *Energy and Buildings* 41(5): 512–520. https://doi.org/10.1016/j.enbuild.2008.11.017.

Pohl, Jens. 2011. "Principles of Thermal Comfort." In *Building Science*, 20–32. Hoboken, NJ: Blackwell. http://dx.doi.org/10.1002/9781444392333.ch2.

Pollio, Marcus Vitruvius. 1999. *Vitruvius: Ten Books on Architecture*. Edited by Ingrid D. Rowland and Thomas Noble Howe. New York: Cambridge University Press.

Sivak, Michael. 2013. "Air Conditioning versus Heating: Climate Control Is More Energy Demanding in Minneapolis than in Miami." *Environmental Research Letters* 8(1): 014050.

Susorova, Irina, Parham Azimi, and Brent Stephens. 2014. "The Effects of Climbing Vegetation on the Local Microclimate, Thermal Performance, and Air Infiltration of Four Building Facade Orientations." *Building and Environment* 76(June): 113–124. https://doi.org/10.1016/j.buildenv.2014.03.011.

U.S. Environmental Protection Agency. 2018. "Emission Factors for Greenhouse Gas Inventories (V2)." Washington, DC: U.S. Environmental Protection Agency. https://www.epa.gov/climateleadership/center-corporate-climate-leadership-ghg-emission-factors-hub.

9 Solid Waste

Solid waste ... is a striking by-product of civilization.

—Hoornweg, Bhada-Tata, and Kennedy 2013

Out of close to 10 million species of animals on Earth,[1] only one creates *solid waste*:[2] humans. And even this statement became true only relatively recently since, for the longest time, humans did not create any solid waste. As the quote above illustrates, it really is civilization, perhaps best manifested by cities—the very thing we have been promoting since the beginning of this book—that causes humans to generate waste and, as a response, to come up with ways to manage solid waste. In the twenty-first century, solid waste management has become a worldwide problem that accounts for about 5% of global greenhouse gas (GHG) emissions (Mohareb and Hoornweg 2017). In the United States in 2016, GHG emissions related to solid waste amounted to 131.45 Mt CO_2e, which was about 2% of all the GHGs emitted (U.S. Environmental Protection Agency 2018d).

While the first efforts to seriously manage solid waste emerged in Greece around 3,000 BCE, we had to wait until the end of the nineteenth century to see any substantial engineering efforts to develop solid waste–management strategies, as we will learn in this chapter. By the beginning of the twenty-first century, sights like the ones shown in figure 9.1 had become common virtually everywhere around the world.

Back to the quote above, what is even more concerning is that in the original document the sentence right after reads: "The average person in the United States throws away their body weight in rubbish every month." It is therefore not necessarily the fact that we create waste that is most remarkable (although it is remarkable); it is that we generate so much of it in such relatively short periods of time that we need an entire science called *solid waste management* to deal with it. It is this science that is at the heart of this chapter.

a) Dumpster b) Open Dump c) Landfill

Figure 9.1
Examples of solid waste management practices. Credits: (a) author; (b) Jonathan McIntosh; (c) Carlisle Energy.

We can even go further and state another concerning fact: solid waste management is mostly an urban problem, as people in rural areas generate about half as much solid waste as urban residents (Hoornweg, Bhada-Tata, and Kennedy 2013). This is perhaps the reason why solid waste management is often the one urban phenomenon that gets cities to talk with one another; as the authors of the quote above wrote in another study: "Solid waste is usually the one service that falls completely within the local government's purview." (Hoornweg and Bhada-Tata 2012) In fact, while solid waste management regulations are often developed at the national or federal level in most countries, it is really the cities and regions themselves that have to work together to come up with strategies to deal with their solid waste. Solid waste management therefore has an important role in urban engineering, and it can be considered at the same level as the infrastructure systems discussed in previous chapters: electricity, water, transport, and buildings.

So, what is solid waste and solid waste management, and are there preferred ways to manage solid waste? Put simply, there are two main categories of solid waste:

- organic (e.g., food waste, paper, plastics)
- inorganic (e.g., glass, metals, construction debris)

Whether the waste is organic or inorganic has a substantial impact on how we manage it. For starters, much energy can be recovered from organic waste (in the chemical sense[3]), while this is not the case for inorganic waste. Even within organic waste, some can be composted (e.g., food waste, yard trimmings) and others cannot (e.g., plastics). In the United States, roughly 80% of the solid waste is organic, and 20% is inorganic (not accounting for construction and demolition debris and some industrial waste, as will be defined later). Despite significant differences in the types of solid wastes that exist, there are only three ways to manage solid waste once it is generated:

- Reuse/recycle
- Burn
- Throw away

Figure 9.2 shows how these three management options were split in the United States in 2015 (U.S. Environmental Protection Agency 2018a). Reuse and recycle represent two of the famous three *R*s of solid waste management—reduce, reuse, recycle—and they accounted for about 35% of all solid waste in 2015 (this figure includes composting). As we will see, a fourth *R* has joined the rank. Burning solid waste is typically referred to as incineration or combustion; in the United States, close to 15% of all solid waste was combusted with energy recovery in 2015. While incineration has many advantages—for example, it does not attract pests, the produced heat can be used to generate electricity and/or heat, and it dramatically reduces the volume of waste—it also has disadvantages, as we will see. Finally, throwing away waste can either be done in a controlled manner—for instance, in sanitary landfills—or it can be done in an uncontrolled manner—for instance, by throwing trash in open air dumps or in bodies of water. In the United States in 2015, more than 50% of all solid waste was sent to sanitary landfills. We will discuss all three ways in section 9.3.

Following the structure used previously, we will first learn about some fundamentals of solid waste management. In particular, we need to define what solid waste actually is. We will also go over a brief history of solid waste management and estimate how much energy we can extract from solid waste. Similar to other chapters, we will then study the "demand" for solid waste management, which here essentially means that we will look

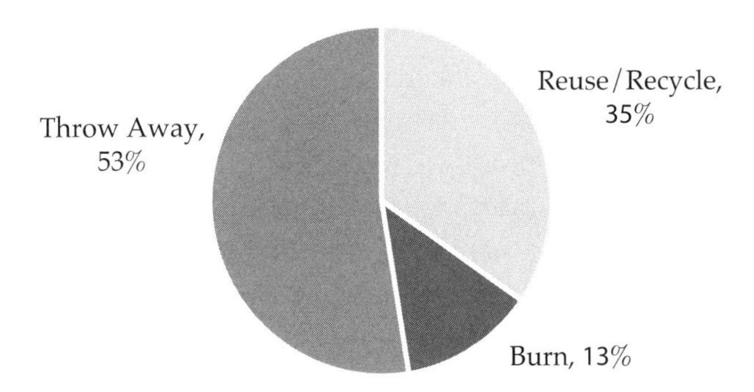

Figure 9.2
Solid waste management strategies in the United States in 2015.

into how solid waste is generated and composed (i.e., types of material). In other words, we will look into patterns of solid waste generation and composition in the United States and the world. Finally, we will focus on the "supply" of management strategies—that is, how solid waste is treated—looking into the three ways in which we can dispose of solid waste, including by learning about how sanitary landfills are engineered.

We should also clear out some problems of semantics before we go on. *Solid waste, trash, rubbish, garbage,* and *refuse* are somewhat synonymous.[4] *Refuse* is an older term that we will use when discussing the history of solid waste management because it used to be used in official documents. We will also use *trash* here and there as a catch-all term, but largely, we will simply use the term *solid waste* (properly defined later) throughout the chapter since it has become the scientific norm, and we will simply not use the term *garbage*.

The science of solid waste management is extensive, and entire textbooks are dedicated to it. While we cannot cover everything here, this chapter is particularly indebted to the seminal book *Integrated Solid Waste Management: Engineering Principles and Management Issues* by Tchobanoglous, Theisen, and Vigil (1993), which should be used as a general guide for matters of solid waste management. Moreover, most of the data shown are for the United States in 2015—the latest available at the time of writing—and they come from the U.S. Environmental Protection Agency ([EPA] 2018a). More recent updates of the database may be available on the EPA's website—note that the website shows units of U.S. short ton, which we convert in this chapter to metric tons.

Let us now learn some fundamentals.

9.1 Fundamentals of Solid Waste Management

Without proper solid waste management strategies, there is absolutely no way that cities could have grown to become as large as they have. The seemingly seamless operation to remove solid waste from cities and to dispose of it has evolved over many centuries, eventually growing into its own science. In this section we will also properly define what solid waste is, and more specifically, we will itemize the various types of solid waste beyond the two categories that we have already seen (i.e., organic and inorganic). Notably, we will also list some important physical, chemical, and biological properties of solid waste that are important in order to determine how much energy can be extracted from solid waste when it is incinerated. The review of solid waste treatment technologies will have to wait until section 9.3. One aspect we will not cover is optimal solid waste collection strategies—for example, the optimized scheduling and routing of garbage trucks.

First, let us go back in time and learn about the history of solid waste.

9.1.1 History

The statement at the beginning of this chapter about the fact that humans represent the only species that generates waste may be a bit unfair. In their textbook, Tchobanoglous, Theisen, and Vigil (1993, p. 3) define *solid waste* as "all the wastes arising from human and animal activities that are normally solid and that are discarded as useless or unwanted." Therefore, all living creatures create some kind of "waste" (if only the remains of the creature once it has died, whether it is a plant, an insect, or an animal). What separates humans, however, is that the waste from other species either gets reused or is naturally biodegradable, and no other species has had to develop a science around managing its waste. And even then, it is only when humans started to settle in relatively large cities that the management of solid waste became an issue. It should therefore not be surprising to learn that the first solid waste management strategies emerged when the first relatively large cities emerged as well. Perhaps this is best articulated by Wilson (1976, p. 123): "Solid waste treatment has necessarily always been more advanced in the largest cities. Disposal problems become difficult with increase of population density. Simultaneously there is a greater production of waste per unit area, and a decreased proportion of land available for its disposal."

As we saw in chapter 4, cities started to develop in the Neolithic era, roughly between 10,000 BCE and 2000 BCE. While it is nearly impossible to date when the first solid waste management practices were adopted in cities—that is, as a deliberate act to deal with solid waste—we should remember that it is often by studying the trash of ancient civilizations that archaeologists have been able to estimate how people lived at the time. The presence and maintenance of municipal dumps must have therefore been reasonably common from the onset of the creation of cities.

Signs of more advanced efforts to deal with solid waste seem to have first emerged in the Bronze Age. Specifically, the Minoan civilization on the island of Crete (Greece) that lasted from roughly 3000 BCE to 1000 BCE developed the practice of dumping its solid waste in large pits, with layers of soil at intervals. By 500 BCE, the Greeks had then developed a fairly rigorous system to deal with solid waste, forbidding the disposal of trash in the streets, organizing trash pickup routines, and managing municipal dumps. These practices were highly advanced for the time.

In general, however, the favored practice used (and kept almost until the end of the nineteenth century, except for a few cases) was to simply throw trash in the streets and allow animals (especially swine) to roam around the cities and eat the trash and sewage. This practice also applies to the Romans, who did not adopt any significant solid waste management practices despite their advances in water resources engineering—although we need to remember that solid waste generation rates were not nearly as high then as they have become by the twenty-first century and that most solid waste consisted of organic produce that animals could easily eat.

Funnily, because the streets were littered with trash, many cities built on top of them, literally "elevating" themselves. Wilson (1976), for instance, recalls that the city of Bath in the United Kingdom is between 3.5 m and 6 m above where it was during the Roman Empire. This is in fact also the case for Rome—and in general in historical cities, the deeper we look, the older the civilization.[5]

When trash could not be thrown in the streets, it was often sent to municipal dumps, as was done in Athens in ancient Greece. This is the case for London, as well, which passed several laws forbidding throwing trash in the city streets. The first of these laws passed in 1297, though with limited success. In 1354, London, after being hit by the bubonic plague that killed roughly half of all the inhabitants of Europe, passed a second law to organize the weekly collection of trash in front of buildings.

Paris initially tried a different strategy. In 1185, Philip II of France, better known as Philip Augustus, had some of the main streets of Paris paved (to make them impermeable) with a gutter in the center. The paved streets facilitated the movement of people and animals in the capital, and the gutter facilitated the management of stormwater and trash and could be washed out with water (Béguin 2013).

Despite significant efforts from local authorities, the management of solid waste remained relatively superficial in most cities. At best, cities organized trash collection practices to carry the trash to municipal dumps. London, for example, used twelve carts to carry out trash. In contrast, Tokyo's network of canals aided significantly in transporting both trash and sanitary sewage out of the city, where they were then partly used as compost. Again, we must remember that the trash in those times was very different from the trash generated in the twenty-first century, and much of what arrived at these dumps was sorted and recycled.

We had to wait until the eighteenth century, also known as the Age of Enlightenment, to see some focus on hygiene. The Age of Enlightenment notably corresponds to formidable progress in science, and two main theories emerged to explain how diseases spread: the miasma theory and the germ theory. The miasma/miasmatic, or *filth*, theory stipulated that diseases emanated from "bad" air originating from decaying organic waste and sewer gas present in the city streets.[6] Advocates of the miasma theory were also called *anticontagionists* because they did not believe diseases were communicable. In contrast, the germ theory stipulates that diseases are spread by microorganisms that can, incidentally, thrive in trash and sewage. The miasma theory was later discredited, and the germ theory was proven correct, but it had a lasting impact on decisions in early solid waste management practices (Louis 2004). The goal therefore became evident—get rid of the trash—but the solution was not obvious.

Aligned with the general focus on hygiene, the eighteenth and nineteenth centuries also saw great advances in water treatment, and we can recall that the very first paragraph of this book tells the story of Dr. John Snow, who found that a single water well in London was the origin of an epidemic of cholera in 1854. In fact, in the nineteenth century most investments went into building water treatment and distribution systems, which is partly why it took some time before seeing significant investment in solid waste management (Louis 2004).

Nonetheless, the use of trash cans slowly started to emerge in some cities to facilitate the pickup of solid waste. In particular, in 1883 the French administrator Eugène Poubelle forced property owners in Paris to use containers with closed lids and to separate their trash into three categories—organic, paper and fabric, and remaining trash—to be picked up regularly by municipal employees. While the practice took some time before it was fully enforced and accepted by the population, eventually both occurred (Béguin 2013), and the word *poubelle* in French became the common term for "trash can."

Along with a better solid waste collection process, various ways of dealing with the trash emerged. In particular, the first incinerator was built in 1874 in England. Subsequent incinerators were built both in England and in Continental Europe. We can recall from chapter 5 that electricity production emerged in the late nineteenth century and early twentieth century, and many incinerators were in fact steam power plants that produced electricity and heat (Wilson 1976).

In the United States, the turning point came from Colonel George E. Waring Jr., who was commissioner of the New York Department of Street-Cleaning from 1895 to 1898. As an ex-military officer who had done extensive work in sewage and drainage engineering, Waring applied a military rigor to the management of solid waste in New York City, setting up strict rules and routines for his crew. In particular, he split solid waste management in five main operations: street cleaning, source separation, trash collection, transport, and resources recovery and disposal. To ensure an effective trash pickup in the winter months, he also added snow removal to the entire operation.[7] In addition, Waring instituted the pickup of trash at curbside, and he required residents to separate their trash similar to Poubelle in Paris. He also established the first recycling center with the goal to recover what could be recovered. For instance, he sold rags to tradesmen and papermills. He also recovered ammonia, grease, glue, and dry residue that could be sold to chemical industries and farmers (Louis 2004).

At the beginning of the twentieth century, the miasma theory was completely discredited, and efforts focused on public health to deal with the spread of disease, as opposed to focusing on sanitation. The emphasis on sanitary engineering (and the importance given to sanitary engineers) therefore fell somewhat. At the same time, the composition of trash changed substantially. Most markedly, ash started to disappear from trash as wood was substituted with electricity, fuel oil, and gas for building heating.

Simultaneously, the economic boom of the 1920s also saw a sudden increase in trash generation, which meant that new trash collection and disposal practices were needed. In particular, municipal dumps could not handle the amount of trash that arrived. In the 1920s in England, the practice of "controlled tipping" was introduced, which is essentially identical to what the Minoan civilization did: dig a pit, toss in the trash that cannot be used, and recover with soil at intervals. Controlled tipping essentially represents the origins of modern landfills. This practice was first implemented in the United States in Fresno, California, by Jean Vincenz, who built the first landfill in 1934.

The practice of landfilling slowly became the dominant practice to handle solid waste. The American Society of Civil Engineers published a guide in 1959 recommending that solid waste be thrown in landfills and covered with dirt at the end of every day to avoid attracting pests (Louis 2004).

Incinerators were also common at the time. In fact, many people had their own incinerators at home that could be used for heating as well. As one can imagine, this practice significantly polluted the air and represented a fire hazard, and it was eventually discontinued.

As the generation of solid waste kept increasing and as cities kept growing, the need to establish regional solid waste management strategies became evident. Eventually, in the United States, the federal government got involved, and President Johnson passed the Solid Waste Disposal Act in 1965 that at first had limited success. If we recall from chapter 2, the National Environmental Policy Act was also introduced in 1969, which is one of the reasons why the Solid Waste Disposal Act was amended in 1970, eventually becoming the Resources Recovery Act that put more emphasis on recycling and energy recovery. In 1970 the Clean Air Act was also passed, and it regulated the emissions from landfills and municipal waste incinerators. Although not related to solid waste management, the Clean Water Act (for wastewater) became law in 1972, and the Safe Drinking Water Act (for drinking water) was passed in 1974, as we saw in chapter 6. Two years after, in 1976, the Resource Conservation and Recovery Act was also passed, and with its 1984 amendment, the Hazardous and Solid Waste Amendment, the U.S. government properly regulated the management of solid waste. Overall, the twenty years from 1965 and 1985, and more particularly the 1970s, saw a significant increase in environmental regulations in the United States to protect the health of the American people and the environment. Thanks to this series of regulations, by the beginning of the twenty-first century sanitary landfills had become complex engineering systems, and we will see how they work toward the end of this chapter.

There is one more use of trash that we have not discussed yet: *land reclamation*. Around the world, many cities have experienced a shortage of land as they have grown. One option, if they are located close to a large body of water, is to reclaim land. Specifically, dirt, rubble, rock, and of course trash have been used to reclaim land.

To name but a few examples, in New York City, a lot of the rubble excavated to build the subway system was used to reclaim land—Ellis Island is mostly reclaimed land. In Toronto (Canada), Front Street used to be on the shoreline, but it is now about 800 m (half a mile) behind the shoreline. Even more dramatically, much of what we know as Boston is actually land that was reclaimed after 1630, as shown in figure 9.3, including most of the Back Bay neighborhood, and much of this reclaimed land is made of trash.[8]

Similar processes are in place in the twenty-first century. For example, figures 9.4a and 9.4b show a map of Singapore in 1984 and in 2016. In total, about one-fifth of the island was reclaimed, mostly from rubble imported from Indonesia. This includes Marina Bay, located on the southern shores of the island and featuring the iconic Marina Bay Sands hotel, and the industrial district, located on the western shores. Figures 9.4c and 9.4d show the development of an island in Tokyo Bay, built purely from trash, that will be used for business and leisure activities once completed. Even using the modern process of converting trash to fill/soil, it is not that easy to avoid contaminating surrounding water bodies. Hornyak (2017) briefly describes the process specifically for the island shown in figures 9.4c and 9.4d.

In conclusion, the general problem of managing solid waste emerged at the same time as cities grew. Early strategies remained mostly limited until the end of the nineteenth century, when solid waste management became a science that further developed in the twentieth century. In the United States, this movement was aided by a series of important regulations passed in the 1970s. To learn more about the history of solid waste management in the United States, Melosi (2008) offers a great account from a historian's perspective.[9]

We now need to be more specific about what we mean by solid waste, and this is the topic of the next section.

9.1.2 Definition of Solid Waste and Solid Waste Management

To be able to discuss solid waste, we first need to properly define it. In fact, we already saw a definition of solid waste in the previous section given by Tchobanoglous, Theisen,

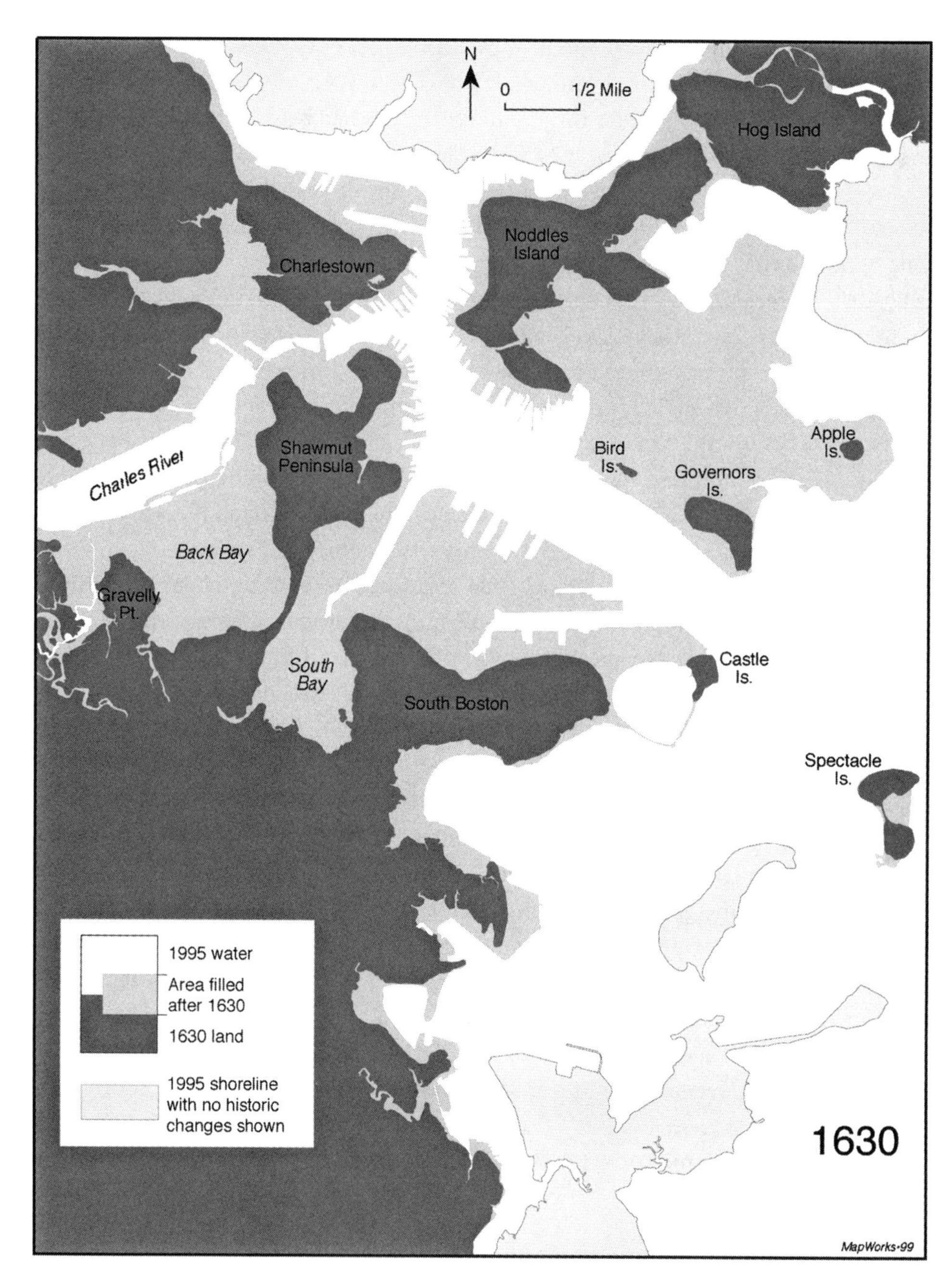

Figure 9.3

Boston land reclamation map. Courtesy of the Norman B. Leventhal Map Center, Boston Public Library. Cartography by Herb Heidt and Eliza McClennen, MapWorks.

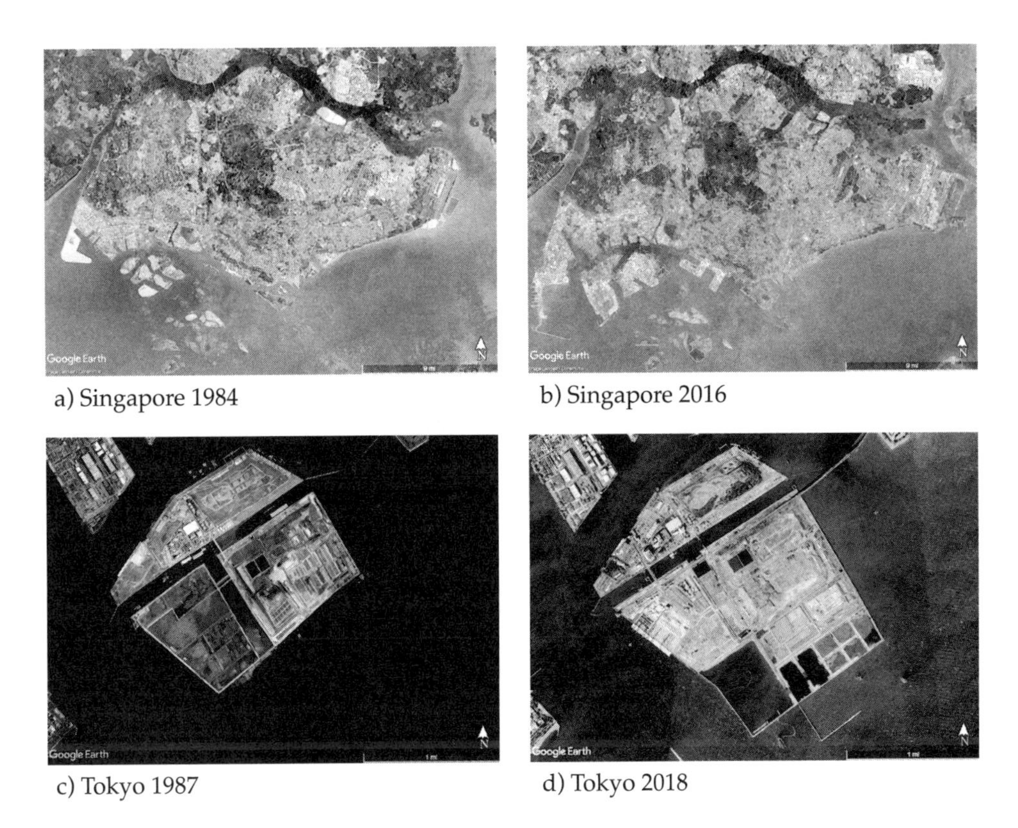

Figure 9.4
Land reclamation in Singapore and Tokyo.
Source: Google Earth.

and Vigil (1993, p. 3): "Solid wastes comprise all the wastes arising from human and animal activities that are normally solid and that are discarded as useless or unwanted." To get a more refined definition, however, we can go to the EPA, which adopts the definition given in the Resource Conservation and Recovery Act that was passed in 1976. The act defines solid waste as "any garbage or refuse, sludge from a wastewater treatment plant, water supply treatment plant, or air pollution control facility and other discarded material, resulting from industrial, commercial, mining, and agricultural operations, and from community activities" (U.S. Environmental Protection Agency 2018c).

Because whether a product can be defined as solid waste or whether it is considered *hazardous waste* has legal implications, the EPA is careful to be specific in its definition of solid waste, and it even offers a list of particular materials to distinguish solid from hazardous waste.[10] Examples of household hazardous waste include paint, herbicides,

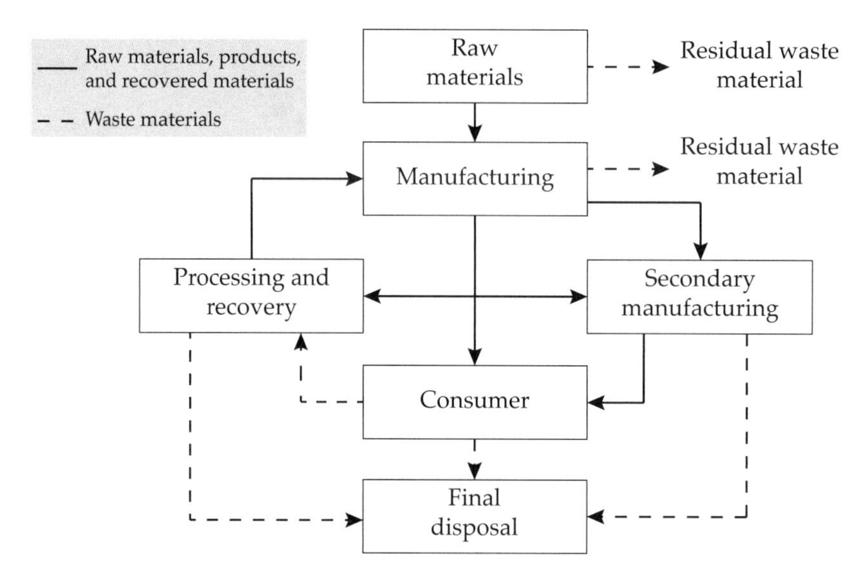

Figure 9.5
Simplified flow chart of solid waste generation. Adapted from Tchobanoglous, Theisen, and Vigil (1993, fig. 1-2).

batteries, and outdated medicines that should not be discarded with nonhazardous waste.[11]

Moreover, the website from which the previous quote was taken also adds: "It is important to note that the definition of solid waste is not limited to wastes that are physically solid. Many solid wastes are liquid, semi-solid, or contained gaseous material" (U.S. Environmental Protection Agency 2018c). The name of "solid" waste is therefore a little misleading, but it essentially refers to any type of material that is discarded.

In addition, the definition from the EPA aptly recognizes that solid waste can be generated by all types of entities and at any point in the life cycle of a product. Figure 9.5 offers a simplified flow diagram of where solid waste can be generated. Virtually every process generates some kind of waste.

Different entities also generate different types of solid wastes. Table 9.1 lists nine sources of solid waste, with typical waste generators and typical types of solid waste. Municipal solid waste (MSW) can comprise the first six categories, but the construction and demolition (often abbreviated C&D) category is often not included in MSW. In this chapter, we will focus mostly on MSW, not including C&D.

From the table, we can see that MSW types vary from food waste, paper, and glass to wood, steel, dirt, and the sludge generated by wastewater treatment plants. In section 9.2,

Table 9.1
Generators and types of solid waste

Source	Typical waste generators	Types of solid wastes
Residential	Single and multifamily dwellings	Food wastes, paper, cardboard, plastics, textiles, leather, yard wastes, wood, glass, metals, ashes, special wastes (e.g., bulky items, consumer electronics, white goods, batteries, oil, tires, and household hazardous wastes)
Industrial	Light and heavy manufacturing, fabrication, construction sites, power and chemical plants (excluding specific process wastes if the municipality does not oversee their collection)	Housekeeping wastes, packaging, food wastes, C&D materials, hazardous wastes, ashes, special wastes
Commercial	Stores, hotels, restaurants, markets, office buildings, and so on	Paper, cardboard, plastics, wood, food wastes, glass, metals, special wastes, hazardous wastes, e-wastes
Institutional	Schools, hospitals (nonmedical waste), prisons, government buildings, airports	Same as commercial
Construction and demolition	New construction sites, road repair, renovation sites, demolition of buildings	Wood, steel, concrete, dirt, bricks, tile
Municipal services	Street cleaning, landscaping, parks, beaches, other recreational areas, water and wastewater treatment plants	Street sweepings; landscape and tree trimmings; general wastes from parks, beaches, and other recreational areas, sludge
All of the above can be included as MSW, although we will mostly not include C&D. Industrial, commercial, and institutional wastes can be grouped together and usually represent more than 50% of MSW. C&D waste is often (but not always) treated separately: if well managed it can be disposed of separately. The items below are usually considered MSW if the municipality oversees their collection and disposal.		
Process	Heavy and light manufacturing, refineries, chemical plants, power plants, mineral extraction and processing	Industrial process wastes, scrap materials, off-specification products, slag, tailings
Medical waste	Hospitals, nursing homes, clinics	Infectious wastes (bandages, gloves, cultures, swabs, blood and body fluids), hazardous wastes (sharps, instruments, chemicals), radioactive waste from cancer therapies, pharmaceutical waste
Agricultural	Crops, orchards, vineyards, dairies, feedlots, farms	Spoiled food wastes, agricultural wastes (e.g., rice husks, cotton stalks, coconut shells, coffee waste), hazardous wastes (e.g., pesticides)

Adapted from Hoornweg and Bhada-Tata (2012, table 2).

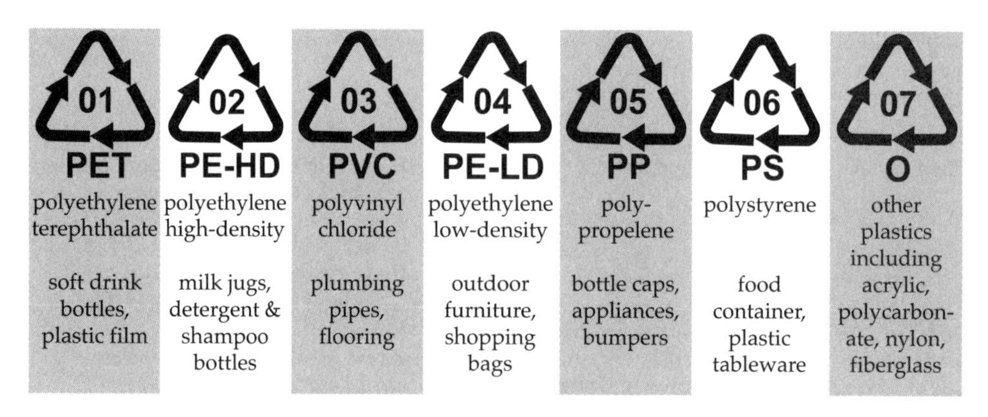

Figure 9.6
Resin identification codes and common uses.

we will see how they are distributed for a typical community—for example, how much paper is typically discarded compared to plastics.

In contrast to MSW, non-MSW solid waste sources usually have to organize the treatment of their own waste, or the pretreatment of their waste, before it is disposed or sent into the MSW loop.

Moreover, the list given in table 9.1 offers broad categories, but most solid waste types also have subcategories. This is the case for plastics, for example, and figure 9.6 shows the familiar resin identification codes and some common uses.

Managing such a variety of solid waste types can be challenging, and strategies must therefore be flexible. While the definition of solid waste management can vary, it usually includes elements of the collection, the treatment, and the disposal of solid waste. In a rather comprehensive manner, Tchobanoglous, Theisen, and Vigil (1993, p. 7) define solid waste management as "the discipline associated with the control of generation, storage, collection, transfer and transport, processing, and disposal of solid wastes in a manner that is in accord with the best principles of public health, economics, engineering, conservation, aesthetics, and other environmental considerations, and that is also responsive to public attitudes." The authors therefore distinguish six main processes that they refer to as *functional elements*.

From this list, what is interesting is that one of these functional elements is *generation*, meaning that reducing the generation of solid waste—for example, at home—is in the purview of solid waste management. Moreover, they directly relate solid waste management to multiple disciplines, thus highlighting the intrinsically interdisciplinary nature of solid waste management, which requires the involvement of multiple parties. Finally, they also recognize the cultural aspect of solid waste management by

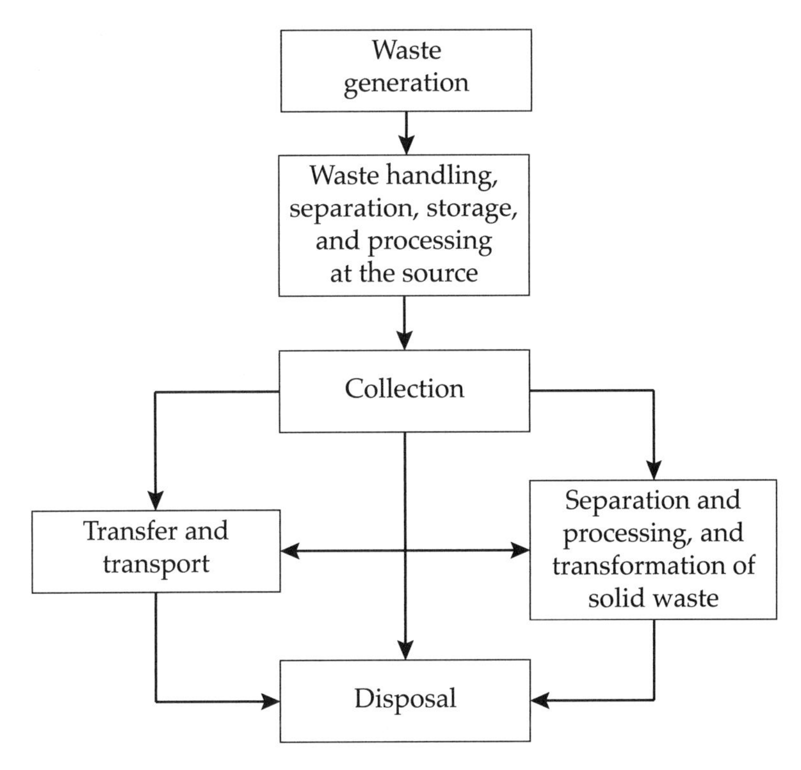

Figure 9.7
Functional elements involved in solid waste management. Adapted from Tchobanoglous, Theisen, and Vigil (1993, fig. 1-5).

adding that strategies should be "responsive to public attitudes." Dealing with solid waste is therefore more complicated than it might seem at first.

Figure 9.7 shows a flow chart of the six functional elements involved in solid waste management. *Waste generation* is quite simply the identification of material that no longer possesses any value to the owner, from cutting out the price tag on a new piece of clothes to discarding an old piece of furniture.

Waste handling, separation, storage, and *processing* at the source represent how solid waste is used or discarded before it is placed in a container ready for pickup—for example, whether one or multiple trash cans are used. This is especially important when it comes to recycling, and many households opt to have multiple trash cans to store different kinds of solid waste, such as food waste, recyclables, and others.

Collection involves not only the physical collection of the solid waste but also its transport to the primary facility where the vehicle is emptied, whether it is a recycling plant, a processing facility, a landfill site, or any other intermediary or final facility where the waste is to be first handled after coming from the original generation site. This step in the process tends to be the costliest, representing about 50% of the total cost.

The *separation, processing,* and *transformation* of solid waste can occur either right after it has been picked up or after it has already been processed and transported (see the next functional element). Possible processes include manual separation, automatic screening of larger items, collection of ferrous metals using magnets, and collection of nonferrous metals, such as aluminum, with eddy current separators. The general goal of the transformation step is to recover potential products and as much energy as possible or to simply reduce the size and weight of the solid waste. For example, organic matter can be combusted (i.e., burned), and the energy produced can be used to generate heat and/or electricity. Food waste can also be used for composting, which is technically called *aerobic composting*—in contrast to *anaerobic digestion*, as we will see in section 9.3.

Transfer and *transport* deal specifically with the aggregation of solid waste, typically from smaller to larger facilities (e.g., from a small municipal to a large regional facility), and its transport to processing facilities and between processing facilities (in conjunction with the previous functional element). For example, during a trip to Barcelona, Spain, I took the picture of the small garbage truck shown in figure 9.8 that has to deal with small medieval streets present in the city center. Once full, the truck discharges its load to a larger truck before it is to be brought to a processing facility, or it may go directly to the city's Waste-to-Energy processing plant (i.e., combustion with energy recovery).

Disposal consists of the final location where the solid waste is taken. This includes not only the solid waste that was not processed along the way but also the residues involved in all processes—for example, the remaining ash from the incineration of waste. In the United States, all residues and remaining solid waste go to sanitary landfills that are often privately owned—see figure 9.1c for an example—and that have been engineered to minimize impacts on public health and the environment.

Overall, these six functional elements often span multiple jurisdictions, which is why solid waste management is most often dealt with at the regional level, thus requiring cooperation between municipalities.

Following the six functional elements involved in solid waste management, we can now think of ways to coordinate between them to achieve specific objectives. This process is generally referred to as *integrated solid waste management* (ISWM), and it has

Figure 9.8
Small garbage truck in Barcelona, Spain.

defined solid waste management practices in many urban regions around the world since the 1990s at least. In general, the objectives that we try to achieve with ISWM follow a particular hierarchy defined by the EPA (and others) and shown in figure 9.9.

The primary objective is to reduce how much waste is generated (e.g., by reducing demand, by using less packaging material through buying in bulk, etc.), to reuse a material instead of discarding it (e.g., through upgrading and repair, or simply by donating it), and also to reduce the toxicity of the waste as much as possible (e.g., by banning certain materials). The second objective is to promote recycling and composting to reduce the amount of solid waste that ends up in landfills. The third objective is to recover as much energy as possible before the solid waste is sent to the landfill. Finally, the fourth objective is to ensure proper treatment, whether physical (e.g., shredding), chemical (e.g., incineration), or biological (e.g., anaerobic digestion), and the disposal of any remaining solid waste.

Solid waste therefore has the potential to be transformed extensively, and to determine how we can transform it, we first need to learn its properties, which is the topic of the next section.

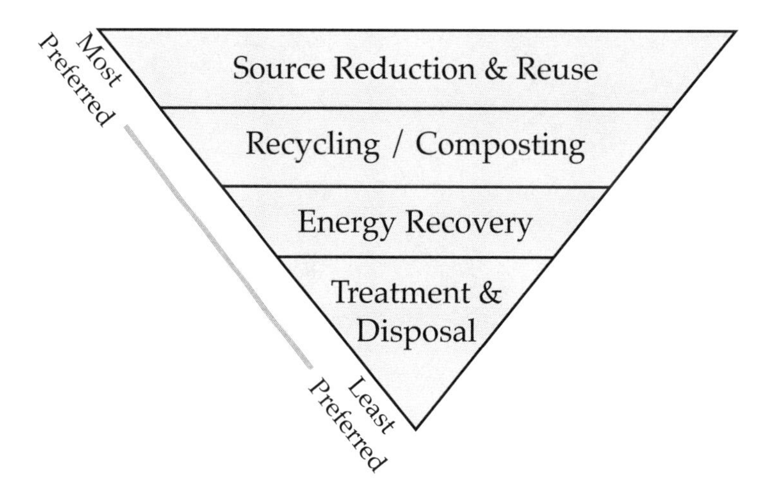

Figure 9.9
The EPA's waste management hierarchy.

Example 9.1
Integrated Solid Waste Management in Civitas

The Civitas government is looking into strategies to reduce the amount of solid waste generated in the city, and it is considering partnering with other cities located in the region. (1) Discuss the ways solid waste generation can be reduced and (2) list the ways Civitas could partner with other cities looking at the six functional elements of solid waste management.

Solution

(1) From figure 9.9, the primary and preferred way to reduce the amount of solid waste generated is to try to reduce the amount generated at the source. This can be done by encouraging people to reuse or repair the objects they use—like repairing furniture or machinery and reusing shopping bags as opposed to discarding them—and by encouraging people not to use disposable products, like plastic cutlery. Moreover, manufacturers can be encouraged to use less packaging and to repair (instead of replace) an object.

(2) Many resources can be put together to handle the management of solid waste at the regional level. Common campaigns can be organized to reduce solid waste generation at the business and household level (see (1)). Moreover, trash pickup routes can include multiple jurisdictions, and sorting and recycling efforts can be shared among multiple jurisdictions as well—for example, a recycle facility can be built to service several cities. A biomass power plant that burns organic solid waste can also be built to provide electricity to the region, and common landfills could be used by several cities to limit the number of trips by heavy trucks.

Solutions to (1) and (2) are not exhaustive, and many solutions can be found to cater to specific local characteristics.

9.1.3 Physical, Chemical, and Biological Properties of Solid Waste

The properties of solid waste can be generally divided into three main categories: physical, chemical, and biological. Physical properties include those related to size and moisture content. Chemical properties include those related to the chemical content of the waste, which is important to estimate how much energy can be recovered by burning the waste. Biological properties include those related to biodegradability and odors. In this section, we will cover the three categories, although the section on biological properties will be particularly short. We start with the physical properties.

9.1.3.1 Physical Properties Solid waste holds many physical properties that partly dictate how it can and cannot be processed. For example, if a material is brittle or soft enough, it can be shredded before being sent to landfills, to reduce its volume. Emphasizing MSW, in this section we will cover properties related to mass and volume (i.e., specific weight), moisture content (i.e., that particularly impacts weight), particle size (i.e., for shredding), and permeability (i.e., for the movement of liquids and gases in solid waste).

The first physical property of solid waste is mass and volume. Because pieces of solid waste can vary greatly in size, we typically look at the *specific weight* of different types of solid waste, which is defined as the weight of solid waste per unit volume and is often expressed in [kg/m^3].[12] Even then, the specific weight of solid waste varies greatly depending on whether it is compacted or not and whether it is wet or not. Table 9.2 shows ranges of specific weights of various types of solid waste, but even these ranges can vary based on individual studies. Typical values are also shown and can be used to get a general estimate of the weight and volume of the solid waste to expect. Moreover, certain types of solid waste that are made of various parts can be considered units—for example, a dishwasher might be expressed in [kg/unit].

From table 9.2, we can see that typical specific weight values range from 50 kg/m^3 for cardboard (uncompacted) to 1,420 kg/m^3 for mixed demolition debris (noncombustible). Moreover, the last set of rows shows the specific weights of general MSW. In particular, we can see that uncompacted solid waste weighs about 160 kg/m^3 compared to 1,100 kg/m^3 for well-compacted landfilled solid waste, which is close to seven times higher—thus showing how compacting solid waste can save a lot of space. An exhaustive list of material can be found in U.S. Environmental Protection Agency (1997), appendix B.

Table 9.2 also shows that some types of waste have an important moisture content. We therefore generally refer to the *wet weight* or the *dry weight* of solid waste. The moisture content M is collected by drying solid waste at 105°C (221°F) for more than twenty-four hours and by using the equation:

Table 9.2
Specific weight of typical MSW

	Specific weight [kg/m^3]		Moisture content (% by weight)	
Type of waste	Range	Typical	Range	Typical
Residential (uncompacted)				
Food wastes (mixed)	130–480	290	50–80	70
Paper	40–130	90	4–10	6
Cardboard	40–80	50	4–8	5
Plastics	40–130	65	1–4	2
Textiles	40–100	65	6–15	10
Yard wastes	60–225	100	30–80	60
Wood	130–320	240	15–40	20
Glass	160–480	200	1–4	2
Steel cans	60–160	90	2–4	3
Aluminum	65–240	160	2–4	2
Other metals	130–1,150	320	2–4	3
Dirt, ashes, and so on	320–1,000	480	6–12	8
Commercial				
Food wastes (wet)	475–950	540	50–80	70
Appliances	150–200	180	0–2	1
Rubbish (combustible)	50–180	120	10–30	15
Rubbish (noncombustible)	180–360	300	5–15	10
Rubbish (mixed)	140–180	160	10–25	15
Construction and demolition				
Mixed demolition (noncombustible)	1,000–1,600	1,420	2–10	4
Mixed demolition (combustible)	300–400	350	4–15	8
Mixed construction (combustible)	180–360	260	4–15	8
Broken concrete	1,200–1,800	510	0–5	—
Industrial				
Chemical sludge (wet)	800–1,100	1,000	75–99	80
Fly ash	700–900	800	2–10	4
Metal scrap (mixed)	700–1,500	900	0–5	—
Oils, tars, asphalts	800–1,000	950	0–5	2
Wood (mixed)	400–675	500	30–60	25
Agricultural				
Agricultural (mixed)	400–750	560	40–80	50
Dead animals	200–500	360	—	—

Table 9.2 (continued)

Type of waste	Specific weight [kg/m^3]		Moisture content (% by weight)	
	Range	Typical	Range	Typical
Fruit wastes (mixed)	250–750	360	60–90	75
Vegetable wastes (mixed)	200–700	360	60–90	75
Manure (wet)	900–1,050	1,000	75–96	94
Municipal solid waste (general)				
Uncompacted	150–180	160	15–40	20
In compactor truck	180–450	300	15–40	20
Landfill normally compacted	700–1,000	830	15–40	25
Landfill well compacted	1,000–>1,200	1,100	15–40	25

Sources: Tchobanoglous, Theisen, and Vigil (1993); U.S. Environmental Protection Agency (1997).

$$M = 100 \cdot \left(\frac{W_{wet} - W_{dry}}{W_{wet}} \right) \tag{1}$$

where W_{wet} and W_{dry} are, respectively, the wet and dry weights of the solid waste. Table 9.2 also shows ranges and typical values of moisture content for different solid waste types. We can see that food wastes have a typical moisture content of 70%, which means that 70% of the weight of food wastes is simply water. In contrast, other solid waste types, such as metals and plastics, have little to no moisture content.

The next important physical property of solid waste is its particle size since it affects how solid waste can be transported and separated (e.g., using screens). There are several ways to define the size of component S_c in solid waste, including:

$$S_c = l \tag{2}$$

$$S_c = \left(\frac{l + w}{2} \right) \tag{3}$$

$$S_c = \left(\frac{l + w + h}{3} \right) \tag{4}$$

$$S_c = (l \times w)^{1/2} \tag{5}$$

$$S_c = (l \times w \times h)^{1/3} \tag{6}$$

where l, w, and h are the length, width, and height of the component, generally expressed in [mm].

Example 9.2

Dry Weight of Solid Waste in Civitas Household

A house in Civitas generated 20 kg of solid waste in a week, with the composition shown in the table below. Calculate (1) the moisture content of the solid waste and (2) the dry weight of the solid waste.

Type of waste	Percent by weight
Food wastes (mixed)	29%
Cardboard	38%
Plastics	16%
Glass	13%
Aluminum	4%

Solution

(1) By first retrieving the typical moisture contents from table 9.2, we can then calculate a proportional moisture content value based on the percent by weight of the solid waste, which gives us a final moisture content of 22.86%.

Type of waste	Percent by weight	Typical moisture	Proportional moisture
Food wastes (mixed)	29%	×70=	20.30
Cardboard	38%	×5=	1.90
Plastics	16%	×2=	0.32
Glass	13%	×2=	0.26
Aluminum	4%	×2=	0.08
Total	100%		22.86%

(2) Using equation (1) and considering the household generated 20 kg of solid waste with a 22.86% moisture content, the dry weight W_{dry} of the solid waste is:

$$W_{dry} = W_{wet} - \frac{M}{100} \cdot W_{wet} = 20 - \frac{22.86}{100} \times 20 = 15.43 \text{ kg}$$

Particle size might also affect how liquids and gases move in solid waste, which is particularly important in landfills. Accordingly, another important physical property of solid waste is its *permeability*, generally expressed by its hydraulic conductivity K, which we already saw in chapter 6 when learning about groundwater engineering. In equation form, K is defined as:

$$K = Cd^2 \frac{\gamma}{\mu} = k \frac{\gamma}{\mu} \tag{7}$$

where C is a dimensionless shape factor, d is the average size of pores, and γ and μ are the specific weight and viscosity of the fluid traveling through the solid waste, which is water for us since we are dealing with moisture.

Let us now look at the chemical properties of solid waste.

9.1.3.2 Chemical Properties Solid waste contains multiple chemical properties, such as the time it takes for the wet solid waste to dry or the fusion point of ash. In this section we will only look at two properties: ultimate analysis and energy content.

Ultimate analysis focuses on identifying the chemical components of solid waste. Namely, it measures the amount of volatile carbon (C), hydrogen (H), oxygen (O), nitrogen (N), sulfur (S), and nonvolatile ash (Ash) yielded by the combustion of a material. Table 9.3 shows typical values (percent by weight) for the combustible components of residential MSW. From the table we can see that the combustion products of organic solid waste contain mostly C and O (primarily as CO_2). This essentially means that much energy can be recovered from burning organic wastes, and little energy can be recovered from burning inorganic wastes (although they are combustible).[13]

Another way to express the chemical property of solid waste is to take into account molar composition—that is, to estimate the number of moles from each chemical element. Table 9.4 shows the atomic weight of the five common chemical elements. This will be particularly useful when we estimate the energy content of solid waste, which is also why we do not include ash (since it does not burn).

To come up with the molar composition, the weights found for each element are divided by their atomic weight. We need to be careful about water, however. We must first find the dry weight of the solid waste, and the molar composition can then be calculated for the solid waste with and without water. The molar composition with water is calculated by assigning the water weight to H and O, precisely by assigning two-eighteenths of the weight (i.e., 2/18) to H and sixteen-eighteenths (i.e., 16/18) to O, as the total atomic weight of a water molecule is eighteen for two Hs and one O—the familiar water molecule, H_2O—and using the individual atomic weights of H

Table 9.3
Ultimate analysis data of MSW

	Percent by weight (dry basis)					
Component	C	H	O	N	S	Ash
Organics						
Food wastes	48.0	6.4	37.6	2.6	0.4	5.0
Paper	43.5	6.0	44.0	0.3	0.2	6.0
Cardboard	44.0	5.9	44.6	0.3	0.2	5.0
Plastics	60.0	7.2	22.8	—	—	10.0
Textiles	55.0	6.6	31.2	4.6	0.2	2.5
Yard wastes	47.8	6.0	38.0	3.4	0.3	4.5
Wood	49.5	6.0	42.7	0.2	0.1	1.5
Inorganics						
Glass	0.5	0.1	0.4	<0.1	—	98.9
Metals	4.5	0.6	4.3	<0.1	—	90.5
Dirt, ash, and so on	26.3	3.0	2.0	0.5	0.2	68.0

Source: Tchobanoglous, Theisen, and Vigil (1993).

Table 9.4
Atomic weight of five chemical elements

Component	C	H	O	N	S
Atomic weight	12	1	16	14	32

and O from table 9.4. Solid waste can then be expressed by its molar composition as $C_xH_xO_xN_xS_x$, where the *x*'s are substituted by their molar value. A practical example is given in example 9.3.

After the chemical composition, another important chemical property of solid waste is energy content. Specifically, we are interested in the energy content of the organic components since we can burn organic matter and use the heat to produce electricity and/or to heat buildings. From a unit perspective, we are looking for energy per unit weight in [Wh/kg], also referred to as *specific energy*—units of [J/kg] (metric) and [Btu/lb] (British) are also used in practice.

Energy content is generally measured by burning the solid waste and by measuring how much energy is released, which is often done with a device called a calorimeter. The term *calor* from Latin means "heat," and the *calorie* has become a unit of energy produced from heat,[14] hence the name *calorimeter*. The principle is fairly simple: a piece

Example 9.3

Ultimate Analysis in Solid Waste of Civitas Household

Using data for the Civitas household from example 9.2, (1) perform an ultimate analysis of the dry and wet solid waste to determine the weight of each chemical component and (2) calculate the molar composition of the dry and wet solid waste. Note that because of rounding issues, differences may appear. Here, the rounded values were taken from the results in the spreadsheet software package used for this example.

Solution

(1) Remembering that the total weight of the solid waste was 20 kg, we can first calculate the dry weight W_{dry} of each waste type and then use table 9.3 to determine the weight of each chemical component by waste type before summing them. We can then add the weight of water—which is $20-15.43=4.57$ kg—to the H and O components by using the proportions given in the text (i.e., 2/18 for H and 16/18 for O). The results are shown in the table below.

Overall, we see that the dry solid waste contains close to 5.95 kg of C, 4.63 kg of O, and 3.99 kg of ash and that the wet solid waste contains, in contrast, 8.69 kg of O.

				Chemical components (kg)					
Type of waste	Percent by weight	Moisture content	Dry weight W_{dry} (kg)	C	H	O	N	S	Ash
Food wastes (mixed)	29%	70	1.74	0.84	0.11	0.65	0.05	0.01	0.09
Cardboard	38%	5	7.22	3.18	0.43	3.22	0.02	0.01	0.36
Plastics	16%	2	3.14	1.88	0.23	0.72	0.00	0.00	0.31
Glass	13%	2	2.55	0.01	0.00	0.01	0.00	0.00	2.52
Aluminum	4%	2	0.78	0.04	0.00	0.03	0.00	0.00	0.71
Total (dry)			15.43	5.95	0.77	4.63	0.07	0.02	3.99
Water			4.57		0.51	4.06			
Total (wet)			20.00	5.95	1.28	8.69	0.07	0.02	3.99

(2) To find the molar composition, we then divide the weight of each chemical component by its atomic weight listed in table 9.4 and shown in the table below. It is then customary to calculate the molar ratio by setting the chemical component with the lowest number of moles to 1 (which is sulfur in our case). This gives us a dry solid waste molar composition of $C_{787}H_{1222}O_{459}N_8S$ and a wet solid waste molar composition of $C_{787}H_{2032}O_{862}N_8S$.

Component	C	H	O	N	S
atomic weight	12	1	16	14	32
Moles (dry)	0.49583	0.77000	0.28938	0.00500	0.00063
Moles (wet)	0.49583	1.28000	0.54313	0.00500	0.00063
Mole ratio (dry)	787	1,222	459	8	1
Mole ratio (wet)	787	2,032	862	8	1

Table 9.5
As discarded inert residue and energy content data of MSW

	Inert residue (%)		Specific energy (Wh/kg)	
Type of waste	Range	Typical	Range	Typical
Organics				
Food wastes (mixed)	2–8	5	1,000–2,000	1,300
Paper	4–8	6	3,200–5,200	4,600
Cardboard	3–6	5	3,800–4,900	4,500
Plastics	6–20	10	7,700–10,350	9,000
Textiles	2–4	3	4,200–5,200	4,800
Yard wastes	2–6	5	650–5,200	1,800
Wood	0.6–2	2	4,850–5,500	5,200[a]
Inorganics				
Glass	96–99+	98	30–65	40
Steel cans	96–99+	98	65–320	200
Aluminum	90–99+	96	—	—
Other metals	94–99+	98	65–320	200
Dirt, ashes, and so on	60–80	70	650–3,200	2,000
Municipal solid waste (general)			2,500–3,900	3,200

Source: Tchobanoglous, Theisen, and Vigil (1993).
Note: [a]Updated numbers suggest values closer to 4,150 Wh/kg for wood.

of material is burned, the heat produced is used to heat water, and the energy released is then measured based on how much the temperature of the water increases.

As a result, we expect solid waste that is made mostly with organic chemical components—especially C and H—to contain a lot of combustible energy.[15] Table 9.5 shows the specific energy of different types of solid waste, and as predicted, organics contain significantly more energy. The table also contains a column for *inert residue*, which is the solid waste left after the combustion process that is not chemically or biologically reactive.

We should note that the numbers shown in table 9.5 are for *as discarded* solid waste, which is therefore wet. To calculate how much energy E_{dry} is contained in dry solid waste, we can use the equation:

$$E_{dry} = E_{asdiscarded}\left(\frac{100}{100 - M}\right) \tag{8}$$

where $E_{asdiscarded}$ is the energy content of the solid waste as discarded. Naturally, $E_{dry} > E_{asdiscarded}$ because E_{dry} does not include water; noting here again that we are referring to *specific energy* here in [Wh/kg] as opposed to energy quantities in [Wh].[16] Similarly, ash does not burn—in fact, it is a residue from the burning process—and the recoverable energy $E_{recoverable}$ contained in solid waste can be calculated as:

$$E_{recoverable} = E_{asdiscarded}\left(\frac{100}{100 - M - Ash}\right) \tag{9}$$

In the absence of specific energy values, as shown in table 9.4, we can also use the modified Dulong formula that estimates how much energy is contained in solid waste based purely on the chemical content. The modified Dulong formula is defined as:

$$E_{Dulong} = 94C + 394\left(H_2 - \frac{1}{8}O_2\right) + 6N + 26S \tag{10}$$

Where E_{Dulong} is in [Wh/kg] and it represents the specific energy of the organic content of the solid waste (i.e., no ash), whether it is dry or wet. We have to be careful because the coefficients given are not dimensionless, and they cannot be used with other units.[17]

This completes our investigation of the chemical properties of solid waste, and we can review a few biological properties before moving on to studying demand.

9.1.3.3 Biological Properties In addition to physical and chemical properties, solid waste also possesses important biological properties. For example, organic solid waste contains sugars, fats, and proteins.[18] Moreover, some types of solid waste can dissolve in water and others cannot. Biological processes in solid waste are the origin of the unpleasant odors that often emanate from trash left unattended for a number of days, which can lead to the breeding of flies and other pests. This partly dictates why solid waste is often handled away from populated areas.[19]

Importantly for us, biological processes are at the origins of *aerobic composting* and *anaerobic digestion*. In essence, both processes are about letting organic solid waste decompose—essentially, letting microorganisms break down (i.e., eat) the organic wastes—but the main difference resides in whether the decomposition involves oxygen or not. *Aerobic* comes from the Greek word for "air" (meant as free air that contains oxygen).

Aerobic composting consists of letting organic solid waste decompose in open air, and when controlled—which we will see in section 9.3—it can produce *compost* that is commonly used as a fertilizer. The general process in aerobic composting is:

$$\begin{aligned}\text{organic matter} + O_2 + \text{nutrients} \rightarrow \text{new cells} + \text{humus} + CO_2 + H_2O \\ + NH_3 + SO_2^{2-} + \text{heat}\end{aligned} \tag{11}$$

Example 9.4

Energy Content of Solid Waste of Civitas Household

Using data for the Civitas household from examples 9.2 and 9.3, calculate the energy content of the wet solid waste using (1) the values given in table 9.5 and (2) the modified Dulong formula.

Solution

(1) Using the typical specific energy values from table 9.5, we get an energy content of the wet solid waste as 3,532 Wh/kg, as shown in the table below. The results are relatively close to the typical value of 3,200 Wh/kg for MSW given in table 9.5.

Type of waste	Percent by weight	Typical energy (Wh/kg)	Energy content (Wh/kg)
Food wastes (mixed)	29%	1,300	377
Cardboard	38%	4,500	1,710
Plastics	16%	9,000	1,440
Glass	13%	40	5
Aluminum	4%	0	—
Total	100%		3,532

(2) To use the modified Dulong formula, we need to determine the contribution of each chemical element in terms of weight. We therefore first need to multiply the molar composition of the wet solid waste from example 9.3, $C_{787}H_{2032}O_{862}N_8S$, by the atomic weight of each element and then calculate the percentage contribution of each element as shown in the table below.

Component	Composition	Atomic weight	Weight contribution	Percentage contribution
C	787	12	9,444	37.16%
H	2,032	1	2,032	8.00%
O	862	16	13,792	54.27%
N	8	14	112	0.44%
S	1	32	32	0.13%
Total			25,412	100.00%

Then, we can apply the modified Dulong formula, and the energy content of the wet solid waste is:

$$E_{Dulong} = 94 \times 37.16 + 394\left(8 - \frac{1}{8} \times 54.27\right) + 6 \times 0.44 + 26 \times 0.13 = 3{,}978 \text{ Wh/kg}$$

Can you guess why the modified Dulong formula gives us a higher energy content? It is because we are only considering the organic content of the solid waste—that is, it does not include ash that does not contribute any energy content in terms of heating value. Again, note that it is the energy content that is higher, not the total energy. The same batch of solid waste has the same energy, but if we omit the ash, then the specific energy (i.e., per kilogram) is higher—the same note was made for dry versus wet energy in the text.

The *humus* is the compost as commonly defined. In contrast, anaerobic digestion limits the presence of oxygen by sealing the solid waste in an airtight container, leading to the following process:

$$\text{organic matter} + H_2O + \text{nutrients} \rightarrow \text{new cells} + \text{humus} + CO_2 + CH_4 + NH_3 + H_2S + \text{heat} \quad (12)$$

Anaerobic digestion also generates humus that can be used as fertilizer, but the big difference is the production of natural gas—that is, CH_4 (methane)—that can be burned to produce heat and/or electricity. Anaerobic digestion can therefore be more desirable to produce energy, but the process is not as simple as aerobic composting since the solid waste needs to be carefully sealed to avoid oxygen intrusion and methane leaks.

This concludes our section on the biological properties of solid waste and our review of the fundamentals of solid waste management. Now that we are equipped with the necessary knowledge, it is time to learn about demand in the context of solid waste, or more appropriately, it is time to learn about solid waste generation.

9.2 Solid Waste Generation and Composition

When it comes to solid waste management, solid waste generation and composition is key. If we look back at figure 9.7, we see that waste generation is only one of the six functional elements of solid waste management, and yet it deserves its own section. First of all, without waste generation, the other five elements would not exist. Moreover, these other five elements depend entirely on the types and quantities of solid waste that are generated. Gaining a deep understanding of waste generation is therefore crucial to properly manage solid waste.

In terms of units, and as we have already seen in this chapter, we will use weight per unit time, generally in metric tons [t] per year or kilograms [kg] per month—as opposed to short tons, which are normally used in U.S.-based studies, such as in the EPA study used in this chapter.[20] For example, figure 9.10 shows the historical evolution of total yearly MSW generation (not including C&D) and monthly MSW generation per person in the United States from 1960 to 2015. In 2015, a total of 238 Mt of solid waste was generated, which represents 61.82 kg per person per month. From the figure, we can see that both the total and per person figures initially increased. Per person MSW generation plateaued around the year 2000 and remained relatively stable after. Total MSW, in contrast, kept increasing. In fact, similar to figure 7.14—that showed the evolution of annual vehicle kilometers traveled in the United States—total MSW generation showed signs of peaking between 2005 and 2010, but the trend then continued to increase. These trends suggest that current efforts to limit MSW

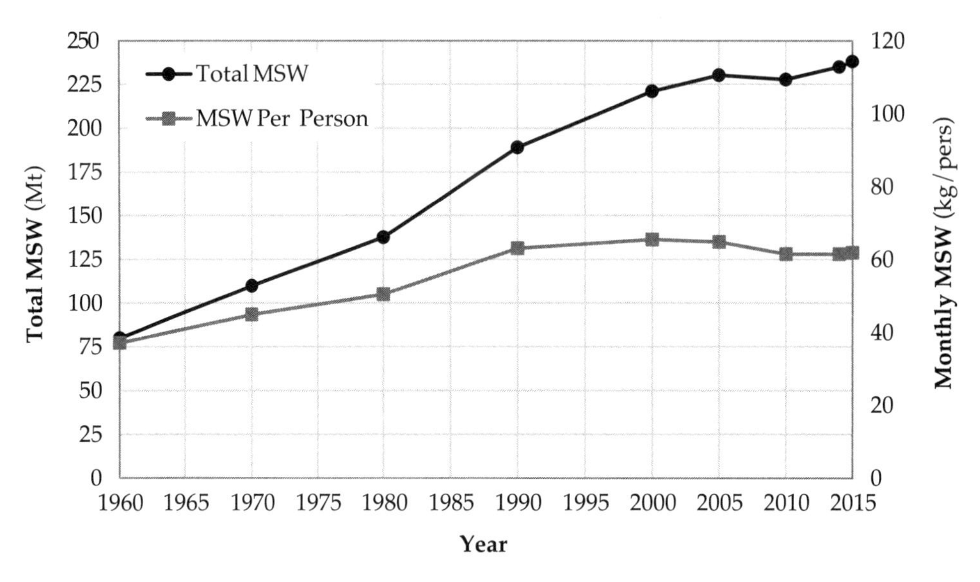

Figure 9.10
Historical trends in MSW generation in the United States.

generation show some success—that is, controlling the demand, our first sustainability principle—but that total MSW generation has increased because the U.S. population has increased.

In this section, we will discuss several important aspects of waste generation and composition. First, we will briefly learn how to perform waste audits so that we can effectively measure the quantities (i.e., generation) and types (i.e., composition) of solid waste generated. Second, we will look at general trends and at the evolution of solid waste generation and composition over time. Finally, we will focus on individual sectors (residential, commercial, etc.) to get a better understanding of how solid waste is generated and composed by sector. Moreover, we will uniquely focus on solid waste types (e.g., paper, glass, etc.) as opposed to products (e.g., major appliances, beer cans, diapers), but data on products are available from the U.S. Environmental Protection Agency (2018a).

One last point needs to be mentioned about data. Solid waste generation is typically reported in two ways: disposal-based or generation-based. Disposal-based refers to the solid waste that is disposed of in facilities located in an area, and generation-based refers to the solid waste that is generated in an area. When we focus on entire countries, the numbers tend to be similar because most countries treat their own solid waste,[21] but we will focus on cities and regions as well in this chapter, and the total weight numbers tend to be different because solid waste can easily be treated outside of the city.

In Chicago, for example, about 7 Mt of solid waste were generated in 2007, but only 4.3 Mt were disposed of in the city of Chicago (CDM 2010).[22] We will focus mainly on generation-based solid waste when it comes to total weight generated. Breakdowns by solid waste composition (e.g., paper, plastics, etc.) tend to be disposal-based, however, since studies are done at disposal facilities—generation-based studies only report how the solid waste was generated (e.g., disposed of in a bin, curbside recycled, curbside organics, etc.). In this section we will therefore break down percentages of solid waste composition from disposal-based studies, and we will use the total weight generated from generation-based studies. At the level of the United States, it will not make a difference—the EPA only reports one set of values for both.

Solid waste studies systematically start with a solid waste audit, and this is the topic of the next section.

9.2.1 Solid Waste Audit

A waste audit or waste assessment is essentially the process of collecting data on solid waste generation and composition. Waste audits can be performed by governing bodies (e.g., a municipality) to get information on the general waste stream generated (i.e., on the solid waste composition of MSW), but they can also be performed by individual companies looking to get information on how much waste is generated in their facility and how this amount can be reduced.

Waste audits are essential to get detailed information about solid waste generation and composition in order to develop effective strategies and policies to reduce waste generation. For example, waste audits directly inform us about the proportion of waste that is reusable, recyclable, and compostable, the proportion that can be incinerated, and even the proportion that is hazardous.

The concept of a waste audit is relatively simple: collect the waste, separate the waste, weigh each waste type, enter the results in a spreadsheet, and analyze the results. In practice, however, waste audits need a lot of preparation and coordination between all members involved. Figure 9.11 shows the general waste audit process, and we can see that it contains three primary phases and two optional phases.

The first phase is the audit setup, in which the goals and priorities are defined. For example, the locations where the audit will take place needs to be agreed upon depending on the boundaries of the study—for example, whether the goal is to survey MSW, C&D, or specific industries. The team also needs to be formed, and the schedules need to be established. The equipment needed to conduct the audit must be secured as well, including containers or bins and scales to weigh the waste. Waste audit setups are lengthy but crucial to ensure the waste audit is implemented successfully.

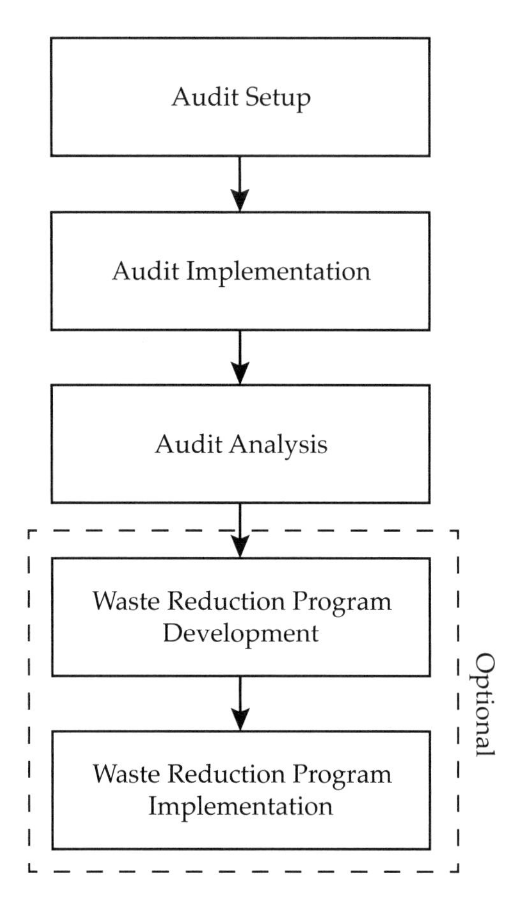

Figure 9.11
The waste audit process.

The second phase is the actual audit implementation. Teams are dispatched over a number of sites identified during the setup phase. All waste is sorted according to distinct categories and sometimes according to distinct subcategories as well, also agreed upon during the setup phase; for example, categories might include paper, plastics, glass, metal, compostable organics, and so on, and subcategories might then differentiate between clear and green glass and between different paper grades (i.e., high grade vs. newspaper). The waste is then placed in containers or bins and weighed (adjusting for the weight of the empty bin), and the results are entered in a spreadsheet. Figure 9.12 shows the snapshot of a spreadsheet given by the EPA—the EPA website also includes other useful spreadsheets. Several spreadsheets are used to account for the solid waste that will be recycled or reused versus the solid waste that will be incinerated or landfilled.

Waste Sort Worksheet				
Organization:				
Function Area(s)/ Department(s):				
Date(s):				
Waste Disposal				
Waste Product/Material	Net component weight (lbs/week)	Weeks per Year	Estimated annual weight (lbs/year)	Percent total sample weight
	0			
	0			
	0			
	0			
	0			
	0			
	0			
	0			
	0			
	0			
	0			
	0			
	0			
	0			
	0			
Total:	0	lbs	0	lbs

Figure 9.12
The EPA waste sorting spreadsheet.
Source: U.S. Environmental Protection Agency (2018b).

Finally, waste audits are typically conducted over several days (generally, a week) to ensure that representative solid waste streams are recorded. For example, if a company discards its paper waste every Friday and the audit is pursued only on a Friday, it would find a disproportionate amount of paper waste. Similarly, if the audit is conducted any other day, it would find that the company does not generate any paper waste. This is why waste audits are conducted over several days, agreed upon during the waste audit setup phase. Some audits are even repeated several times per year to capture seasonal differences (e.g., with yard trimmings).

The third phase is the audit analysis, which essentially consists of compiling all spreadsheets and analyzing the data. This may be started during phase two in case some adjustments are needed. The results measured during the audit implementation phase are then scaled up to estimate the quantity of solid waste generated for longer periods of time (e.g., a year) and/or for similar users (e.g., scaling up the results from the waste collected by several garbage trucks to an entire city). This step must be done carefully since a small error during the audit process may become a large error when scaled up. Therefore, several characteristics must be considered, such as the volume of the vehicle that brought the solid waste (e.g., the volume of a garbage truck and the number of garbage truck trips) and whether the solid waste is compacted or not. Moreover, the results from this phase include both a detailed analysis of the solid waste composition as well as information on the amount of solid waste reused, recycled, incinerated, and landfilled.

The fourth and fifth phases are optional and are typically applied when the waste audits are conducted with the purpose of establishing waste reduction work plans—for example, when a company orders the waste audit. Based on the results from phase three, a waste reduction work plan is first developed in phase four, and it is then implemented in phase five. Phase four can include both strategies to reduce the amount of waste generated as well as to divert some of the waste that is incinerated and landfilled so that it is reused or recycled instead. In particular, in phase four several criteria are considered, including the volume/weight of certain solid waste types, the ability to easily separate the waste, the economic value of separated waste, the impact of solid waste separation on daily activity, and whether certain solid waste types need to follow specific government regulations.

This is all we will cover on solid waste audits. Although the concept of a waste audit appears simple in principle, many elements need to be taken into consideration during the audit, making it a fairly complicated process. The EPA website offers many great resources to pursue a successful waste audit. Another good resource is the *Waste Audit*

Example 9.5
Solid Waste Audit in Civitas

The Civitas government has decided to conduct a solid waste audit to learn the solid waste generation patterns of its residents. (1) List the steps needed to conduct a waste audit and (2) discuss three important items that must be considered for the audit to be successful.

Solution

(1) As shown in figure 9.11, the three main phases of a waste audit are the following: audit setup, audit implementation, and audit analysis. Two more phases can be added if Civitas is looking for recommendations to reduce the amount of solid waste generated.

(2) Many important items must exist for an audit to be successful, including the following

a. Audit duration: For an entire city, an audit should be done over multiple days (e.g., a week) and over multiple seasons. For example, two one-week audits could be conducted, one in the summer and one in the winter.

b. Selection of disposal facility: The audit should carefully select the disposal facilities that will be analyzed to ensure the solid waste received is representative of the typical waste discarded. As a result, facilities that receive solid waste from commercial areas (e.g., a central business district) and from more residential areas should be selected.

c. Scaling factors: The number of garbage trucks arriving at a facility should be counted, and details about which areas the trucks service should be recorded. For example, 50% of the trucks might service commercial areas, and the other 50% might service residential areas, but if Civitas is mostly composed of residential areas, then the scaling factor for residential areas should be larger.

Users Manual: A Comprehensive Guide to the Waste Audit Process by Fenco MacLaren and Angus Environmental (1996).

Once the audit is completed, it is time to analyze the solid waste composition.

9.2.2 Solid Waste Trends and Composition

Solid waste composition reports the amount of solid waste (generally in terms of weight) by category and sometimes by subcategory. The categories and subcategories are selected during the waste audit process, but we have already seen typical categories reported in table 9.1 and in subsequent tables in section 9.1. In the United States, the EPA reports MSW composition in nine main categories: paper (that includes cardboard); glass; metals; plastics; rubber, leather, and textiles; wood; food wastes; yard trimmings; and other. Here, to keep the analysis simple, we will not discuss subcategories, but they

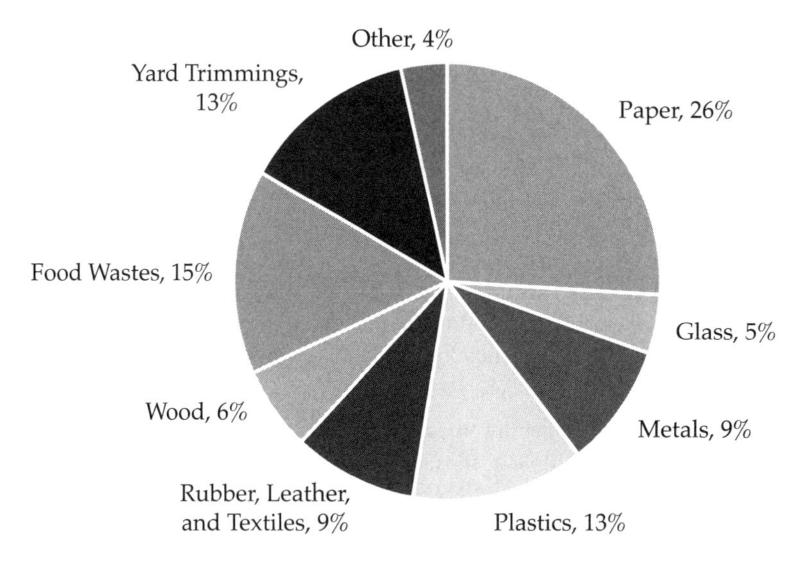

Figure 9.13
2015 U.S. MSW composition.

are important in practice—for example, to establish an adequate sorting process (e.g., ferrous vs. nonferrous metals) and to establish recycling strategies.

From figure 9.10, we recall that in 2015 a total of 238 Mt of MSW was generated. Figure 9.13 shows the composition of the 2015 MSW. We can see that paper leads the way with slightly more than one-quarter of all MSW generated; paper includes cardboard and all paper-based packaging products. Food wastes come in second, with 15%, followed by yard trimmings and plastics, both at 13%. These four categories alone represent two-thirds of all MSW generated.

These numbers are difficult to put in perspective, however—after all, it is hard to get a grasp of what 238 Mt means. Instead of looking at yearly percentages, table 9.6 shows the per person monthly MSW composition in [kg/pers], as reported in figure 9.10 as well. First of all, and as already reported, we can see that the average person generates 61.82 kg (135 lb) of solid waste per month. Essentially, this corresponds to the weight of an average person, as already discussed in chapter 1, which is significantly high.

The good news is that much of this solid waste is recyclable and compostable, but as we saw in figure 9.2, only about 35% of all solid waste gets recycled or reused. Moreover, if we recall the EPA's waste management hierarchy (figure 9.9), recycling and composition is the second-best strategy. The best strategy is reduction and reuse. When we think that humans used to generate barely any waste before the 1800s—and even

Table 9.6
2015 U.S. monthly MSW composition

Material	Monthly weight (kg/pers)
Paper	16.03
Glass	2.70
Metals	5.65
Plastics	8.13
Rubber, leather, and textiles	5.77
Wood	3.84
Food wastes	9.36
Yard trimmings	8.18
Other	2.16
Total	61.82

then, it was mainly in the form of food wastes—it is amazing to see how we have managed to increase our amount of waste in such a short time period. Put differently, it is a strong indicator that our first sustainability principle can be applied here—that is, control the demand or, more appropriately, reduce the amount of solid waste generated.

We can bring this analysis a little further and analyze how solid waste composition has evolved over time. Put differently, we can visualize the historical MSW generation trends from figure 9.10 while taking into account the nine main categories of MSW. This is what figure 9.14 shows by plotting the historical trends in total MSW composition in the United States from 1960 to 2015.

First, figure 9.14 reiterates that the amount of solid waste generated has significantly increased between 1960 to 2015. Precisely, the total amount of MSW has increased from 80 Mt to 238 Mt, which represents roughly a threefold increase. In terms of per person, the monthly MSW generated increased from 37 kg to 62 kg, which represents close to a factor of 1.7 and which is also significant.

Second, figure 9.14 shows that paper wastes have always dominated the MSW composition. In terms of percentages, however, the proportion of paper wastes decreased from about 35% in the 1960s and 1970s to 25% in 2015, likely because many paper transactions were substituted with electronic transactions. Moreover, we can see that yard trimmings used to be the second-largest type of waste, but they were overtaken by food wastes slightly before the year 2000. This either means that the per person quantity of yard trimmings has decreased or that the per person quantity of food wastes has increased. To determine which one, we can plot the same information shown in figure 9.14 but per person, which is exactly what figure 9.15 shows.

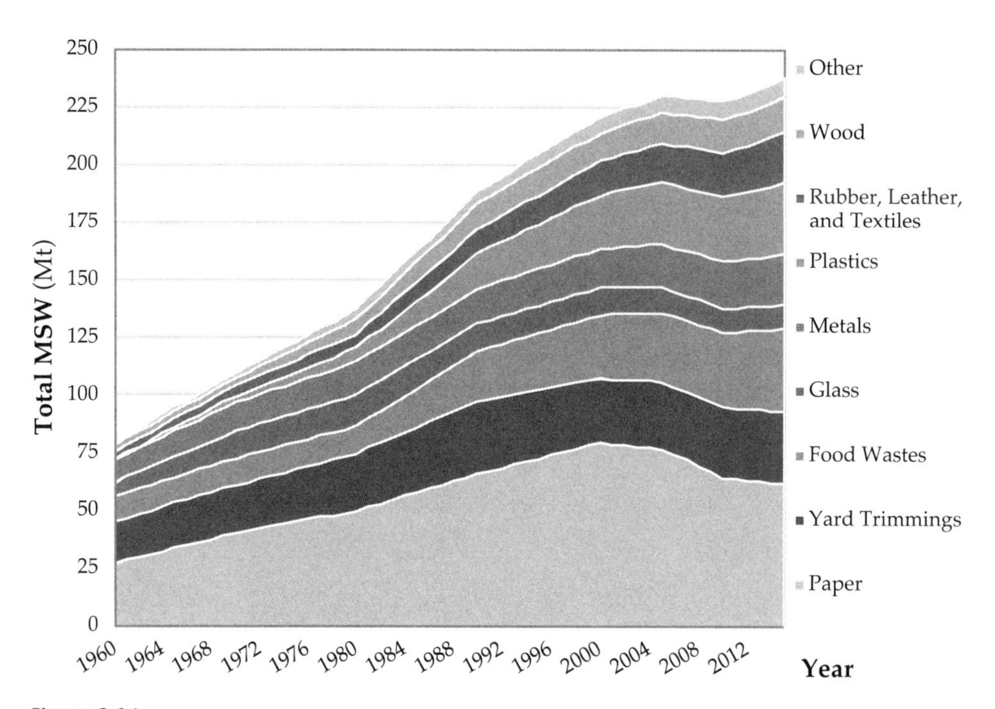

Figure 9.14
Historical trends in total MSW composition in the United States.

Figure 9.15 reiterates the fact that the proportion of paper waste has significantly decreased since the year 2000. Moreover, we can also see the amount of yard trimmings increased a little between 1960 and 1990, but it then decreased as well. In fact, the proportion of yard trimmings hovered around 9 kg per person per month during the period studied. Food wastes, however, increased fairly significantly. In fact, the amount of food wastes in 1960 was 5.1 kg per person per month, but it increased to 9.4 kg by 2015. These results suggest that we buy too much food and that a significant amount ends up in the trash.

Both figure 9.14 and figure 9.15 show that the proportion of plastics has increased significantly between 1960 to 2015. Both in terms of total and per person MSW, the proportion of plastics increased from about 0.4% in 1960 to more than 13% in 2015. This is again enormous, and in 2015 the average American threw away close to 8.1 kg of plastics per month. Remembering that plastics are made from oil and that oil is available in finite quantities—thus by nature unsustainable, as we saw in chapter 2—and looking at recent trends, we can see that our current MSW generation trends are simply not sustainable.

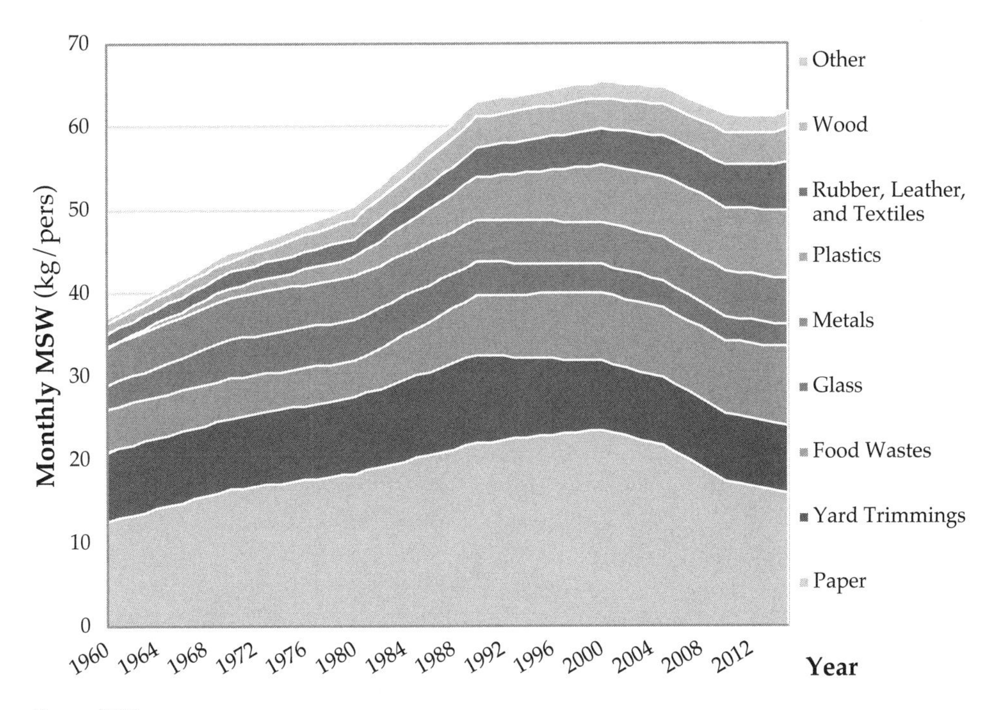

Figure 9.15
Historical trends in per-person MSW composition in the United States.

Finally, we can also see that the amount of rubber, leather, and textiles discarded has also increased, which largely represents the amount of clothes discarded. In terms of total MSW, the proportion of rubber, leather, and textiles has increased from 4.1% to 9.3%, and in terms of per-person MSW, it has increased from 1.5 kg to 5.8 kg per month. Again, this is significant. It means that in 2015 we discarded, on average, close to four times more rubber, leather, and textiles than we did in 1960.

The generation and composition figures we have seen so far apply to the United States, and they are not valid for all places. Table 9.7, for example, shows total waste generated for the city of Saint Pierre (Saint Pierre and Miquelon) in 2017 in monthly [kg/pers].[23] The Saint Pierre municipal government had been impressively proactive when it comes to solid waste management. Waste was either discarded at the curb and collected (38%) or it was self-hauled (62%) (i.e., waste directly taken to the disposal facility). In total, the average person in Saint Pierre generated 48.28 kg of waste per month in 2017, which included C&D debris.

The table further provides a breakdown of what was recycled and what was not from what was collected. In particular, we can see that 57% of the waste collected was

Example 9.6

Solid Waste Generation in Civitas

The waste audit from example 9.5 in Civitas was conducted in 1990, 2000, and 2010, and the gross MSW values are shown in the table below. (1) Recalling that the Civitas population was 38,500, 48,000, and 60,000 in 1990, 2000, and 2010, calculate the percentage of each waste type and calculate the per person monthly MSW generated. (2) Analyze the trends from the data calculated in (1).

Waste type (t)	1990	2000	2010
Paper	836	999	1,051
Glass	203	235	238
Metals	158	235	305
Plastics	136	294	407
Rubber, leather, and textiles	136	235	305
Wood	158	176	170
Food wastes	203	324	441
Yard trimmings	339	353	339
Other	90	88	136
Total	2,259	2,939	3,392

Solution

(1) The percentages and per person monthly values can easily be calculated, and they are shown below.

	Percentages			Monthly (kg/person)		
Waste type	1990	2000	2010	1990	2000	2010
Paper	37%	34%	31%	21.72	20.82	17.52
Glass	9%	8%	7%	5.28	4.90	3.96
Metals	7%	8%	9%	4.11	4.90	5.09
Plastics	6%	10%	12%	3.52	6.12	6.78
Rubber, leather, and textiles	6%	8%	9%	3.52	4.90	5.09
Wood	7%	6%	5%	4.11	3.67	2.83
Food wastes	9%	11%	13%	5.28	6.74	7.35
Yard trimmings	15%	12%	10%	8.81	7.35	5.65
Other	4%	3%	4%	2.35	1.84	2.26
Total	100%	100%	100%	58.70	61.23	56.50

(2) In terms of weight, we can see that the monthly per person amount of MSW increased from 58.70 to 61.23 kg between 1990 and 2000, which is a 4% increase. Generation

Example 9.6 (continued)

numbers then decreased to 56.50 kg by 2000, representing an 8% decrease from 2000 numbers. This trend follows the general trend in the United States, especially as the total MSW generated increased systematically between 1990 to 2010 to close to 3.4 kt in 2010. Nonetheless, Civitians tend to generate less solid waste than the U.S. average; for example, in 2015, while the average American generated 61.82 kg of solid waste, the average Civitian generated 56.50 kg, which is 8.6% lower.

In terms of percentages, Civitas experienced trends similar to the United States, as the amount of paper waste generated generally decreased, and both plastics and food wastes numbers increased. Nonetheless, we can see that the amount of yard trimmings generated decreased as well, which might be indicative of some land use changes, as we will see in the next section.

Table 9.7

Total waste generation in Saint Pierre, Saint Pierre and Miquelon, in 2017

	Total waste (kg/month/person)		Total waste (%)	
Waste collected	18.47		38%	
Recycled				
Glass		3.85		36.34%
Plastics		0.38		3.63%
Steel		0.15		1.45%
Aluminum		0.05		0.44%
Plastic-lined paper products		0.15		1.45%
Cardboard (noncorrugated)		0.62		5.81%
Food wastes		5.39		50.87%
Total recycled		10.59		57.33%
Nonrecycled		7.88		42.67%
Waste self-hauled	29.81		62%	
Corrugated fiberboard		5.23		17.55%
Metals		2.62		8.77%
Large items		11.54		38.71%
Wood (treated)		2.32		7.79%
Wood (untreated)		3.08		10.32%
Appliances		1.92		6.45%
Hazardous waste		0.31		1.03%
Batteries		0.79		2.66%
Textiles		0.92		3.10%
Paper		1.08		3.61%
Total	48.28		100%	

Table 9.8
Monthly MSW plus C&D generation in world countries

Cont.	Country	Monthly MSW (kg/pers)	Cont.	Country	Monthly MSW (kg/pers)
Africa	Algeria	25.00	Asia	Israel	53.33
	Benin	10.00		Japan	29.17
	Ghana	13.33		Korea, South	30.00
	Kenya	11.67		Malaysia	35.83
	Mali	10.00		Nepal	5.00
	Nigeria	15.00		Pakistan	13.33
	South Africa	30.00		Philippines	11.67
	Zambia	15.00		Singapore	115.00
The Americas	Brazil	32.50		Vietnam	9.17
	Canada	59.17	Europe	Denmark	65.83
	Chile	32.50		France	41.67
	Colombia	21.67		Germany	51.67
	Cuba	20.00		Italy	40.83
	Ecuador	27.50		Netherlands	43.33
	Mexico	35.00		Poland	24.17
	Trinidad & Tobago	45.83		Spain	35.83
	United States	67.50		Switzerland	60.00
Asia	Bangladesh	7.50		United Kingdom	40.00
	China	12.50	Oceania	Australia	46.67
	India	13.33		New Zealand	60.83

Source: Kaza et al. (2019).

recycled, most of which was glass (36%) and food wastes (51%). Most of the self-hauled waste was also recycled, with the exception of wood (treated) and large items that make up about 50% of all self-hauled waste.[24] The goal in Saint Pierre is to divert as much as possible from landfill, and we will discuss more disposal strategies in section 9.3.

At the country level, table 9.8 shows monthly per person MSW generated for forty countries in the world. The same data for many more countries are available from Kaza et al. (2019). We can note that the value reported for the United States is 67.50 kg, which is more than the 61.82 kg that we saw earlier. This is possibly because the original data set includes C&D debris in the MSW, among others.

In table 9.8 the highest value reported is for Singapore, with 115 kg per person per month, and the lowest value reported is 5 kg for Nepal. This difference is staggering.

These values suggest that an average person in Singapore generates over twenty-three times more MSW than a person in Nepal. We should be a bit careful here, however. It is extremely hard to acquire consistent data for world countries, and some values shown might include or omit some waste categories. Countries may also have a different definition of MSW (as was the case for the United States). That being said, there are undoubtedly large differences. Referring back to U.S. trends, it is not hard to imagine that people in lower-income countries discard less paper, food wastes, plastics, and yard trimmings than people in higher-income countries.

Table 9.8 also groups countries by continent. If we look within each continent, we do not see systematic trends. For example, in Africa, values range from 10 (Benin and Mali) to 30 kg (South Africa), and in North and South America, values range from 20 (Cuba) to 67.50 (United States). This suggests that it is not geography that affects MSW generation. If we look a little closer at the values, we can quickly see that income level seems to have an impact. To further investigate this hypothesis, we can group world countries by income level and calculate average trends. This is exactly what table 9.9 does by grouping countries based on four income levels: high, upper middle, lower middle, and lower.[25]

First, table 9.9 shows that as income level increases, monthly MSW generation systematically increases as well, which we already expected. What is more surprising, however, is that the change in average monthly MSW values are not even. Values increase from 12.17 to 16.12 kg between lower- and lower-middle-income countries, which is a 32% increase. Values then increase from 16.12 to 21 kg between lower-middle to upper-middle-income countries, which is a 30% increase. Finally, values increase from 21 to 48.06 kg between upper-middle and high-income countries, which is a 130% increase. These percentages keep getting higher as we move to countries with higher income levels, which suggests that as countries get richer, they disproportionally generate more MSW, and this is a problem. Applying the first sustainability principle is therefore a priority when it comes to solid waste management.

Table 9.9
Monthly MSW plus C&D generation in world by income level

Income level	Monthly MSW (kg/person)
High	48.06
Upper middle	21.00
Lower middle	16.12
Lower	12.00

Source: Kaza et al. (2019).

Before we list strategies to deal with all the MSW generated, we should also account for the fact that the values shown here are averaged over multiple sectors. Put differently, the monthly 61.82 kg of waste generated per person in the United States includes commercial and some industrial wastes as well as residential waste. In the next section, we will differentiate MSW generation by sector.

9.2.3 Solid Waste Composition by Sector

So far, we have looked at MSW as a whole, irrespective of individual sectors. First, we should be aware that in the United States, the residential sector is responsible for roughly 50% of all MSW, and the solid waste composition of this 50% is different than the remaining 50% attributed to the commercial sector (U.S. Environmental Protection Agency 2013). Moreover, even within sectors, we might see differences. The residential sector, for example, can be further divided into single-family residential (i.e., houses) and multifamily residential (i.e., apartments and condominiums). Similarly, the commercial sector can be further divided into offices, restaurants, retail stores, and so on; put simply, think about the different types of commercial buildings we listed in table 8.10 in chapter 8.

The problem is that this type of data is not as available as general MSW data. In this section we will go over the results from a large disposal-based study conducted for California that reports the 2014 values available in Cascadia Consulting Group (2015a).[26] Values are bound to change depending on city size, climate, and income level, as well as other factors, but these values will offer us a taste of what to expect in general. California conducted similar studies in 1999, 2004, and 2008, all of which are available online.

The study defines three waste sectors: franchised residential, franchised commercial, and self-hauled (i.e., waste directly taken to the disposal facility). Moreover, it further divides the residential sector into two subsectors: single-family and multifamily, and it further divides the self-hauled sector into two subsectors: residential and commercial. The contributions of each sector are shown in table 9.10. In particular, we can see that the commercial sector (franchised and self-hauled) accounts for $38.6+11.3=49.9\%$, which is close to 50%, as mentioned above. The next step is to analyze the solid waste composition by sector.

The report differentiates between sixty-two material types, but it then aggregates them into ten major categories: paper, glass, metal, electronics, plastic, other organic (e.g., food, yard trimmings, textiles), inerts and other (e.g., concrete, lumber), household hazardous (e.g., paint, batteries), special waste (e.g., medical, tires), and mixed residue.

Table 9.10
2014 California solid waste contribution by weight by sector

Sector	Proportion of disposed waste by weight (%)	
Franchised commercial	38.6%	
Franchised residential	47.0%	
Single-family residential		35.4%
Multi-family residential		11.6%
Self-hauled	14.4%	
Commercial self-hauled		11.3%
Residential self-hauled		3.1%
Total	100.0%	

The full sixty-two material types can be found directly in the report. Table 9.11 shows the solid waste composition by sector and subsector, but the results are easier to analyze in the form of graphs.

Figure 9.16 shows pie charts of solid waste composition by sector in California, and figure 9.17 shows the same information by subsector. Some waste types with negligible values were removed from the graphs for clarity.

The first thing we notice from figure 9.16 is that except for self-hauled, the other organic category dominates. Other organic includes food wastes, yard trimmings, and textiles, which account for a significant portion of all solid waste in the United States, as we saw before. Moreover, other organic accounts for 45% of residential solid waste, compared to 35% for commercial. This difference is not surprising because households are bound to discard more food wastes, yard trimmings, rubber, leather, and textiles.

In addition, the figure also shows that the inerts and other category is significant, especially for commercial and self-hauled. This is again not surprising considering inerts and other are composed of concrete, asphalt, lumber, rock, and so on—that is, essentially C&D debris. The commercial sector is bound to generate more inerts. Moreover, it is logical for self-hauled to be composed primarily of inerts—most direct trips to a waste disposal facility are to discard inerts. Finally, figure 9.16 also shows that paper and plastics comprise about 20% and 13% of solid waste, which agrees with the U.S. trends discussed in the previous section.

Figure 9.17 shows the same information for subsectors. Focusing on the residential sector for now, we see that both single-family and multifamily residential share similar characteristics. The main difference is in the proportion of inerts that make up 12% of

Table 9.11
2014 California solid waste composition by weight by sector and subsector

Waste type	Overall	Franchised commercial	Franchised residential	Self-hauled
Paper	17.4%	20.4%	19.2%	3.3%
Glass	2.5%	3.3%	2.2%	1.1%
Metal	3.1%	3.3%	2.9%	3.4%
Electronics	0.9%	0.8%	1.1%	0.5%
Plastic	10.4%	12.5%	10.2%	5.4%
Other organic	37.4%	34.8%	45.2%	19.0%
Inerts and other	19.9%	17.9%	10.8%	54.9%
Hazardous	0.4%	0.4%	0.5%	0.0%
Special waste	5.0%	4.8%	3.2%	12.0%
Mixed residue	3.0%	1.8%	4.8%	0.4%

	Residential		Self-hauled	
Waste type	Single-family	Multifamily	Commercial	Residential
Paper	17.7%	23.9%	3.5%	2.7%
Glass	1.9%	3.0%	0.6%	2.6%
Metal	2.7%	3.3%	3.2%	4.3%
Electronics	1.0%	1.4%	0.4%	1.0%
Plastic	10.0%	11.0%	5.5%	4.9%
Other organic	45.7%	43.8%	19.7%	16.5%
Inerts and other	12.3%	6.1%	56.4%	49.3%
Hazardous	0.6%	0.1%	0.0%	0.1%
Special waste	2.9%	3.9%	10.5%	17.4%
Mixed residue	5.1%	3.6%	0.1%	1.1%

the waste of single-family residential compared to 6% for multifamily residential. This is likely due to the fact that single-family residential represents houses, as opposed to multifamily residential that represents apartments and condominiums and in which fewer activities occur that require material belonging to the inerts category such as home remodeling and landscaping.

Similarly, single-family residential has a higher proportion of other organic, which suggests that this category may generate more yard trimmings, for example, since single-family residential often includes outdoor property (e.g., a backyard), unlike multifamily residential.

As a consequence, we can see that multifamily residential units discard proportionally more paper waste, but this is probably simply to compensate for the difference

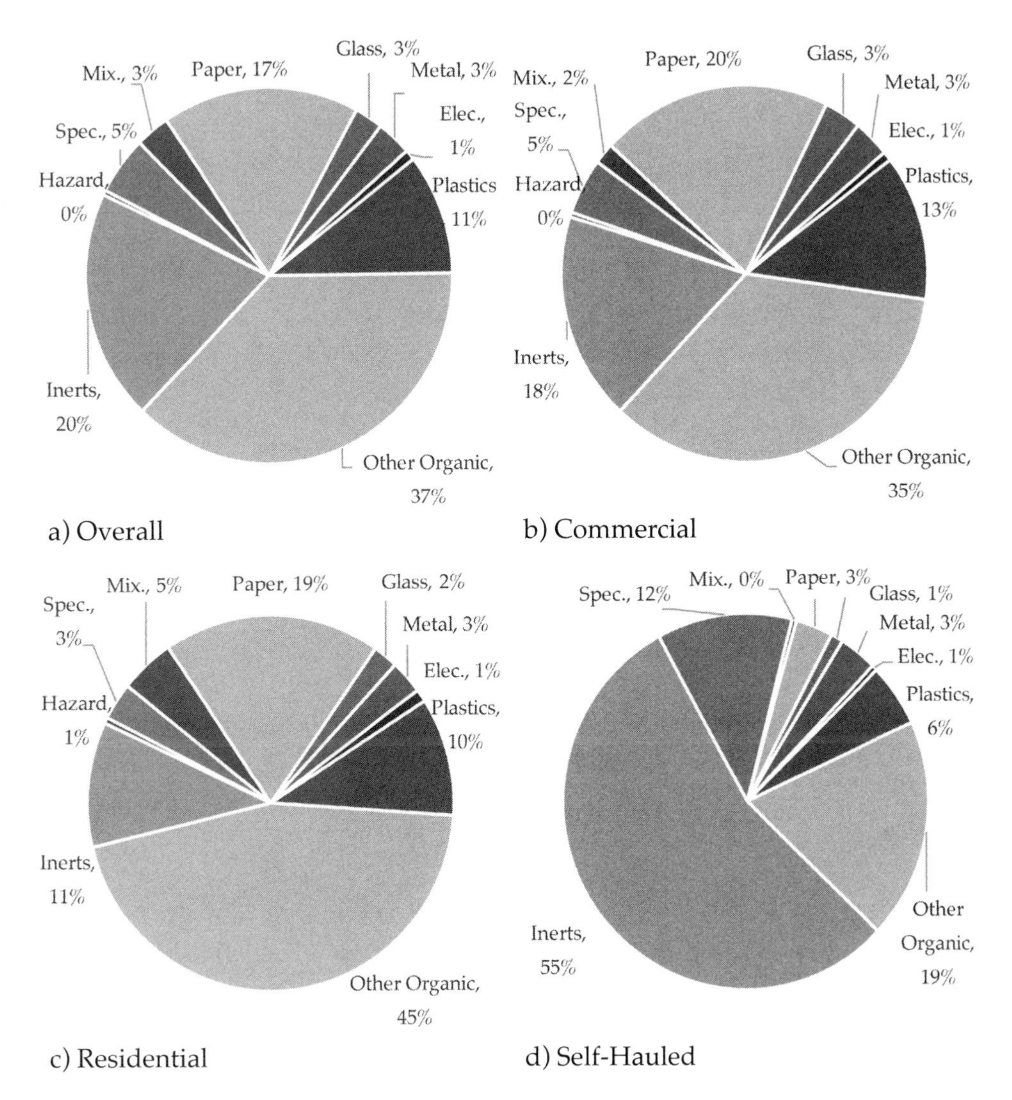

Figure 9.16
2014 California solid waste composition by sector.

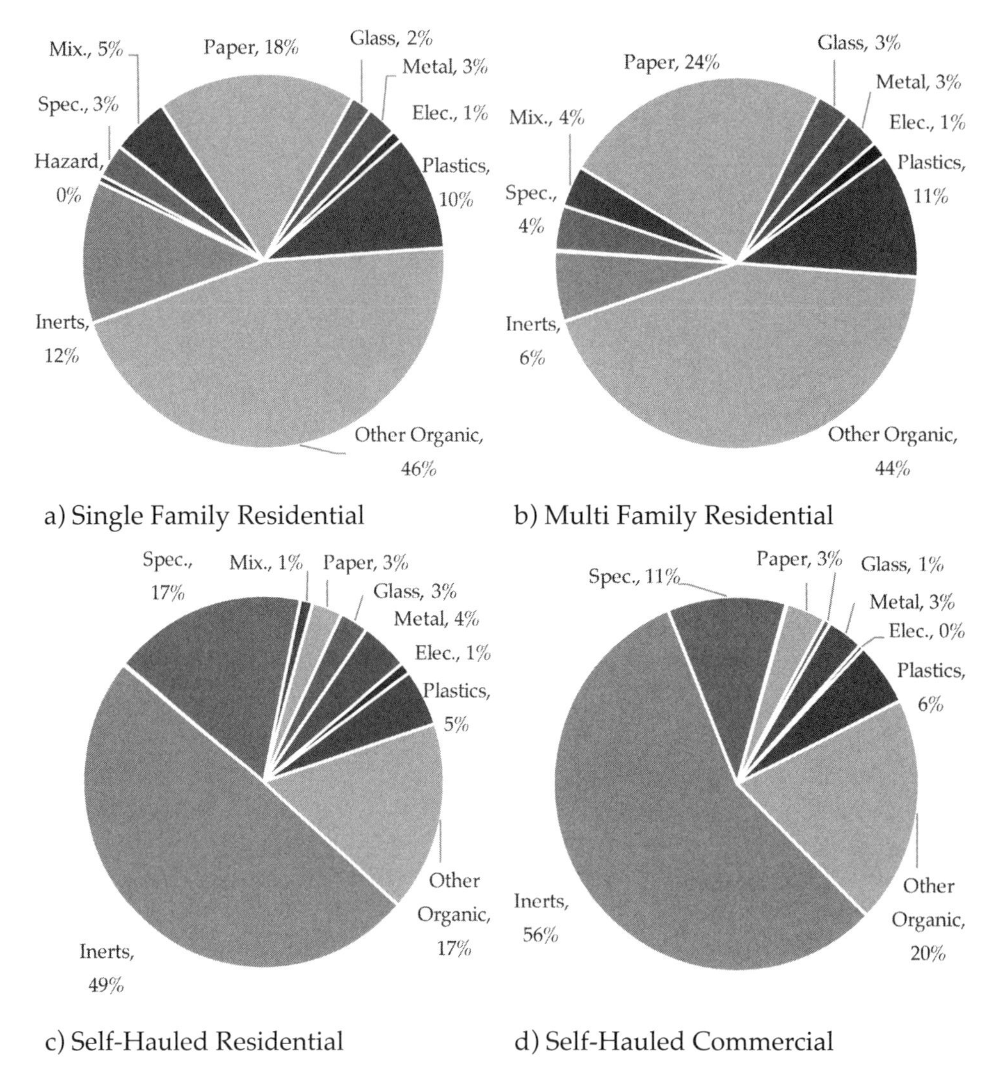

Figure 9.17
2014 California solid waste composition by subsector.

in inerts—that is, more paper is not necessarily discarded, but proportionally, out of 100%, a higher percentage is attributed to paper because the proportion of inerts and other organic is lower.

Finally, both self-hauled subsectors show a high percentage of inerts, which is again due to the fact that the primary purpose of a direct trip to a disposal facility is to discard inerts.

Overall, we can conclude that the main difference between the residential and commercial sectors resides in the fact that the residential sector has a higher proportion of organics, and the commercial sector has a higher proportion of inerts. Moreover, the self-hauled sector is dominated by inerts. Unfortunately, no distinction is made within the commercial sector, which means we will not be able to differentiate between offices, restaurants, stores, and other commercial activities.

The next question is what to do with all this waste once it is generated and discarded. In other words, how is waste disposed of in disposal facilities and where does it go afterward? This is the topic of the next section.

Example 9.7

Solid Waste Generation by Sector in Civitas

Based on the results calculated in example 9.6 and the main observations made in this section, (1) discuss how you expect different sectors to be represented in Civitas.

Solution

(1) The categories given in example 9.6 are different from the categories given in this section. Nonetheless, we can sum rubber, leather, and textiles with wood, food wastes, and yard trimmings to get an estimation of the "other organic" category. When doing this for 1990, 2000, and 2010, we get a constant percentage of 37%, which is identical to the percentage given for overall MSW in figure 9.16. This suggests that Civitas has a similar representation of the residential and commercial sectors to California.

The other apparent change in Civitas is the reduction in yard trimmings. As we saw in this section, yard trimmings are generally attributed to single-family residences because individual houses often include outdoor property, unlike multifamily residences. A decrease in yard trimmings therefore may suggest that the proportion of Civitians who live in multifamily residential buildings has increased, which would go along with the significant population increase discussed in chapter 3.

9.3 Solid Waste Disposal

Once solid waste has been generated, it needs to be disposed of, and disposing of solid waste properly is important—in particular, about 2% of the United States' and 5% of global GHG emissions are attributed to solid waste management (Mohareb and Hoornweg 2017; U.S. Environmental Protection Agency 2018d).[27]

As discussed at the beginning of the chapter, there are really three ways to manage solid waste: reuse/recycle, burn, and throw away. Figure 9.18 shows how solid waste was managed in the United States from 1960 to 2015. Precisely, instead of three strategies, the figure shows four strategies (i.e., recycling was split into recycling and composting[28]):

- Composting
- Recycling
- Combustion with energy recovery
- Landfilling and other disposal

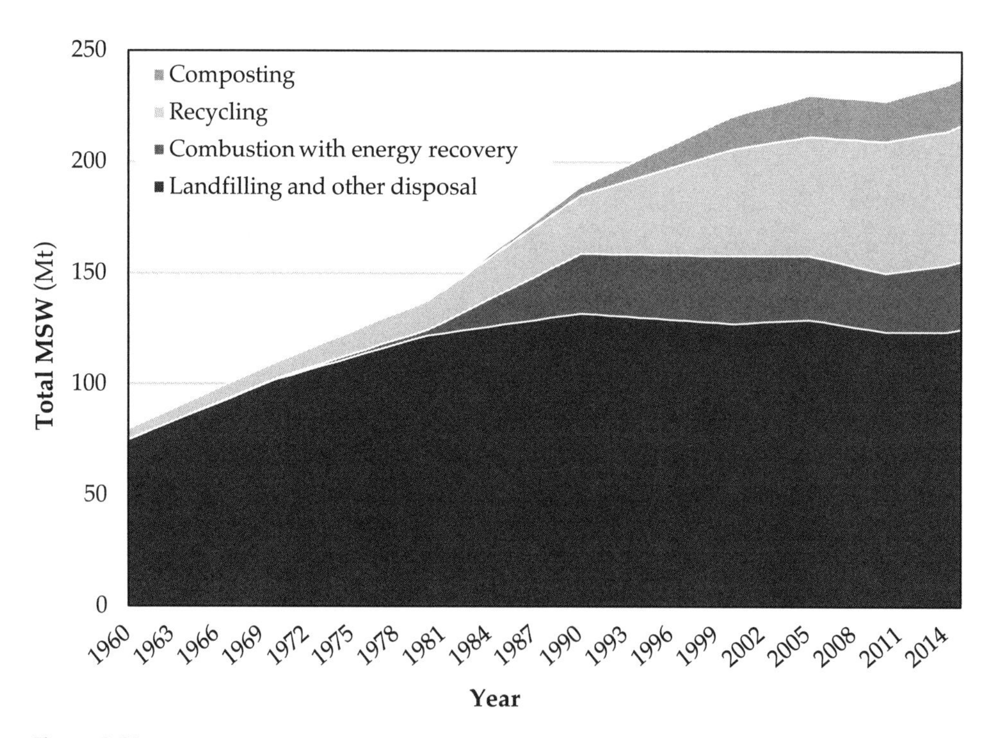

Figure 9.18
Historical trends in MSW management in the United States.

Combustion with energy recovery includes all solid waste that is incinerated to produce energy, whether in the form of electricity and/or heat. Finally, landfilling belongs to the throw-away category, which we will discuss in this section. This last category also includes "other disposal" methods such as incineration without energy recovery.

Differentiating between the categories is also important because they do not generate the same amount of GHG emissions per ton of solid waste. As a cautionary warning, coming up with GHG emission factors for solid waste is notoriously difficult because emissions depend on waste types (e.g., paper vs. food wastes), local contexts (e.g., distances traveled), and what is done with the energy recovered (e.g., electricity vs. heating). The EPA has developed the Waste Reduction Model (WARM), which can be used to estimate the amount of GHG emissions avoided from reducing, recycling, digesting, and recovering solid waste in the United States. The model is available on their website,[29] and it should be used for a proper GHG assessment. We will use it here to estimate GHG emission values for recyclables.[30]

To make things simpler, in this chapter, for all strategies except recycling, we can use the 2015 U.S. GHG inventory and divide the GHG emissions attributed to each category by the amount of solid waste processed by each category; both data sets come from the EPA. The results are shown in table 9.12 and are discussed in individual sections. These numbers cannot directly be used for a proper GHG assessment, however—WARM can be used for example—but they will suffice to give us a general idea of the impact of each solid waste disposal strategy.

For recycling, table 9.12 shows the GHG emission factor avoided by mixed recyclables instead of producing new products—that is, not only instead of landfilling, but instead of producing new products from raw materials.[31] This is why there is a negative sign in front of the factor. Applying a similar approach for composting and combustion with energy recovery, WARM gives us emission factors of –0.176 and –0.077 t CO2e per wet ton respectively.

Table 9.12
GHG emission factor by solid waste management strategy

Management strategy	GHG emission factor (t CO_2e per wet ton)
Composting	0.188
Recycling	–3.114[a]
Combustion with energy recovery	0.361
Landfills	0.894

Note: [a]Different method used. See section 9.3.2.2.

In the early 2000s, landfilling tended to be the most common means of solid waste disposal around the world. High-income countries operated sanitary landfills (43%) that were highly engineered, as we will see, compared to low-income countries that mostly used open dumps (49%). Combustion with energy recovery was also specific to high-income countries. Upper- and lower-middle-income countries mostly used sanitary landfills (60%) (Hoornweg and Bhada-Tata 2012).

Figure 9.18 shows the historical breakdown for the United States. We can see that solid waste management was largely dominated by landfilling and other disposal methods in the 1960s and 1970s, and the trend changed in the 1980s, partly thanks to all the regulations that were passed, as discussed in section 9.1. By 2015, only about 50% of the solid waste was landfilled, but there is no reason to believe that this number cannot further decrease. We now need to apply our second sustainability principle: increase supply within reason. Therefore, once the amount of solid waste generated has been reduced as much as possible (i.e., the first sustainability principle), efforts can be put into increasing how much solid waste is reused and recycled and how much is incinerated with energy recovery. In this section we will discuss the four methods shown in figure 9.18, as well as some others.

Going back to figure 9.7, we have already covered the first functional element—waste generation. Once solid waste has been generated, it is first handled at the source. There are several ways to apply our second sustainability principle at the source—for instance, by having separate bins to sort out trash. Solid waste is then collected, often by garbage trucks, but it can also be self-hauled to a disposal facility. Once it arrives at a disposal facility, solid waste is then processed and transported before finally being disposed of.

In this section we will focus on the separation, processing, and transformation of solid waste, as well as on the disposal of solid waste. More specifically, strategies like screening, recycling, and composting belong to the separation and processing and transformation of solid waste functional elements. In contrast, landfilling and other disposal belong to the disposal element. To differentiate between the various functions of each element, we will divide the section into three parts: (1) the separation and processing of solid waste, (2) the transformation of solid waste, and (3) the disposal of solid waste.

9.3.1 Solid Waste Separation and Processing

Solid waste can be separated and processed in various ways. In section 9.1 we covered the physical, chemical, and biological properties of solid waste, and based on these properties, various strategies can be adopted.

Figure 9.19
Waste sorting posters. Courtesy of the University of Illinois at Chicago Office of Sustainability.

In general, the first strategy to pursue is to separate solid waste. Most of us are somewhat familiar with this process when we separate our trash that is recyclable versus nonrecyclable. Posters like the ones shown in figure 9.19, for example, have become ubiquitous. This is a typical example of separation at the source.

Solid waste is often further sorted at the disposal facility, automatically and/or manually. The glass/metal/plastic category, for example, is often sorted by a combination of manual and automatic sorting with an eddy current separator (figure 9.20). For this, the solid waste is put on a conveyor belt equipped with magnets that produce an eddy current. At the end of the belt, the magnets attract ferrous metals, which end up in a bin on the left-hand side, while nonmetals fall into another bin or onto another conveyor belt, and nonferrous metals are expelled by the eddy current to another bin on the right-hand side. Nonmetals can then be sorted manually if desired—for example, to sort glass from plastics.

Many other solid waste types are separated. Electronics are often separated for recycling; for example, figure 9.21c shows discarded monitors that have been packed on a pallet, ready to be transported to a specialized recycling facility. Naturally, hazardous products like batteries and paints also tend to be separated. Some solid waste facilities also conduct a screening process to separate large items that might be processed in a different way.

Once the separation process is complete, solid waste often goes through size-reduction processes. The simplest kind of size reduction is compacting (also called *baling* for most materials and *crushing* for cans). Essentially, categories of solid waste are compacted

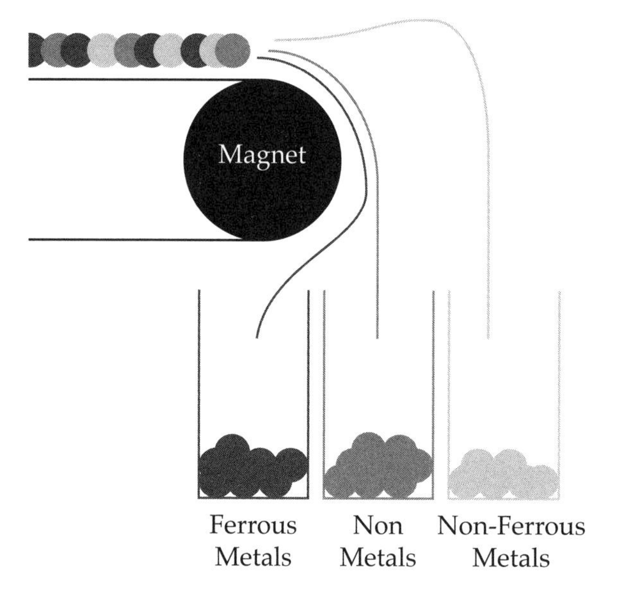

Figure 9.20
Eddy current separator.

Figure 9.21
Solid waste size reduction and recycling.

and attached so they can then be transported to other facilities. Figures 9.21a and figure 9.21b show two examples, one for cardboard and one for plastics. Once the size is reduced, the solid waste can be sent to a recycling facility or to an incinerator to be further processed or disposed of.[32]

In terms of processing, some solid waste is shredded, sometimes before sorting and sometimes after, depending on the final size desired. Shredding is helpful to reduce the size of solid waste so it can be compacted/baled more easily before being sent to a recycling facility, a Waste-to-Energy processing plant, or to a landfill. As an extreme case of shredding, glass (if not reused) is often crushed into sand that can be used for a variety of purposes (including to replenish beaches). Similarly, while wood can be burned to produce electricity and/or heat, if it is untreated, it can also be chipped and used for composting, for example.

After being separated and processed, solid waste needs to be transformed.

9.3.2 Solid Waste Transformation

Solid waste transformation often involves processes that give new life to solid waste. As the old idiom goes, "one man's trash is another man's treasure," and this is exactly what solid waste transformation is about. Solid waste transformation is notably about leveraging the chemical and biological properties of solid waste.

Again, the famous three *R*s of solid waste management are to reduce, reuse, and recycle. As mentioned at the beginning of the chapter, a fourth *R* has joined the rank, and it stands for recover. We have already covered the *R* that stands for reduce in the solid waste generation section, but we can discuss the three other *R*s in this section.

In urban engineering, we particularly like solid waste transformation because it often also leads to a reduction in solid waste that ends in landfills, thus contributing to both sustainability principles, as discussed in chapter 2. For example, reusing an item means a new one is not needed.

9.3.2.1 Reuse Reuse can be as simple as reusing an item for the same function it originally had to reusing an item for a different function. The premise for reuse is that the item has not been physically, chemically, and biologically transformed in a significant way.

We have all discarded an item that has been reused or collected a discarded item, keeping the original function of the item, if only by reusing a disposable bag from a store. Other examples of reused items include furniture, clothes, and even electronic equipment (e.g., giving/selling an electronic device to get a new one), whether they were donated, sold, or exchanged.

Moreover, we have all used an item for a slightly different function than that for which it was made. For example, containers and boxes bought at a grocery store (such as coffee cans) can be used for a different purpose—to store nonfood items.

At a larger scale, the field of industrial symbiosis is about finding reuses for the waste of an industry. The most famous example of industrial symbiosis is Kalundborg symbiosis in Kalundborg, Denmark, where industries exchanged materials, energy, and water.[33] At a smaller scale, facility-scale industrial symbiosis is also possible when multiple businesses colocate in a single facility (Mulrow et al. 2017). The Plant—a collection of food businesses collocated in one single facility in Chicago—offers a great case study (Chance et al. 2018). For example, CO_2 produced from a brewery is used by an indoor farm, melted water from an ice maker goes to an indoor wetland, and wastewater from fish tanks goes to beds growing spirulina algae. There are countless ways to reuse solid waste, and it should be the first disposal strategy to consider.

The next *R* stands for recycle.

9.3.2.2 Recycle Recycling can have different meanings. Colloquially, many people use the term *recycling* instead of *reusing*, although reuse may be the more appropriate term. For us, recycling implies a transformation of the solid waste, whether physically, chemically, or biologically, to the same or a different product.

We have already seen some forms of recycling. In fact, from the beginning of solid waste management, recycling has held an important place. Colonel George E. Waring Jr., who we recall from section 9.1, established the first recycling center in the United States, essentially to generate some revenue. He notably recovered ammonia, grease, glue, and dry residue that he then sold to chemical industries and farmers.

In the twenty-first century, recycling has often been associated with efforts to sort out what can be recycled from what cannot (see figure 9.19). Typical items that are recycled include paper and paper-based products, drink containers (whether made of aluminum, plastic, or glass), and steel cans. Once collected, these products are processed to make new products such as aluminum cans, cereal boxes, egg cartons, laundry detergent bottles, newspapers, and even trash bags. We will not go into the processes to recycle these products, but the practice has become common, and often, the reason for not recycling resides in the sorting of products to identify recyclable material instead of the recycling process itself.

In the end, recycling these products means new products do not have to be produced, thus contributing to fewer GHG emissions, which is why table 9.12 shows −3.114 t CO_2e per wet ton of solid waste for recycling, with a negative sign in front of the value as mentioned previously for table 9.12. Table 9.13 lists some GHG emission factors for common recyclable materials, collected from the EPA WARM. The table, notably, shows us that aluminum is particularly carbon-intensive, and therefore recycling aluminum products is highly desirable. For more international values see Turner, Williams, and Kemp (2015).

As an important note, the value of −3.114 t CO_2e per wet ton from tables 9.12 and 9.13 is described as "mixed recyclables" in EPA WARM. It should not be used for

Table 9.13
GHG emission factor of common recyclable materials

Waste material type	GHG emission factor (t CO_2e per wet ton)
Aluminum cans	–10.040
Steel cans	–1.998
Mixed metals	–4.784
Glass	–0.305
Wood (dimensional lumber)	–2.713
Mixed paper	–3.892
Mixed plastics	–1.127
Concrete	–0.009
Asphalt concrete	–0.090
Mixed recyclables	–3.114

Source: EPA WARM, https://www.epa.gov/warm (accessed March 11, 2019).

individual recycled products, but in the context of this book, it offers us a general indicator.

Although food wastes are not in table 9.12, they can also be recycled, but in a different way. As we saw when we discussed the biological properties of solid waste, there are two ways to leverage the biodegradability of solid waste: *aerobic composting* and *anaerobic digestion*. The technical difference is that aerobic composting produces compost, which is essentially a fertilizer that can be used for farming. Anaerobic digestion produces a solid digestate that can also be used for farming, but in addition, it produces methane, which can be captured and burned to produce electricity and/or heat.

In practice, aerobic composting is extremely easy to do—and most people who own a vegetable garden do some sort of composting. Food wastes are first separated and piled together and mixed with other organic wastes like yard trimmings or wood chips (figure 9.22a). Every once in a while, air has to be added to the compost, essentially fueling the aerobic respiration of microorganisms that are breaking down the waste. This is either done manually—generally called *turned composting*—by mixing the compost (by hand, with a shovel, or even with a front loader for industrial composting facilities), or it is done by forcing air through the compost—generally called *aerated static pile composting*. In general, the more air is added, the less time it takes to produce the compost. Mixing the compost continuously, weekly, three to five times per year or one time per year results in a finished product that is ready in three to four months, four to six months, fourteen to eighteen months, or twenty-four to thirty-six months, respectively (Tchobanoglous, Theisen, and Vigil 1993).

Example 9.8

Recycling Solid Waste in Civitas

The table below shows the gross MSW values for 2010 in Civitas as shown in example 9.6. (1) Assuming that 50% of the paper, glass, plastics, and wood are recycled and assuming that 40% of the metals consist of steel cans and aluminum cans (20% each) and that 100% of all the cans are recycled, calculate how many GHG emissions were avoided thanks to recycling in 2010 in Civitas. (2) Assuming a general GHG emission factor of −3.114 t CO_2e per wet ton, calculate how many GHG emissions were avoided thanks to recycling and compare your results with (1).

Waste type	Wet weight (t)
Paper	1,051
Glass	238
Metals	305
Plastics	407
Rubber, leather, and textiles	305
Wood	170
Food wastes	441
Yard trimmings	339
Other	136
Total	3,392

Solution

(1) The solution is shown in the table below. The total wet weight of solid waste is shown in the second column (separating steel and aluminum cans from metals). The weight of the recycled solid waste is then shown in the third column. The fourth column shows the GHG emission factors from table 9.13, and the last column shows the GHG emissions avoided. As an example for paper, the procedure is $GHG_{paper} = 1{,}051 \times 0.5 \times -3.892 = -2{,}045$ t CO_2e.

Overall, recycling in Civitas directly contributed to 3,275 t CO_2e of GHG emissions avoided.

Waste type	Wet weight (t)	Recycled (t)	GHG emission factor (t CO_2e per wet ton)	GHG emissions (t CO_2e)
Paper	1,051	525.5	−3.892	−2,045
Glass	238	119	−0.305	−36
Steel cans	61	61	−1.998	−122
Aluminum cans	61	61	−10.040	−612
Plastics	407	203.5	−1.127	−229
Wood	170	85	−2.713	−231
Total	1,988	1,055		−3,275

Example 9.8 (continued)

(2) The total weight of recycled solid waste is 1,055 t. Assuming an average GHG emission factor of −3.114 t CO_2e per wet ton, the number of GHG emissions avoided is $GHG_{recycling} = 1{,}055 \times -3.114 = -3{,}285$ t CO_2e.

The value found in (2) is close to the value from (1), which is likely because we are considering a mixed combination of recycled waste types, but as discussed, it does not apply to individual materials. Refer to EPA WARM or other sources to get GHG emission factors from other recyclable materials.

a) Aerobic Composting

b) Anaerobic Digestion Plant
(Courtesy of HoSt Bioenergy Installations)

Figure 9.22
Aerobic composting and anaerobic digestion.

The story is different for anaerobic digestion. As we saw, the term *anaerobic* means the food wastes are deprived of oxygen so that anaerobic microbes proliferate, producing methane and carbon dioxide as gaseous by-products. This means that food wastes must be placed in a solid airtight container, and a mechanism must be installed to capture and process the methane. Figure 9.22b shows an example of an industrial anaerobic digestion plant. In addition, to maximize the activity of anaerobic microbes, moisture and nutrients are added to the food wastes, and the mixture is kept at temperatures between 55°C and 60°C and mixed continuously. At the time of this writing, anaerobic digestion was relatively uncommon in the United States, which is why it was not reported in figure 9.18.

Aerobic composting and anaerobic digestion offer the best ways to recycle food wastes, but they do emit GHG emissions—although fewer than if they were landfilled and much fewer than the emissions from the production of fertilizers that the compost replaces; in fact, EPA WARM reports a GHG emission value of –0.176 t CO_2e per wet ton as mentioned previously. In terms of direct emissions, in 2015 in the United States, composting was responsible for 4 Mt CO_2e. Considering 21.22 Mt of solid waste was composted, we get a GHG emission factor of 0.188 t CO_2e per wet ton of solid waste.

It is time to discuss the last *R* for recover.

9.3.2.3 Recover Recover might not be part of the original three *R*s, but it has been practiced forever. Essentially, if solid waste cannot be reused or recycled, the last option is to recover anything that can be recovered instead of sending it to a landfill. The simplest type of recovery is arguably incineration if energy is recovered from the combustion process, but we will discuss incineration in the next section on solid waste disposal.

We have already encountered a popular means to recover material. Indeed, we have seen how solid waste is often used to reclaim land—for instance, by using rubble excavated from construction sites. Instead of discarding it, C&D debris can be used for other purposes. Table 9.12 shows that recycling rubble can save 2 kg CO_2e per ton of rubble, which might not look impressive compared to 8,143 for aluminum cans. When we scale it up to the actual number of tons of rubble used, however, the savings are actually impressive.

The general field of recovering useful resources from cities in particular is called *urban mining*, and many materials can be mined, from wood and concrete to copper, steel, zinc, and many others (Graedel 2011; Koutamanis, van Reijn, and van Bueren 2018). As many cities face problems with their infrastructure, recovering existing materials to produce new materials has become an attractive option.

Overall, there exist countless ways to reuse, recycle, and recover materials. In 2015 in the United States, there were close to 600 materials recovery facilities typically focusing on recycling products (U.S. Environmental Protection Agency 2018a), but there will no doubt be many more in the future.

When solid waste cannot be transformed, it needs to be disposed of, and this is what we will focus on next.

9.3.3 Solid Waste Disposal

The final stage of solid waste management is disposal. Essentially, this should be the last resort to deal with solid waste, and there exist two primary strategies: incineration or landfill. We will discuss both strategies here, starting with incineration.

9.3.3.1 Incineration Incineration consists of burning the solid waste. As seen previously, only organic waste can actually burn—that is, food wastes, plastics, yard trimmings, textiles, and so on. Inorganic waste can be thrown in the fire as well, but it will not burn. Once the incineration process is completed, most of what is left is inert residue (e.g., ash) that even organic wastes contain, as detailed in table 9.5.

There are two main types of incineration plants: mass-fired combustors and RDF-fired combustors (where RDF stands for refuse-derived fuel). In mass-fired combustors, the solid waste goes through little to no treatment before it is incinerated; in other words, what comes out of a garbage truck is incinerated. In RDF-fired combustors, the solid waste is separated (e.g., to only include organic waste) and processed (e.g., shredded) to meet specific conditions to ensure that the amount of energy desired is produced.

Incineration has many benefits. First, incineration burns all organic wastes, and the inert residue does not attract pests—many of the major epidemics that occurred in the past (e.g., cholera, plague) came from microorganisms that pests first contracted from piles of solid waste. Second, as we will see, piles of solid waste can produce methane and leachate—in the same fashion as anaerobic digestion—which, if not captured, is a public health concern; methane is highly combustible, and GHG and leachate can contaminate groundwater aquifers, which is again a health hazard.

While early incineration plants simply burned the waste, most modern plants (at least in high-income countries) use the heat produced to generate electricity and/or heat. These plants are called municipal Waste-to-Energy processing plants—like the one in Barcelona that might receive the solid waste collected by the small truck shown in figure 9.8.

In 2015 in the United States, a total of seventy-seven incineration plants, mostly located in the Northeast, produced energy. Figure 9.23 shows the general functions of a Waste-to-Energy processing plant. Essentially, the heat created from the combustion of the waste is used to create steam, which turns a turbine to produce electricity and/or be distributed in a district heating system (*top right*). In the summer, electricity generation may be favored for air conditioning, and in the winter, heat distribution may be favored for building heating. The inert residue left is then separated. Whatever metallic residue that can be recovered is collected, and whatever ash/residue that can be used as aggregate is separated. The rest is then landfilled.

Considering that the incineration of waste was responsible for close to 11 Mt CO_2e in the United States in 2015 and that 30.45 Mt of solid waste were combusted with energy recovery, we get a GHG emission factor of 0.361 t CO_2e per wet ton, as shown in table 9.12, which is roughly twice as much as composting. But again, if we compare to landfilling—similar to the process we followed for recycling—we get an emission factor of –0.077 t CO_2e per wet ton.

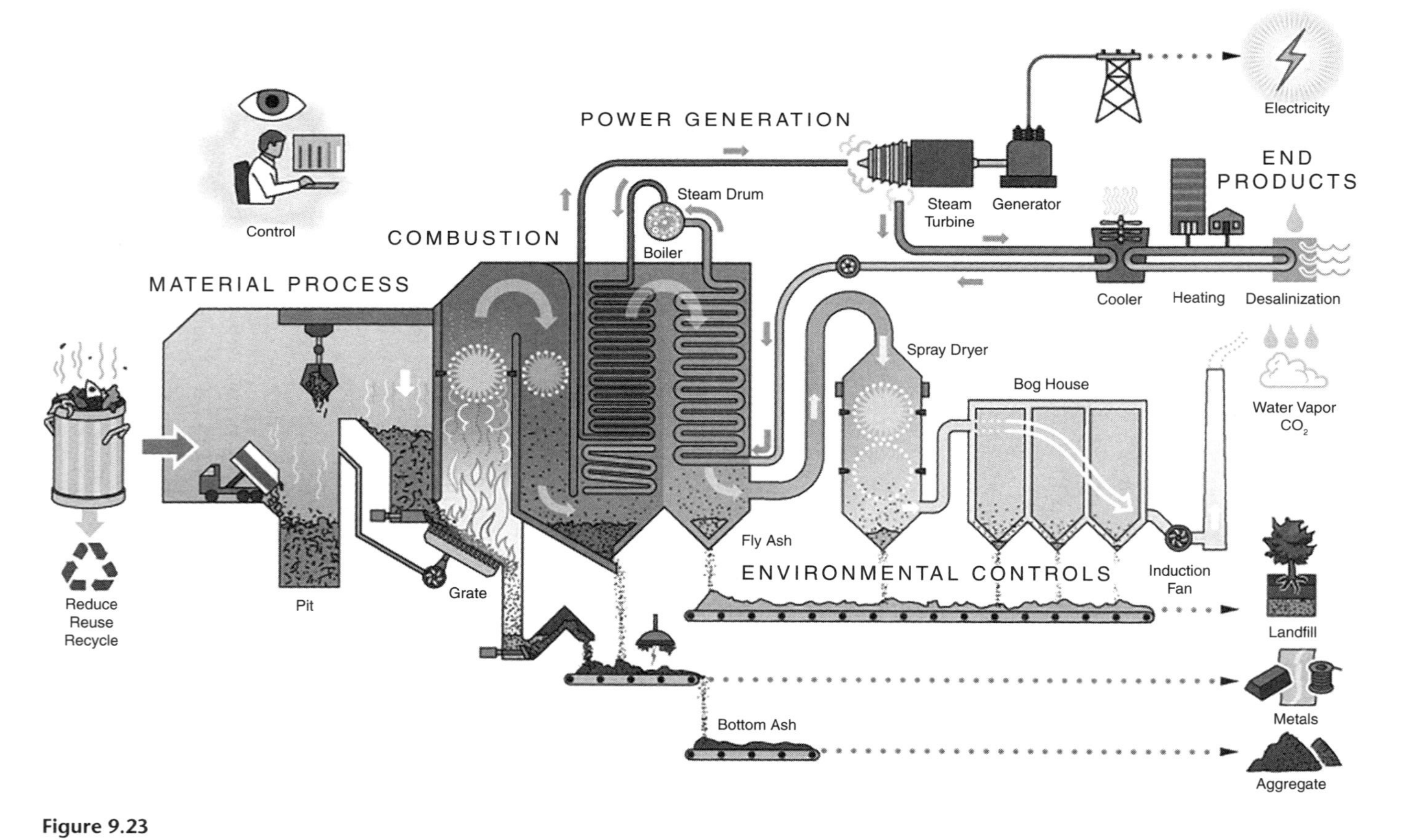

Figure 9.23
Waste-to-Energy processing plant. Adapted with permission from Deltaway Energy. For more about Deltaway Energy, visit http://www.deltawayenergy.com/.

We actually already discussed the burning of solid waste in chapter 5 on electricity. Incineration plants are essentially biomass power plants that focus on burning solid waste. In chapter 5, we saw that biomass power plants have a GHG emission factor of about 200 g CO_2e/kWh. In addition, table 9.5 shows that the energy content of general MSW is 3,200 Wh/kg, which is 3,200 kWh/t. Using these numbers, we get a GHG emission factor of $200 \times 3{,}200 = 640{,}000$ g CO_2e per ton = 640 kg CO_2e per ton = 0.64 t CO_2e per ton of wet solid waste, which is twice what we see in table 9.12. Can you guess why? There are at least two reasons. First, the value of 200 g CO_2e/kWh from chapter 5 includes the life cycle of the power plant (i.e., construction, operation, and demolition) as well, while the value from table 9.12 only includes the burning of solid waste. Moreover, the value of 200 g CO_2e/kWh from chapter 5 only includes the generation of electricity and omits the generation of heat, and we remember that generating electricity is far less "efficient"—natural gas power plants have efficiencies of about 42%—while furnaces producing heat have efficiencies of about 90%—because the major losses in natural gas power plants are heat losses. This is why it is important to properly account for how the energy is recovered.

The final disposal strategy is landfilling.

9.3.3.2 Sanitary Landfill Landfills are essentially plots of lands where solid waste is discarded. An open-air dump is a form of landfill, but modern landfills are highly engineered, which is why they are called *sanitary landfills*. Ideally, landfilling is the final strategy that is adopted when no more energy or resources can be recovered from solid waste, but this is often not the case in practice. Landfills, therefore, frequently receive high volumes of solid waste from which larges volumes could have been reused, recycled, and recovered. In fact, this is the reason why landfills have to be engineered. If everything that could be recovered were recovered, the only thing left would be inert residue, which possesses little value.

The principle of landfilling seems relatively easy. Solid waste is discarded and covered with soil at the end of the day to prevent pests from coming into contact with it—and that is the way it used to be, as we saw in section 9.1. The problem is that with time the organic wastes decompose, and because the waste is not exposed to air, it produces methane and leachate. As seen, methane is a GHG with a high global-warming potential. Leachate is the liquid that trickles down the solid waste pile, similar to how stormwater infiltrates the ground. In fact, the production of leachate can be estimated with standard groundwater-engineering techniques (e.g., Darcy's law) using the hydraulic conductivity defined in equation (7). Leachate is heavily polluted and can contaminate groundwater aquifers located under the landfill if it is not properly collected.

Example 9.9

Incinerating Solid Waste in Civitas

The table below shows the gross MSW values for 2010 in Civitas as shown in examples 9.6 and 9.8. (1) Focusing purely on organic waste, using the inert residue values given in table 9.5, and assuming that organic waste has a specific weight of 300 kg/m^3 and that the inert residue has a specific weight of 600 kg/m^3, determine the volume reduction in solid waste if 50% of all organic waste was incinerated. (2) Calculate the GHG that would be emitted with incinerating 50% of all organic waste.

Waste type	Wet weight (t)
Paper	1,051
Glass	238
Metals	305
Plastics	407
Rubber, leather, and textiles	305
Wood	170
Food wastes	441
Yard trimmings	339
Other	136
Total	3,392

Solution

(1) The total weight of the inert residue left is shown in the table below, where 50% of all organic waste was kept, and the typical percentage of inert residue left was taken from table 9.5.

Waste type	Wet weight (t)	Wet weight incinerated (t)	Typical inert residue (%)	Inert residue (t)
Paper	1,051	525.5	6	31.5
Plastics	407	203.5	10	20.4
Rubber, leather, and textiles	305	152.5	3	4.6
Wood	170	85	2	1.7
Food wastes	441	220.5	5	11.0
Yard trimmings	339	169.5	5	8.5
Total	2,713	1,356.5		77.7

Since the original organic waste has a specific weight of 300 kg/m^3, this means that the original volume of the organic waste $V_{organic}$ was:

Example 9.9 (continued)

$$V_{organic} = 1000 \times \frac{1{,}356.5}{300} = 4{,}521.7 \text{ m}^3$$

Similarly, and considering the inert residue has a specific weight of 600 kg/m^3, the volume of the inert residue left, V_{inert}, is:

$$V_{inert} = 1000 \times \frac{77.7}{600} = 129.5 \text{ m}^3$$

These results give us a volume reduction V_{reduc} of:

$$V_{reduc} = \left(\frac{4{,}521.7 - 129.5}{4{,}521.7}\right) \times 100 = 97\%$$

(2) Considering that the per ton GHG emission factor linked with the incineration of MSW with energy recovery is 0.361 t CO_2e per ton of wet solid waste, the GHG emissions linked with burning 50% of all organic waste $GHG_{organic}$ is:

$$GHG_{organic} = 1{,}356.5 \times 0.361 = 489.7 \text{ t } CO_2\text{e.}$$

Based on these considerations, landfill design generally follows this process:

1. A hole is often dug in the ground that will contain the bottom part of the landfill.
2. The bottom of the landfill is covered with a liner so that the leachate cannot percolate into the groundwater aquifer. Pipes are added to the bottom of the landfill to collect the leachate.
3. Environmental monitors are also installed to ensure the area surrounding the site does not get contaminated, both from methane and from leachate.
4. Compacted solid waste is laid in the landfill and is covered with soil or dirt at periodic intervals (usually at the end of the day). A layer of solid waste and soil is typically called a *cell.*
5. A series of cells that cover the entire surface area of a landfill is called a *lift.* Once a lift is completed, new cells are started on top of it.
6. Structural elements called *benches* can be added when needed to ensure the landfill does not collapse.
7. Once the final lift is installed, the landfill is covered with a liner to ensure that methane cannot escape—figure 9.1c shows a sanitary landfill right before it is about to be covered with its final liner. Pipes are also installed to collect the methane that will be burned to produce energy.

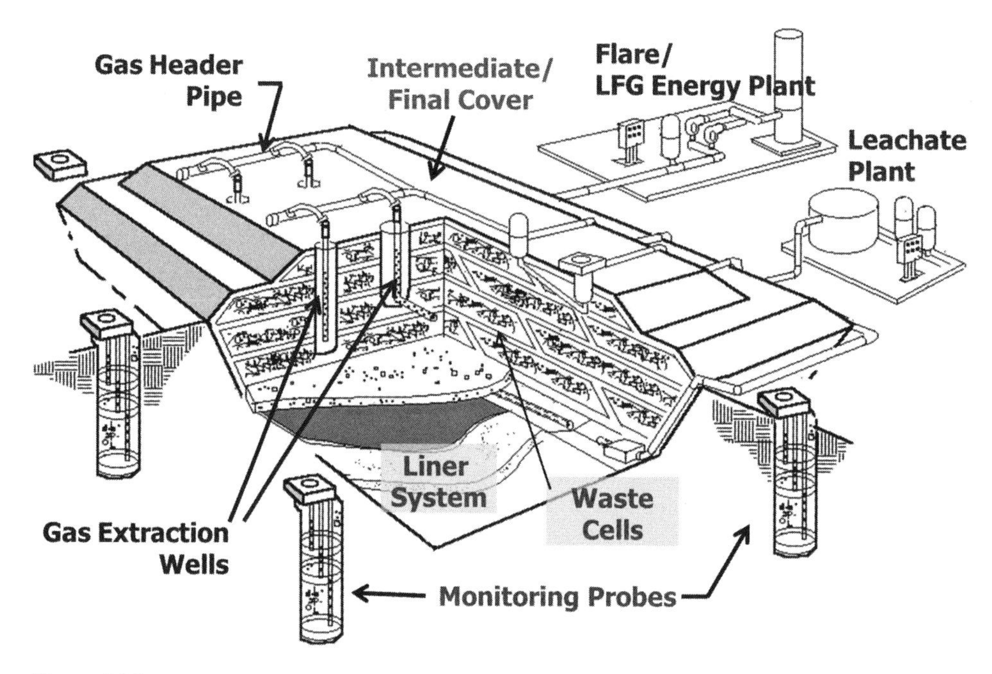

Figure 9.24
Diagram of a sanitary landfill.

Figure 9.24 shows the diagram of a sanitary landfill, further demonstrating that they are complex engineering systems. In the United States in 2015, there were 1,738 landfill facilities, mostly located in the West and in the South.

There are several types of landfills. Common landfills accept any type of MSW, but some landfills only accept shredded solid waste (which means such landfills can therefore hold more solid waste). Other types of landfills accept only one or few material types (e.g., combustion ash or asbestos), and they are called *monofills*.

The life of a landfill generally ranges from thirty to fifty years. This means that electricity and/or heat can be generated for about thirty to fifty years, and it also means that environmental monitoring must be pursued for at least thirty years.

Moreover, akin to incineration, the methane collected from landfills is generally used to produce electricity and/or heat. So which one is better: incineration or landfill? Let us go through some numbers first.

Considering that landfills were responsible for about 112 Mt CO_2e in the United States in 2015, and that 137.7 Mt of solid waste was landfilled, we get a GHG emission factor of 0.894 t CO_2e per wet ton, as shown in table 9.12. This value is a factor of 2.5 and 4.75 times higher than combustion with energy recovery and composting, respectively.

Looking at these numbers, incineration therefore seems to be the preferred choice—in addition, we remember the EPA WARM negative emission factor for incineration. The story is actually a bit more complicated, and it is actually a big debate in the solid waste community.

First, the solid waste community acknowledges that both incineration and landfilling should be the last strategies to adopt. As shown in the EPA's waste management hierarchy (figure 9.9), solid waste reduction strategies come first, followed by reuse, and then by recycling. It is therefore clear that once solid waste is generated, if it cannot be reused, recycling and composting are preferable to incineration and landfilling.

The choice between incineration and landfilling highly depends on what is incinerated. Some experts argue that plastic products emit a significant amount of GHG emissions if they are burned and that they should therefore be landfilled. Nonetheless, as we saw in table 9.5, plastic is the waste material with the highest specific energy, and therefore incinerating plastics produces a lot of energy. In the end, the debate is not over, and more research is needed to identify the preferable strategy.

This concludes our section on sanitary landfills and on solid waste disposal in general.

9.4 Conclusion

Solid waste management has evolved tremendously over time, in particular since the late 1800s. The very issue of having to deal with solid waste only emerged when cities started to form, and as the quote at the beginning of this chapter reminded us, solid

Example 9.10

Landfilling Solid Waste in Civitas

The solid waste audit conducted in Civitas revealed that in 2010, a total of 3,392 t of wet solid waste was generated. (1) Assuming that 50% of all the solid waste is landfilled, calculate the GHG that would be emitted with landfilling.

Solution

(1) Considering that the per ton GHG emission factor linked with landfilling is 0.894 t CO_2e per ton of wet solid waste (table 9.12), the GHG emissions linked with landfilling 50% of all solid waste $GHG_{landfill}$ is:

$$GHG_{landfill} = 0.5 \times 3{,}392 \times 0.894 = 1{,}516 \text{ t } CO_2e.$$

waste is a by-product of civilization. Solid waste management is therefore intrinsically an urban problem that holds an important role in urban engineering. Akin to other chapters, this chapter had three main parts. First, we covered some fundamentals of solid waste management, then we studied patterns and trends in solid waste generation and composition, and finally, we studied solid waste disposal.

In the fundamentals section, we first focused on the history of solid waste. We learned that initially, solid waste was simply thrown into streets, and since most of it consisted of organic waste, animals often ate it. By the 1600s and 1700s, cities had grown so much that concerns over general hygiene became a large focus for local authorities. It is not before the 1800s, however, that solutions were formed—notably, by Colonel George E. Waring Jr. in New York City. By the early 1900s, solid waste management had become a science in itself, as we discovered throughout the remainder of the chapter.

After reviewing the history of solid waste management, we then properly defined solid waste and solid waste management. Specifically, we saw that solid waste management contains six functional elements, starting with waste generation and ending with waste disposal. We also learned about integrated solid waste management (ISWM) and about the EPA's waste management hierarchy. Subsequently, the physical, chemical, and biological properties of solid waste were examined. We learned properties of the specific weight, moisture content, ultimate analysis, and energy content of solid waste.

In the second section, we dove into studying solid waste generation and composition trends, with a focus on the United States, although we analyzed data from Saint Pierre and Miquelon and from other places in the world as well. But first, we briefly covered how to conduct a waste audit, learning about the three main and two optional phases of an audit. We then saw that the average American generated 61.82 kg of solid waste per month in 2015, most of which was paper (26%), food wastes (15%), yard trimmings (13%), and plastics (13%). By looking at generation rates around the world, it became apparent that solid waste generation is largely a problem of high-income countries.

Finally, we differentiated waste composition between sectors using survey data for California and learned that the main difference between the residential and commercial sectors was in the generation of organics (i.e., the residential sector generates more organic waste) and inerts (i.e., the commercial sector generates more inerts).

In the last section, we focused on solid waste disposal. Specifically, we first learned how solid waste is separated and processed, then how it can be transformed, and finally how it can be disposed of. In terms of separation and processing, the ubiquitous recycling posters were used as an example of separation at the source, and eddy current separators were used as an example of separation at a disposal facility. We also discussed waste compacting, glass crushing, and wood grinding. In terms of transformation, we looked at three of the four *R*s: reuse, recycle, and recover. We saw that significant amounts of GHG emissions could be avoided thanks to recycling (especially by recycling aluminum), and the processes for aerobic composting and anaerobic digestion were detailed. In the recover section, we discussed urban mining that has in fact been practiced for a very long time—for example, with the practice of land reclamation.

Finally, the disposal section focused on two major strategies: incineration and landfilling. Waste-to-Energy processing plants essentially burn solid waste and use the resulting heat to create steam that can then generate electricity or distribute heat in a district heating system for building heating. Sanitary landfills are highly engineered facilities with a fairly intricate design process to capture the methane and leachate produced by solid waste. While the use of sanitary landfills should be avoided as much as possible in favor of other solid waste management methods (e.g., recycling), their use is common in the world.

We have come a long way since the beginning of this book. Having reviewed the five big types of urban infrastructure systems—electricity, water, transport, buildings, and solid waste—it is now time to combine them. This is important because urban infrastructure systems are highly codependent and interdependent. We will now discuss how they fit together by looking through the lens of *urban metabolism*.

Problem Set

9.1 The main purpose of this problem is to estimate your solid waste generation and composition habits and carbon footprint. (1) Conduct a waste audit by weighing how much waste and determining the composition of waste you generate for one week. Make sure to separate your solid waste by material following the categories from table 9.6. Report the results in a table. (2) Analyze your solid waste generation and composition habits and compare your numbers with an average person in America in 2015 (table 9.6). (3) Come up with three ways to reduce how much solid waste you generate. (4) Estimate your solid waste carbon footprint by assuming all of your solid waste is sent to a sanitary landfill.

9.2 In about 500 words, describe the history of solid waste management.

9.3 Similar to Boston, New York City, Singapore, Tokyo, and Toronto, identify a city that has practiced land reclamation and show a map of the city before and after land was reclaimed.

9.4 In your own words, define *solid waste* and *solid waste management.*

9.5 In your own words, describe the difference between organic and inorganic waste.

9.6 Define *household hazardous waste* and list at least ten types of household hazardous waste.

9.7 Select any solid waste category other than resins (plastics) and identify the subcategories of solid waste types and common uses, akin to figure 9.6.

9.8 In your own words, define the six functional elements of solid waste management.

9.9 In your own words, define the waste management hierarchy shown in figure 9.9.

9.10 Identify how solid waste is managed in your city and develop an ISWM strategy to enhance it.

9.11 By searching the web or any other source, look up and report the specific weight of the items shown in the table below:

Waste type	Volume	Specific weight (kg)
Dishwasher	1 unit	
Clothes dryer	1 unit	
Stove	1 unit	
Refrigerator	1 unit	
Clothes washer	1 unit	
Desktop	1 unit	
Laptop	1 unit	
TV	1 unit	
Cellular phone	1 unit	
Aluminum cans (uncompacted)	m^3	
Steel cans (whole)	m^3	

9.12 A house generated 120 kg of solid waste in a month with the composition shown in the table below. Based on table 9.2, calculate (1) the moisture content of the solid waste and (2) the dry weight of the solid waste, and (3) estimate the volume of solid waste generated by the house in that one month.

Waste type	Percent by weight
Food wastes (mixed)	12%
Paper	33%
Plastics	15%
Textiles	13%
Glass	12%
Steel cans	9%
Aluminum cans	6%

9.13 Using table 9.2, calculate (1) the moisture content and (2) the volume of the solid waste generated by an average person in the United States in 2015. Clearly list any assumption made.

9.14 Based on the moisture content (table 9.2) and ultimate analysis data (table 9.3) given in the chapter and using the same data as problem 9.12, (1) perform an ultimate analysis of the dry and wet solid waste generated by the house to determine the weight of each chemical component and (2) calculate the molar composition of the dry and wet solid waste generated by the house. Calculate the energy content of the wet solid waste using (3) the values given in table 9.5 and (4) the modified Dulong formula. (5) Explain the difference between the two values.

9.15 Using tables 9.2 and 9.3, (1) perform an ultimate analysis of the dry and wet solid waste generated by an average person in the United States in 2015 to determine the weight of each chemical component and (2) calculate the molar composition of the dry and wet solid waste generated by an average person in the United States in 2015. Clearly list any assumption made. Calculate the energy content of the solid waste using (3) the values given in table 9.5 and (4) the modified Dulong formula. (5) Explain the difference between the two values. Clearly list any assumption made.

9.16 In your own words, define and compare *aerobic composting* and *anaerobic digestion.*

9.17 By doing your own research and by going beyond the description in the chapter, describe the biological processes of aerobic composting and anaerobic digestion.

9.18 By visiting the EPA website, download the most up-to-date data on solid waste management trends in the United States and reproduce figure 9.10. Update the figure if later data are available.

9.19 In your own words, describe the process to perform a successful solid waste audit for a municipality.

9.20 In your own words, describe the process to perform a successful solid waste audit for a company.

9.21 By visiting the EPA website, download the most up-to-date data on solid waste composition in the United States and reproduce figure 9.13 and table 9.6. Update the figure if later data are available.

9.22 By visiting the EPA website, download the most up-to-date solid waste generation data for one particular category and plot a pie chart of the subcategories of solid waste types (e.g., for paper and paperboard, show the breakdown between newspapers, books, magazines, and so on).

9.23 By visiting the EPA website, report the total weight in [Mt] of all major and small appliances and furniture and furnishings generated in the municipal waste stream.

9.24 Performing your own research, collect solid waste generation data for any city / country not reported in the chapter and analyze the results.

9.25 A waste audit was conducted in a community in 1970, 1990, and 2010, and the gross MSW values are shown in the table below. Considering that the population in this community was 347,639 in 1970, 408,476 in 1990, and 487,312 in 2010, (1) calculate the percentage of each waste type and calculate the per person monthly MSW generated. (2) Analyze the trends from the data calculated in (1).

Waste type (t)	1970	1990	2010
Paper	4,655	7,512	7,724
Glass	1,790	2,332	2,861
Metals	1,613	2,332	2,861
Plastics	1,255	2,590	3,431
Rubber, leather, and textiles	897	1,814	2,573
Wood	1,790	2,071	1,433
Food wastes	1,432	2,332	2,573
Yard trimmings	3,045	3,627	3,431
Other	1,432	1,295	1,715
Total	17,909	25,905	28,602

9.26 A waste audit was conducted in a community to differentiate solid waste composition trends between the residential and commercial sectors, and the results are shown in the table below. Discuss the main differences between the two sectors.

Waste type (t)	Residential	Commercial
Paper	26%	28%
Glass	10%	11%
Metals	7%	15%
Plastics	9%	15%
Rubber, leather, and textiles	12%	7%
Wood	7%	3%
Food wastes	12%	7%
Yard trimmings	15%	5%
Other	2%	9%
Total	100%	100%

9.27 In your own words, explain what solid waste disposal is and discuss the different strategies available to dispose of solid waste.

9.28 By visiting the EPA website, download the most up-to-date data on solid waste disposal trends in the United States and reproduce figure 9.18. Update the figure if later data are available.

9.29 Identify and plot the solid waste disposal strategies used by any city/country other than the United States.

9.30 By visiting the EPA website, download the most up-to-date data on U.S. GHG inventory and (1) compile the latest results for table 9.12 (do not include recycling). (2) Discuss any differences with the values shown in table 9.12.

9.31 Take pictures of a local recycling initiative (e.g., posters by a trash can) and report on it.

9.32 Stop a garbage truck and ask where they take the solid waste (i.e., which disposal facility) and whether they know where the solid waste is to be disposed of (i.e., a Waste-to-Energy processing plant or a sanitary landfill).

9.33 In your own words, explain how solid waste can be separated.

9.34 By searching the web or any other source, report on a solid waste separation practice other than those discussed in the chapter. Include pictures to illustrate the method.

9.35 By searching the web or any other source, report one example of industrial symbiosis other than that of Kalundborg and The Plant discussed in the chapter.

9.36 Assuming that all of the recyclable solid waste generated by the house from problem 9.12 is recycled (do not include food waste and textiles), calculate how many GHG emissions are avoided in that one month using (1) the values from table 9.13 and (2) the general recycling value from table 9.12. (3) Compare the results.

9.37 Assuming that all of the paper, glass, and plastics generated by the community from problem 9.25 in 2010 is recycled, (1) calculate how many GHGs are avoided in that one month using the values from table 9.13. In addition, a new campaign pushes for all of the steel and aluminum cans to be recycled, which make up 30% of all metals each (i.e., 15% steel and 15% aluminum). (2) Calculate how many GHG emissions would be avoided with the campaign and compare the results.

9.38 Assuming that all of the paper and food wastes generated by the house from problem 9.12 are composted, calculate how many GHGs are emitted in that one month.

9.39 Assuming that all of the food wastes generated by the community from problem 9.25 in 2010 are composted, calculate how many GHGs are emitted.

9.40 In your own words, explain how solid waste can be recovered.

9.41 In your own words, explain how a Waste-to-Energy processing plant works.

9.42 Based on the inert residue values given in table 9.5, using the same data as problem 9.12, and focusing on organic waste only, (1) determine the volume reduction in solid waste if all of the organic waste was incinerated; assume that organic waste has a specific weight of 300 kg/m^3 and that inert residue has a specific weight of 600 kg/m^3. (2) Calculate the GHGs that would be emitted by incinerating all the organic waste.

9.43 Assuming that the community from problem 9.25 has always incinerated all of its organic waste, calculate the per capita GHG emissions associated with the incineration of organic waste in 1970, 1990, and 2010.

9.44 In your own words, explain how a sanitary landfill is designed and how it works.

9.45 Assuming that all of the solid waste from the house in problem 9.12 goes to a sanitary landfill, calculate how many GHGs are emitted.

9.46 Assuming that all of the solid waste from the community in problem 9.25 goes to a sanitary landfill, calculate how many GHGs are emitted per person in 1970, 1990, and 2010.

Notes

1. The exact number is surprisingly hard to estimate since most species have not been discovered yet, but a 2011 study put the number at 8.7 million ± 1.3 million (Mora et al. 2011).

2. Let us be loose on the definition for now. We will formally define what solid waste is later.

3. We have to be careful here and look at how organic waste is defined, since some databases only include food waste and yard trimmings (i.e., readily biodegradable) in organic waste. In this chapter we roughly define organic waste in a chemical sense, as in combustible.

4. Although some experts may talk to you at length about how they are different.

5. The construction of the Athens metro, for example, unearthed many important archaeological finds.

6. Some academics at the time even thought that one could gain weight simply by smelling food.

7. Waring also believed in the miasma theory, despite the fact that it had been disproven several decades before, which partly explains why he was so adamant to rid the New York streets of their trash as effectively and quickly as possible.

8. We should also note that many reclaimed lands are considered brownfields, which means, for example, that they should not be used to grow food.

9. His book also discusses the history of water and wastewater systems, as seen in chapter 6 of this book.

10. Hazardous waste is a specific type of waste that we will not discuss in this book. Instead, we will focus on municipal solid waste (defined later).

11. Search online to see how your city deals with household hazardous waste.

12. We can see that the unit used is the same as density, but specific weight does not measure density since solid waste can be compacted and contains moisture—all of which have an impact on the specific weight but not the density of the raw material.

13. This is also why we use wood to make a fire and not glass or metal—much more energy is released from burning wood.

14. This is also why we often say "burning calories" when we engage in physical activity.

15. Oxygen is also important, but it is present in many molecules (including water), and it is therefore primarily C and H that give us an idea of energy content. In fact, this is why methane (CH_4) yields a significant amount of energy.

16. The same batch of solid waste, therefore, does not have more energy if it is dry. Its energy per unit weight is higher when it is dry, but when it is dry it is also lighter, so the overall quantity of energy does not change.

17. The same four coefficients for C, ($H_2 - 1/8\ O_2$), N, and S are 337, 1419, 23, and 93 for units in [kJ/kg] and 145, 610, 10, and 40 for units in [Btu/lb].

18. Like food, partly because food waste is a part of organic waste. The exceptions here include plastic, rubber, and leather.

19. Flies develop in less than two weeks after eggs have been laid.

20. The conversion factor between a short ton (i.e., 2,000 pounds) and a metric ton is 1 short ton = 0.907184 metric tons. Other conversion factors can be found in chapter 1 and in the appendix.

21. But this is not always the case. Switzerland, for example, exports a lot of solid waste to Germany, which then incinerates it to produce electricity and heat.

22. The document actually reports a total of 7,668,097 generated tons and 4,745,685 disposed tons, but the unit is expected to be in short tons.

23. The data was generously provided by Christophe Caignard, who works for the Ville de Saint Pierre (i.e., the City of Saint Pierre).

24. Although at the time of this writing, there were plans to build a Waste-to-Energy processing plant to produce energy from burning the waste, either for electricity or to heat. Moreover, wood (untreated) was used to make compost.

25. In their book *Factfulness*, Rosling, Rönnlund, and Rosling (2018) make a great argument to discard the old dichotomy of the world based on two categories (developed and developing countries) and instead suggest the adoption of these four categories based on income level.

26. A generation-based study is also available to learn how solid waste was discarded/generated (Cascadia Consulting Group 2015b).

27. The values are from 2015.

28. The figure does not include composting from individual activities, like backyard composting, simply because these types of compositing are not disposed of in the general MSW stream. In other words, they do not make it to a disposal facility, and no data are available. Moreover, it does not include anaerobic digestion simply because the use of anaerobic digestion in the United States at the time of this writing was rare and thus negligible.

29. Available at https://www.epa.gov/warm (accessed March 11, 2019).

30. We have to be a bit careful when looking at WARM directly, since it reports values in metric tons of CO_2e avoided per U.S. short ton. Since we only use metric units in the book, the WARM values were converted to metric tons of CO_2e per metric ton.

31. Which does not fully consider the life cycle since material extraction and other processes are not included.

32. The pictures in figure 9.21 were taken in Saint Pierre (Saint Pierre and Miquelon). From there, cardboard and plastics are sent to Canada to be recycled. Monitors, appliances, and other electronic devices that follow European regulations are sent to France to be recycled.

33. The website of the Kalundborg symbiosis contains a great animation to visualize the types of exchanges happening. Visit http://www.symbiosis.dk/ (accessed March 11, 2019).

References

Béguin, Marine. 2013. "L'histoire des ordures: De la préhistoire à la fin du dix-neuvième siècle." *VertigO* 13(3). https://doi.org/10.4000/vertigo.14419.

Cascadia Consulting Group. 2015a. *2014 Disposal-Facility-Based Characterization of Solid Waste in California*. DRRR 2015-1546. Sacramento: California Department of Resources Recycling and Recovery.

———. 2015b. *2014 Generator-Based Characterization of Commercial Sector Disposal and Diversion in California*. DRRR 2015-1543. Sacramento: California Department of Resources Recycling and Recovery.

CDM. 2010. *Waste Characterization Study*. Chicago: City of Chicago Department of Environment.

Chance, Eva, Weslynne Ashton, Jonathan Pereira, John Mulrow, Julia Norberto, Sybil Derrible, and Stephane Guilbert. 2018. "The Plant—An Experiment in Urban Food Sustainability." *Environmental Progress & Sustainable Energy* 37(1): 82–90. https://doi.org/10.1002/ep.12712.

Fenco MacLaren and Angus Environmental. 1996. *Waste Audit Users Manual: A Comprehensive Guide to the Waste Audit Process*. Winnipeg, MB: Canadian Council of Ministers of the Environment.

Graedel, T. E. 2011. "The Prospects for Urban Mining." *Bridge* 41(1): 43–50.

Hoornweg, Daniel, and Perinaz Bhada-Tata. 2012. "What a Waste: A Global Review of Solid Waste Management." Working Paper 68135. Urban Development Series. Washington, DC: World Bank.

Hoornweg, Daniel, Perinaz Bhada-Tata, and Chris Kennedy. 2013. "Waste Production Must Peak This Century." *Nature* 502(7473): 615–617. https://doi.org/10.1038/502615a.

Hornyak, Tim. 2017. "Wasteland: Tokyo Grows on Its Own Trash." *Japan Times*, February 18. https://www.japantimes.co.jp/life/2017/02/18/environment/wasteland-tokyo-grows-trash/ (accessed March 14, 2019).

Kaza, Silpa, Lisa Yao, Perinaz Bhada-Tata, and Frank Van Woerden. 2019. *What a Waste 2.0: A Global Snapshot of Solid Waste Management to 2050*. Urban Development Series. Washington, DC: World Bank.

Koutamanis, Alexander, Boukje van Reijn, and Ellen van Bueren. 2018. "Urban Mining and Buildings: A Review of Possibilities and Limitations." *Resources, Conservation and Recycling* 138: 32–39. https://doi.org/10.1016/j.resconrec.2018.06.024.

Louis, Garrick E. 2004. "A Historical Context of Municipal Solid Waste Management in the United States." *Waste Management & Research* 22(4): 306–322. https://doi.org/10.1177/0734242X04045425.

Melosi, M. V. 2008. *The Sanitary City: Environmental Services in Urban America from Colonial Times to the Present*. Pittsburgh: University of Pittsburgh Press.

Mohareb, Eugene, and Daniel Hoornweg. 2017. "Low-Carbon Waste Management." In *Creating Low Carbon Cities*, edited by Shobhakar Dhakal and Matthias Ruth, 113–127. Cham, Switzerland: Springer International. https://doi.org/10.1007/978-3-319-49730-3_11.

Mora, Camilo, Derek P. Tittensor, Sina Adl, Alastair G. B. Simpson, and Boris Worm. 2011. "How Many Species Are There on Earth and in the Ocean?" Edited by Georgina M. Mace. *PLoS Biology* 9(8): e1001127. https://doi.org/10.1371/journal.pbio.1001127.

Mulrow, John S., Sybil Derrible, Weslynne S. Ashton, and Shauhrat S. Chopra. 2017. "Industrial Symbiosis at the Facility Scale." *Journal of Industrial Ecology* 21(3): 559–571. https://doi.org/10.1111/jiec.12592.

Rosling, H., A. R. Rönnlund, and O. Rosling. 2018. *Factfulness: Ten Reasons We're Wrong about the World—and Why Things Are Better than You Think*. New York: Flatiron Books.

Tchobanoglous, George, Hilary Theisen, and Samuel Vigil. 1993. *Integrated Solid Waste Management: Engineering Principles and Management Issues*. New York: McGraw-Hill.

Turner, David A., Ian D. Williams, and Simon Kemp. 2015. "Greenhouse Gas Emission Factors for Recycling of Source-Segregated Waste Materials." *Resources, Conservation and Recycling* 105: 186–197. https://doi.org/10.1016/j.resconrec.2015.10.026.

U.S. Environmental Protection Agency. 1997. *Measuring Recycling: A Guide for State and Local Governments*. EPA530-R-97-011. Washington, DC: U.S. Environmental Protection Agency.

———. 2013. *MSW Residential/Commercial Percentage Allocation—Data Availability*. Washington, DC: U.S. Environmental Protection Agency. https://www.epa.gov/sites/production/files/2016-01/documents/rev_10-24-14_msw_residential_commercial_memorandum_7-30-13_508_fnl.pdf (accessed March 14, 2019).

———. 2018a. *Advancing Sustainable Materials Management: 2015 Tables and Figures*. Washington, DC: U.S. Environmental Protection Agency. https://www.epa.gov/facts-and-figures-about-materials-waste-and-recycling/advancing-sustainable-materials-management (accessed March 14, 2019).

———. 2018b. *Best Practices for WasteWise Participants*. Washington, DC: U.S. Environmental Protection Agency. https://www.epa.gov/smm/best-practices-wastewise-participants#wastesort (accessed March 14, 2019).

———. 2018c. *Criteria for the Definition of Solid Waste and Solid and Hazardous Waste Exclusions*. Washington, DC: U.S. Environmental Protection Agency. https://www.epa.gov/hw/criteria-definition-solid-waste-and-solid-and-hazardous-waste-exclusions (accessed March 14, 2019).

———. 2018d. *Inventory of U.S. Greenhouse Gas Emissions and Sinks: 1990–2016*. EPA 430-R-18-003. Washington, DC: U.S. Environmental Protection Agency.

Wilson, D. G. 1976. "A Brief History of Solid-Waste Management." *International Journal of Environmental Studies* 9(2): 123–129. https://doi.org/10.1080/00207237608737618.

III Urban Metabolism and Novel Approaches

10 Urban Metabolism and Infrastructure Integration

> The essence of freedom ... is something that is able to overcome its own boundaries. The question is not only to be able to define things, but also to have the boundaries be felt in the proper way—they are defining but not limiting.
>
> —Frank Stella

We have now reached a significant step. After having learned about the five big types of infrastructure, it is now time to start integrating them. This is far from being trivial, but it must be achieved if we truly aspire to build a sustainable world. The quote above comes from Frank Stella, one of the most renowned American artists of the second half of the twentieth century. As one of the pioneers of the minimalist movement in the late 1950s, Stella painted stripes and played with the shape of canvases. His style later changed, however. In the 1970s his paintings emerged from the canvas, almost becoming sculptures and getting much larger in size. Initially bounded by the canvas, he ventured into a new space that has allowed him to produce amazing works of art.

Engineers now need to follow a process similar to Frank Stella's. Since the beginning of the twentieth century, the engineering profession has siloed itself into distinct fields, forming boundaries that can be difficult to cross. While these boundaries do indeed exist—that is, electricity and transport are two different things—they are "defining but not limiting," to reiterate Stella's quote. Therefore, in this chapter we will try to understand how the various infrastructure systems interact with one another and how they can be better integrated. After all, infrastructure systems are interdependent, and it simply makes sense to study them together. We actually saw this a little in chapter 1. Go back and have a look at figure 1.2, which showed the concept of infrastructure ecology between water, energy, transport, and land use. Does the figure make more sense now?

As importantly for us, all infrastructure systems produce or require energy, and all emit greenhouse gases (GHGs). It therefore makes sense to study them together. We

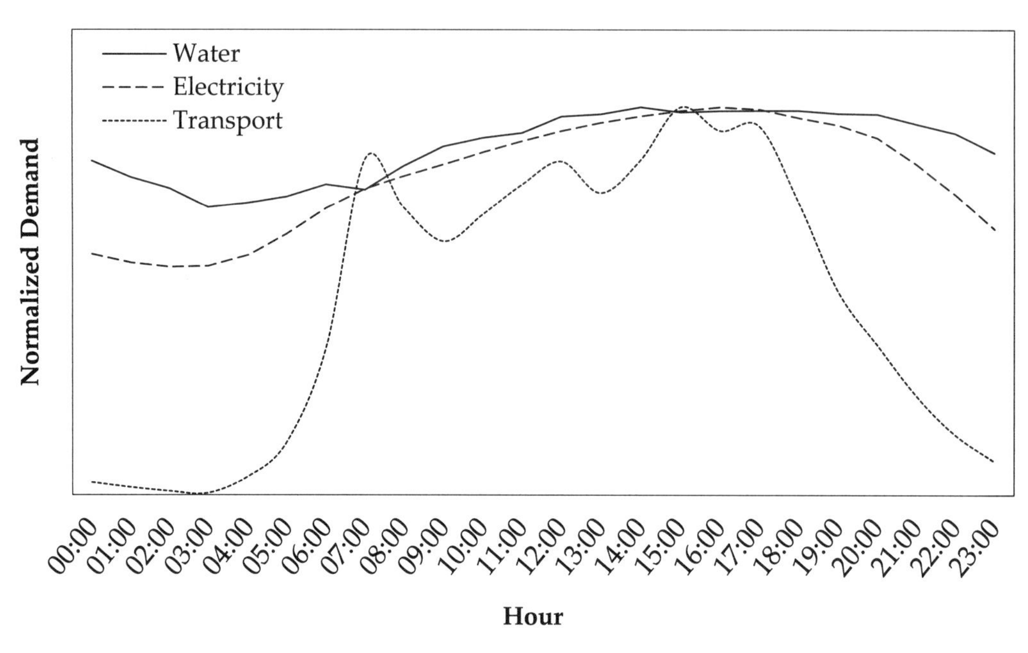

Figure 10.1
Hourly demand for water, electricity, and transport.

should not let the boundaries limit us and should view infrastructure as an interconnected and interdependent system.

Instead of starting with some fundamentals of infrastructure interdependencies and integration—which we cannot do since they did not yet exist at the time of this writing—we will first focus on the demand for infrastructure. Throughout the past five chapters, we have learned to express the demand for energy, water, transport, buildings, and solid waste in terms of energy in [Wh] and GHG emissions in [g CO_2e]. Figure 10.1 shows the daily demand curves that we have seen in chapters 5–7 in one single graph. To make the data comparable, every value was simply divided by the maximum of each data set, which is why labels on the y-axis are missing. Even though the data have been collected from different places (New England for electricity, Chicago for water, and the U.S. average for transport), we can clearly see that the three demands are connected. The demand for water and electricity is relatively low at night and picks up in the morning, when people wake up, take a shower, use their toasters, and so on. The demand for transport then picks up very quickly and decreases after the morning rush hour, while demand for water and electricity increases further. In the late afternoon, the demand for transport increases again, when people travel back home from school/work. The demand for electricity and water remains high as well since this is when people return home, when the sun sets, and when more households turn their lights on, use their electric appliances, and need water, while the demand for transport decreases significantly. From the figure, the demand for water is relatively high at night, which is possibly because the data was collected in the summer; many people water their lawns at night in the summer so the plants and grass do not burn from the heat.

One field is particularly good at integrating these demands together, and it is called *urban metabolism*. We will therefore learn first about urban metabolism in this chapter. After that, we will perform an inventory of interdependencies between infrastructure systems. This will notably take the form of a large matrix showing how infrastructure systems affect one another. Many of these interdependencies should be obvious now that we have covered infrastructure systems separately. We will finally learn about a conceptual approach to think about planning future infrastructure systems so that they are not only more integrated but also more decentralized. Decentralization is something we have discussed a little throughout the chapters but have yet to fully explore. This last section will give us a chance to do just that.

But first, we start with urban metabolism.

10.1 Urban Metabolism

To understand what urban metabolism is, we first need to recall what *metabolism* is, and we discussed it a little in chapter 8. Metabolism is the process of burning calories to supply the human body with energy and heat. Urban metabolism is therefore similar but is applied to sections of cities (like a neighborhood) or even to entire cities. More formally, Kennedy, Cuddihy, and Engel-Yan (2007) define urban metabolism as "the sum total of the technical and socio-economic processes that occur in cities, resulting in growth, production of energy, and elimination of waste." From a technical perspective, it is about achieving a mass and energy balance between what comes into a city, what goes out of a city, what stays in a city, and what is produced/consumed within a city.

This mass balance approach to large systems first became popular in the field of *industrial ecology* with a technique called material flow analysis that—as the name suggests—only looks at the flow of materials (for example, between various industries).[1] For cities, we extend the method to include energy and simply call it material and energy flow analysis (MEFA). MEFA has become the most common technique applied to measure urban metabolism, but this could change as much more data was becoming available at the time of this writing.

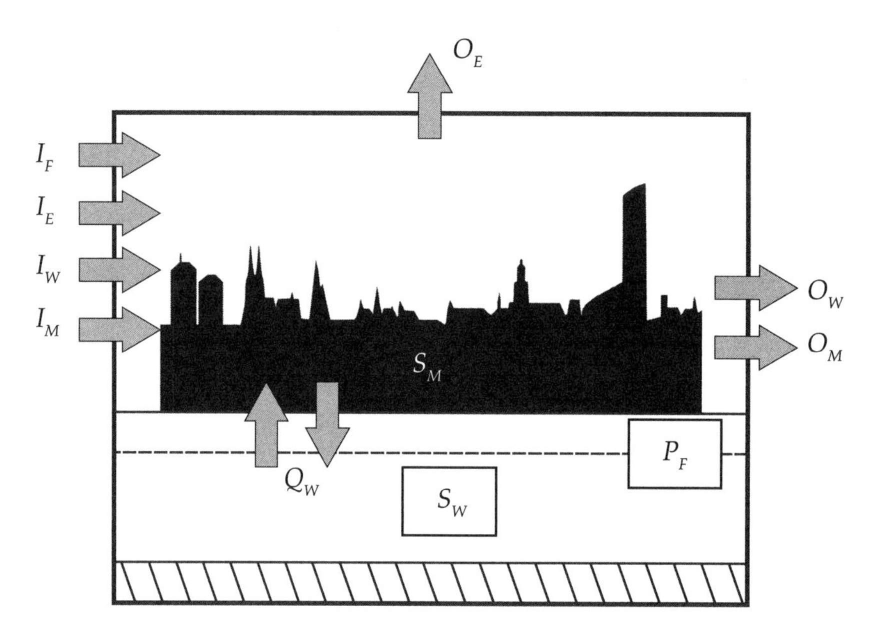

Figure 10.2
Illustration of urban metabolism, accounting for inputs (*I*), outputs (*O*), internal flows (*Q*), storage (*S*), and production (*P*) of water (*W*), energy (*E*), material (*M*), and food (*F*).

Although we can find traces of urban metabolism in the works of many famous writers like Patrick Geddes,[2] many place the beginning of urban metabolism to Abel Wolman's 1965 study. In this work, Wolman (1965) essentially conducted a large-scale mass and energy balance of everything going in and out of a hypothetical city of 1 million people in the United States. He notably focused on water, food, coal, oil, natural gas, and motor fuel (gasoline and diesel). Urban metabolism has been studied extensively since then, including by Chris Kennedy,[3] who has several publications that compare the urban metabolism of world cities.

What is included in the mass and energy balance depends on the scope of what we are studying and on the data that are available. Figure 10.2 shows inputs *I* for food *F*, energy *E*, water *W*, and materials *M*. It shows output *O* for energy *E*, water *W*, and materials *M*. Moreover, within a city there is a production *P* of food *F*. Finally, water *W* and materials *M* stay in the city in the form of storage *S*, and internal flows *Q* between the groundwater *W* and the city are measured.

All of these flows can include a number of things. For example, energy can include only the electricity and gas that is being imported into a city, but it could also include the energy received from the sun and the wind. Figure 10.3 shows an early example of

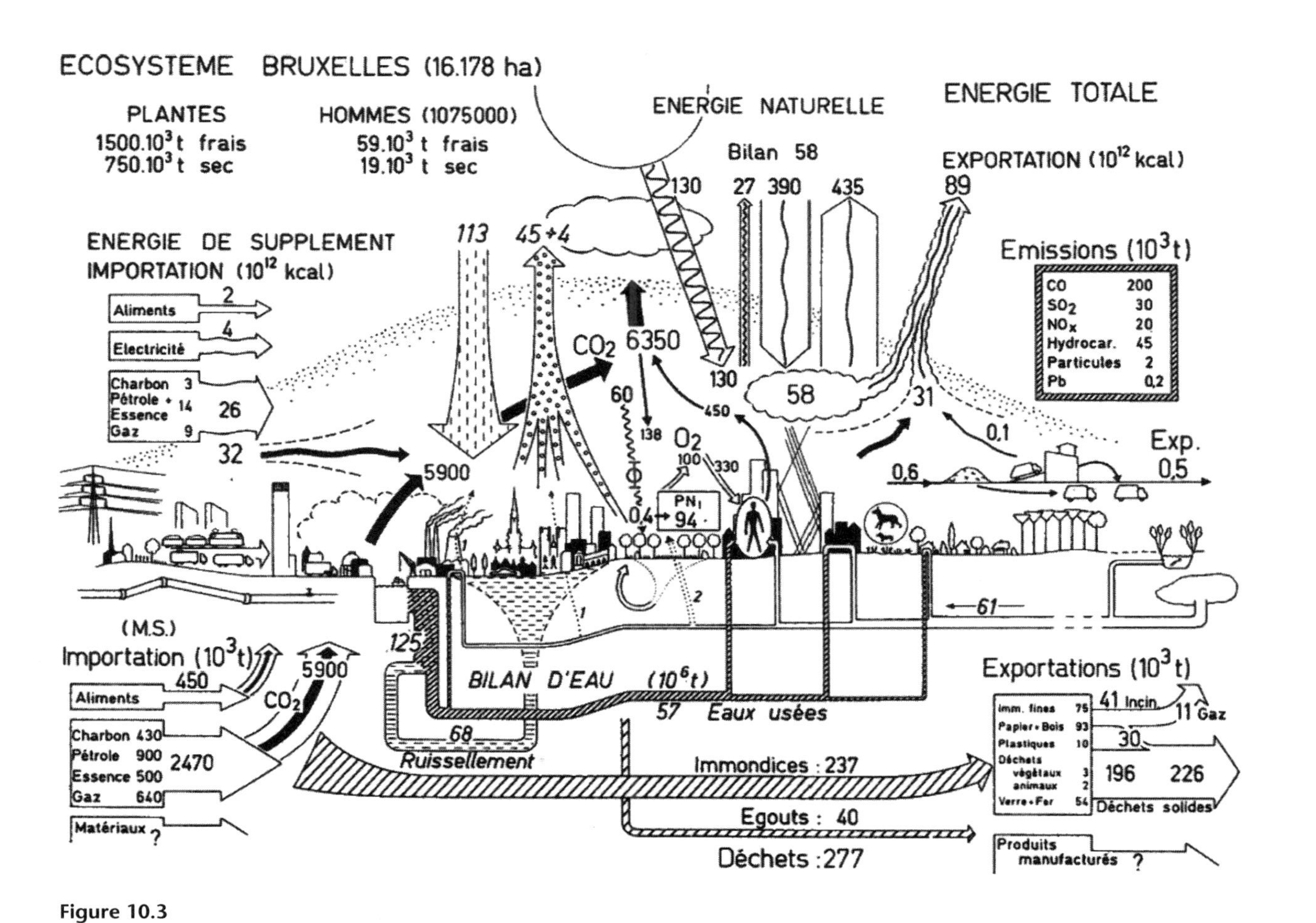

Figure 10.3
Urban metabolism of Brussels, Belgium. Adapted from Duvigneaud and Denaeyer-De Smet (1977).

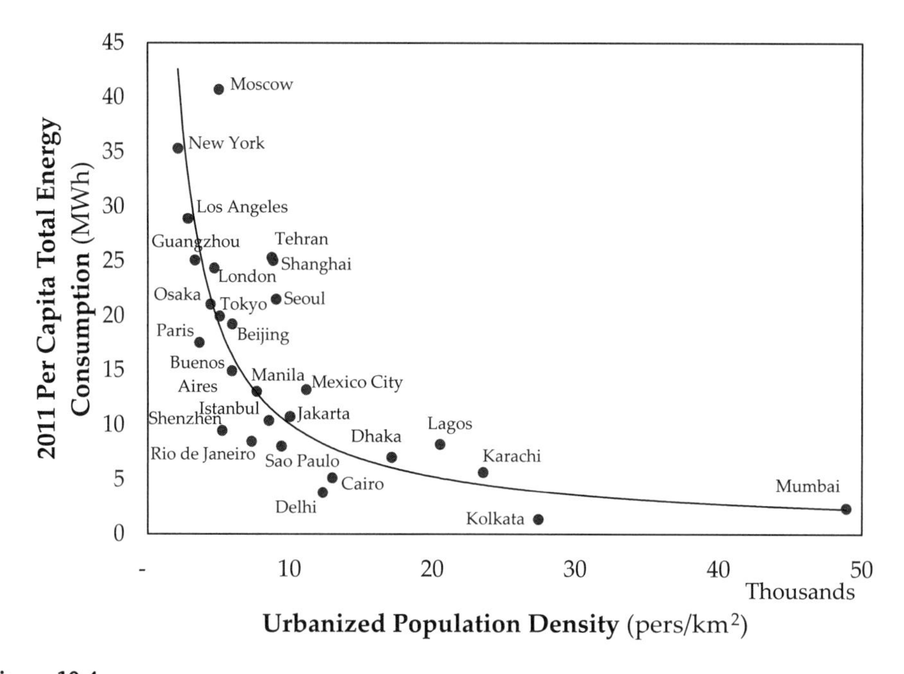

Figure 10.4
Energy use versus population density for world cities.
Source: Facchini et al. (2017).

urban metabolism applied to the city of Brussels, the Belgian capital. The figure is in French, but most words can be translated easily. The figure is also very thorough. In fact, it even accounts for pets—can you spot them in the figure? Most urban metabolism studies are not as detailed.

Naturally, many aspects can have an impact on the urban metabolism of cities. From what we saw in chapter 8, we should expect that climate has a large impact since it takes more energy to heat up a home in colder climates. We can also relate energy consumption with space use. Similar to Newman and Kenworthy's work in chapter 7, Facchini et al. (2017) found a power law trend between per capita total energy use and population density. Figure 10.4 shows this trend; the units were simply converted from [GJ] (gigajoules) to [MWh]. This means that the denser a city, the less energy the residents will consume on a per capita basis. Moreover, they found that the exponent of the power law was –0.75, which has important implications, as we will see in the next chapter.

What we can do now is dive into some of these flows and come up with equations for all of them. We can do this for materials, food, energy, and water. The equations were mostly taken from the book chapter "A Mathematical Description of Urban Metabolism" by Kennedy (2012).

10.1.1 Materials

Materials is a broad term and can include almost everything, from aggregates to make concrete to all of the things that we consume daily—like the computer, tablet, phone, or book from which you are reading right now—and we literally consume tons and tons of materials every year. Figure 10.5 shows the annual world production of twenty-seven materials. The figure was taken from Michael Ashby's (2013) book *Materials and the Environment*, which is a great resource to learn more about the consumption and environmental impact of materials.

The figure includes oil and coal as a benchmark. We have to be careful since the figure uses logarithmic scaling. Assuming the numbers given are from 2012, we can note that we consumed about 30 billion tons of concrete and 1 billion tons of asphalt in that year. That is enormous! Roughly, 1 billion tons of asphalt represents a little more than 1 million kilometers (620,000 miles) of new or resurfaced roads per year,[4] which then translates into about 8 Mt CO_2—but we will learn more about GHG emissions and materials in a section to come.

Generally, too many types of materials would need to be included if we were to complete a full MEFA for an entire city. This is why what we include in the mass and energy balance depends on the scope of our study, as mentioned before.

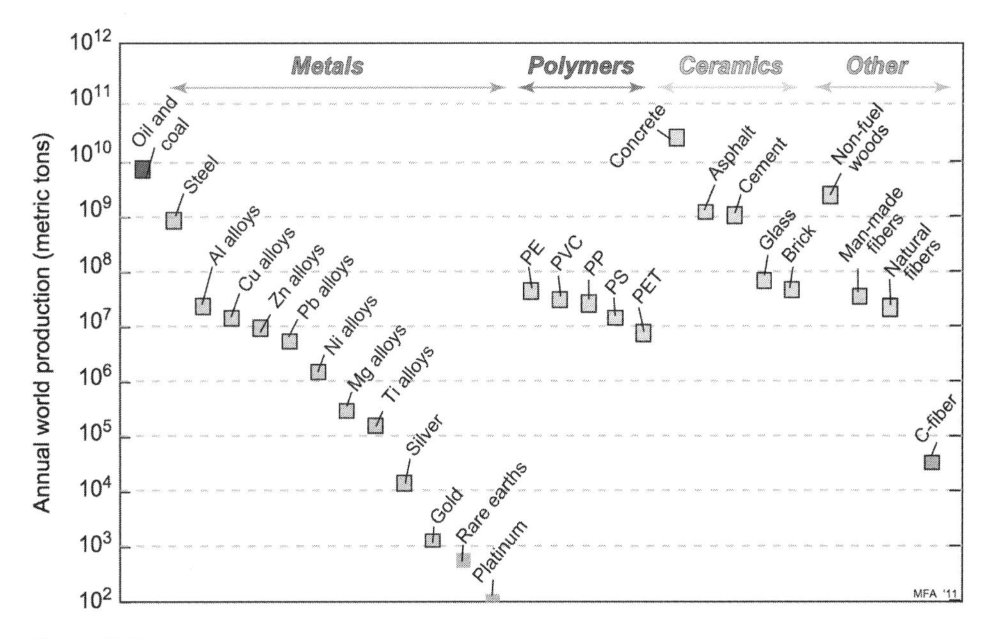

Figure 10.5
Annual world production of twenty-seven materials in metric tons. Note that the scale is logarithmic. *Source*: Ashby (2013, fig. 2.3). Courtesy of Michael Ashby.

While we might think that the more we include the better, data are rarely available for everything. Moreover, it is often difficult to find a total flow of a certain material. What we can do instead is measure or estimate the average rate or density of something—for example, floor space per capita—and scale it up. This type of approach should now sound familiar since it is essentially the IPAT equation and the Kaya identity. The main difference is that at first we are not necessarily trying to quantify the impact but instead are trying to determine the total stock of materials *S*—for example, in tons.

Let us apply this logic to two major types of infrastructure in cities that require a lot of materials: buildings (e.g., concrete) and roads (e.g., asphalt). For building type *i*, the stock *S* of materials *m* can be calculated as:

$$S_{i,m} = P \cdot A_{i,m} \cdot M_{i,m} \quad (1)$$

where *P* is the population of a city in [pers], *A* is the floor space per capita in [m^2/pers], and *M* is the material intensity of the building in [t/m^2]. When we apply equation (1), we therefore get a final result as a weight in [kg] or [t].

For roads, we may want to take a slightly difference approach. We usually talk about kilometers of roads as opposed to square meters of roads. We can therefore adapt *A* and *M* from equation (1). Moreover, instead of population, we may prefer to consider the dimensions *D* of a city in terms of area, such that:

$$S_{i,m} = D \cdot A_{i,m} \cdot M_{i,m} \quad (2)$$

where *S* is still in [t], *D* is in [km^2], *A* is in [km/km^2], and *M* is in [t/km]. We note that equation (2) can easily be applied to railways and canals. In fact, it can be applied to any "linear" infrastructure that is defined by its length (in [km]) rather than by its area (in [km^2]).

We can now go further and try to estimate the total energy use or the total emissions related to a material. The impact/emissions *Y* (e.g., in [g CO_2e]) can be calculated as:

$$Y_{i,m} = S_{i,m} \cdot T_{i,m} \quad (3)$$

where $T_{i,m}$ is the emission factor or energy intensity of material *m* for building/road type *i*. Emission factor can be expressed in [g CO_2e/m^2] or [g CO_2e/km], and energy intensity can be expressed in [Wh/m^2] or [Wh/km] depending on the infrastructure being considered. From the original IPAT equation and the Kaya identity, we simply changed the symbol of impact *I* or emissions *E* to *Y* because *I* refers to inputs, and *E* refers to energy in this chapter.

The energy intensity of a material is sometimes referred to as its *embodied* energy, which accounts for all the energy that has been required for the extraction of the raw material, the transport, the manufacturing, the construction, and even the end of life of the material. Embodied energy is therefore similar to life-cycle assessment except that it does not include the energy used to maintain the material during its operating life. A group of researchers in the United Kingdom called Circular Ecology,[5] based in Bristol, estimated the embodied energy of more than 200 materials. A small sample is listed in table 10.1. The two energy columns are identical; the units are simply different, as one is in [MJ], and the other one is in [kWh]. The two carbon columns show the value of a low carbon intensity and a high carbon intensity.

As a small caveat, only the "general" types of materials as reported by Hammond and Jones (2011) are shown, but they have data for many different subtypes of materials, and the data are absolutely free. Everyone is recommended to download their report. Moreover, the data are reported per unit weight. Density values therefore have to be used to get volumes, but an average thickness has to be estimated since equations (1) and (2) require energy intensities per square meter and per kilometer.

From the table we can see that aluminum has the largest energy footprint of the materials given. It has, in fact, over two orders of magnitude more embodied energy than concrete. This should not be a surprise, as we already saw how carbon-intensive aluminum is when we discussed recycling in chapter 9. Moreover, this is not to say the concrete is not energy-intensive, because it is. We should really account for how much a material is used, as well. Concrete, for example, has an average density of 2,400 kg/m^3, and therefore the embodied energy and carbon intensity of 1 m^3 of concrete are 504 kWh and 240 kg CO_2, respectively. A small house foundation with a 12.5 cm thick (5 in.) concrete slab of 8×12 m (about 26×40 feet) has a volume of 12 m^3, thus representing more than 6 MWh and about 2.8 t CO_2. Considering the millions of cubic meters of concrete that are poured every day, the carbon emissions from concrete use are substantial.

If we look a little bit further, however, we can see that it really is the Portland cement that has a high embodied energy, with 1.53 kWh/kg. The aggregates are comparable to stone, with 0.35 kWh/kg. The key would therefore be to produce concrete without using Portland cement, and significant effort is put to find alternatives (using fly ash, for example).

In the table we also see that the low and high carbon intensities for timber/lumber—that is, wood especially used as building material—are divided into two parts: *fos* and *bio*. The first part, *fos*, is attributed to all fossil fuels needed to process wood until it can be used. The *bio* part relates to whether the wood itself was sustainably grown—that is, whether it comes from a sustainably managed forest.[6] In the original data set, we are therefore told that if the wood comes from a sustainably managed forest, we can simply omit the *bio* part.

Table 10.1
Embodied energy and carbon intensities of several common materials

	Embodied energy and carbon intensities			
	Energy		Carbon (kg CO_2/kg)	
Materials	*MJ/kg*	*kWh/kg*	*Low*	*High*
Aluminum (general)	155	43.06	8.24	9.16
Asphalt road (hot construction)	2,509 MJ/m^2	696.95 kWh/m^2	93 kg/m^2	99 kg/m^2
Common brick	3	0.83	0.23	0.24
Average Portland cement	5.5	1.53	0.93	0.95
Concrete (general)	0.75	0.21	0.1	0.11
Lead (general)	25.21	7.00	1.57	1.67
Paint (general)	70	19.44	2.42	2.91
Plastics (general)	80.5	22.36	2.73	3.31
Steel (UK average recycled)	20.1	5.58	1.37	1.46
Stainless steel	56.7	15.75	6.15	
Stone (general)	1.26	0.35	0.07	0.08
Timber/lumber (general)	10	2.78	$0.30_{fos}+0.41_{bio}$	$0.31_{fos}+0.41_{bio}$
Vinyl flooring (general)	68.6	19.06	2.61	3.19

Source: Hammond and Jones (2011).

Going back to our original equations, we can also see that it is easy to combine equation (1) and (3) directly or equations (2) and (3).

If we now want to sum all the materials to get a total weight for the entire stock of materials, we can simply use the equation:

$$S = \sum_i \sum_m S_{i,m} \tag{4}$$

If instead we prefer to assess the total impact of the stock of materials, we can use the equation:

$$Y = \sum_i \sum_m Y_{i,m} \tag{5}$$

Another aspect that we may be interested in is time. In this case we need to account for all inputs I and all outputs O of materials. Flows of material m can therefore be captured by using:

$$\frac{dS_m}{dt} = I_m(t) - O_m(t) \tag{6}$$

$I(t)$ and $O(t)$ can be modeled from past data. $I(t)$, for example, can be linked with population or economic growth. Typically, $I(t)$ is quite large when a city is expanding, and $O(t)$ tends to be low since the life of a building is generally long—which is why it is important to consider S. Moreover, many projects try to recycle materials from existing structures, within a framework called the *circular economy* that has an obvious parallel with the *circular ecology* group from which we collected data for table 10.1.

When we have data on imports and exports for specific years, we can simplify equation (6) to:

$$\Delta S_m^{t_2-t_1} = [I_m(t_2) - I_m(t_1)] - [O_m(t_2) - O_m(t_1)] \tag{7}$$

Equation (7) is particularly interesting to study the flow of materials into and out of a city. In terms of units, we can think of [t/year]. Presumably, a city that is doing well economically and that is growing may have a high ΔS for several years and a low ΔS when it is not doing well.[7]

A last point should be made about data. As geographic information systems (GIS) technology becomes more popular, it becomes increasingly easy to get data on the total floor area of buildings and the total length of roads and rails. The variables P and A from equations (1) and (2) can therefore be easily collapsed into one. Moreover, in GIS, we can easily study urban metabolism at different scales, from a single building, to a neighborhood, and to an entire city. The equations given here should therefore be used more as a guide rather than rigid urban metabolism laws.

Example 10.1

Embodied Energy and Carbon Footprint in an Apartment Complex in Civitas

An apartment building in Civitas is home to forty people, and a brief investigation suggests it has a total volume of concrete of 440 m^3. Considering an average concrete density of 2,400 kg/m^3 and an average concrete carbon footprint of 0.1 kg CO_2/kg, (1) calculate the total and per resident CO_2 emissions of the building. In addition, the building has a small parking lot behind it that measures 100 m^2. Assuming an asphalt carbon footprint of 96 kg CO_2/m^2, (2) calculate the total and per resident CO_2 emissions of the parking lot. Considering the apartment building has an estimated lifetime of forty-five years and the parking lot has to be repaved every fifteen years, (3) calculate the yearly total and per resident CO_2 emissions linked to the apartment complex.

Solution

(1) A total volume of 440 m^3 of concrete translates to a mass M_c of $M_c = 2{,}400 \times 440 = 1{,}056{,}000$ kg. With an average carbon footprint of 0.1 kg CO_2/kg, we get a total carbon emissions for concrete: $Y_c = 1{,}056{,}000 \times 0.1 = 105{,}600$ kg $= 105.6$ t CO_2.

Dividing the total carbon emissions for concrete by the number of residents, we get $Y_{c,resident} = 105.6 / 40 = 2.64$ t CO_2.

(2) The total surface area of the parking lot is 100 m^2. Considering an asphalt carbon footprint of 96 kg CO_2/m^2, we get $Y_a = 100 \times 96 = 9{,}600$ kg $CO_2 = 9.6$ t CO_2.

Accordingly, the per resident carbon emission linked with the parking lot is $Y_{a,resident} = 9{,}600 / 40 = 240$ kg $CO_2 = 0.24$ t CO_2.

(3) With a lifetime of forty-five years for the building and a lifetime of fifteen years for the parking lot—that is, it has to be paved three times at years 0, 15, and 30—the total carbon footprint Y of the apartment complex that accounts for the concrete used and asphalt is $Y = 105.6 + 3 \times 9.6 = 134.4$ t CO_2.

The yearly carbon emissions Y_y therefore are $Y_y = 134.4 / 45 = 2.99$ t CO_2e.

Assuming the apartment complex systematically has forty people, the per resident yearly carbon footprint $Y_{y,resident}$ attributed to the concrete and asphalt used in the apartment complex is $Y_{y,resident} = 2.99 / 40 = 0.075$ t CO_2.

Although this number looks small, when we sum all the buildings in the world, it becomes significant—remember figure 10.5 and the fact that we consumed about 30 billion tons of concrete and 1 billion tons of asphalt in a single year.

This is it for the materials section. Again, the extent of what is included is purely based on the scope of the study being done. One could only look at the impact of bricks, for example, or include many different types of building materials. Moreover, for transport, one can look purely at roads or include rails as well. It is therefore important to know what we want to measure and what we can measure based on the data that we have.

We will now move on to food.

10.1.2 Food

Applying the IPAT equation or the Kaya identity for food requires more elements. After all, food involves both liquids and solids. The general equation proposed by Kennedy (2012) is:

$$I_F + P_F + I_{W,Kit} = O_{F,FW} + O_{F,Met} + O_{F,S} \quad (8)$$

where

- I_F: food and packaged drinks imported to a city,
- P_F: food and packaged drinks produced in a city for internal consumption,
- $I_{W,Kit}$: kitchen water used,
- $O_{F,FW}$: food waste going to landfill, compost, or organic waste collection,
- $O_{F,Met}$: carbon and water lost via respiration and transpiration in residents' metabolism, and
- $O_{F,S}$: feces and urine going to a sanitary sewer.

Equation (8) can either be expressed in terms of weight in [t] or in terms of energy in [Wh] or [J].

We note that equation (8) does not include a variable for liquid waste, such as any excess water, but it can be included.

Although we will not get into it, to investigate I_F (food and packaged drinks imported to a city), we can look at freight imports and exports, which are also important to estimate the GHG emissions linked with freight transport. In the United States, the Freight Analysis Framework offers some data.[8]

This is it for food. Although the section for food is short, it is easy to see that equation (8) can be extremely difficult to measure if we try to include all types of food and drinks (especially those that are locally produced). Like we did for materials, rates and densities (e.g., the average food intake per person) can be used to quantify some of the variables in equation (8), but we will not reiterate the process.

10.1.3 Energy

Energy is also a broad term and can include all types of energy being consumed in a city. This includes not only electricity and gas consumption but also fuel consumption from transport and the energy used to pump and distribute water. Naturally, we have to be a bit careful here. If we include gasoline in the materials process of urban metabolism, then it should not be included here.

When it comes to input energy I_E, Kennedy (2012) includes six types of anthropogenic energies (as opposed to natural energy like sunrays). They are:

$$I_E = I_{E,\ buildings} + I_{E,\ transport} + I_{E,\ industry} + I_{E,\ construction} + I_{E,\ water\ pumping} + I_{E,\ waste} \quad (9)$$

The subscripted labels should be self-explanatory. We notably see that we have only inputs here since none of the six types of energy uses generate any output of energy. For a city that outputs energy in the form of exports, it can be added to equation (9) as a negative energy—that is, $O_{E,\ exports}$.

We have already discussed some of the variables in equation (9). Specifically, we have discussed $I_{E,\ buildings}$ in chapter 8, $I_{E,\ transport}$ in chapter 7, $I_{E,\ waste}$ in chapter 9, and $I_{E,\ water\ pumping}$ in chapter 6 (that includes water collection and treatment and wastewater collection). We have not directly discussed $I_{E,\ industry}$ and $I_{E,\ construction}$, and although they are significant, we will not discuss them here. We simply need to remember that we can adapt the IPAT equation and the Kaya identity based on the data that are available if we want to estimate them.

In chapter 8 we specifically looked at space conditioning. Here, $I_{E,\ buildings}$ should account for all energy uses, and we can therefore define it as:

$$I_{E,\ buildings} = I_{E,\ heating} + I_{E,\ cooling} + I_{E,\ lights\ and\ appl.} + I_{E,\ water\ heating}. \quad (10)$$

Out of the four elements, we see that three relate to the generation or displacement of heat—as we saw in chapter 8, space conditioning and water heating tend to require the most energy.

While large-scale averages may be sufficient for $I_{E,\ lights\ and\ appl.}$ and $I_{E,\ water\ heating}$, energy use related to space conditioning needs to account for temperature (i.e., climate). In chapter 8 we developed equation (46), which uses the concept of degree days to estimate the energy consumption related to the space conditioning of a building. In line with the approach that we have followed for materials, we can estimate total energy use for heating and cooling from:

$$I_{i,\ heating} = \sum_i P \cdot A_i \cdot HDD \cdot T_{i,\ heating} \quad (11)$$

$$I_{i,\ cooling} = \sum_i P \cdot A_i \cdot CDD \cdot T_{i,\ cooling} \quad (12)$$

where P is the population of a city in [pers], and A_i is the floor space per capita for a building type i, like equation (1), or the exposed surface area per capita in [m^2/pers] depending on the value given for $T_{i,\ heating}$ and $T_{i,\ cooling}$ (see below). HDD and CDD are the heating degree days and cooling degree days, respectively, in [Kd] or [Kh] (kelvin days or kelvin hours). Finally, $T_{i,\ heating}$ and $T_{i,\ cooling}$ are the energy intensities for space heating and space cooling, respectively, for building type i.

The energy intensities $T_{i,\ heating}$ and $T_{i,\ cooling}$ can be expressed in [W/m^2K] as a power or [Wh/m^2K] or [Wd/m^2K] as an energy.

We should note that the only difference between $T_{i,\,heating}$ and $T_{i,\,cooling}$ is the technology used. For both intensities, we should use the building conductance U_{sys} (from chapter 8) and then apply the heating and cooling efficiencies—that is, the annual fuel utilization efficiency (AFUE), the energy efficiency ratio (EER), the seasonal energy efficiency ratio (SEER), and the coefficient of performance (COP).

Moreover, we have to be careful with the units. Most often, we should multiply HDD and CDD by twenty-four to get [Kh]. Moreover, we can use actual energy intensities—for example, as estimated with eQuest—or we can use estimated energy intensities per floor area, depending on the data that are available—we saw ways both in chapter 8. Finally, we have to be careful with the values that we pick as well. Equations (11) and (12) require per capita values as opposed to per building values. Per building values should first be divided by household occupancy.

Building types can range from comparing residential and commercial buildings only, to dividing building types based on the number of floors, to using one single average for all buildings.

Back to equation (9), about transport energy use, we remember that we need to account for distance traveled, travel modes, and emission factors by modes. In chapter 7 we developed equations (36) and (37), but we will not repeat them here.

With regard to solid waste management, we need to estimate the total weight of the solid waste generated and then calculate the associate GHG emissions, potentially using the values given in tables 9.12 and 9.13 in chapter 9.

Regarding energy use for water pumping, we remember equation (11) in chapter 6, which calculates the shaft power needed to run a pump and depends on the specific weight of water, the flow rate, the pump head, and the pump efficiency. Because energy use for water pumping also heavily depends on land topography (gradient vs. flat terrain for pump head), we normally need detailed numbers from the local water management department, for instance. In the absence of power consumption data, we can use the value 1 Wh/L that accounts for water collection, treatment and distribution (0.45 Wh/L) and for wastewater treatment (0.55 Wh/L) as seen in chapter 6. For the flow rate, we can also use average county-level public use water withdrawal data from the U.S. Geological Survey if more detailed data are not available. We actually did this before when we first introduced the IPAT equation and the Kaya identity in chapter 2. But we can also estimate flow rate another way, as we will learn in the next section.

Finally, we may or may not want to account for "natural" energy inputs and outputs. Kennedy (2012) refers to the entire process as surface urban energy balance and defines it as:

Example 10.2

Energy Consumption in an Apartment Complex in Civitas

The apartment building from example 10.1 has a total floor area of 1,200 m^2 (including common areas) for forty people, and it has a total of ten apartments. The building has a floor area heating intensity of 0.61 W/m^2K and a floor area cooling intensity of 0.23 W/m^2K. From chapter 8, we also know that Civitas has 2,902 HDD and 104 CDD. Moreover, a separate investigation found that the average energy use per apartment is 1.6 MWh for lights and appliances and 2.4 MWh for water heating. Calculate the (1) building, (2) per apartment, and (3) per resident heating and cooling energy use and total energy use.

Solution

(1) Using equation (11) and (12) but directly using the floor area, the total energy used for heating and cooling is:

- $I_{E,\ heating} = 1{,}200 \times 2{,}902 \times 0.61 = 2.124 \times 10^6$ Wd $= 50.98$ MWh
- $I_{E,\ cooling} = 1{,}200 \times 104 \times 0.23 = 28{,}704$ Wd $= 0.69$ MWh

Moreover, considering there are ten apartments, the total lights and appliances and water heating energy use are:

- $I_{E,\ lights\ and\ appl.} = 1.6 \times 10 = 16$ MWh
- $I_{E,\ water\ heating} = 2.4 \times 10 = 24$ MWh

This means that the total energy use I_E in the building is:

- $I_E = 50.98 + 0.69 + 16 + 24 = 91.67$ MWh

(2) To obtain the per apartment energy use values, we simply divide all the results from (1) by 10:

- $I_{E,\ heating\ per\ apt.} = 50.98 / 10 = 5.098$ MWh
- $I_{E,\ cooling\ per\ apt.} = 0.69 / 10 = 0.069$ MWh
- $I_{E,\ per\ apt.} = 91.67 / 10 = 9.167$ MWh

We do not need to add the energy used for lights and appliances and for water heating since it was given to us to start with.

(3) To obtain the per resident energy use values, we simply divide all the results from (1) by 40:

- $I_{E,\ heating\ per\ res,} = 50.98 / 40 = 1.27$ MWh
- $I_{E,\ cooling\ per\ res.} = 0.69 / 40 = 0.013$ MWh
- $I_{E,\ lights\ and\ appl.\ per\ res.} = 16 / 40 = 0.4$ MWh
- $I_{E,\ water\ heating\ per\ res.} = 24 / 40 = 0.5$ MWh
- $I_{E,\ per\ res.} = 91.67 / 40 = 2.30$ MWh

$$I_{E,S}+I_{E,F}+I_{E,I}=O_{E,I}+O_{E,G}+O_{E,E} \quad (13)$$

where

- $I_{E,S}$: radiant energy from the sun,
- $I_{E,F}$: generation of heat due to combustion and dissipation in machinery,
- $I_{E,I}$: heat from the earth's interior,
- $O_{E,I}$: heat loss by evapotranspiration,
- $O_{E,G}$: heat loss by conduction to soil, buildings, roads, and so on, and
- $O_{E,E}$: heat loss by radiation.

We will not get into the details of energy (13), but we should remember that we discussed some of the variables listed. In particular, we covered the radiant energy from the sun in equations (37) in chapter 8 and with figure 5.19 in chapter 5 (for the United States). Moreover, we discussed heat loss by radiation in equations (25) to (36) in chapter 8.

Although we will not discuss it, equation (13) can help account for a phenomenon called the *urban heat island* effect. This phenomenon essentially describes the fact that cities tend to be warmer than the rural areas around them because they have many more dark surfaces (thus they receive more radiation from the sun), and because of heat loss from buildings (partly linked with the generation of heat for space heating). The urban heat island effect may not seem important, but it has multiple (negative) impacts, including on rainfall (the temperature difference with the surrounding areas may be enough for water to condensate), air quality (an increase in pollutants), and on water quality (warmer water flowing may impact the entire ecosystem).

Being able to quantify the impact of energy consumption is therefore quite complicated but essential since it has the largest effect on GHG emissions. Instead of considering energy as one category, however, we can split it into multiple categories such as electricity, gas, transport, solid waste, and water. We can also choose to separate space conditioning from water heating, or not. Moreover, we need to add consumption to the water category, which is the goal of the next section.

10.1.4 Water

Even though water only deals with one substance (i.e., water), quantifying water flow is far from trivial. Like energy, we can start here with the anthropogenic use of water—we will see natural flows of water, like precipitation, later. As opposed to inputs, here we will work in terms of flow Q, which can then serve as an input to estimate $I_{E,\text{ water pumping}}$, for example.

We divide anthropogenic water use Q_W into only two types:

$$Q_W = Q_{W,D} + Q_{W,L} \tag{14}$$

where $Q_{W,D}$ is the water demand (i.e., consumed), and $Q_{W,L}$ is the water lost because of leaks. If desired, we can further divide $Q_{W,D}$ by consumer type: residential, commercial, industrial, and so on.

Although we did not discuss it much in chapter 6, water consumption varies depending on the season. The base water demand, mostly for indoor use, remains relatively constant throughout the year, but water demand tends to be higher in the summer for outdoor use and increased cooling needs (e.g., in thermal power plants). We can therefore define water demand as:

$$Q_{W,D} = Q_{W,D,\ base} + CDD \cdot T_{W,\ cooling} \tag{15}$$

where $Q_{W,D\ base}$ is the base water demand in [m^3/d], CDD are in [Kd], as we have already seen, and $T_{W,\ cooling}$ is intensity of water demand based on the temperature in [m^3/K]. If the data are available, they can be easily approximated using average daily demand $Q_{W,\ avg\ daily}$ and maximum daily demand $Q_{W,\ max\ daily}$ in a year, as:

$$T_{W,\ cooling} = \frac{Q_{W,\ max\ daily} - Q_{W,\ avg\ daily}}{CDD} \tag{16}$$

As we saw in chapter 6, losses because of leaks generally account for between 6% and 16% of the total water withdrawal in the United States, but we can be a little more specific here. If we know the total length L of the pipe system and if we can measure or approximate an average leak rate by kilometers of pipe, then $Q_{W,L}$ takes the form:

$$Q_{W,L} = L \cdot r_{leak} \tag{17}$$

where r_{leak} is the average leakage rate in [m^3/km]. However, if we do not know the total length of the system, we can try to estimate it based on an average length per unit area ρ_W and the total area D of the city, such that:

$$Q_{W,L} = D \cdot \rho_W \cdot r_{leak} \tag{18}$$

where ρ_W can be estimated from GIS maps. For example, we can use the length of streets per unit area as an indicator or even the total length of streets to estimate $D{\cdot}\rho_W$ altogether—although we have to be careful since many streets, like bike alleys and expressways, for example, do not generally carry water pipes.

Moving to wastewater, the total flow of wastewater Q_{WWT} needs to account for three separate flows:

$$Q_{WWT} = Q_{WW,D} + Q_{WW,R} + Q_{Inf} \tag{19}$$

Example 10.3

Estimating Water Consumption in a Civitas Neighborhood

A new neighborhood is being built in Civitas, and we are trying to estimate how much water will be consumed to correctly size the water pipes. The neighborhood is projected to have about 180 people living in fifty townhouses. The typical base water demand is 190 L/day per person, and in general the water demand intensity, accounting for warm weather, is 1.40 m^3/K per house (note that CDD=104, as we were reminded in example 10.2). In addition, 400 m of new pipes will be laid, and as the pipes get older, we are told to account for a leakage rate of 4 m^3/km per day. (1) Calculate the total, per house, and per resident annual and average daily water consumption. Subsequently, compute (2) the total water lost due to leakage and (3) the total water that will be consumed per year and the average water flow rate in [m^3/s].

Solution

(1) We note that the water demand intensity is given per house, while the base water demand is given per person. Let us start with per house. The annual water intensity per house is $CDD \cdot T_{W,\ cooling\ per\ house} = 104 \times 1.40 = 145.6\ m^3$.

Moreover, since we have 180 people in fifty townhouses, this means that the average number of residents per house is 180 / 5 = 3.6. Therefore, the base water demand per year per house is $Q_{W,D,base} = 3.6 \times 190 \times 365 = 249{,}660\ L = 249.66\ m^3$.

Thus, the annual and average daily water consumption per house is:

- Annual: $Q_{W\ per\ house} = 249.66 + 145.6 = 395.26\ m^3$ per year
- Daily: $Q_{W\ per\ house} = 395.26\ /\ 365 = 1.083\ m^3 = 1{,}083$ L per day

As a result, the annual and average daily water consumption per resident is:

- Annual: $Q_{W\ per\ res.} = 395.26\ /\ 3.6 = 109.89\ m^3$ per year
- Daily: $Q_{W\ per\ res,} = 1{,}083\ /\ 3.6 = 300.8$ L per day

Finally, the annual and average daily water consumption for the neighborhood are:

- Annual: $Q_W = 395.26 \times 50 = 19{,}763\ m^3$ per year
- Daily: $Q_W = 1.083 \times 50 = 54.15\ m^3$ per day

(2) Using equation (17), the total water lost due to leakage $Q_{W,L}$ is:

- Daily: $Q_{W,L} = 400 \times 10^{-3} \times 4 = 1.6\ m^3$ per day
- Annual: $Q_{W,L} = 1.6 \times 365 = 584\ m^3$

(3) Finally, the total water use for the neighborhood is:

- Annual: $Q_W = 19{,}763 + 584 = 20{,}347\ m^3$
- Flow rate: $Q_W = 20{,}347\ /\ (365 \times 24 \times 60 \times 60) = 0.000645\ m^3/s$

where $Q_{WW,D}$ is the flow of wastewater that comes from the water demand, $Q_{WW,R}$ is the flow of wastewater from surface runoff, and Q_{Inf} is the groundwater that infiltrates the sewer system. Not all water demanded $Q_{W,D}$ is included in $Q_{WW,D}$ since a significant amount is used to water plants and lawns and is evaporated.[9] Kennedy (2012) found that 20% to 25% of the water demand in Toronto did not end up as wastewater.

Accounting for natural flows of water, the total balance of water coming in and out of a city can be calculated from:

$$I_{W,\,precip} + I_{W,\,pipe} + I_{W,\,sw} + I_{W,\,gw} = O_{W,\,evap} + O_{W,\,out} + \Delta S_W \tag{20}$$

where

- $I_{W,\,precip}$: natural inflow from precipitation,
- $I_{W,\,pipe}$: inflow from pipes into the city,
- $I_{W,\,sw}$: net surface water flow into the city (e.g., streams),
- $I_{W,\,gw}$: net groundwater flow into the city,
- $O_{W,\,evap}$: water loss through evapotranspiration,
- $O_{W,\,out}$: outflow from pipes out of the city, and
- ΔS_W: annual change in water stored within the city (normally close to zero unless there are significant changes in groundwater level—i.e., over pumping or major drought).

The first variable, $I_{W,\,precip}$, is usually collected from past data—that is, using data from nearby weather stations and adjusting it to the location as needed.[10] The second variable, $I_{W,\,pipe}$, depends completely on the city where the study is being performed. For example, New York City imports most of its water from upstate New York. Similarly, Los Angeles gets most of its water from Central California. In contrast, Chicago gets most of its water from Lake Michigan.[11] The third variable, $I_{W,\,sw}$, and the fourth variable, $I_{W,\,gw}$, can be estimated from applying the Natural Resources Conservation Service (NRCS) method and using Darcy's law, for instance, which we covered in chapter 6. Several equations exist to estimate $O_{W,\,evap}$ that can be incredibly simple (from collected measurements) or more complex. Outflow $O_{W,\,out}$ also depends on the city, but it will be linked to all inputs. Finally, ΔS_W should be zero or close to zero unless a well is being overpumped and is drying out.

This concludes the section on urban metabolism. Again, we must remember that these equations are used as a means to perform a mass and energy balance. They are not physical laws. They can therefore easily be adapted based on the type of data that are available to us. With *Big Data* and *open data portals*, the availability of high resolution and accurate data should be able to help us calculate the urban metabolism of neighborhoods in a much more accurate way.

Example 10.4

Urban Metabolism of a Civitian Neighborhood

The table below shows some properties of a neighborhood in Civitas. Calculate the per capita and total materials stocked in the neighborhood for the buildings and the roads in terms of (1) weight and (2) carbon emissions. Based on the energy values given, calculate the per capita carbon footprint of the neighborhood related to (3) buildings, (4) transport (calculate carbon footprint only), (5) solid waste, and (6) water (assume all energy is provided from the power grid). From the results computer, calculate the (7) per capita and total carbon footprint of the neighborhood.

Category	Value	Unit
General properties		
Population *P*	10,000	pers
Area *D*	0.7	km^2
Floor area *A*	80	m^2/pers
Exposed area *A*	120	m^2/pers
Road density *A*	15	km/km^2
Grid emission factor	125	g CO_2e/kWh
Materials		
Building intensity (per floor area)	1,000	t/m^2
	500	kg CO_2e/m^2
Road intensity	3	Mt/km
	1,000	t CO_2e/km
Energy		
Buildings		
Average *U* value (exposed area)	0.4	W/m^2K
HDD	2,902	Kd
CDD	104	Kd
AFUE (electric heating)	1	
SEER (cooling)	3	
Lights and appliances	22	kWh/m^2
Water heating (electric)	9.6	kWh/m^2
Transport		
Passenger km traveled (PKT) (year)	6,837	km
Average emission factor	151	g CO_2e/km
Solid Waste		
Waste generated (monthly)	56.50	kg
Average emission factor	0.432	t CO_2e per wet ton
Water		
Average consumption	300	L/d
Energy intensity	1	Wh/L

(continued)

Example 10.4 (continued)

Solution

(1) Equation (1) can be used for buildings, and equation (2) can be used for transport.

Buildings:

- total $S_{building} = 10{,}000 \times 1{,}000 \times 80 = 800{,}000{,}000$ t $= 800$ Mt
- per capita $S_{building} = 800 \times 10^6 / 10{,}000 = 80{,}000$ t

Roads:

- total $S_{road} = 0.7 \times 15 \times 3 = 31.5$ Mt
- per capita $S_{road} = 31.5 / 10{,}000 = 0.00315$ Mt $= 3{,}150$ t

(2) Adopting the same procedure as question (1), we simply need to use the emission factors given.

Buildings:

- total $I_{building} = 10{,}000 \times 500 \times 80 = 400{,}000{,}000$ kg $CO_2 = 0.4$ Mt CO_2
- per capita $I_{building} = 500 \times 80 = 40{,}000$ kg $CO_2 = 40$ t CO_2

Roads:

- total $I_{road} = 0.7 \times 15 \times 1{,}000 = 10{,}500$ t CO_2
- per capita $I_{road} = 10{,}500 / 10{,}000 = 1.05$ t CO_2

(3) From equation (10), the energy and carbon footprint of the buildings is:

$$I_{E,\ buildings} = I_{E,\ heating} + I_{E,\ cooling} + I_{E,\ lights\ and\ appl.} + I_{E,\ water\ heating}$$

From equations (11) and (12), both $I_{E,\ heating}$ and $I_{E,\ cooling}$ require energy intensities. Using the U value, AFUE, and SEER given, we get:

$$T_{i,\ heating} = \frac{U}{AFUE} = \frac{0.4}{1} = 0.4$$

$$T_{i,\ cooling} = \frac{U}{SEER} = \frac{0.4}{3} = 0.133$$

Therefore, from equations (11) and (12):

- per capita $I_{i,\ heating} = 120 \times 2902 \times 0.4 \times 24 = 3.34 \times 10^6$ Wh $= 3.34$ MWh
- per capita $I_{i,\ cooling} = 120 \times 104 \times 0.133 \times 24 = 39{,}836$ Wh $= 0.039836$ MWh

Since the energy intensities of the lights and appliances and water heating are given in kWh/m^2, we can simply calculate:

- per capita $I_{i,\ lights\ and\ appl.} = 22 \times 80 = 1{,}760$ kWh $= 1.76$ MWh
- per capita $I_{i,\ water\ heating} = 9.6 \times 80 = 768$ kWh $= 0.768$ MWh

The per capita building energy consumption of the neighborhood, therefore, is:

- per capita $I_{E,\ buildings} = 3.34 + 0.039836 + 1.76 + 0.768 = 5.91$ MWh

Moreover, because all the energy consumed came from electricity, we can simply calculate the carbon footprint from building energy use as:

- per capita $I_{E,\ buildings} = 5.91 \times 0.125 = 0.74$ t CO_2e

Example 10.4 (continued)

(4) Since the data provided already take travel mode and number of trips into account, the carbon footprint from transport is:

- per capita $I_{E,\ transport} = 6{,}837 \times 151 = 1.03 \times 10^6$ g $CO_2e = 1.03$ t CO_2e

(5) As we are given a monthly value in kilograms, we need to multiply our result by 12 and convert the weight from [kg] to [t]. Therefore, the carbon footprint from waste is:

- per capita $I_{E,\ waste} = 12 \times 56.50 \times 10^{-3} \times 0.432 = 1.03 \times 10^6$ g $CO_2e = 0.29$ t CO_2e

(6) Based on elements learned in chapter 6, the average yearly energy use for water is:

- per capita $I_{E,\ water} = 365 \times 300 \times 1 = 0.1095 \times 10^6$ Wh $= 0.1095$ MWh

Moreover, because all the energy used in water came from electricity, we can use the power grid emission factor given such that:

- per capita $I_{E,\ water} = 0.1095 \times 0.125 = 0.014$ t CO_2e

(7) Based on the results collected, the per capita footprint of the residents of the Civitas neighborhood is:

- per capita $I_E = 0.74 + 1.03 + 0.29 + 0.014 = 2.074$ t CO_2e

This value is particularly low, which suggests that Civitians have a particularly low carbon footprint. This is partly thanks to the relatively low power grid emission factor. Nonetheless, we need to remember also that we did not account for any of the energy used for commercial and industrial activities, which would likely significantly increase the carbon footprint of the neighborhood residents.

Based on the per capita results, the total carbon footprint of the neighborhood is:

- $I_E = 2.074 \times 10{,}000 = 20{,}740$ t $CO_2e = 20.74$ kt CO_2e

To learn more about urban metabolism, the book *Sustainable Urban Metabolism* by Ferrão and Fernández (2013) is recommended. The book *Understanding Urban Metabolism: A Tool for Urban Planning* by Chrysoulakis, de Castro, and Moors (2014) can also be a valuable source.

Overall, the future of urban metabolism is bright and, in fact, essential if we are to successfully integrate urban infrastructure together. It is therefore now time to go through some of the interdependencies that intricately connect urban infrastructure systems.

10.2 Infrastructure Interdependencies

Think about what you do on a regular day and which infrastructure systems enable you to do these things. Chances are you wake up with an alarm, perhaps from a smartphone, thus integrating the telecom infrastructure with the power grid. Naturally, the power grid requires a lot of water to be able to generate electricity, and it also depends on the transport system to supply it with equipment and fuel. You then take a shower and brush your teeth. The water system requires a lot of energy to collect, treat, and distribute the water, and for this it depends on the power grid. Moreover, whether your city has a water pump that runs on natural gas or whether your water heater uses gas to heat water, the water system also relies on the gas system. Even coffee itself requires both water and electricity. Once all of the water you used goes down the drain, it must be treated by a wastewater treatment plant.

Moreover, once you are finished with your tube of toothpaste and once you are done with all the packaging used to store the food you ate for breakfast, all is thrown in the trash, which is picked up by a garbage truck and taken to a disposal facility, where it may be recycled, incinerated, or landfilled. Even some of the sludge that came from treating the wastewater that you generated might end up in a landfill. The electrical and water/wastewater systems therefore depend on solid waste facilities.

When you go out you may walk/bike, use transit, or drive. Both transit and driving directly require electricity—if only to account for the battery needed to start the engine and the electricity needed to run the pumps at gas stations. Transport is also reliant on the water system, since flooded streets cannot be used. We can go on like this for pages and find that almost everything we do requires a complex chain of actions involving many, if not most, infrastructure systems simultaneously. These chains of actions are typically called *interdependencies*, and we discussed them a little in chapter 1.

Rinaldi, Peerenboom, and Kelly (2001) define four types of interdependencies: physical, cyber, geographic, and logical. Physical means that two systems are interdependent if the state of one changes the state of the other (i.e., two systems are physically connected). Cyber means that two systems are interdependent in terms of "information" in the computational sense (i.e., an error or lack of information transmission can affect a system). Geographic means that a system might feel an impact from a system geographically close to a system that was affected (e.g., flooding affects roads). Logical means that two systems are interdependent by mechanisms that are not physical, cyber, or geographical, such as two systems that are interdependent based on a human decision.

In this section we will not separate the types of interdependencies, but we will focus on seven types of infrastructure systems:

- Transport
- Water
- Utility (gas, district heating and cooling)
- Electricity
- Telecom

- Solid waste
- Buildings

A nonexhaustive infrastructure interdependency matrix between the types is shown in table 10.2.

Technically, electricity, water, and telecom can be considered as utilities, but we will separate them here because they are important in their own way. In the utilities, we include the gas system and district heating and cooling systems—we discussed those at the end of chapter 8.

The way to read the table is as follows. For each row, each cell specifies how an infrastructure system is being influenced by another infrastructure system. For instance, if we look at the second cell from the left at the top, we see that transport is affected by water in part because water conduits/pipes are buried in streets. Another way to look at this, especially when linking these interdependencies with demand, is to think in terms of *infrastructure ecology* (figure 1.2 in chapter 1). Here we are expanding from figure 1.2 to include all interactions across infrastructure systems.

Naturally, the "degree" of interdependency varies across infrastructure systems. One system can be "more" dependent than another system, but defining what "more" means is not obvious. For example, buildings use electricity, and many people are significantly impacted during power outages, which is why electric utilities spend so much time and effort ensuring reliable services.[12] Comparatively, the dependence of the water system on transport infrastructure does not seem as high, and yet if the right chemicals do not get to the water treatment plant in time, an entire water system can get contaminated. Therefore, which of these two examples is "more" dependent on the other? One may be more frequent, but the other may be more severe. A formal and objective way to measure the "degree" to which two systems are interdependent did not exist at the time of this writing.

What we can do here is to go over each infrastructure system individually and discuss its impacts on other infrastructure systems.

This chapter greatly borrows from the articles "Urban Infrastructure Is Not a Tree: Integrating and Decentralizing Urban Infrastructure Systems" (Derrible 2016) and "An Approach to Designing Sustainable Urban Infrastructure" (Derrible 2019).

Let us start with transport.

10.2.1 Transport

Although transport may at first seem to be a little less interdependent than other systems, we really need to take into account all aspects of an infrastructure system. In particular, here, transport is geographically highly interdependent with other infrastructure

Table 10.2

Infrastructure interdependency matrix

	Transport	Water	Utility	Electricity	Telecom	Solid waste	Buildings
Transport		• Underground water conduits in streets • Leaks and runoff leading to street flooding • Overflowing of stormwater channels leading to flooding	• Underground utility lines in streets • Occasional construction and maintenance of infrastructure leading to traffic disruption	• Raw material transport for electricity generation • Electricity needed for electric vehicles, electric rail and bus modes, and for operations (e.g., traffic signals, streetlights)	• Underground telecom lines in streets • Transmission of real-time information	• Bins/cans located on sidewalks, back alleys, roads, etc. • Solid waste collection and transfer vehicles use roads • Land reclamation creates space for transport infrastructure	• Conflict for land • Buildings as location where people go to or depart from
Water	• Restricted right-of-way • Hard-to-reach water infrastructure when located underground • Impermeable surfaces leading to flooding		• Competition for underground space • Gas-run pumps for water distribution • Gas leak can contaminate groundwater wells	• Competition for underground space • Electricity to treat and distribute water (Energy-Water Nexus)	• Competition for underground space • Information to manage water distribution systems (e.g., SCADA) • Increasing reliance on telecom with smart meters	• Contamination of surface water bodies and aquifers with incineration and landfilling • Ability of waste facilities to receive solid waste from treatment plants	• Force water conduits to be below streets • Impermeable surfaces leading to flooding • Buildings as places of water consumption

Utility	• Restricted right-of-way • Hard-to-reach gas lines as well as steam and chilled water pipes when located underground	• Competition for underground space		• Competition for underground space	• Competition for underground space • Information transmission for real-time monitoring	• Predictable generation of methane for natural gas and district heating systems	• Buildings as places of gas consumption • Buildings as places of steam and chilled water consumption for space heating
Electricity	• Restricted right-of-way • Hard-to-reach distribution infrastructure when located underground • Movement of raw material for electricity generation	• Competition for underground space • Thermal power systems require significant amounts of water (Energy-Water Nexus)	• Competition for underground space • Electricity generation from natural gas		• Competition for underground space • Similar to water, increasing reliance on telecom with smart meters	• Predictable generation of electricity • Ability of waste facilities to receive solid waste from power plants (e.g., nuclear waste)	• Partially directs how distribution lines are installed • Buildings as places of electricity consumption • Hazard with tree branches next to buildings
Telecom	• Restricted right-of-way • Hard-to-reach telecom lines when located underground • Many Internet cables are located next to rail tracks	• Competition for underground space • Large amounts of water are needed for cooling, especially in data centers	• Competition for underground space	• Competition for underground space • All telecom devices require electricity • Data centers require a significant amount electricity for cooling		• Ability of waste facilities to receive solid waste from telecom (e.g., wires)	• Buildings as end points where telecom lines are installed

(continued)

Table 10.2 (continued)

	Transport	Water	Utility	Electricity	Telecom	Solid waste	Buildings
Solid waste	• Roads must be accessible for solid waste collection and transport vehicles • Space must be dedicated to solid waste infrastructure	• Some processes require stable supply of water • Heavy rains to impact landfilling activities • Facilities use water	• Natural gas needed to initiate / aid combustion • Heating / cooling solid waste facilities	• Some processes require stable supply of electricity (e.g., eddy current separators) • Facilities use electricity	• Environmental monitoring of landfills • Increasing reliance on telecom (e.g., GPS in garbage trucks)		• Solid waste generated in buildings • Periodic service of solid waste collection • Buildings host solid waste facilities
Buildings	• Conflict for land • Building location (e.g., in real estate)	• Presence / availability of water • Water problems lead to flooding (e.g., basement) • Conflict for land for larger water infrastructure	• Presence / availability of gas • Systems' size for district heating/ cooling • Conflict for land for larger gas infrastructure	• Presence / availability of electricity • Conflict for land for larger transmission lines	• Presence / availability of telecom lines • Buildings are sometimes strategically located to be near a main telecom hub	• Accommodating solid waste generation (e.g., trash chute, dumpster at back) • Ability of waste facilities to receive solid waste from buildings	

Source: Derrible (2016, 2019).

systems. After all, water pipes, gas lines, telecom cables, and power distribution lines are located in streets and back alleys, whether underground, at grade, or elevated. Moreover, solid waste collection and transfer vehicles use streets, back alleys, and other transport infrastructure. Geographically and from a network perspective, the way roads are arranged largely dictates how the other infrastructure systems are planned. Put differently, transport dictates the spatial properties of the other infrastructure systems. This is important because a desirable network configuration is key when it comes to reliability and resilience.

Moreover, because streets host these other infrastructure systems, transport is affected by their performance. Most noticeably, each time a water pipe bursts or a gas line leaks, streets have to be opened up so that the damaged infrastructure can be repaired properly. These events can significantly and negatively affect traffic. Specifically, and as discussed many times, transport infrastructure is vulnerable to flooding, and it therefore depends on the water system. In particular, stormwater sewer systems get backed up during heavy rains. This is particularly serious because transport infrastructure needs to be operational during severe weather events, like heavy rainfalls, so that emergency services can function properly.

In addition, transport infrastructure is used to supply other infrastructure systems. Water treatment requires chemicals that are shipped in. Thermal power plants require fuel to run and produce electricity. Even hydroelectric power plants and solar and wind farms require transport infrastructure to ship in spare parts and maintenance equipment. Similarly, transport infrastructure is used to dispose of the stuff we do not want, from radioactive waste from nuclear power plants to biosolids from wastewater treatment plants.

Naturally, transport is greatly dependent on the power grid. Many transit modes, such as heavy rail, tramway, and streetcars, use electricity to run. Moreover, as electric vehicles become more common, it will not be long before transport depends more heavily on the power grid. This is why we need a safe and reliable power grid, and decentralization can help here. We will say more on this later.

Increasingly, the safe and efficient operation of transport infrastructure relies on real-time data collection, and these data are transmitted by telecom lines. In fact, the telecom infrastructure is becoming increasingly important for all infrastructure systems, and we seldom discuss it beyond matters of cybersecurity. We will learn more about telecom in section 10.2.5.

Transport also depends on solid waste. First, the construction of transport infrastructure generates waste. Moreover, the material used for transport might have been recovered from solid waste. Even the space itself when transport infrastructure is built might have come from solid waste (e.g., land reclamation).

Furthermore, although this is not shown in the matrix, transport depends on itself as well. All gas stations depend on the transport infrastructure to keep them supplied with gasoline.

Finally, we must remember from chapter 7 that land use and transport are highly interdependent. To some extent, buildings and transport infrastructure "compete" for space. After all, there would probably be no transport infrastructure if there were no buildings, and vice versa. These types of interdependencies are important, and they have a direct impact on the livability of a neighborhood, as we learned in chapter 4.

We should note that most of these interdependencies were not taken into account at the design phase. As we better integrate infrastructure systems in the future, we need to ensure that future interdependencies help strengthen a system, as opposed to making it more vulnerable.

We can now go through a similar exercise for water infrastructure.

10.2.2 Water

Here, water includes drinking water and wastewater infrastructure, as well as infrastructure related to the "control" of water, such as floodgates and green infrastructure. Unsurprisingly, water infrastructure depends on all other infrastructure, and it also has an impact on all other infrastructure systems.

We have discussed the energy-water nexus several times already, in particular in chapters 5 and 6. The water system is indeed highly dependent on the power grid. Electricity is needed to pump water to a treatment plant. More importantly, massive amounts of power are required to run the pumps that ensure we get water from the tap at a sufficient pressure.[13] The wastewater treatment process also requires significant amounts of energy, and so do control devices such as floodgates, to operate as desired. From a different perspective, water and electricity also "compete" for space since they are laid in the streets. In North America, power lines tend to be elevated, often in back alleys, but in Europe they tend to be buried underground. In these cases it is important not to bury them next to one another since a water leak can have dramatic impacts on power lines, and the current that flows in the electric wires can affect water pipes that are made of iron.

The water system also depends heavily on transport infrastructure. We have already seen that one of the interdependencies relies on the ability of transport to supply water treatment plants with the chemicals needed for the treatment process. The way streets are designed also has a direct impact on the amount of runoff that ends up in the sewer systems. After all, gutters are small open channels. Streets tend to be curved across their width (a little like a frown), with the top part at the median of the street and the sides acting as the gutters. Poor street designs can cause flooding. Moreover, green infrastructure and low-impact development strategies, like rain gardens, are part of both the transport and the water infrastructure. This is why some cities, like Copenhagen, are completely redesigning some of their streets so that roads are designed not only to facilitate traffic but also to carry runoff to strategically located green areas that can handle them (i.e., retention basins).

In addition, and at a more physical level, water infrastructure is heavily dependent on transport because pipes are buried in streets—that is, if a street is damaged, a water conduit can be damaged too.

Although interdependencies between the utilities and water are a little less obvious, they still exist. First off, many water pumps still ran on natural gas at the time of this writing. This is actually not a bad thing in regions where the power grid has a high emission factor. Gas lines also compete for space with water conduits, as we have discussed already. Gas leaks can also contaminate groundwater if the leaks are severe. On the positive side, wastewater treatment produces methane, which can be used by the treatment plant itself or sold to a different entity for heating or to produce electricity.

Most water systems are monitored constantly, notably through Supervisory Control and Data Acquisition (SCADA) systems. These SCADA systems help detect leaks in a water distribution system by monitoring water consumption at various points across a city. The telecom infrastructure is also used to collect many different types of data in real time, from the stage (i.e., depth) of a river and the level of water behind a floodgate to the precipitation at a weather station, for example. Importantly for the future, smart meters are being deployed massively around the world. Similar to the smart grid, smart meters are being used to monitor water consumption in real time. This will be increasingly relevant as we move to a more decentralized infrastructure.

Naturally, water infrastructure is heavily reliant on buildings. Water is treated and consumed in buildings. Wastewater is also treated in buildings. Beyond these obvious aspects, buildings also represent impermeable surfaces, therefore contributing to surface runoff, linking back to geometric road design.

Finally, water also depends on solid waste. First of all, many of the processes needed to treat water and wastewater generate solid waste that needs to be collected and treated. Moreover, historically, solid waste used to be dumped in bodies of water.[14] In most places (at least in high-income countries), large volumes of solid waste are disposed in landfills that produce leachate, as we saw in chapter 9, which can easily contaminate surface water bodies and groundwater aquifers if they are not properly collected.

Considering the extent of water infrastructure, we can think of many more interdependencies, but this should be enough for us. Next is the utility infrastructure.

10.2.3 Utility

In this book, utility includes gas lines as well as district heating and cooling systems as defined in chapter 8, since they mainly deal with space conditioning. Like water, vast gas networks are present below nearly every street of a city. Gas systems therefore rely heavily on transport infrastructure. Damages to underground infrastructure may also have an impact on gas lines.

Surprisingly, gas systems rely on electricity. Although gas pumps use gas as a source of power, compressors used to compress the gas after the initial extraction run on electricity. A short report by Judson (2013) offers a great resource to learn more about the topic.

Moreover, and similar to water, gas networks are monitored by SCADA systems. Gas systems therefore rely on the telecom infrastructure. Here again, interdependencies between the gas and the telecom infrastructure systems are likely to increase in the future. A little like the smart grid, gas systems will have to adapt. At the same time, the gas system is the only infrastructure that can be completely substituted by another infrastructure system. Indeed, gas is primarily used for space heating, water heating, and cooking, all of which can be substituted with electricity. There is therefore quite a bit of uncertainty about how the gas system will evolve in the future. In fact, as natural gas is a fossil fuel—and thus available in finite quantities—the gas system will likely either disappear or adopt a much smaller role in the future (since gas is also produced by anaerobic digestion plants and sanitary landfills).

In complete contrast, district heating and cooling systems are likely to become more important in the future. These systems essentially benefit from economies of scale. On colder days, large boilers (that can use gas) are used to create steam, which is then distributed through a network of pipes. On warmer days, a separate network of pipes is used to distribute cold water to the system. District heating and cooling systems tend to be relatively small in size (about the size of a neighborhood[15]). Moreover, because they require an agreement between all the tenants, they exist mainly in large complexes, campuses, and central business districts, but they could become more common in residential neighborhoods as well. The extent of their success will depend partly on the price of gas and electricity, since there needs to exist a measurable economic incentive to build the system in the first place. An alternative, and perhaps preferable, option will be to switch to air-source or ground-source heat pumps for residential buildings.

10.2.4 Electricity

As discussed extensively in chapter 5, nearly every activity we engage in requires electricity. The list of interdependencies is long, and we have already listed several of them.

This is actually quite a big concern. As we saw, the current structure of the power grid cannot be maintained. Heavily centralized power plants that are supposed to be able to accommodate perfectly a certain demand in real time do not make sense from a resilience perspective. So instead of adding new interdependencies to the power grid, we should really step back and spend some efforts into making it more flexible—notably, by decentralizing some of the power generation and, as importantly, by finding ways to effectively store electricity.[16] Only then should we consider increasing our demand for electricity.

At the moment, electricity relies on transport for the supply of fuel used to generate electricity as well as to bring spare parts and equipment to power plants, substations, transmission and distribution lines, and so on. Natural gas–fired power plants depend on the gas system to supply them with a constant and reliable flow of gas, which includes anaerobic digestion plants and sanitary landfills. Electricity also depends on Waste-to-Energy processing plants and, more generally, biomass plants that burn solid waste to produce electricity. Electricity also relies on (lots of) water for cooling in thermal power plants. Solar thermal plants require large amounts of water to be able to produce electricity as well. Even hydroelectric power plants rely heavily on precipitation patterns and surface runoff to ensure the level of water at the dam does not decrease significantly. In addition, and as part of the smart grid, electricity is increasingly dependent on telecom infrastructure.

Naturally, power lines are also located along streets, typically in back alleys, elevated on poles. They are therefore "competing" in terms of space with the other infrastructure systems. Moreover, because they are elevated they are particularly susceptible to severe weather events; wind can knock off a tree branch and create a significant power outage. In some cities, electricity transmission and distribution lines are located along rail corridors.[17]

Power plants also produce solid waste that needs to be managed. The most infamous type of waste is, arguably, nuclear waste, but this is not the only type of solid waste generated by the electrical system.

As we found out toward the end of chapter 5, the electrical grid will have to change significantly in the future. There is no doubt that the smart grid and microgrids will have major roles to play in this endeavor. They will probably be key to ensuring a safe and reliable supply of electricity. The goal will be to leverage these advances in the power grid to make urban infrastructure in general more reliable, and the telecom infrastructure will play a vital part.

10.2.5 Telecom

We have not discussed the telecom infrastructure in this book. Although telecom has become the nervous system of our society, transmitting information from one side of the globe to the other side of the globe in less than one second, telecom is not typically planned by urban engineers. This task is left to communications and computer engineers.

This should really change in the future, however. After all, fiber optic cables transmit information—small cables transmit less information, and large cables can transmit more information, just like water, electricity, and gas. And again just like water, electricity, and gas, most telecom lines are buried in the streets. They are therefore "competing" for space with one another and are dependent on the transport infrastructure.

Moreover, similar scientific fundamental principles exist to calculate how many bits of data can be transmitted in a cable—the *bit* is the unit for information. In urban engineering, we could actually learn from the structure of the Internet, which is famous for being impressively reliable—even if a significant proportion of the system fails, the network will likely work almost as well.

The telecom system does not only consist of the fiber optic cables and telephone lines that run underneath the street (often in old Western Union conduits) and on ocean floors. It also includes data centers and Internet exchange points. Exchange points are physical locations where cables are connected, and service providers exchange Internet traffic. The telecom infrastructure therefore depends on the buildings where those Internet exchange points are located.

Data centers tend to be massive buildings with servers that store enormous amounts of information (a little like a gigantic hard drive). These servers run constantly, producing excessive heat that requires excessive cooling. Data centers therefore rely heavily on water. Google and Facebook strategically decided to place their main data centers in Oregon since the yearly temperature tends to be mild, and they have easy access to large sources of water, but it was Ashburn, Virginia (close to Washington, DC), that contained the highest number of data centers at the time of this writing. Most likely, these data centers currently store your e-mails, your online shopping lists, and a copy of this book.

The popular book *Tubes: A Journey to the Center of the Internet* by Blum (2012) offers a great introduction to the world of "physical" telecommunications.[18] For example, in the book we learn about an interesting interdependency between the telecom and transport infrastructure. Can you guess where the fiber optic cables that connect cities

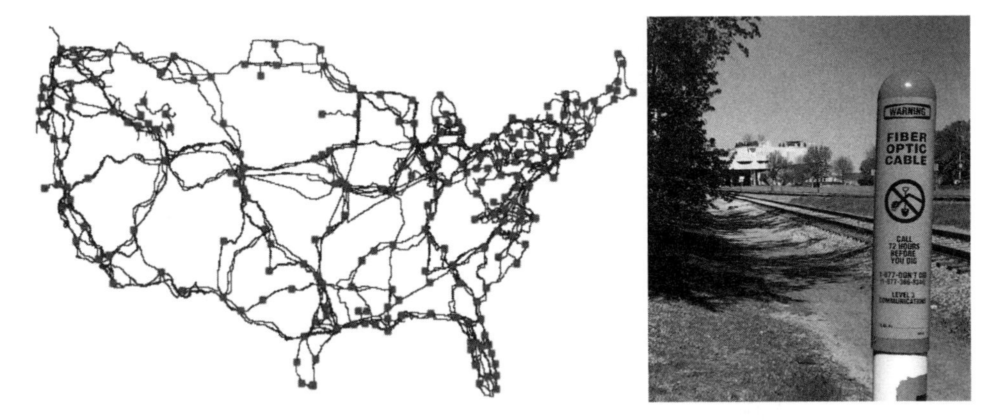

Figure 10.6
Left, map of the physical Internet in the continental United States.
Source: Durairajan et al. (2013, 2015).
Right, warning post close to railroad tracks in Madison, Wisconsin.

with one another are located in the United States? Most people by far think they are located next to phone lines or even electric transmission lines, but fiber optic cables actually tend to be located next to rail lines, thus creating a geographical interdependency with transport. This is because rail infrastructure is linear, with an exclusive right-of-way that is perfect for telecom lines. Figure 10.6 shows a map of the physical Internet in the continental United States (*left*), and it shares many features with a map of the railways in the United States. Figure 10.6 (*right*) also shows a post in Madison, Wisconsin, with a warning sign not to dig close to the railroad tracks since fiber optic cables are located there.[19]

Blum (2012) also tells us how every continent is connected with fiber optic cables laid on the ocean floors. In 2018, even Saint Pierre and Miquelon—where I grew up, as discussed in chapter 1—was hooked up physically to the Internet. Figure 10.7 shows where the cable arrives in Saint Pierre. The sight is fairly unimpressive despite its vital importance.

Furthermore, in 2013 the city of Chicago opted to locate fiber optic cables in sewer systems to connect an area of the city to the Internet (City of Chicago 2013). This is possible because sewer systems and telecom lines share similar physical properties, as opposed to sharing operational properties. This physical aspect of infrastructure has been greatly overlooked when it comes to integration.

The telecom system is highly interdependent with all other infrastructure systems. Telecom depends on the transport infrastructure for space, on electricity (and thus

Figure 10.7
Arriving point of fiber optic cable in Saint Pierre, Saint Pierre and Miquelon.

water and gas) for operations (i.e., to "pump" bits in wires), on buildings that "consume" information and house Internet exchange points, and on solid waste management, which handles all the solid waste generated (e.g., discarded wires).

The telecom infrastructure is also quickly becoming vital for other infrastructure to run. New interdependencies are being created every day. Here again, it is paramount to design these interdependencies so that they strengthen urban infrastructure in general, as opposed to making it more vulnerable.

10.2.6 Solid Waste

Virtually every activity, directly or indirectly, generates some kind of solid waste in one form or another, which is partly why the average American generated about 62 kg of solid waste per month in 2015. Therefore, all infrastructure systems depend directly on solid waste management. This is a two-way relationship, however, and solid waste also depends on all infrastructure systems.

Remembering the six functional elements of solid waste management, we are reminded that solid waste is first and foremost often generated in buildings, and some space in buildings must therefore be dedicated to solid waste, if only with the presence of one or multiple trash bins. Moreover, waste disposal facilities are themselves physical buildings, and as such, they require water, electricity, gas, telecom service, and sometimes district heating and cooling systems to be able to operate properly.

In addition, many of the separation and transformation processes discussed in chapter 9 require energy. For example, conveyor belts require electricity to function, whether used to screen, compact, crush, or grind solid waste or to separate solid waste—for instance, with eddy current separators. Waste-to-Energy processing plants often require gas to start the combustion process. Some separation and transformation processes also require water, if only to clean items that might be eligible for reuse and recycling.

Furthermore, garbage trucks are increasingly being outfitted with GPS devices to monitor the solid waste collection operation in real time. Sanitary landfills, Waste-to-Energy processing plants, and other solid waste facilities are also equipped with environmental monitors to ensure no public health issues arise. All of these sensors depend directly on the telecom infrastructure and on the electricity infrastructure. While some of this electricity may be generated on-site (e.g., by burning the methane produced by sanitary landfills), such facilities are generally connected to the power grid and to the Internet.

Solid waste naturally depends on the transport infrastructure as well, which serves as the physical backbone for garbage trucks that collect trash from buildings and take it to waste disposal facilities. Transport infrastructure is also used to transport solid waste after it has been transformed—for example, from a disposal facility to a landfill. Effectively managing the transport infrastructure is critical for solid waste, and this is why Colonel Waring wanted snowplowing to be in the purview of the Department of Street-Cleaning in New York City when he became commissioner in the late 1800s. Solid waste, including nuclear waste, is also transported by rail in many places.

Considering the amount of solid waste generated in the world, solid waste management is unlikely to become less interdependent with other infrastructure systems in the future.

The last infrastructure system that we will study is buildings.

10.2.7 Buildings

Buildings essentially act as the locations where the services that the other infrastructure systems provide are consumed. Put differently, they are end points, or demand points. Electricity, water, and gas are consumed in buildings, and solid waste is generated in buildings. Moreover, even though transport is technically "consumed"—for example, on the roads and rails—a trip is typically made between two buildings. By their very nature, buildings are therefore dependent on all infrastructure systems.

In parallel, all infrastructure systems also involve buildings in one form or another. Power plants, water treatment plants, wastewater treatment plants, utility plants, Internet exchange points, waste processing facilities, and so on are physically located in buildings themselves.

Beyond these aspects, buildings "compete" with transport for space. This is naturally linked to the transport and land use relationship that we discussed in chapter 7. In fact, we should also remember the paradox of parking that we discussed in chapter 7. It may be desirable to build more parking spaces in a congested area, but if we build too many, there will no longer be any buildings to visit in the first place.

We should also remember that buildings create impermeable surfaces. This means that sewer systems may not be able to handle heavy surface runoff, which not only floods streets but basements as well.

In terms of real estate, buildings compete to some extent with all other infrastructure systems. Buildings may compete with power transmission lines, aqueducts, and rail tracks, for example. They also compete with power substations, rail yards, and water reservoirs (often located in parks).

Because district heating and cooling systems are limited in size (to minimize thermodynamic losses), they can only service certain buildings within a given radius. Again, this is why they tend to exist mainly for campuses, hospitals, central business districts, and other large complexes.

Buildings also depend heavily on solid waste management to properly and effectively rid them of their solid waste in a timely manner—for example, in order to avoid attracting pests.

Although many more interdependencies exist between infrastructure systems, this section has offered an overview of some of the main interdependencies. The next step is to start better integrating infrastructure, which is the topic of the next section.

10.3 Integrating and Decentralizing Urban Infrastructure Systems

Seeing as how all urban infrastructure systems are intricately and intrinsically interconnected with one another, it would make sense to control how they are integrated. This is generally not the case, and it will probably not be the case until engineers start to look beyond their disciplinary boundaries and design integrated systems. As mentioned in chapter 1, this is the strongest message this book has to give.

In this section we will learn about a simple conceptual matrix to think about how we can better integrate and decentralize infrastructure systems. But before we do that, we should discuss the general way infrastructure systems provide their services. Power

Example 10.5

New Pedestrian Bridge to Design Interdependencies in Civitas

The Civitas government aspires to build a new pedestrian bridge over the river Vita, and some people are wondering whether the bridge could be used for different purposes. (1) Elaborate on which other infrastructure systems could be integrated with the bridge.

Solution

(1) Considering the pedestrian bridge is a transport infrastructure, the remaining infrastructure systems are water, utility, electricity, telecom, solid waste, and building (that we will not consider here).

As the number of HDD in Civitas is fairly high, some days can be below freezing, which is detrimental for water pipes that are normally buried below the frost line (unless they are insulated). In terms of surface runoff, the bridge is not likely to create any in significant amounts. The bridge can still be integrated with the water system by placing sensors under the bridge to monitor the river stage and discharge (i.e., depth and flow) and possibly water quality.

To communicate these water data, the telecom systems should be integrated with the bridge. In fact, the bridge can be used as an additional connecting point to link the two shores of Civitas, and telecom companies can use it to expand their system and make it more robust. Telecom cables can therefore be placed under the bridge.

Similar to telecom, the bridge could be used to transfer electricity between the two shores. Depending on the precautions taken (to minimize any chances of electrocution), small distribution lines could be placed under the bridge or on one of the sides, which may improve the robustness of the system.

In terms of utility, it may not be a good idea to use the bridge for natural gas, in case of a leak. However, depending on how much energy it requires, a small *snow-melting* system may be installed. A snow-melting system places electric wires or small pipes that distribute warm water below a surface (typically a sidewalk, but here a bridge) to melt the snow in the winter. This way, salt is not needed, which would corrode the bridge and possibly pollute the water flowing under the bridge.

While the pedestrian bridge should not be used to transport solid waste, trash cans may be added on both ends of the bridge and maybe even on the bridge. Moreover, the bridge itself could be made from reused, recycled, or recovered material.

Many more purposes can be imagined, but we will stop here. The goal is to think beyond the first and obvious purpose for the infrastructure being built. We will talk more about this in the conclusion (chapter 12), in which we will be introduced to a four-step urban infrastructure design process.

plants and water treatment plants tend to be centralized, but not all infrastructure systems are, and there is a simple and effective way to look at the design patterns of infrastructure systems.

10.3.1 The Design Patterns of Infrastructure

Whenever two entities are connected, it is a good idea to look at how they are connected. This is fairly well known in the systems analysis realm, and the different types of connections are usually referred to as *design patterns*. Quite simply, there are four types of design patterns, as illustrated in figure 10.8.

Specifically, what we want to look at here instead is how an infrastructure system provides its service. In other words:

- How is transport provided?
- How is water supplied?
- How is wastewater collected?
- How is gas supplied?
- How is electricity supplied?
- How is telecom provided?
- How is solid waste disposed of?
- How are buildings provided?

Most infrastructure systems at the moment follow a one-to-many type of design pattern. A service is generated in a centralized location and supplied to many consumers. This is the case for electricity, water, and natural gas. In contrast, wastewater treatment plants and solid waste processing plants have a many-to-one type of design pattern.[20] This is because wastewater treatment and solid waste processing plants do not "supply"

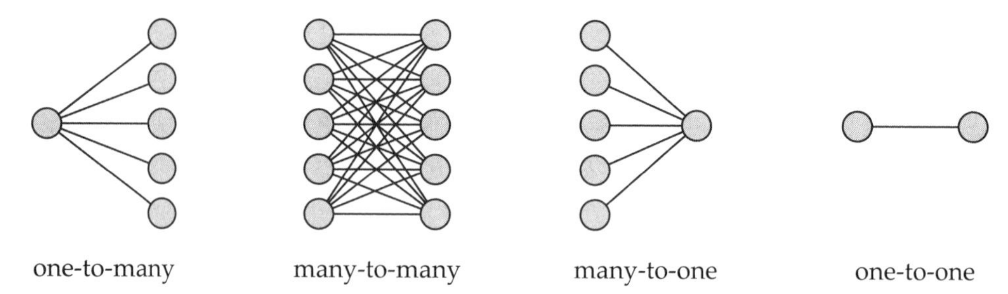

Figure 10.8
Four types of urban infrastructure design patterns.

something but instead "collect" something. Inherently, both one-to-many and many-to-one design patterns rely heavily on centralized infrastructure.

One the one hand, this may be desirable because centralized facilities benefit from increasing economies of scale—that is, as a facility gets bigger, a process may be easier and cheaper to do. On the other hand, these types of design patterns are vulnerable because the failure of a centralized facility can negatively affect the entire system. Moreover, when we expand our focus a little, we may find that these increasing economies of scale are actually not that significant after all. For example, flooding is a major concern in most cities in the world, costing enormous sums of money in congestion and repairs. If stormwater runoff ended up in rain gardens and bioswales as opposed to the sewer system, these costs would be avoided. The benefits of having a larger treatment plant may therefore be offset by the cost of flooding. A similar observation can be made with electricity. Centralizing power production may make it easier to generate electricity, but each blackout has a cost that may have been avoided with a more decentralized power grid—for example, with solar panels, wind turbines, and microturbines.

At some point in the second half of the twentieth century, we reached the limit on how many more benefits we could gain from building larger infrastructure systems. For example, Chicago hosts the Jardine Water Purification Plant, which started operating in 1968 and is reportedly the largest water treatment plant in the world. Have we had no opportunity to build a bigger treatment plant since? Of course we have, but we know that relying on one centralized facility is no longer the preferred option. Cities now prefer to build smaller systems but have more of them. We can still do better, however, which is why we need more urban engineers.

Transport is different since transport infrastructure is used to link buildings. It therefore has more of a many-to-many type of design pattern. Even if transit riders gather together at semicentral locations to take the train and hop on a bus, the general design pattern remains many-to-many. This is different from one-to-many and many-to-one design patterns. For starters, except for transit there is no "central" command. In contrast, power is typically generated by one or few companies. Similarly, water, wastewater, and solid waste are treated by one or few agencies. In transport, every trip is made by a different person. To some extent, telecom is the same way. Whenever we are connected to the Internet, we are both sending and receiving information. Every computer can be seen as a small data center. Moreover, even if large data centers are centrally located, because the scale of the Internet is so large (i.e., the entire world), the design pattern remains many-to-many. This is in fact the reason why the Internet network is so reliable and resilient. If the information cannot go one way, it will simply go another way.

One-to-one design patterns are not as frequent in infrastructure systems because buildings (or more accurately, people in buildings) tend to be the consumers of whatever service is being provided. Thermal power plants offer a good example since they require large amounts of water to produce electricity. This water is most often pumped from a nearby river, hence the one-to-one design pattern. Waste-to-Energy processing plants and sanitary landfills offer another good example—for example, the solid waste processed by a disposal facility tends to serve only one Waste-to-Energy processing plant (that produces heat / electricity) or only one sanitary landfill (that can burn onsite the methane released to produce heat / electricity). To some extent, buildings have a one-to-one design pattern as well in the context of their main service: offering space to people. Even if some people live and work in high rises, there is no one centralized building where everyone lives and works.

Finally, we should highlight the case of district heating and cooling systems that combine both a one-to-many and a many-to-one design pattern. Steam or cooled water is distributed through pipes to buildings from a central location. Heat is then exchanged with the building, and the fluid goes back to the central location to be heated or cooled again. The scale of these systems is different, however. As mentioned several times, district heating and cooling systems are used in large complexes as opposed to entire cities. Although the failure of a system may have important negative impacts, the damages are relatively limited. This is also why microgrids are desirable. If one fails, a single neighborhood might be impacted only as opposed to an entire city or an entire region. Neighborhoods affected by a failed microgrid can also rely on the microgrids of their neighbors, making power generation and distribution particularly resilient.

Taking into account these different design patterns will be important in future cities. To help us on the way, we can use a simple matrix, as we are about to see.

10.3.2 Integration-Decentralization Matrix

Future urban infrastructure systems might become more integrated in multiple ways. One of the other dimensions we might want to consider is whether this integration happens only in one location—that is, in a centralized way—or whether it is more decentralized. To make it simple, we can view it as an integration-decentralization (ID) matrix. The matrix is shown in figure 10.9. Illustrations to showcase each of the four quadrants are shown in figure 10.10.

In the bottom-left quadrant, we have the current paradigm. Even though they are interdependent, infrastructure systems are seldom planned and designed as integrated systems. They are therefore seen and planned more like tree networks, going back to the work of Christopher Alexander in chapter 4. Power plants are built to produce

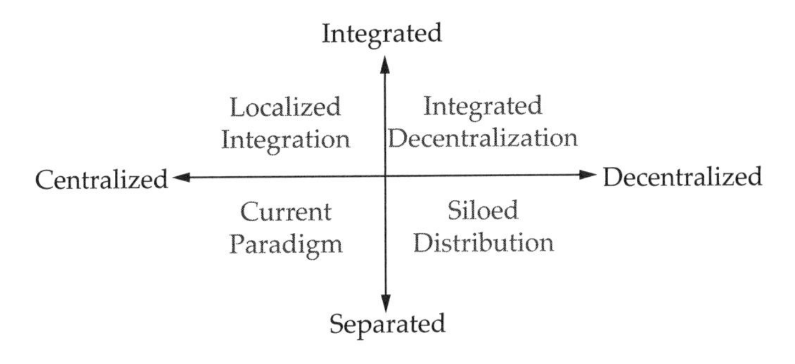

Figure 10.9
ID matrix. Adapted from Derrible (2016).

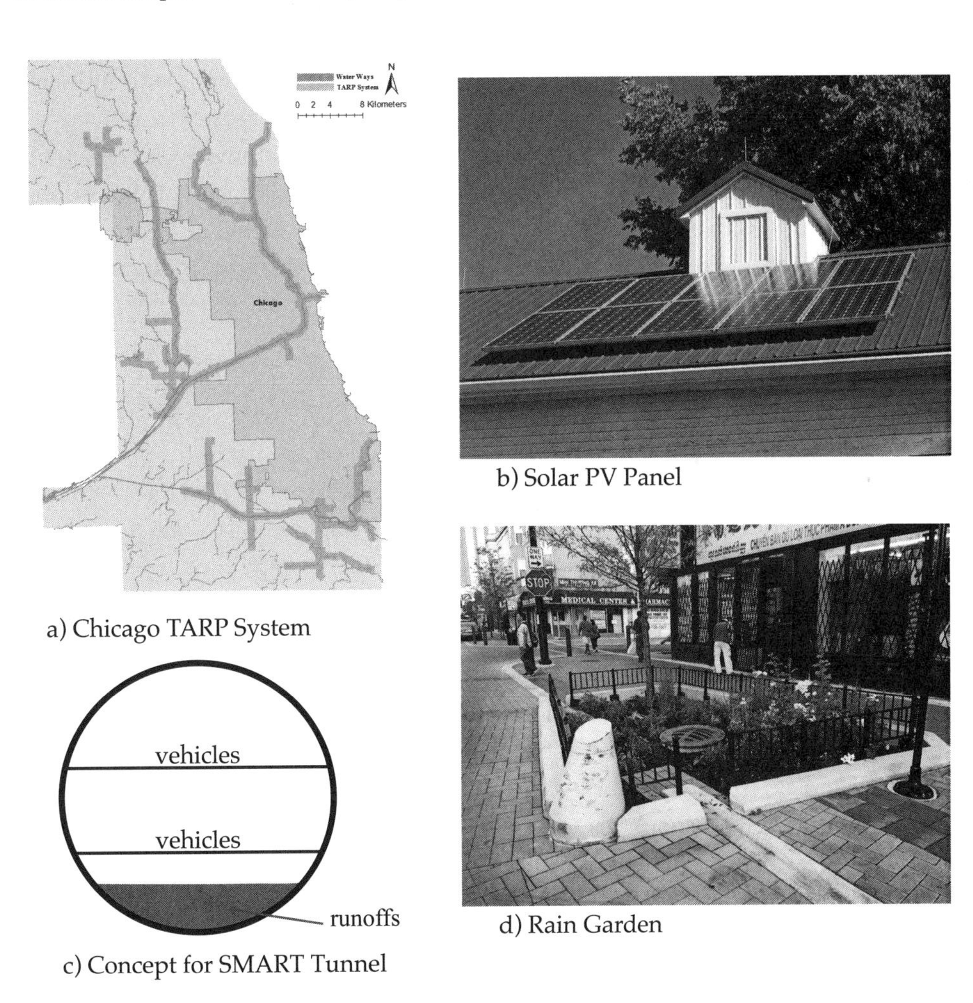

Figure 10.10
Illustrations for the ID matrix. Adapted from Derrible (2016).

electricity only. They do not try to think about the specific needs of the water system or the transport system, for instance. In microgrids, this would have to change since the demand of a specific neighborhood fluctuates a lot more.

Other good examples include the large underground reservoirs that some cities have built to handle excess stormwater during heavy rain events. Chicago has one called the Chicago Tunnel And Reservoir Plan (TARP), shown in figure 10.10a. Whenever the wastewater treatment plant cannot handle the amount of sewage arriving or the sewer system cannot accommodate the large intake flows, some of sewage is stored in massive underground tunnels and old quarries to prevent sewer backups and minimize the number of sewer overflows (i.e., dumping of raw sewage in the river). Once the rain stops, the sewage is pumped back up and sent to the wastewater treatment plant. Although there is logic to the process, we should remember that the sewer system is full because of impermeable surfaces. The right thing to do is not to build massive underground tunnels but to make surfaces more permeable. This is a classic case of diminishing marginal returns, going back to chapter 1. In addition, these tunnels are located underneath rivers. Rivers were selected partly because the untreated sewage is dumped in rivers. They were also selected because the authority in charge also handles the waterways, making it easier to stay in the silo.

In the bottom-right quadrant, we have siloed distribution. Infrastructure systems are still separated, but they start to be decentralized. The best examples of this are solar panels, as shown in figure 10.10b. Solar panels can easily be installed on roofs, thus producing electricity in a decentralized fashion, but they are not integrated with any other infrastructure systems. Another example could be building-scale water treatment facilities. If every building could treat its own water, there would no longer be a need to maintain a minimum pressure in water distribution systems since any contaminated water would simply be purified once it gets to the building. In this case, water infrastructure would be decentralized but still separated.

In the top-left quadrant, we have localized integration. Here, infrastructure systems are still centralized, but they start to be integrated. Figure 10.10c shows quite an interesting example. Like many cities in the tropics, Kuala Lumpur, the Malaysian capital, has a rainy season (monsoon) that can dramatically flood the city. At the same time, the city was suffering from serious congestion problems. The government therefore decided to build an underground tunnel to handle traffic, but the tunnel transforms itself into a massive open channel during heavy rain events. This infrastructure project is aptly called the Stormwater Management and Road Tunnel (SMART). Waste-to-Energy processing plants and sanitary landfills also offer good examples since they integrate solid waste management with the electricity and/or gas system in a localized manner.

Another good example of localized integration involves district cooling. We already saw this example in chapter 8, but we can repeat it here. In Toronto (Canada), the water at the bottom of Lake Ontario has a temperature of about 4°C (39°F). In the summer, the water is used for district cooling. Quite simply, the water is pumped from the lake and goes through a heat exchanger to cool down water that is used to cool down buildings. After the heat exchanger, the pumped water is simply sent to the water treatment plant to become drinking water. In this case, district cooling and water treatment are integrated.

In the top-right quadrant, we have integrated decentralization. Here, infrastructure systems are integrated in a decentralized way. The best example of this type of integration is green infrastructure. Figure 10.10d shows the example of a rain garden on Argyle Street in Chicago. Can you see the permeable pavers that can store runoff before water reaches the rain garden? Can you also see how the stormwater is naturally channeled from the street to the rain garden? Finally, if both the permeable pavers and the rain garden are full, excess water is then sent to the sewer system—notice that the sewer inlet is slightly elevated, which is better than the example we saw in figure 6.23 in chapter 6. Rain gardens can be installed in every street in a city, and they contribute both to water and transport infrastructure. Rain gardens offer a particularly fitting example for our purpose because they are low tech. Preferable infrastructure designs do not have to be high tech. In fact, high tech may be a hindrance since it might make it more complicated to implement. A good example of that may be the solar roadways that we already discussed a little in chapter 2. The rationale there is to put solar panels on roads. Although this would also offer an integrated-decentralized solution, how effective could it be? After all, road traffic can easily damage solar panels and block exposure to the sun; they do not offer the best place to install solar panels (certainly not better than roofs and empty fields for example). Moreover, they require the addition of a lot of "complexity" in a system, but they are unlikely to produce increasing marginal returns. Even though they are not integrated with other infrastructure systems, microgrids offer many more benefits than solar roadways. In contrast, installing solar panels and small wind turbines on streetlights and traffic signals makes a lot more sense as an integrated-decentralized solution. Not only do the streetlights and traffic signals require less electricity to run (contributing to our first sustainability principle)—and they may even produce some electricity—but they also become less dependent on the grid (i.e., back to our interdependencies).

William Gibson has a quote that applies well here: "The future is already here, it's just not very evenly distributed." During a trip to Okinawa (Japan) in 2015, I saw the house shown in figure 10.11, and it struck me as a house from the future.

Figure 10.11
A house in Okinawa, Japan.

At first sight, the house looks absolutely normal. If you look a little closer, however, you will see solar panels on the roof of the house. The house therefore produces its own electricity, and because Okinawa enjoys warm weather, we can speculate that by storing excess energy in a battery the owners can self-sustain their electricity consumption. Right above the solar panels, we can also see a large metallic drum. This is a solar water heater. Water is stored in the tank during the day, heating up the water and thus greatly reducing the need for a conventional water heater. We may also speculate that if a conventional water heater is used in addition, it is electric and therefore also powered by the solar panels. Above the water tank, we can see power distribution lines. The goal for this house is not necessarily to be completely self-sufficient but perhaps to be able to generate a significant proportion of the electricity it consumes. The house can still be connected to the grid, however. In fact, it really should be connected to the grid. This way any excess electricity generated during the day can be sold back to the grid, and power is then drawn from the grid at night. Moreover, if the solar panels fail or need to be disconnected, the house can be fully supplied by the grid. At the bottom of the picture, we can also see a car in the garage. The people who live in the house

Example 10.6
Decentralizing Water Treatment in Civitas

Civitas currently has one water treatment plant and there is a growing concern to make the system more resilient. (1) Elaborate on strategies that can improve the resilience of the Civitas water treatment system.

Solution

(1) An obvious solution would be to build a second water treatment plant somewhere else in the city-state. This would be a typical siloed distribution strategy. If this is possible, then it could be a desirable option, and the plant could be equipped with solar panels (giving us an integrated and decentralized strategy). A limiting factor may be where to find the freshwater for the new treatment plant. In Civitas, the river Vita is currently used, and the water is taken upstream of the river. A second water treatment plant could be built either even more upstream or downstream but preferably upstream of the location where the wastewater treatment discharges its treated wastewater (which is still fine, but more energy would have to be put into the water treatment process).

Another concern may be the elevation of the water treatment plant. Generally, it is preferable to have a water treatment plant located on a higher elevation to avoid excessive costs for water pumping.

If a second water treatment plant cannot be built, water reservoirs could be installed in several locations around the city, again equipped with solar panels or other another type of infrastructure (e.g., Internet exchange point or blue / green roofs) to contribute to an integrated and decentralized strategy. These reservoirs would improve the "buffer" capacity of water system. In addition, water quality could be monitored at these reservoirs and a small secondary water treatment process could be implemented, which would further improve the resilience of the system.

Many other solutions can also be imagined that would both improve the resilience of the water system and help improve current interdependencies with other infrastructure systems.

therefore own at least one car (that could also be electric), but they may take the bus that drives by every day. We may also speculate that the house has an air-source heat pump at the back or has a ground-source heat pump to provide heat in the winter and cool air in the summer. Finally, we could imagine that the solid waste generated in the house is reused or recycled in the house (e.g., to make compost for a small garden) or that it is recycled in a larger facility, and the little solid waste that remains is used by a Waste-to-Energy processing plant to generate the electricity that runs in the transmission lines above the house.

We often assume that our GHG emissions are so high because we are missing a key technology that will solve all of our problems. But by and large, these technologies already exist. It is more a matter of being able to harness their potential. Inevitably, this requires a change in the standard practice. This is not to say that any new technology will not be useful. But we must realize that with what we currently have at our disposition, we can substantially cut our carbon footprint.

What is certain, however, is that we need to think more in terms of systems in order to integrate urban infrastructure better. The solution might have been given a long time ago by Minett (1975), who wrote an article in response to Christopher Alexander's article "A City Is Not a Tree." In his article, Minett (1975) argues for a *supraplanning* agency that is not responsible for the planning of a city but for the coordination and the communication of the various departments within a city. In other words, he argues that the departments of transport should work more with the departments of water management and the electricity and gas utility companies. Put simply, he argues for a system's view of cities, and he has that quote that is quite telling: "It is the links that create the system." In other words, it is not the power plants, water treatment plants, roads, data centers, gas pipes, and so on that create the system but how they are connected.

10.4 Conclusion

We have a long way to go before infrastructure systems become better integrated, but we are starting to have the right elements to pursue this goal. In this chapter we covered many aspects of cities and infrastructure to help us reach this goal. First, we looked at the daily demand curves for electricity, water, and transport that we had seen in chapters 5, 6, and 7. Clearly, we saw an obvious parallel between the three curves.

To study urban infrastructure together, we first learned about urban metabolism. Urban metabolism is about measuring and accounting for all the materials and the energy that enter, exit, or are processed within a city. The main method to measure urban metabolism at the time of this writing was to essentially perform a large-scale mass and energy balance—notably, by applying a method from the field of industrial ecology called Material and Energy Flow Analysis (MEFA). We saw several equations that can be used to measure the stocks of materials in a city and learned about embodied energy—which is captured in a concept called the *circular economy*. We then quickly covered the flows of food in a city before moving on to energy. There, we saw an equation that takes into account and further details the energy from buildings, transport, industry, construction, water pumping, and waste. After that, we looked at flows of water, paying particular attention to those related to our demand for water for everyday use (and accounting for seasons). None of the equations that we developed were fundamental laws of physics, however. Instead, all were derived from simple mass and energy balance principles. As more and more data are becoming available to us, these equations may not be as relevant to measure the urban metabolism of a neighborhood or of a city. Nonetheless, this also means that we should be better able to calculate the urban metabolism of a neighborhood or of a city, which is significant and a definite step forward to being able to better integrate infrastructure.

In the second section of this chapter, we discussed some of the current interdependencies between infrastructure systems. For this, we used a relatively large infrastructure interdependency matrix that consisted of a seven-by-seven matrix for transport, water, utility, electricity, telecom, solid waste, and building infrastructure. From this exercise we learned that all infrastructure systems are deeply and intricately interconnected and interdependent. Nonetheless, these interdependencies were not purposefully planned for at the design stage. By interdependencies, we did not limit ourselves to operating characteristics—that is, electricity needed to treat water. Markedly, we discussed physical and geographical interdependencies, such as streets and back alleys, that are used not only for transport purposes but by the water, utility, electricity, and telecom infrastructure. This is important to realize because some of these physical and geographical interdependencies offer avenues for higher integration. For example, fiber optic cables that connect cities with one another are often located next to railroad tracks. Moreover, in Chicago, fiber optic cables were laid out in sewers to provide Internet service to neighborhoods previously disconnected from the virtual world. These types of solutions are actually not new. In Paris, this has been the case since the 1860s. Baron Haussmann (whom we met in chapter 4) built large tunnels underneath roads with canals to handle sewers, but these tunnels also have large water mains, and they now carry fiber optic cables. This was possible because Baron Haussmann was responsible for all infrastructure systems, not just a single one.[21]

In the third section, we discussed how infrastructure systems can not only be further and better integrated but also more decentralized. For this, we first learned about four typical design patterns that are used in systems analysis: one-to-many, many-to-many, many-to-one, and one-to-one. We paid particular attention to the one-to-many and many-to-one design patterns that rely on centralized infrastructure and make urban infrastructure quite vulnerable if they fail. The logical response was to develop an ID matrix that can help us come up with new designs. The ID matrix had four quadrants. The current paradigm quadrant has separated and centralized infrastructure, and it was illustrated with the Chicago TARP system. The siloed distribution quadrant still has separated infrastructure, but the infrastructure is also decentralized, which was illustrated with solar panels. The localized integration quadrant is more centralized, but infrastructure systems start to be integrated, and it was illustrated with the SMART project in Kuala Lumpur, where a tunnel is used to handle both traffic and stormwater. Finally, the integrated decentralized quadrant integrates infrastructure systems but in a decentralized way, and this was illustrated by low-impact development strategies (including rain gardens).

While developing new designs that are more integrated and decentralized is challenging on its own merit, convincing the authorities that run these infrastructure systems can be even more challenging. This will have to be done, however, and a possible solution proposed by Minett (1975) is to create a new supraplanning organization to enhance communication and coordination across the departments that plan, design, build, and operate infrastructure systems.

We have now come to the end of a long journey that started with the diminishing marginal returns of Joseph Tainter. Before we conclude, however, we should spend some time discussing the science and algorithmic tools created in the 1960s and 1970s that mostly emerged later partly thanks to the profusion of data and computing power made possible in the early 2000s. This science is boldly called the Science of Cities, and the algorithmic tools belong to a family of computational techniques called Machine Learning. They are the topic of our next chapter.

Problem Set

10.1 Based on all your calculations from problem 1 in each chapter, compute your overall yearly carbon footprint, similar to problem 1 of chapter 1, and compare your results. Results must be reported in [kg CO_2e]. Make sure to add emissions that have not been discussed in the various homework but are accounted for by the online calculator of your choice (e.g., clothes).

10.2 Akin to figure 10.1, track your consumption of transport, water, and electricity for a typical day and sketch them in a single graph.

10.3 Redraw figure 10.2 and include at least one type of flow for all the arrows shown.

10.4 Name three ways the climate has an impact on urban metabolism.

10.5 Name three ways land use has an impact on urban metabolism.

10.6 Select any building that you know and estimate its energy and carbon footprint using the values given in table 10.1.

10.7 The front and back of a small two-story store made of concrete measures 8 m high × 10 m wide. The sides are 14 m long. Assuming the roof is flat and made of concrete as well and that all concrete surfaces are 12.5 cm thick, determine the (1) energy and (2) carbon footprint of the store. Make any necessary assumptions.

10.8 A three-story house made of timber (density of 600 kg/m^3) measures 9 m high × 6 m wide. The sides are 12 m long. Assuming the roof is flat, no other timber is used in the house, and all timber used is 5 cm thick, calculate the carbon footprint of the house if the wood is (1) sustainably grown and (2) not sustainably grown. Make any necessary assumptions.

10.9 Look up the total road length of your city or a specific neighborhood in your city, and based on a simple estimation of the road width and thickness, calculate the (1) total and (2) per capita energy and carbon footprint of the roads. Make any necessary assumptions.

10.10 A six-lane highway made of asphalt is being built in an area. The width of one lane is 3.7 m, and the highway will span 10 km. Calculate the total (1) energy and (2) carbon footprint of the highway.

10.11 A community of 50,000 people with an average density of 800 pers/km^2 has an average road density of 9 km/km^2. Calculate (1) the surface area of the community and estimate (2) the total length and surface area of all the roads in the community (assume an average road width of 8 m). Based on this information, calculate the (3) total and (4) per capita carbon footprint of all the roads in the community. Assuming all roads are repaved every twelve years, (5) calculate the yearly per capita carbon footprint of all the roads.

10.12 A house with a floor area of 175 m^2 was found to have floor-area heating and cooling intensities of 0.55 and 0.14 W/m^{2}K, respectively. The area where the house is located has 2,549 HDD and 185 CDD. The total electricity consumption for lights and appliances was estimated to be 1.16 kWh/m^2 per month, and the water heater consumes, on average, 0.65 kWh/m^2 per month. Based on this information, calculate the yearly energy consumption of the house attributed to (1) space conditioning, (2) lights and appliances, and (3) water heating, and (4) calculate the total year energy consumption of the house.

10.13 A house has an exposed area of 220 m^2, and it had a total yearly electricity bill of \$688.92 last year. The electricity cost \$0.12 per kWh. The house is located in an area that had 565 HDD and 829 CDD last year. Both heating and cooling are provided through an air-source heat pump (ASHP) with a SEER of 4. The water is also heated with the ASHP, and it is approximated that water heating required 0.62 MWh last year, while the ASHP consumed a total of 1.89 MWh. Based on this information, calculate (1) the electricity consumed by the house and (2) the part attributed to lights and appliances. Moreover, (3) calculate how much electricity was used for space conditioning and (4) determine the floor-area system conductance of the house.

10.14 A city of 30,000 people is trying to estimate how much water is lost through leakage every year in its system. The average water consumption per inhabitant is estimated to be 555 L/day (accounting for residential and commercial use). The water treatment plant pumps out 6,624,200 m^3 of water per year. The city has 420 km of water pipes laid out throughout the city. Calculate (1) the amount and (2) the percentage of water that is lost due to leakage per year in the city and compute (3) the average leakage rate per kilometer of water pipe.

10.15 A water SCADA system in a neighborhood recorded an average water flow rate of 17.8 m^3/day and a maximum flow rate of 21.4 m^3/day. The neighborhood has 182 CDD, and it is composed of twenty families who are estimated to consume a total of 850 L/day per family. Calculate (1) the amount and (2) percentage of water that was lost due to leakage in a year and comment on the results. Moreover, (3) calculate the intensity of water demand based on the temperature and (4) the base water demand per family.

10.16 Based on the information provided in the table below, calculate the carbon footprint of this person for (1) electricity consumption (no space conditioning), (2) water consumption, (3) transport, (4) space conditioning (both heating and cooling), and (5) solid waste. Then (6) calculate the total for the year.

Household property	Value	Unit
Occupants	4	
House size	205	m^2
Electricity (no space conditioning)	3.6	MWh/year
Energy use (space conditioning)	2.1	W/m^2K
Heating AFUE	0.9	
Cooling COP	3	
Personal consumption		
Water	340	L/day
Transport (auto)	42	km/week

Household property	Value	Unit
Transport (transit)	102	km/week
Solid waste	37	kg/month
Local property		
Heating degree days	2,867	Kd
Cooling degree days	111	Kd
Natural gas emission factor (heating)	0.181	t CO_2e/MWh
Electricity emission factor	0.539	t CO_2e/MWh
Water energy use	1	Wh/L
Transport emission factor (auto)	231	g CO_2e/km
Transport emission factor (transit)	51	g CO_2e/km
Solid waste emission factor	0.543	t CO_2e per wet ton

10.17 Based on the information provided in the table below, calculate the total materials stocked in the neighborhood for the buildings and the roads in terms of (1) weight and (2) carbon emissions. Based on the energy values given, calculate the per capita carbon footprint related to (3) the buildings, (4) transport (calculate carbon footprint only), (5) water (assume all energy is provided from the power grid), and (6) solid waste (calculate carbon footprint only). From the results computed, calculate the (7) per capita and total carbon footprint of the city.

Category	Value	Unit
General properties		
Population *P*	30,000	pers
Area *D*	3	km^2
Floor area *A*	67	m^2/pers
Exposed area *A*	108	m^2/pers
Road density *A*	9	km/km^2
Grid emission factor	321	g CO_2e/kWh
Materials		
Building intensity	1,000	t/m^2
	500	kg CO_2e/m^2
Road intensity	3	Mt/km
	1,000	t CO_2e/km
Energy		
Buildings		
Average *U*-value	0.42	W/m^{2}K

(continued)

Category	Value	Unit
HDD	1,582	Kd
CDD	238	Kd
AFUE (electric)	1	
EER	3	
Lights and appliances	20	kWh/m^2
Water heating (electric)	9	kWh/m^2
Transport		
PKT	9,990	km
Average emission factor	195	g CO_2e/km
Water		
Average consumption	350	L/d
Energy intensity	1	Wh/L
Solid waste		
Average generation	49.5	kg/month
Average emission factor	0.771	t CO_2e per wet ton

10.18 The table below shows properties for a city. Calculate the per capita and total materials stocked in the neighborhood for the buildings and the roads in terms of (1) weight and (2) carbon emissions. Based on the energy values given, calculate the per capita carbon footprint of the neighborhood related to (3) the buildings, (4) transport (calculate carbon footprint only), (5) water (assume all energy is provided from the power grid), and (6) solid waste (calculate carbon footprint only). From the results, calculate the (7) per capita and total carbon footprint of the neighborhood.

Category	Value	Unit
General properties		
Population *P*	640,000	pers
Area *D*	370	km^2
Floor area *A*	68	m^2/pers
Exposed area *A*	123	m^2/pers
Road density *A*	7.3	km/km^2
Grid emission factor	57.41	g CO_2e/kWh
Materials		
Building intensity	1,000	t/m^2
	500	kg CO_2e/m^2
Road intensity	3	Mt/km
	1,000	t CO_2e/km

Category	Value	Unit
Energy		
Buildings		
Average *U*-value	0.45	W/m^2K
HDD	3,172	Kd
CDD	45	Kd
AFUE (electric)	1	
EER	3	
Lights and appliances	23	kWh/m^2
Water heating (electric)	11	kWh/m^2
Transport		
PKT (year)	12,512	km
% private vehicle	43.3	%
Emission factor private	272	g CO_2e/km
% public transport	43.1	%
Emission factor transit	34	g CO_2e/km
% active transport	13.6	%
Emission factor active	0	g CO_2e/km
Water		
Average consumption	380	L/d
Leakage rate	6	%
Energy intensity	1	Wh/L
Solid Waste		
Average generation	45.3	kg/month
Emission from recycling (15%)	–3.114	t CO_2e per wet ton
Emission from incinerating (50%)	0.361	t CO_2e per wet ton
Emission from landfilling (35%)	0.894	t CO_2e per wet ton

10.19 Draw a flow diagram and give specific examples to illustrate how the seven infrastructure systems—transport, water, utility, electricity, telecom, solid waste, and buildings—are interdependent.

10.20 Select two infrastructure systems from the following seven systems and comprehensively assess their interdependencies: transport, water, utility, electricity, telecom, solid waste, and buildings.

10.21 Select any one of the following seven infrastructure systems and comprehensively report on how it is interdependent with all other systems: transport, water, utility, electricity, telecom, solid waste, and buildings.

10.22 A new wastewater treatment plant is being built in an area. Elaborate on other infrastructure systems that could be integrated with the building.

10.23 A new rail line is being implemented in a city. Elaborate on other infrastructure systems that could be integrated with the rail line.

10.24 A series of solar panels are being dispatched and installed throughout a city. Elaborate on other infrastructure systems that could be integrated with the solar panels.

10.25 A new aerobic-composting facility is being built outside of a city. Elaborate on other infrastructure systems that could be integrated with the facility.

10.26 In your own words, describe the four design patterns introduced in the chapter: one-to-many, many-to-many, many-to-one, and one-to-one.

10.27 For each of the four design patterns, cite one example that has not been used in the chapter.

10.28 Give some examples of how electricity, water, and natural gas systems could acquire a many-to-many design pattern.

10.29 Find four examples (other than the one given in the chapter) for the four quadrants of the ID matrix shown as figure 10.8.

10.30 Through an example, describe how a system that is currently centralized and separated could be modified to become any of the other three configurations shown in the ID matrix.

10.31 Similar to the house in Okinawa, describe how a house in the region where you live could be retrofitted to become more independent from other infrastructure systems.

10.32 In your own words, elaborate on Minett's (1975) quote, "It is the links that create the system," and give an example in which looking at the links between two systems is paramount in urban engineering.

Notes

1. We discussed industrial ecology a little bit in chapter 9 when we introduced the idea of industrial symbiosis in section 9.3.2.1.

2. An amazing person and one of the first to study the city as a biological system (as opposed to a physical system). We really should have talked about him in chapter 4. He wrote a famous book titled *Cities in Evolution* (Geddes 1915).

3. Chris Kennedy is an amazing researcher who also happened to be my PhD advisor.

4. That is a rough calculation. Assuming an average asphalt density of 2,300 kg/m^3, an average road width of 8 m (26.2 ft), and an average road thickness of 5 cm (2 in.), we get $2{,}300 \times 8 \times 0.05 \times 1000 = 920{,}000$ kg/km, and 1 billion tons of asphalt therefore gives us 1,086,957 km of roads.

5. To learn more, visit http://www.circularecology.com/ (accessed March 25, 2019).

6. The definition of a sustainably managed forest is not obvious, but it is a forest that is well managed in which new trees grow at a similar rate as they are cut and where biodiversity is respected. Put simply, the forest is sustainably managed if it does not change significantly—unlike massive deforestation, as has happened in the Amazon rain forest, for example.

7. Another indicator, used more anecdotally than scientifically, is the number of cranes in a city. The next time you visit a city, count the number of cranes you see and compare it with your city, then try to see if there is a correlation with how well the two cities are doing economically.

8. To learn more, visit http://faf.ornl.gov/ (accessed August 22, 2018).

9. Both naturally and not so naturally. Power plants and industrial boilers can be a significant source of water vapor.

10. You might have learned or heard of the Theissen polygon method, which separates an area into different tessellations and attributes the rainfall of a weather station to all the points that are nearest to it. Using GIS data, a kriging approach, based on distance, can easily be used to interpolate rainfall data from multiple weather stations.

11. But the wastewater goes to the Chicago River, which serves a completely different watershed. That is actually quite significant since water from Lake Michigan that was supposed to end up in the Great Lakes watershed and then in the Saint Lawrence River now actually ends up in the Mississippi watershed and then in the Gulf of Mexico. This is because the flow of the Chicago River was reversed in the early twentieth century so that wastewater flows away from the lake and does not contaminate it, as the lake is the main source of drinking water for Chicago.

12. If you want to know what a world without electricity would look like, I recommend the 2016 movie *The Survival Family*. The movie traces the story of a family in Japan after everything that had anything remotely electric stopped working because of a solar flare (including all cars, phones, water pumps, and so on).

13. Unless the system works purely with gravity.

14. In fact, this is how New York City used to deal with its solid waste until Colonel Waring revolutionized the way solid waste is collected, transported, and treated in cities.

15. The one major exception is New York City, which possesses a large steam system that services a large portion of Manhattan.

16. This is actually exactly what Tesla Energy was doing at the time of this writing.

17. In Chicago in the 1920s, the electricity mogul Samuel Insull (the one from chapter 5) also bought many railway lines since they sometimes use electricity, and he could also install new transmission lines next to railroad tracks—for example, next to the former Chicago North Shore Line.

18. A particularly great quote from the book says that "a data center doesn't merely contain the hard drives that contain our data. Our data has become the mirror of our identities, the physical embodiment of our most personal facts and feelings. A data center is the storehouse of the digital soul. I liked the idea of data centers tucked away up in the mountains like wizards" (Blum 2012, p. 229).

19. We can also see Frank Lloyd Wright's Monona Terrace in the background.

20. Technically, solid waste follows a many-to-few design pattern and then a few-to-one design pattern as solid waste from buildings is first taken to a few disposal facilities and then what is left over is taken to one Waste-to-Energy processing plant or sanitary landfill.

21. We recall that Mumford (1961) calls these chief architects/engineers "regimenters of human functions and urban space."

References

Ashby, Michael F. 2013. "Chapter 2—Resource Consumption and Its Drivers." In *Materials and the Environment*, 2nd ed., 15–48. Boston: Butterworth-Heinemann. https://doi.org/10.1016/B978-0-12-385971-6.00002-6.

Blum, Andrew. 2012. *Tubes: A Journey to the Center of the Internet.* New York: HarperCollins.

Chrysoulakis, N., E. A. de Castro, and E. J. Moors. 2014. *Understanding Urban Metabolism: A Tool for Urban Planning*. Abingdon, UK: Routledge.

City of Chicago. 2013. "City Council Approves Expansion of Fiber Optic Cables as Part of Broadband Initiative." Cityofchicago.org, September 11. https://www.cityofchicago.org/content/dam/city/depts/mayor/Press%20Room/Press%20Releases/2013/September/9.11.13AppFiber.pdf.

Derrible, Sybil. 2016. "Urban Infrastructure Is Not a Tree: Integrating and Decentralizing Urban Infrastructure Systems." *Environment and Planning B: Urban Analytics and City Science* 44(3): 553–569. https://doi.org/10.1177/0265813516647063.

———. 2019. "An Approach to Designing Sustainable Urban Infrastructure." *MRS Energy & Sustainability* 5(E15). https://doi.org/10.1557/mre.2018.14.

Durairajan, Ramakrishnan, Paul Barford, Joel Sommers, and Walter Willinger. 2015. "InterTubes: A Study of the US Long-Haul Fiber-Optic Infrastructure." *ACM SIGCOMM Computer Communication Review* 45(5): 565–578. https://doi.org/10.1145/2829988.2787499.

Durairajan, Ramakrishnan, Subhadip Ghosh, Xin Tang, Paul Barford, and Brian Eriksson. 2013. "Internet Atlas: A Geographic Database of the Internet." In *Proceedings of the 5th ACM Workshop on HotPlanet—HotPlanet '13*, 15. Hong Kong: ACM Press. https://doi.org/10.1145/2491159.2491170.

Duvigneaud, P., and S. Denayeyer-De Smet. 1977. "L'Ecosystéme Urbain Bruxellois." In *Productivité Biologique En Belgique*, edited by P. Duvigneaud and P. Kestemont, 608–613. Paris: Duculot.

Facchini, Angelo, Christopher Kennedy, Iain Stewart, and Renata Mele. 2017. "The Energy Metabolism of Megacities." *Energy and Urban Systems* 186(2) (January): 86–95. https://doi.org/10.1016/j.apenergy.2016.09.025.

Ferrão, P., and J. E. Fernández. 2013. *Sustainable Urban Metabolism*. Cambridge, MA: MIT Press.

Geddes, Patrick. 1915. *Cities in Evolution: An Introduction to the Town Planning Movement and to the Study of Civics*. London: Williams & Norgate.

Hammond, G., and C. Jones. 2011. "Inventory of Carbon & Energy (ICE)." Bath, UK: University of Bath. http://www.circularecology.com/embodied-energy-and-carbon-footprint-database.html.

Judson, Nicholas. 2013. *Interdependence of the Electricity Generation System and the Natural Gas System and Implications for Energy Security*. Technical report 1173. Lexington, MA: Massachusetts Institute of Technology. http://www.dtic.mil/dtic/tr/fulltext/u2/a584764.pdf.

Kennedy, Christopher. 2012. "A Mathematical Description of Urban Metabolism." In *Sustainability Science*, edited by Michael P. Weinstein and R. Eugene Turner, 275–291. New York: Springer.

Kennedy, Christopher, John Cuddihy, and Joshua Engel-Yan. 2007. "The Changing Metabolism of Cities." *Journal of Industrial Ecology* 11(2): 43–59. https://doi.org/10.1162/jie.2007.1107.

Minett, John. 1975. "If the City Is Not a Tree, nor Is It a System." *Planning Outlook* 16(1–2): 4–18. https://doi.org/10.1080/00320717508711494.

Mumford, Lewis. 1961. *The City in History: Its Origins, Its Transformations, and Its Prospects*. New York: Harcourt, Brace & World.

Rinaldi, S. M., J. P. Peerenboom, and T. K. Kelly. 2001. "Identifying, Understanding, and Analyzing Critical Infrastructure Interdependencies." *IEEE Control Systems Magazine* 21(6): 11–25. https://doi.org/10.1109/37.969131.

Wolman, A. 1965. "The Metabolism of Cities." *Scientific American* 213: 179–190.

11 Science of Cities and Machine Learning

> The problem with a systems theory of cities is that it tends to view systems as being well-behaved, in the sense that external shocks to the system tend to work themselves out, restoring the previous equilibrium or at least evolving to something that is close to the pre-existing state. What has been realized in the last fifty years, however, is that this notion of systems freely adjusting to changed conditions is no longer valid, and in fact never was. Cities admit innovation—indeed they are crucibles of innovation; they generate surprise; they display catastrophe.
>
> —Batty 2013, p. 25

All around the world, a myriad of cities were born and grew substantially in the twentieth century. This means that the impressive population growth we first saw in chapter 3 has happened mostly in cities. In fact, *urbanization*, as the process of building cities and people moving to cities, has grown at such a phenomenal pace that since 2008, and for the first time in the history of mankind, the urban population has overtaken the rural population. From figure 11.1 we can also see that the rural population will remain flat in the future, while the urban population will keep increasing quite rapidly. This is a big deal. In particular, it means that most of the future population growth will happen in cities. By 2050, cities will house more than 6 billion people, which was about the world population in 2000.

Moreover, cities are not only growing in number (i.e., there are many more cities in the world), they are also growing in size. The term *megacity* describes cities with more than 10 million inhabitants. The first city to ever get this title was New York City in the 1950s. Thanks to rapid economic growth, the number of megacities increased to ten in the 1990s. With urbanization, the number of megacities further increased, and by 2015 there were more than thirty megacities in the world. Megacities like Tokyo and Shanghai housed more than 35 million inhabitants at the time of this writing. This was

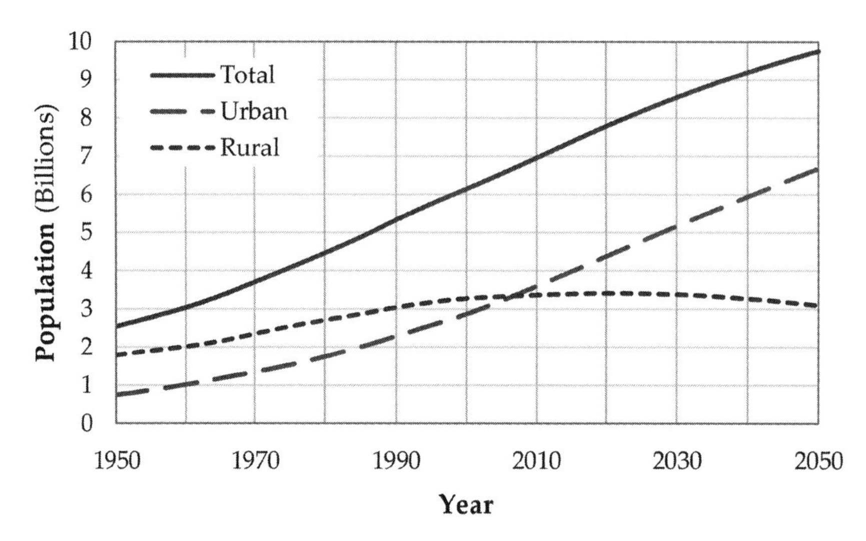

Figure 11.1
World population.
Data source: United Nations (2018).

about the same as the entire population of Canada living in one city.[1] It is therefore not surprising that we need to revisit the way we plan, design, engineer, and operate cities.

In addition, most of the growth is happening in low-, lower-middle-, and upper-middle-income countries from India and China to the Philippines, Indonesia, and Pakistan, where the per capita greenhouse gas (GHG) emissions tend to be low. As new cities are virtually built from scratch, we should really make sure they have better designs than existing megacities and are built to be sustainable and resilient.

Expectedly, understanding how cities are distributed and how they form and grow has also become paramount. In fact, this has led to the emergence of a science, sometimes called *Urban Science* or the *Science of Cities*. This is not the same as urban planning, which we covered in chapter 4. It is also different from all the chapters that we have seen, and it was not particularly developed to look at infrastructure, or even sustainability for that matter. Instead, it has emerged thanks to new abilities provided by computers since the 1970s, and it really started to become a science in the 2000s—for example, computers enabled us to test bottom-up processes that show emergent properties from simple rules.

The Science of Cities is roughly made up of other sciences that provide the technical abilities to study cities, and we will study these other sciences in the next section. In particular, *complexity science* enables us to study systems that show apparent order without the presence of a central command designing this order; these systems are called *complex systems*. To further explore cities, we will then learn about *scaling laws* in cities, which can be quite relevant for us. After that, we will see an impressive underlying phenomenon that seems to capture how city size is distributed, called *Zipf's law*, which we will try to reproduce by going over several simple population models. We will then get quickly acquainted with *network science*, which has become a dominant science to study complex systems.

In the second half of the chapter, we will get a quick introduction to *Machine Learning* (ML), which will undoubtedly revolutionize the way we do things in cities. In addition to a general introduction, we will specifically see three techniques: *K-means clustering*, *decision tree learning*, and *neural networks*. It will, unfortunately, be a very short section, but it should whet your appetite to learn more about the subject. Importantly, ML is likely going to substantially change the way we work as urban engineers, and it directly contributes to making sure we use a different kind of thinking than we used when we created some of the problems we face today, echoing the quote from Einstein that we saw in chapter 1.

In this chapter we will not learn any fundamentals and/or analyze the demand for electricity, water, transport, buildings, or solid waste, as we have done previously. Nevertheless, many of the techniques we are about to learn can be applied firsthand to better understand trends and patterns in the demand for something, and there is no doubt that these techniques will become a big part in how we design and operate cities.

But first, let us learn about the Science of Cities.

11.1 The Science of Cities

11.1.1 Complexity Science

The quote at the beginning of the chapter from Michael Batty, one of the fathers of the Science of Cities, sums up the familiar way of thinking about how cities evolve quite well. As human beings, we tend to seek an equilibrium point, and we are taught that the economy does the same by matching supply and demand. This kind of thinking has led many (if not most) models to perform some kind of optimization, hoping that once optimized, a city will become stable. This has obviously never happened since cities are in constant evolution, either improving or declining; there is just no other way. Although he predated the Science of Cities, a quote by Patrick Geddes captures this point well: "A city is more than a place in space, it is a drama in time."[2] This Science of Cities therefore takes an "evolutionary" approach, much closer to the way of thinking in biology, rather than an equilibrium approach, which is more common in traditional physics (and, incidentally, in engineering). To quote Batty (2013, p. 14) once again: "The image of a city as a 'machine' has been replaced by that of an 'organism.'"

We therefore need a completely new tool kit to study cities as organisms.[3] This is where complexity science, or *complexity theory*, becomes important. Complexity science is a relatively new field of study,[4] although it goes back to the middle of the twentieth century.[5] To showcase a quick example that shows complex behavior, we consider the following simple equation:

$$x_{n+1} = r \cdot x_n(1 - x_n) \quad (1)$$

where x is just any chosen variable (think of it as the consumption of a good, for instance). While being "just" a constant, the other variable r has a critical impact on the whole process. Figure 11.2 shows how this equation behaves for five different values of r. For values lower than 2, the equation slowly goes to 0; at a value of 2, it stays at 0.5; for values higher than 2 but lower than 4, the equation goes up and

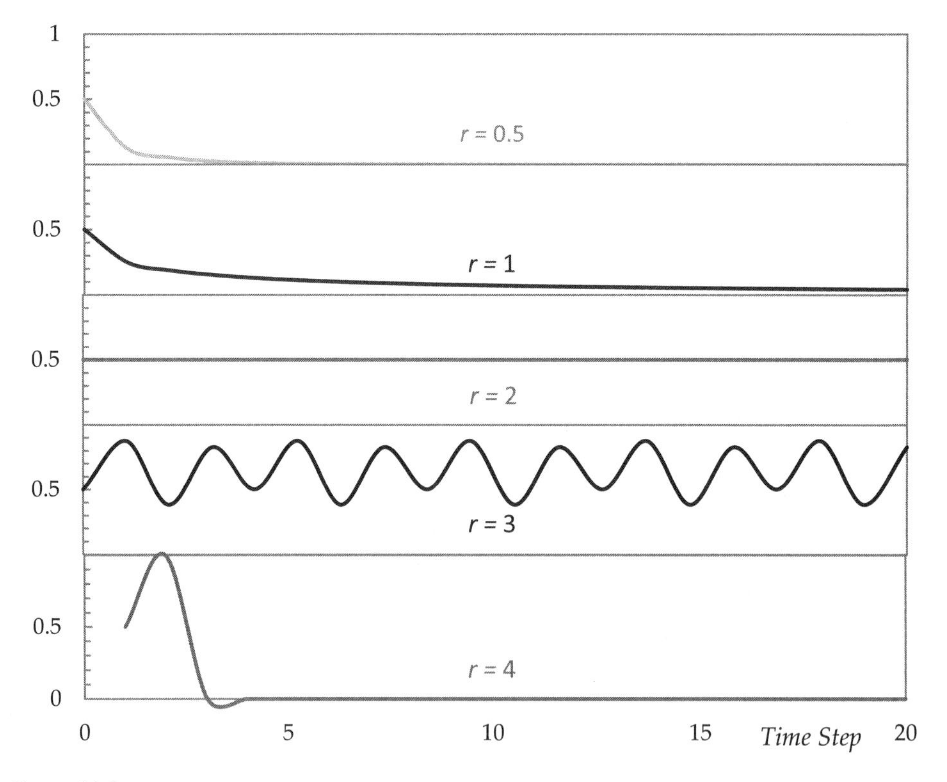

Figure 11.2
Complexity in a simple equation.

down in a sinusoidal fashion; and for values higher than 4, it starts to embark on a sinusoid and then crashes at 0. This is quite incredible for such a simple equation. Now, imagine countless equations like this, interacting in the same system: that is exactly what is happening in a city. No wonder we need a science to understand what happens in cities.

More generally, complexity science is the science that studies complex systems. While there is no official definition of a complex system, most complex systems exhibit one or many of the following four properties (Holland 2014, pp. 5–6):

1. Self-organization into patterns, as occurs with flocks of birds or schools of fish
2. Chaotic behavior, in which small changes in initial conditions ("the flapping of a butterfly's wings in Argentina") produce large changes ("a hurricane in the Caribbean")
3. *Fat-tailed* behavior, in which rare events (e.g., mass extinctions and market crashes) occur much more often than would be predicted by a normal (bell-curve) distribution
4. Adaptive interaction, in which interacting agents (as in markets or the prisoner's dilemma) modify their strategies in diverse ways as experience accumulates

Property 1 is typical, and we have discussed it already. Property 2 is not as typical, especially when it comes to cities, but the idea of "cascading" failure is present, especially when we think of the power grid, for example, in which one failure may propagate and break down a large part of the grid (e.g., the Northeast blackout of 2003). Property 3 has become increasingly relevant in urban engineering, as extreme events are not as rare as we would like them to be—and as we have discussed a little, urban infrastructure will have to be more resilient in the future. Finally, property 4 is very important since rules do change and people do adapt—as mentioned, cities do not seek an equilibrium and are in constant evolution.

Thanks to computers, it has become relatively easy to study and even sometimes model complex systems to identify/simulate these four properties. Say, for instance, that we model a mock city with a certain number of people, and we give simple equations/rules that people have to follow. We can then run the model and see what happens over time. These types of models have become common in research, and they are called *agent-based models*; in the case presented, people are agents, but an agent could really be anything. Agents simply have to follow some rules. The software package NetLogo is extremely user friendly and has many tutorials,[6] thus offering a great way to get into agent-based modeling.

Sometimes, we may also want to model more complicated phenomena, or even add some randomness in our models. This is typically done by performing a Monte Carlo analysis. Essentially, a Monte Carlo analysis means drawing random numbers from a distribution and rerunning the model we are using each time. For instance, let us assume that human height follows a normal distribution, with a mean of 1.70 m (about 5 ft,7 in) and a standard deviation of 40 cm (about 1 ft, 4 in). We are now running a model, and we are asked to generate the height of 10,000 individuals. We cannot give the average height to everyone; it would not make sense. Instead, we randomly pick 10,000 different heights from a normal distribution with a mean of 1.70 m and a standard deviation of 40 cm. This is called a Monte Carlo analysis, and it is very easy to do with a computer. But we are digressing a little bit here. The main point is that computers have enabled us to do a lot more simulation, which has partially enabled the Science of Cities.

Back to our main point, this complex feature of cities may greatly help us to become more sustainable. Indeed, some evidence seems to suggest that people consume less energy in cities (Brand 2009), therefore showing some economies of scale. This *economies-of-scale* aspect is in fact important, and some argue that the only reason people decided to concentrate in cities in the first place is because the benefits outgrow the costs (e.g., transport congestion). One simple way to look at it is to count the possible number of interactions. In a city with a population P, the number of people that one person is able to meet is $P-1$. Consequently, if we sum all possible interactions for everyone, we get $P\times(P-1)$. Of course, if person A knows person B, then person B knows person A, and the actual number of possible relationships is half that: $\frac{1}{2}P\times(P-1)$. The general order of magnitude is therefore P^2, which is quadratic. Put differently, this means that the increase in the number of possible interactions is not linear, and we benefit more by being in a larger city. This nonlinear aspect has strong implications, and we are about to learn some of them.

11.1.2 Scaling Laws in Cities

If we start from the principle that people benefit from being in larger cities, then surely this should manifest itself in some ways that we can measure. There has been quite a bit of work on measuring these benefits, which tends to appear in the form of scaling laws. Mathematically, scaling laws are simple power laws that take the form:

$$Y=\alpha X^{\beta} \qquad (2)$$

where Y is a dependent variable (e.g., kilometers of urban road), X is the independent variable (e.g., city population), and α and β are constants. The value of α changes depending on what we measure, and we will not spend time discussing it. The value of the exponent β, which we call the *scaling exponent*, shows great interest. There are four cases for β:

- $\beta < 0$ gives us a decreasing power law that we will encounter in the next section, and we will therefore skip it for now.
- $0 < \beta < 1$ gives us a sublinear relationship; as X increases, the new addition of Y becomes smaller (i.e., ΔY decreases). For example, as population increases, per capita values decrease. This is similar to the concept of decreasing marginal returns that we saw in chapter 1.
- $\beta = 1$ gives us a linear relationship with which we are familiar.
- $\beta > 1$ gives us a superlinear relationship; as X increases, the new addition of Y becomes larger (i.e., ΔY increases), and we get increasing marginal returns. This is the case for the number of possible connections—for instance, that has $\beta = 2$.

Ideally, we would like to have desirable urban properties, such as income, to be superlinear, which would definitely show increasing economies of scale. In contrast, we prefer not so desirable urban properties, such as energy use, to be sublinear—that is, the larger a city is, the smaller per capita energy use becomes.

Directly inspired by Bettencourt (2013), figure 11.3 shows this scaling for two urban systems. On the left, the 2010 total road length in miles for all U.S. urbanized areas follows a sublinear pattern with $\beta = 0.82$. On the right, the 2015 gross metropolitan

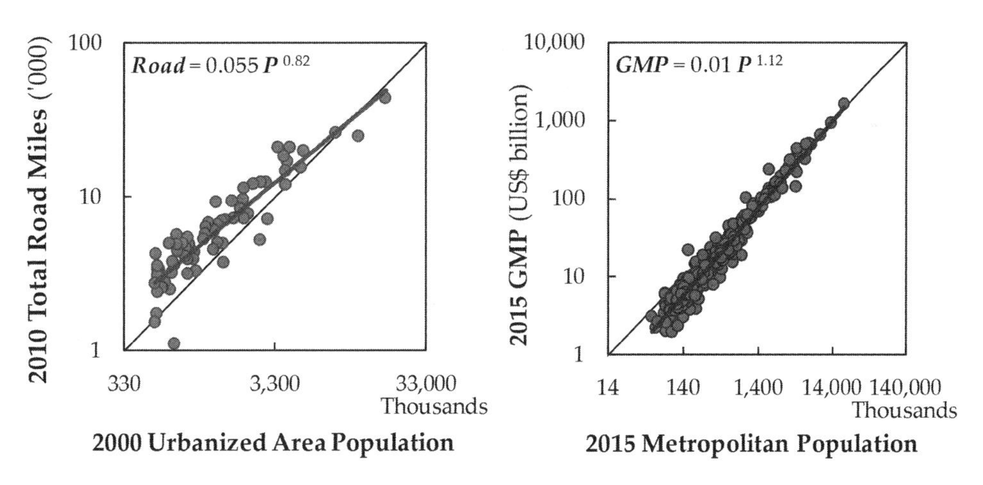

Figure 11.3
Scaling laws and cities. *Left*, 2010 U.S. total road length versus urbanized area population.
Source: Federal Highway Administration (2014).
Right, 2015 U.S. GMP versus 2015 metropolitan population.
Source: Bureau of Economic Analysis (2016). Population data were collected from the U.S. Census Bureau for the respective years and administrative units.

product (GMP)—essentially like the gross domestic product (GDP) but for cities—for all U.S. metropolitan statistical areas follows a superlinear pattern.

In both cases, we plotted the data on a log-log scale so that a power law appears as a straight line. To see this mathematically, we simply need to take the log on both sides of equation (2):

$$\log(Y) = \log(\alpha X^{\beta})$$

$$\log(Y) = \log(X^{\beta}) + \log(\alpha)$$

$$\log(Y) = \beta \log(X) + \log(\alpha) \tag{3}$$

From equation (3), we see that the exponent β simply became the slope of a line that is easy to calculate—although this technique possesses some problems.[7] The presence of scaling is typically the type of emerging order that characterizes a complex system. After all, cities do not build their roads based on what other cities do. There is therefore no "central command" dictating how road length and GMP should be attributed, yet we find the presence of inherent scaling. This is similar to the concept of the "invisible hand" by Adam Smith, the famous eighteenth-century Scottish economist. Despite the lack of a central command, complex systems behave in specific (and sometimes even predictable) ways.

And what about GHG emissions? Well, it is not that easy to tell. Even these two results above are controversial for two main reasons. First, cities are composed of various governments, including municipal and county governments, and it is surprisingly difficult to determine the "area of influence" to be included to define the boundaries of a city; even in the examples given in figure 11.3, one graph used urbanized areas, while the other used MSAs. Second, some of these data are approximations and may therefore be unreliable. GHG emissions from transport, for instance, depend on congestion, and we do not have accurate measures of congestion. Depending on what we include, we can find conflicting results.[8]

Interestingly, however, this type of scaling is not only present between cities but also appears within cities. The scaling of rent in cities (higher in the center, lower in the periphery) has been captured for a while, dating back to von Thünen in 1826,[9] and many more interesting works can be read that measure virtually anything about cities.

That is enough about this type of scaling for now. We can instead look at one more interesting phenomenon that compares cities with one another and that captures the complexity of how they are distributed.

Example 11.1
Distribution of Water Pipes in Civitas

The table below shows the total length of the water distribution system (i.e., length of pipes in kilometers) per zone in Civitas. (1) Determine whether scaling laws appear in the water distribution system of the city-state.

Zone	Population P	Pipe length L (km)
1	24,356	654
2	12,586	357
3	6,789	199
4	7,264	214
5	9,005	265

Solution

(1) To answer the question, we can plot pipe length against population, fit a power law to the data, and analyze the results. The figures below show that some scaling is in fact happening. The figure on the left has a normal scale, and the figure on the right has a log-log scale. More specifically, the trend is sublinear with power 0.93. The first constant also suggests that there is, on average, about 0.057 km = 57 m of pipes per capita.

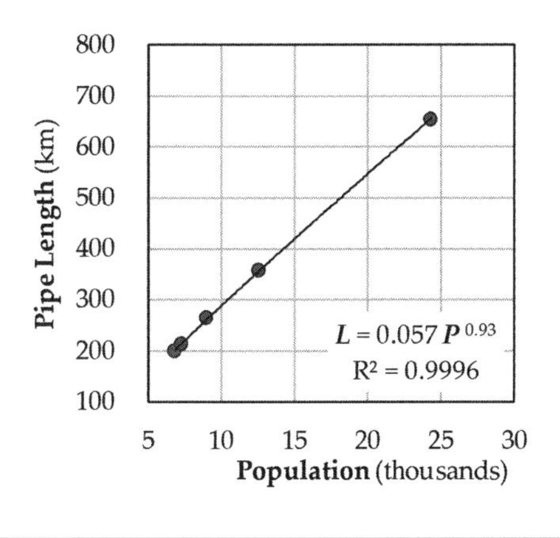

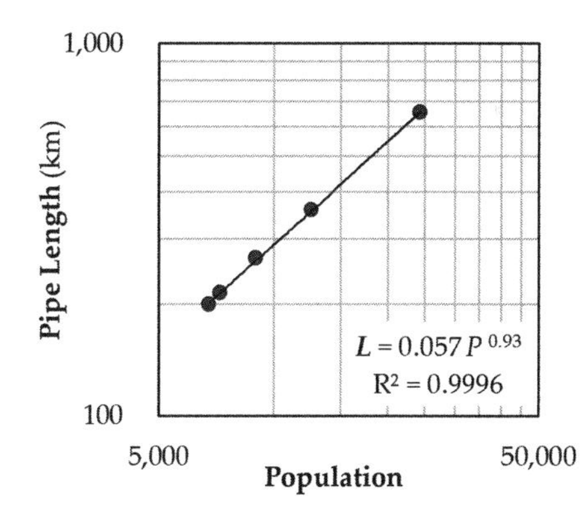

11.1.3 Zipf's Law

Instead of looking at income, energy use, and road length, let us simply look at the size of the cities in terms of their population. Put simply, what happens if we count how many small, medium, and large cities there are?

To start, we can look at the population of all U.S. cities from the 2010 U.S. census. The census actually only gives us the list of cities with more than 50,000 people, and we count 727 cities, totaling 115 million people, which is a little more than one-third of the U.S. population. This number is quite low since the United States had about 310 million people in 2010, more than 80% of whom lived in cities. That means that about half the U.S. population lived in cities of fewer than 50,000 people (and about 15% live in rural areas). This actually perfectly illustrates our earlier problem of defining cities. Many of these cities containing fewer than 50,000 people are actually located in the suburbs of larger cities and are therefore in the area of influence of those large cities. This is why the U.S. Census Bureau has the category of MSAs that we used to compile figure 11.3 (*right*), but we will not discuss them here.

Once we have downloaded the data, we need to "bin" them to determine what small, medium, and large actually means. One way to do it is to double the size of the bins. We will therefore start with all cities below 100,000 people, then between 100,000 and 200,000; 200,000 and 400,000; 400,000 and 800,000; 800,000 and 1.6 million; 1.6 million and 3.2 million; 3.2 million and 6.4 million; and finally, 6.4 million and 12.4 million. The results are plotted in figure 11.4. The figure on the left shows the data in normal scale, and we see that the number of cities drops dramatically when we count

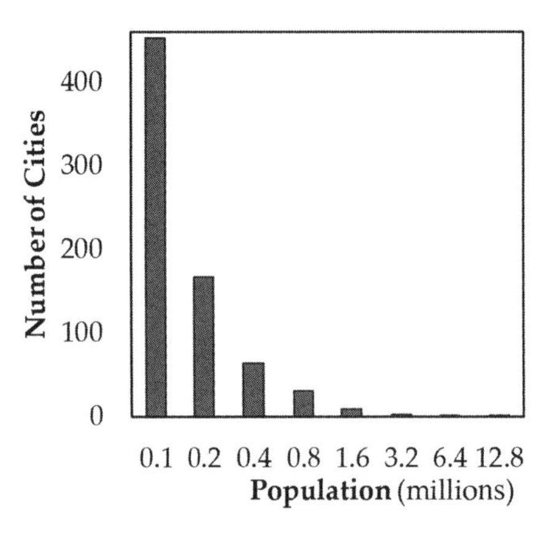

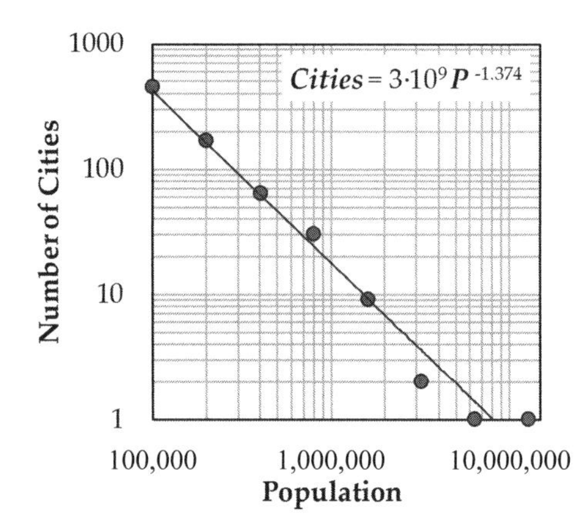

Figure 11.4
Frequency of cities by size in the United States.

larger cities. We therefore switch to a log-log plot on the right, and immediately, we see something that looks like a straight line. Calculating the slope using the technique we learned in equation (3), we get a scaling exponent of −1.374, which is nonlinear.

Despite these impressive results, it is hard to find the right bin size. We could increase some bins and reduce others, and this coefficient may change. What George Kingsley Zipf (1949) did in the middle of the twentieth century was to rank the values instead of binning them. You therefore take the largest value and place it at $x=1$, then the second-largest value and place it at $x=2$, and so on and so forth. Performing the ranking and plotting the results, we find the presence of a power law again, as shown in figure 11.5. The equation captured here is:

$$P = \frac{P_0}{r^{\beta}} \tag{4}$$

where P is population, P_0 is the largest population, r is the rank (i.e., the values on the x-axis), and β is the scaling exponent. This is exactly like equation (2), with β being systematically below 0. This is not just a random observation; Zipf (1949) actually provided quite a sophisticated theory behind it,[10] partially captured by the title of his book *Human Behavior and the Principles of Least Effort*. This general scaling pattern when ranking properties is now typically referred to as Zipf's law. In the pure case of Zipf's law, $\beta = -1$, and Zipf often rewrote the equation as:

$$P \cdot r = \text{constant} \tag{5}$$

where P is our population here again, although it can be any variable under consideration, and r is again the rank. In other words, equation (5) suggests that when we multiply the value (e.g., population) by its rank, we get a constant. In our case of U.S. cities, the scaling exponent is 0.726, as shown in figure 11.5, and it is therefore not a pure Zipf's law. The constant still holds, however, when we multiply the population of a city by its rank to the power 0.726.

Looking more into Zipf's law and assuming a scaling exponent of one for now, it means that the second-largest entity should be half the size of the largest entity and that the third-largest entity should be one-third as large as the largest. We see that our scaling exponent is lower than one, however. This means that the decay (expressed by the slope) is not as fast as we expected. In other words, cities are larger than expected. While we cannot comprehensively answer why, this lower scaling may be due to the nature of the United States, which is divided into fifty states, each of which has a rank-size distribution. Some cities like New York City, Chicago, and Los Angeles are world cities with significantly large populations, and very few states have cities of this

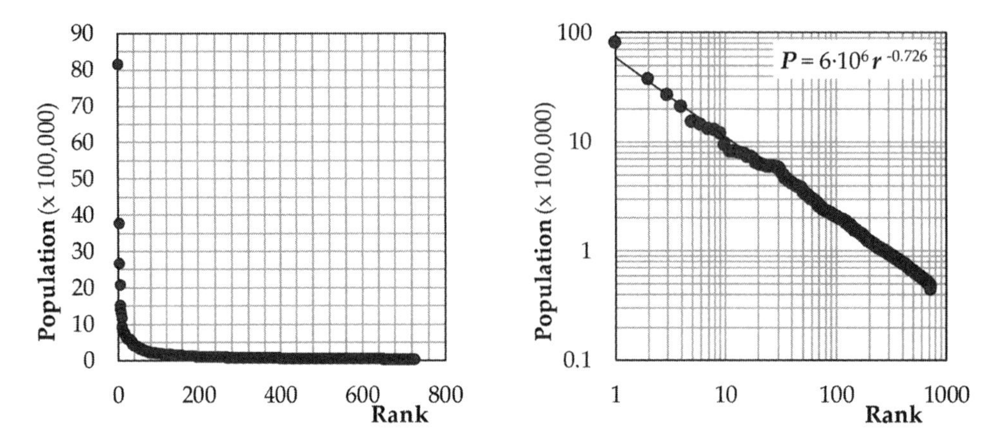

Figure 11.5
Rank-size distribution of cities in the United States.

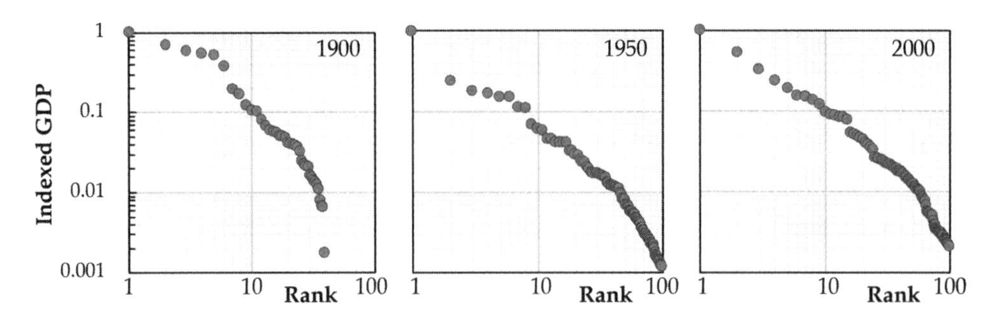

Figure 11.6
Rank-size distribution of country GDP in 1900, 1950, and 2000.

size. Most states, however, have large cities that rival one another, like Milwaukee, Kansas City, Tampa, or Memphis. We can even see it in figure 11.5 (*right*), just around the 500,000 people mark, where we can see a small "step."

This observation is actually quite important. Due to the scaling in Zipf's law, it means a subsample of one data set may not show any signs of the law. Therefore, not seeing the presence of Zipf's law may mean either that the phenomenon is not present in that data set or simply that the data set is incomplete.

As interestingly, one data set may not the show the presence of Zipf's law at one point in time but show it later. This is actually what has happened with the GDP of world countries. In their study, Cristelli, Batty, and Pietronero (2012) showed the rank-size distribution of GDP over time, and the line is getting straighter. Figure 11.6 shows

Example 11.2

Distribution of Population in Civitas

The table below shows the population per zone in Civitas and the zones' respective ranks. (1) Determine whether Zipf's law applies here.

Zone	Population P	Rank R
1	24,356	1
2	12,586	2
3	6,789	5
4	7,264	4
5	9,005	3

Solution

(1) Plotting population versus rank, the two figures below show that indeed population at the zone level seems to follow a power law with the zones' rank. In this case, the scaling exponent calculated is 0.816, which diverges from the theoretical value of 1, although we must remember that we are only dealing with five data points.

The value of 23,223 represents the theoretical value for rank 1 that is in zone 1 and that actually houses 24,356 people.

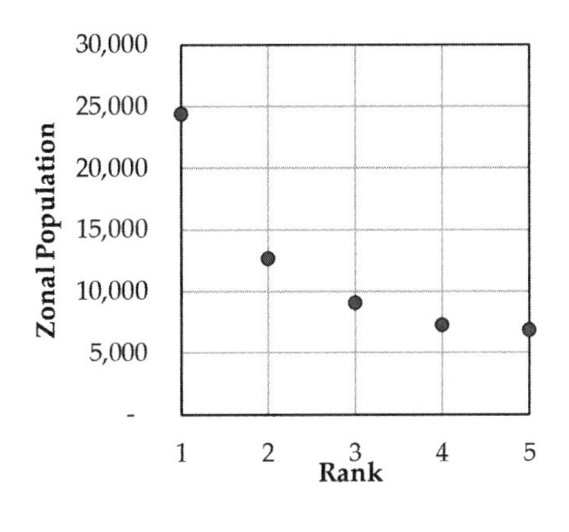

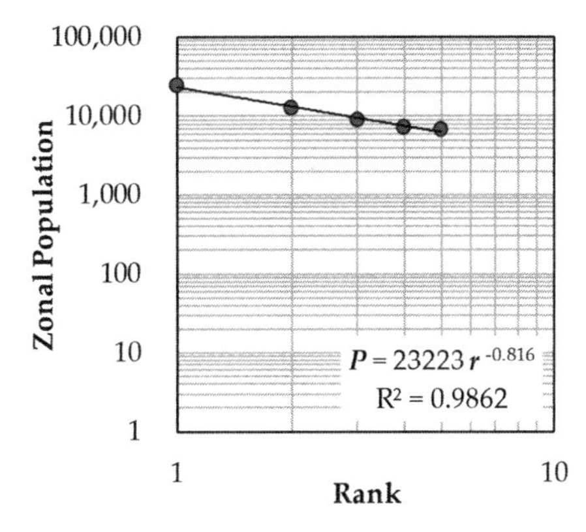

the GDP of world countries (indexed by the maximum value) versus their rank for the years 1900, 1950, and 2000.[11] This phenomenon seems to be related to globalization, with market forces more recently applying at the global scale.

Now that we have looked at the distribution of cities and populations, let us see whether we can model this ourselves.

11.1.4 Simple Population Models

We have gained some significant insights so far into the Science of Cities. Most importantly, we have learned that, in complex systems, a measurable and distinct order emerges despite the absence of a central force. The question is therefore: Can we develop models that can replicate this order? In other words, can we find the rules and equations that when applied at the bottom level result in the same order at the top level? Batty (2013, chap. 4) shows several models that attempt to do just that, and we will recall some of them here.

First, we can try to model how population is distributed between cities. Naturally, people are moving to cities because they are attracted to them, and there are some types of "forces" (like market forces) in play. The next logical step is for us to try to detect and reproduce these forces. Although we do not comprehensively know why these things happen with such order, let us try to investigate population growth, starting with a population P at time t, $P(t)$, which grows or declines by some fixed factor every year, following this model:

$$P(t+1) = (1+\lambda)P(t) \tag{6}$$

Taking the example of Civitas, with its 60,000 people, what if it grows by 1% ($\lambda=+0.01$), stays constant ($\lambda=0$), or declines by 1% ($\lambda=-0.01$) in the next few years? Figure 11.7 (*left*) shows the result assuming 100 time steps (i.e., after 100 years). Even here, with such a simplification, we see that the growth and decline scenarios are nonlinear, which is simply explained by the recursion $(1+\lambda)^T$ for a period T (e.g., 100 years)—essentially like compound interest in economics.

Now, instead of having a fix λ, we can assume that the growth rate changes randomly, $\lambda(t)$, between -0.01 and $+0.01$ every year. Figure 11.7 (*right*) shows that for four different random scenarios. We can see that there is a lot of noise but that three out of four scenarios end up with fewer than the initial 60,000 people. Although it may seem odd, getting less than the initial population is actually an artifact of the model itself since it is *multiplicative*. Declining factors ($\lambda<0$) have a larger impact than growth factors ($\lambda>0$)—again, because of the compound nature of the model.

Eventually, if we go on for more time steps or increase the variability of λ, most curves will go toward 0. To avoid this scenario, we can use a model with an *additive*

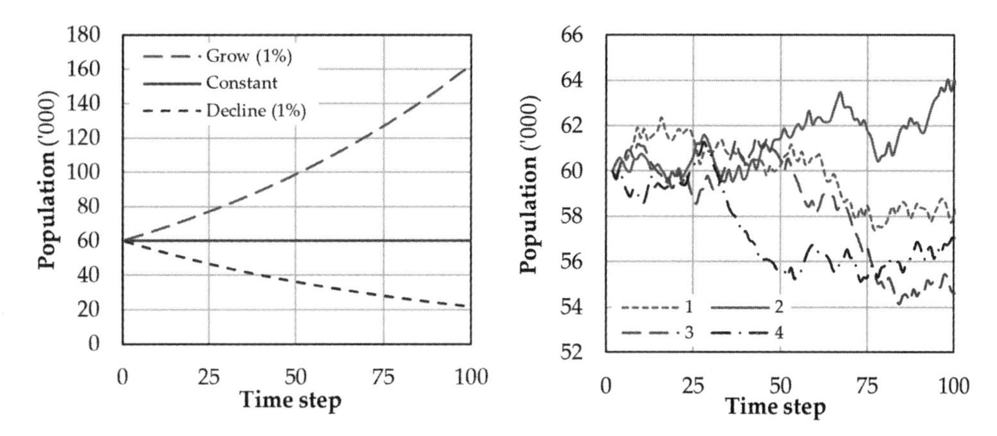

Figure 11.7
Evolution of Civitas population by fix (*left*) and random (*right*) amounts.

property—that is, that adds or subtracts a fixed population at every time step instead of a population that is proportional to the current population. For this, we can use the following model: $P(t+1)=P(t)+(1+\lambda)P_0$, where P_0 is some previously chosen population, but we will not go over it.

Keeping the multiplicative model, let us see what happens when we simulate a high number of "cities" over time. Because we are dealing with multiple cities, we will need to add a subscript i to P, giving us P_i, where P_1 is the population of city 1, P_2 is the population of city 2, and so on. If we say we have 1,000 cities, for instance, then we can go from P_1 to P_{1000}. Similarly, let us assume that our 1,000 cities evolve over 1,000 years, and each city can grow, remain the same, or decline randomly by rates within $-0.01<\lambda(t)<+0.01$. We simply have to assign an initial population to these 1,000 cities—say, one person—and stipulate that the population cannot go lower than one person—that is, we set a lower threshold, and if we get a value lower than 1, then we set it back to 1. Once we go through the 1,000 runs, we can look both at frequencies, as we did in figure 11.4—now, we use bins of 2, 4, 8, 16, 32, 64, 128—and at ranks, as in figure 11.5. The results, shown in figure 11.8, unmistakably show a power law relationship identical to Zipf's law. After 1,000 years and for 1,000 cities, the maximum population we get for the 1,000 cities is 119.5 people.

This result is actually not totally unexpected. Indeed, remember that we did set a lower bound at one person per city, and considering the nature of the multiplicative model, we are bound to have many more cities that have close to one person and very few cities with many more people. Regardless, the fact that it closely follows a power law remains interesting.

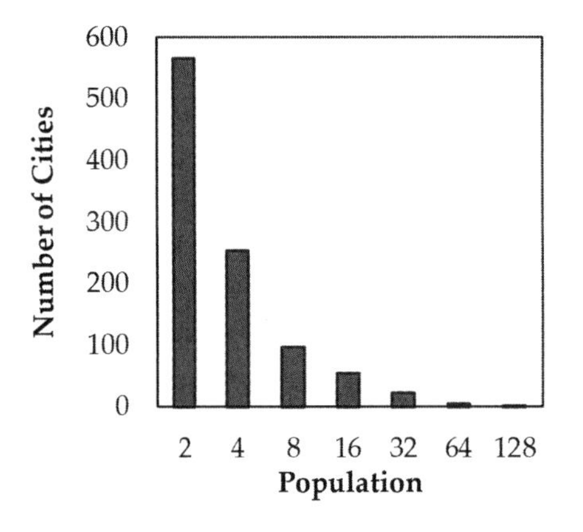

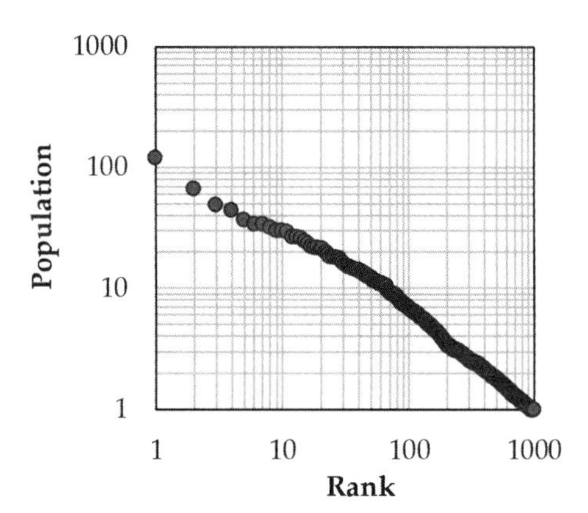

Figure 11.8
Frequency and rank plots for 1,000 simulated cities using the multiplicative model.

Remembering chapter 3, these results should echo what we saw with the geometric and arithmetic growths. The geometric series resembles our multiplicative approach with the recursion $(1+\lambda)^T$, while the arithmetic series resembles our additive approach. In fact, our models are identical to our findings from chapter 3. There is therefore some truth in using these two simple models here, and it is not so surprising that we are collecting the same emerging properties characteristic of complex systems.

We can look at one more simple model. Instead of choosing random rates, we now assume there is some competition between cities for a fixed population, and therefore people can only migrate between cities. Put more simply, we will swap people between cities, making sure we never get below some defined lower bound and that our total population remains constant. Mathematically, for every randomly selected couple of cities i and j, we add a defined population P_0 to one and subtract from the other as follows:

$$\left.\begin{aligned} P_i(t+1) &= P_i(t) + P_0 \\ P_j(t+1) &= P_j(t) - P_0 \end{aligned}\right\} \quad (7)$$

For instance, let us assume we start with 1,000 cities, all of which have a population of 100 people, therefore giving us a total of 100,000 people. We then pick two random cities and add 5 people to one and subtract 5 people from the other as long as the remaining population is not below 0. The total population therefore remains 100,000, but people can move between cities. Instead of doing this systematically for all cities, we just loop over this process 1 million times and see how big cities get. Figure 11.9

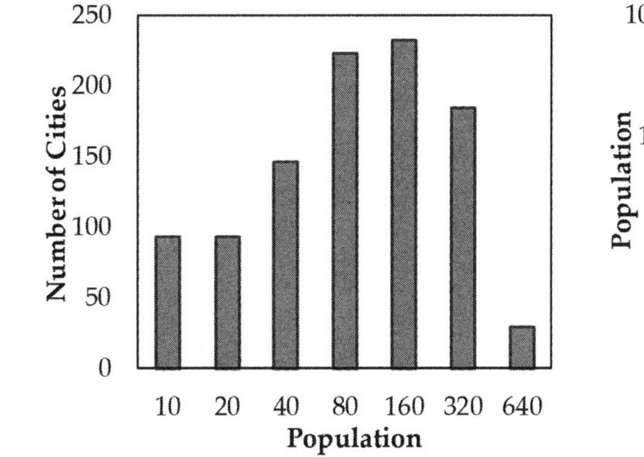

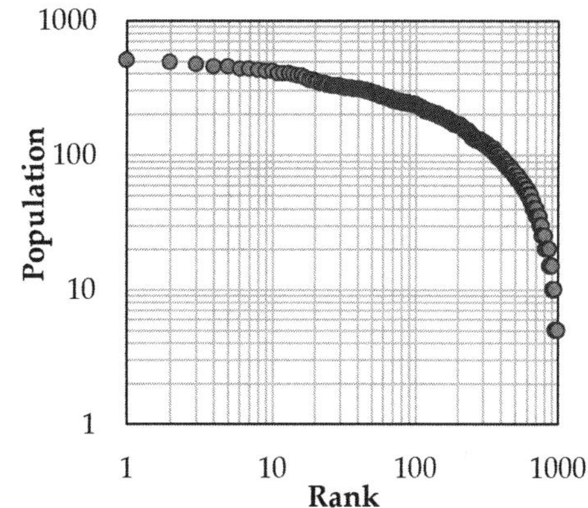

Figure 11.9
Frequency and rank plots for 1,000 simulated cities using the swapping model.

shows the results, and we can see that they are completely different from what we are used to.

We can see that many cities are below their initial population of 100 people—in fact, 93 cities have below 10 people, and another 93 cities have between 10 and 20 people. The high number of small cities is again due to our lower bound. On the rank plot, the largest city has 505 people, followed by a city with 485 people, then 465 and so on and so forth, which definitely does not look like a power law. In fact, the resulting curve is much better fitted with an exponential function, which can be observed if only the x-axis is put on a log scale (figure 11.10)—take equation (2) and take the log only on the right-hand side. This exponential fit can actually be explained theoretically (see Batty 2013, p. 125).

Overall, the fact that these simple models can replicate what we observe in real life is encouraging. It makes us think that we may be capturing real phenomena. In fact, many argue that these scaling laws and power laws express the "signature" of complex phenomena. We are therefore coming up with models that exhibit the same signature. This can be dangerous, however. Getting the same signature does not mean that we have found the right phenomena. First of all, our models are purely mathematical, and cities are much more complicated. More importantly, we can easily be duped by developing designs that have the same signatures. We therefore need to look at more things within complex systems to better understand them. For this, one of the dominant sciences to study complex systems is network science, and it is the topic of the next section.

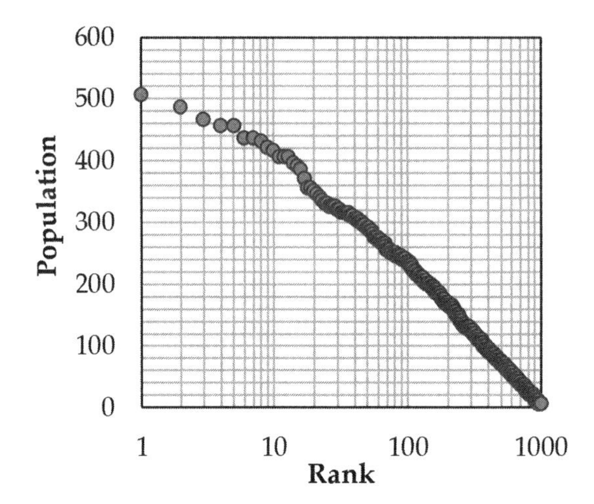

Figure 11.10
A rank plot of figure 11.9 on a semilog scale.

11.1.5 Network Science

As the name suggests, network science focuses on networks. The precursor to network science is graph theory, which originated in the eighteenth century with the famous mathematician Euler. In fact, graph theory was really born thanks to a simple transport problem called the Seven Bridges of Königsberg (Euler 1741). The town of Königsberg (now Kaliningrad in Russia) had seven bridges to reach four different parts of the city, and the dilemma was to prove that it was impossible to cross all of the bridges consecutively and only once—that is, if you want to cross all of them, one will necessarily have to be crossed twice. Euler famously solved the problem by inventing a new branch of mathematics: graph theory.

Graph theory tends to look at relatively simple and well-ordered graphs. The networks that have been collected, especially since the 1990s, thanks to computers, can be extremely large, however, and rarely possess discernable patterns. New tools from graph theory were needed to analyze them, hence the new *network science*.

The thing about networks is that many of their properties are governed by the way the links and nodes are wired (i.e., their topological structure) as opposed to flows happening on the networks. For example, two cities with the same road length and the same travel demand (i.e., traffic) can experience different kinds of congestion levels based purely on their network properties. Network science therefore focuses mostly on measuring the structural properties of networks, and this is what we are going to do here.

In this section we will look at only a few basic concepts of graph theory and network science, but it can really be expanded to large networks. This is all the more relevant as most urban infrastructure systems are physical networks. From transport systems, water conduits, wastewater channels, and electric, gas, and telecom lines, all are physical networks. It is therefore quite surprising that we are not using network science extensively to study them. One of the biggest advocates of using network science in the Science of Cities is Marc Barthelemy (2011), who has studied spatial networks extensively.[12]

First of all, a network/graph G is defined as a set of two things: its nodes N and its links L, which give us the set $G=\{N, L\}$. In graph theory, nodes are called vertices, and links are called edges, or arcs, but they really mean the same thing, so we will stick with nodes and links. Moreover, links can have a direction or not. If they do, the network is said to be *directed*, and if they do not, the network is said to be *undirected*. Here, we will deal only with *undirected* networks. Figure 11.11 shows us an example of four undirected networks with five nodes, and the number of links ranges from three to six. We will define each of the metrics shown in the figure one by one.

Before we start to measure properties of networks, we need to realize that networks do not necessarily have to be fully connected. Instead, a network can have a number of subnetworks. Network 1, for example, is one network but it has a number of subnetworks S of 2 (i.e., you can see two subnetworks: {A, B, C} and {D, E}). In figure 11.11, N, L, and S represent the number of nodes, links, and subnetworks in the general properties column.

The next indicator, Δ, is the diameter of the network. The diameter is the smallest number of connections needed to connect the two nodes that are furthest apart to each other. For example, in network 1, Δ is 2 to connect nodes A to C (i.e., A to B to C). In network 2, Δ is 3 to connect nodes A or C to node E (i.e., A to B to D to E).

The next indicator, μ, is the cyclomatic number. It basically counts the number of cycles/loops in the network. For example, networks 1 and 2 have none, but network 3 has one and network 4 has two. Mathematically, μ is defined as:

$$\mu = L - N + S \tag{8}$$

Essentially, $N-S$ is the number of links in a tree network. Do you remember what tree networks are? We discussed them when we covered Christopher Alexander's article "A City Is Not a Tree" in chapter 4. By definition, a tree network is a network without loops, and by definition a tree network cannot have more than $N-S$ links, which is exactly what we have for networks 1 and 2. So, μ calculates the number of "extra" links—that is, in addition to the links needed to form a tree network.

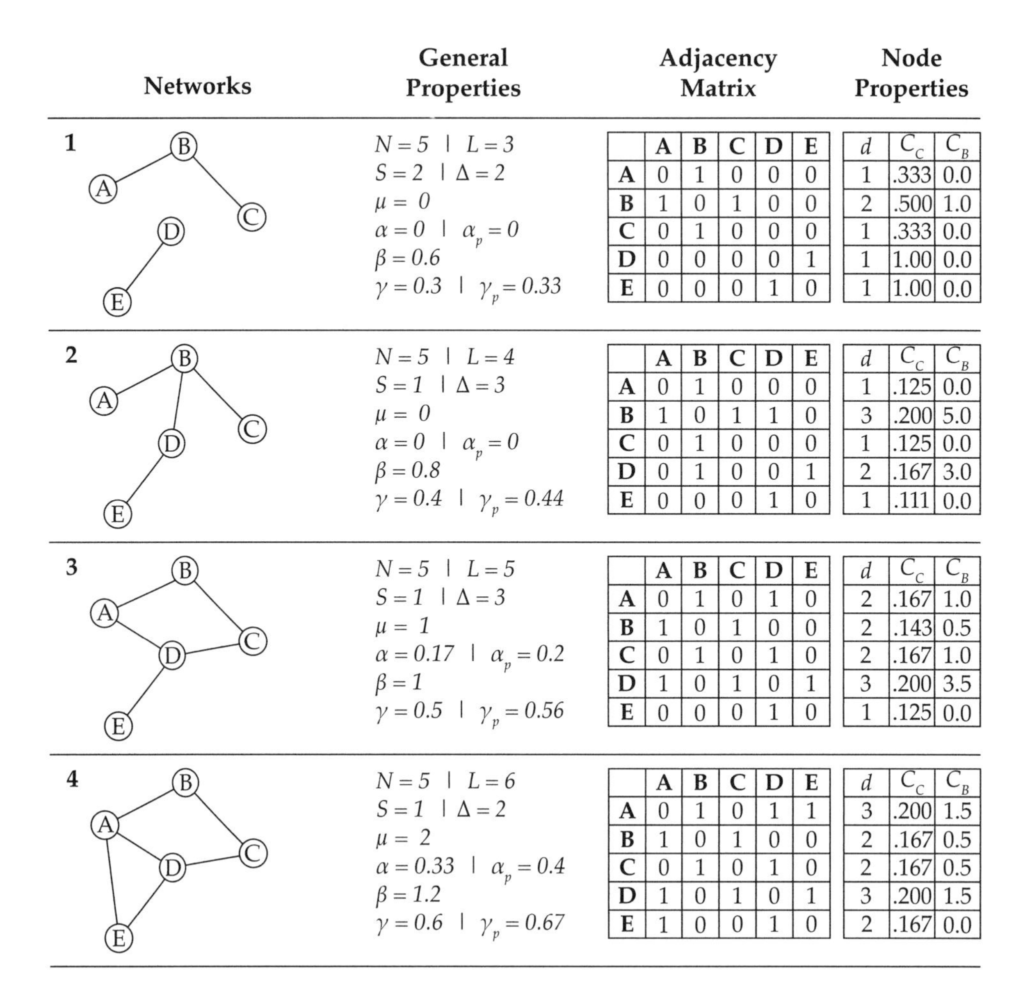

Figure 11.11
Four sample networks.

We can extend the idea of the cyclomatic number by measuring the number of cycles a network has compared to how many it could have in the maximum case. Let us first start by identifying the maximum number of links that a network can have. If we have N nodes and we want to connect all of them to every other, then the maximum number of links is $N \times (N-1)$. This is the case for directed networks. For undirected networks (i.e., connecting A to B or B to A is the same thing), we simply take half of it, so the maximum number of links is ½ $N \times (N-1)$. Since the maximum number of links in a tree network is $N-S$, the maximum number of cycles in a network is ½ $N \times (N-1) - (N-S)$. We can now define the ratio of the actual to potential number of cycles, α, as:

$$\alpha = \frac{L - N + S}{\frac{1}{2}N(N-1) - (N-S)} \tag{9}$$

One feature of infrastructure systems is that many of them are *planar*. This means that if two links cross, they necessarily create a node, like road systems with no underpasses or overpasses. All networks in figure 11.11 are planar. For planar networks, the maximum number of links is $3N-6$. This means that the maximum number of cycles becomes $3N-6-(N-S)=2N-6+S$. In the planar case, α_p is defined as:

$$\alpha_p = \frac{L - N + S}{2N - 6 + S} \tag{10}$$

Since networks 1 and 2 have no cycles, their values of α and α_p are 0. With one cycle, α and α_p of network 3 are 0.17 and 0.2, and they increase again for network 4 since it has two cycles.

The next measure is the Betti number β that simply calculates the ratio of the number of links to nodes:

$$\beta = \frac{L}{N} \tag{11}$$

The symbol is the same as the exponent from sections 11.1.2 and 11.1.3, but the variables are completely different.

The Betti number β is a simple indicator that informs us about the "complexity" of a network. In other words, the more links a network has, the more complex it must be. It is therefore not bound between 0 and 1, unlike α. At the same time, it is half the average number of connections per node, but we will cover the number of connections later. From figure 11.11, values of β range from 0.6 to 1.2.

Another useful indicator is the degree of connectivity γ. It is a little simpler than α since it calculates the ratio of the actual to potential number of links in a network. Since we have already covered how to calculate the maximum number of links in non-planar and planar networks, γ and γ_p are defined as:

$$\gamma = \frac{L}{\frac{1}{2}N(N-1)} \tag{12}$$

$$\gamma_p = \frac{L}{3N-6} \tag{13}$$

From figure 11.11, γ ranges from 0.3 to 0.6, and γ_p ranges from 0.33 to 0.67. In general, having more connections tends to be more desirable, especially for infrastructure systems, because it means there is more than one way to go from one node to another, although we have to account and control for interdependencies. This is actually related to matters of resilience, but we will discuss that a little later when we get to centrality.

A network can be represented as a drawing, as we can see in figure 11.11, but it can also be represented by a matrix, called an adjacency matrix A. The adjacency matrix is a square matrix, in which each node has one row and one column. If node i is connected to node j, then the cell ij has a value of 1; otherwise, the value is 0. More formally, the adjacency matrix A is defined as

$$A_{ij} = \begin{cases} 1 & \text{if nodes } i \text{ and } j \text{ are connected} \\ 0 & \end{cases} \tag{14}$$

Each network in figure 11.11 has its adjacency matrix shown on the right-hand side. We can actually do quite a few things with these matrices. To start with, we can easily calculate the number of connections for each node. In network science, this is called the degree of a node. In network 1, node A has 1 connection and therefore $d_A = 1$. From the adjacency matrix, the degree d_i of node i is defined as:

$$d_i = \sum_j A_{ij} \tag{15}$$

One of the most important features of a network is how its links are distributed, which we call *degree distribution*. In other words, we count how many nodes have one link, two links, three links, and so on, and we plot a frequency diagram. Some networks have degree distributions that follow a normal distribution (often called random networks or Erdős–Rényi networks), while others follow a power law (often called scale-free networks[13]). These properties have a significant impact on the performance of networks, but we will not get into them.

Moreover, sometimes links are given weights—for instance, to represent a volume (like traffic on a road). If we use equation (15), we do not have the degree of a node anymore, but we have its *strength*. To some extent, the degree or strength of a node is also a measure of *centrality*, in the sense that nodes with more connections (or higher strength) tend to be more *central* in a network.

Beyond degree and strength, there are many more ways to characterize the centrality of a node. Another way is to measure how close a node is to another in terms of the shortest number of connections between them. This is called a path,[14] or more accurately, it is a shortest path. From network 1, the shortest path between nodes A and C is 2. Based on this path, we can then define the diameter, Δ, of a network, which we saw earlier, as the largest of all the shortest paths.

Following this idea, we can also define the shortest paths between nodes i and j as $p(i,j)$, and the closeness of a node is the sum of all its shortest paths to all other nodes. The problem with this is that the larger closeness would be, the less central it would be—in fact, the sum of shortest paths for a node is instead referred to as its *farness*. Instead, we prefer to use one over the sum of shortest paths, or in equation form the closeness centrality C_{Ci} of node i is defined as:

$$C_{Ci} = \frac{1}{\sum_j p(i,j)} \tag{16}$$

Nodes with high C_{Ci} values are closer to all other nodes and are therefore more central. We can also see that closeness rapidly becomes a small number for large networks, and thus the normalized version of closeness C_{CN} is often used:

$$C_{CNi} = \frac{N-1}{\sum_j p(i,j)} \tag{17}$$

where we simply multiply equation (16) by $N-1$ as the number of nodes a node i can reach.

We can now expand this concept a little. Instead of considering the distance of a node to other nodes, we can look at how many times it is used to connect two other nodes. Indeed, a node may be off-center a little and have a relatively small closeness centrality, but it may still be used heavily to connect other nodes together. In network science, this is called *betweenness centrality*. The best analogy is traffic volume in transport. Small roads in city centers may be closer to all other roads in a city, but arterials are used heavily and should therefore be considered more *central* in the sense that we adopt here. Mathematically, letting p_{jk} be the number of shortest paths that connect nodes j and k, and letting $p_{jk}(i)$ be the probability of node i to link them, the betweenness centrality C_{Bi} of node i is defined as:

$$C_{Bi} = \sum_{j,k} \frac{p_{jk}(i)}{p_{jk}} \tag{18}$$

For example, in network 4, there are two ways to connect nodes A and C, either through node B or through node D. We assign each a probability of being used of 0.5. Since this is the only shortest path for which node B is used, its betweenness centrality is 0.5. Similarly, node C is one of two nodes that can be used to connect B and D, and therefore its betweenness centrality is 0.5. Node E is never used to connect two other nodes, hence its betweenness centrality of 0. In contrast, node D is used to link both

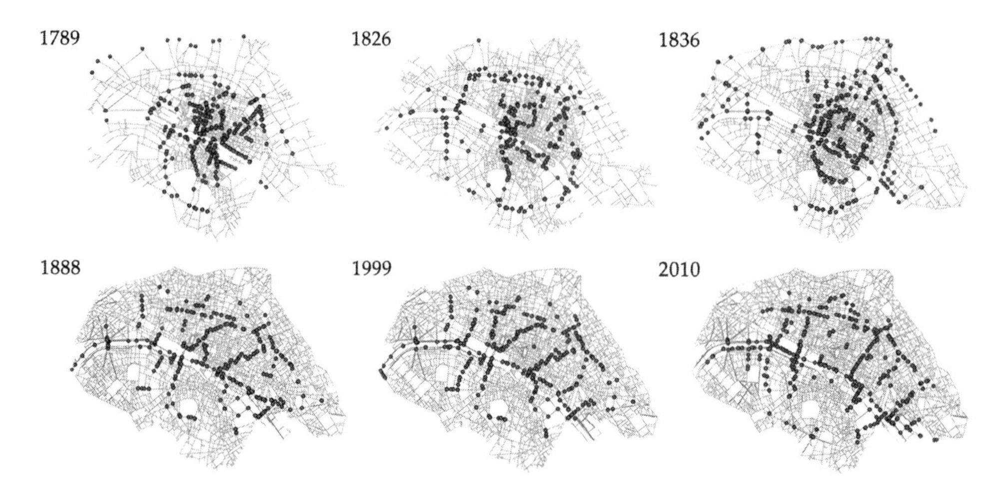

Figure 11.12
The evolution of betweenness centrality in the Paris Road Network. The highlighted nodes represent those with a betweenness centrality at least 10% of the value of the node with the highest betweenness. Courtesy of Patricia Bordin.

C to A (with a 50% chance) and C to D (with a 100% chance), giving it a betweenness centrality of 1.5. Similarly, node A is used to link nodes D and B (with a 50% chance) and B to E (with a 100% chance), hence its betweenness centrality of 1.5.

In practice, network science can be of significant help to identify the properties of urban engineering systems. For example, figure 11.12 shows how betweenness centrality evolved in the Paris road network from 1789 to 2010. The highlighted nodes represent the nodes with a betweenness centrality at least 10% of the value of the node with the highest betweenness—mathematically, all nodes i with $C_{Bi} > (C_{Bmax}/10)$ are highlighted. What can you tell from the figure? Can you identify major changes in the road system? In chapter 4 we briefly talked about Baron Haussmann, who changed Paris significantly in the second half of the nineteenth century. In particular, Haussmann is known for developing large boulevards, and from the figure, we can see that betweenness shifted from being mostly around the center of the city to being distributed along major roads—this is a typical example of top-down planning.

Naturally, many more concepts and metrics exist, some of which are relevant for infrastructure systems. We even talk about multiplex or multilayer networks (i.e., networks of networks) to analyze networks that are connected with one another—for example, transport and water. For this, we use the same bases of nodes, links, and adjacency matrices. Figure 11.13 shows an example of three separate but interconnected

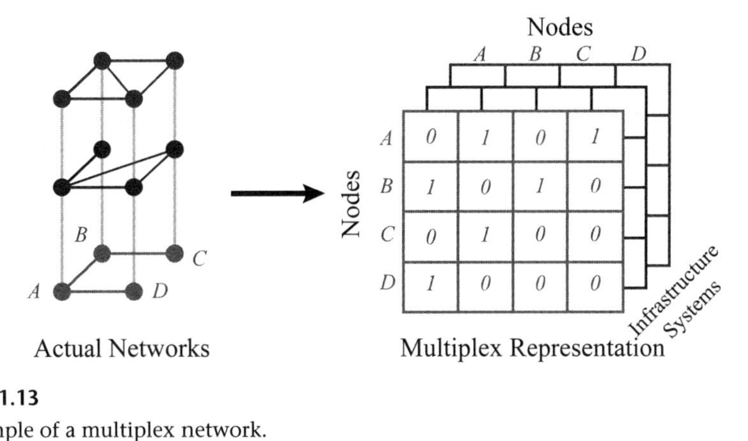

Figure 11.13
An example of a multiplex network.

networks. This is also sometimes called a *tensor* representation because it looks like a tensor from computational mechanics.

Most of the metrics we have seen can easily be calculated by hand or by using the adjacency matrix in any commercial statistical software like Microsoft Excel. Others are more complicated, though, like closeness and betweenness centrality. In general, using software to calculate network properties makes the whole process much easier.

Fortunately, as network science has become more mainstream, many options exist. An easy way to calculate these metrics is to use NodeXL, which is a Microsoft Excel add-on; a basic version was available for free at the time of this writing. It calculates the metrics that we have seen, as well as others, but it can really only be used for relatively small networks, however. A slightly more advanced option is offered by the freeware Gephi.[15] Gephi not only calculates most relevant metrics from network science but is also a visualization software package, and it therefore creates nice layouts (as we will see in example 11.3). Although a little better than NodeXL, large networks still cannot be processed easily in Gephi.

To be able to study large networks, we need to do some coding. Many free libraries exist that calculate just about any network properties.[16] The Python and R libraries igraph and NetworkX were particularly popular at the time of this writing and incredibly easy to use. Naturally, many other tools and software packages exist, most of which are free to use, and a Wikipedia page contains a list of those available.[17] The address of the page indicates that network science really emerged to look at social networks, but the types of networks that are studied have vastly outgrown the realm of social network analysis.

Example 11.3

Network Analysis of Main Arterials in Civitas

The figure below shows the network of the main arterials in Civitas. Based on the figure, (1) construct the adjacency matrix of the network. Based on the adjacency matrix, (2) calculate N, L, S, μ, α_p, β, and γ_p. Using any software, (3) calculate the network diameter Δ and determine the degree d, normalized closeness C_{CN}, and betweenness centrality C_B of each node, and (4) visualize the network in various layouts.

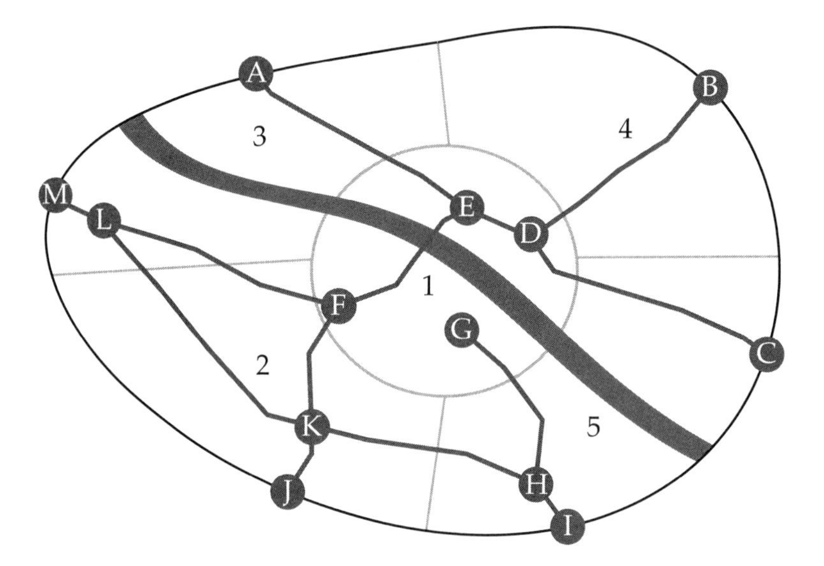

Solution

(1) From the figure, we can easily construct the adjacency matrix of the network. Omitting all the 0s to see more clearly all the links, we get:

Example 11.3 (continued)

	A	B	C	D	E	F	G	H	I	J	K	L	M
A					1								
B				1									
C				1									
D		1	1		1								
E	1			1		1							
F					1						1	1	
G								1					
H							1		1		1		
I								1					
J											1		
K						1				1		1	
L						1					1		1
M												1	

(2) From the figure and the adjacency matrix, we can easily calculate:

- $N=13$
- $L=13$
- $S=1$
- $\mu=1$
- $\alpha_p=0.048$
- $\beta=1$
- $\gamma_p=0.39$

(3) For this question, Gephi was used. The network diameter Δ is 6 to go from nodes B or C to node I. Moreover, the degree d, closeness C_{CN}, and betweenness centrality C_B of each node are

Example 11.3 (continued)

Nodes	d	C_{CN}	C_B
A	1	0.308	0
B	1	0.261	0
C	1	0.261	0
D	3	0.343	21
E	3	0.429	35
F	3	0.480	35
G	1	0.273	0
H	3	0.364	21
I	1	0.273	0
J	1	0.324	0
K	4	0.462	35
L	3	0.414	11
M	1	0.300	0

In particular, we can see that node K has the highest degree, closeness, and betweenness centrality. Although it is not located in the center of Civitas, K must be critical, and effort should be put into making sure that it is always operational.

Moreover, we can see that both nodes H and L have a degree of 3. However, node L has a higher closeness while having lower betweenness than node H. This is interesting. One measure of centrality tells us that the two nodes are equal, while two other methods contradict each other. This dilemma can be seen from the figure. Node L is important to connect to node M, and it also enables the one cycle that the network has (nodes L-K-F). In contrast, node H connects to two nodes G and I that would not be connected to the network otherwise. This simple example shows us the power of network analysis.

(4) Gephi offers a variety of different network layouts. The figure below shows three options. Nodes with higher degrees are darker.

Note that there is no "optimal" layout. The most sensible option is to play with several layouts and determine which one best conveys the sort of information that is sought after. In our case, the Yifan Hu layout seems to best convey which nodes are more central in our network and thus perhaps more important. With traffic volume, we could also redraw these networks by adjusting the size of each link based on traffic volumes.

Example 11.3 (continued)

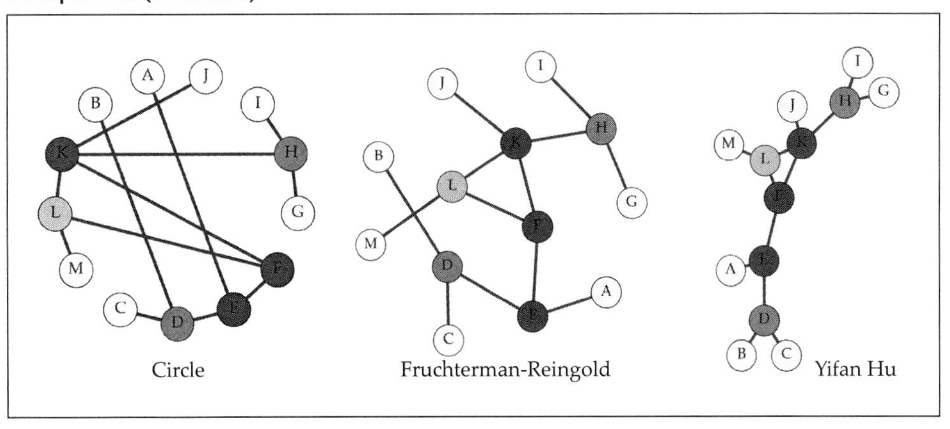

This is all that we will cover on network science. To learn more about network science specifically, the free online book *Network Science* by Barabási (2016) and the textbook *Networks: An Introduction* by Newman (2010) are recommended. To learn more about how the Science of Cities has been used to study cities and urban infrastructure, another work by Marc Barthelemy (2016) is recommended titled *The Structure and Dynamics of Cities: Urban Data Analysis and Theoretical Modeling.*

This overview of network science completes our introduction to the Science of Cities. Next, we will focus on a family of data mining techniques that has proven to be extremely powerful for processing and analyzing data thanks to the use of algorithms, and there is no doubt that these techniques will be used heavily in the future in urban engineering. This family of techniques is commonly grouped under the term Machine Learning (ML), and it is the topic of our next and final section.

11.2 Machine Learning

Developed originally in computer science, ML can essentially be used on just about any data set. The terms used in ML can sometimes sound a little complicated—ML is partly based on algorithms as opposed to the traditional statistics commonly used in engineering—but once we look deeper, we rapidly see that ML techniques are really not that difficult to understand, and more importantly, they are really easy to use. ML will likely get its own chapter in the future, but for now this section will suffice.

As the name suggests, ML is about letting an algorithm learn by itself as opposed to telling it what to do. This is relevant for driverless cars, for instance. Telling a car what to do in all possible cases is virtually impossible. Driving around, collecting data, and letting the "machine learn by itself"—especially how to "see" (i.e., image processing)—is much easier and much more effective.[18] In particular, we sometimes use simple rules that may not be "optimal" but that work much faster and are often (though not always) preferable.[19]

For us, the real strength of ML is its ability to process just about any kind of data. In engineering, we tend to prefer *continuous* data sets that we can easily plot on x- and y-axes, just like the data we used for scaling laws and Zipf's law. What do we do when we have *ordinal* data, however, such as small, medium, and large? Or *categorical/nominal* data such as the name of a city where a person lives?

In particular for us, energy and resource consumption may change by climate zones, which are ordinal (1 is very warm and humid, such as Miami, and 8 is very cold, such as Alaska). Do we first need to manually divide our data by climate zone and then try to plot it? Moreover, what if we have other types of categorical data, like residential versus commercial use for a building? Clearly, we need better tools, and this is where ML thrives. In fact, ML can even deal with text (i.e., text mining), but we will not get into that.

In this section we will first learn about basic concepts in ML. In particular, we will learn that we do not necessarily need to have a dependent variable to use ML—that is, what if we simply want to discover whether patterns exist in our data? We will then learn and apply three powerful ML techniques: K-means clustering, decision tree learning, and neural networks.

Since it is generally easier to learn mathematical and physical concepts by giving practical examples, we will use the Concrete Slump Test (CST) data set from Yeh (2007) in all the example sets of this section. The CST data set is stored at the University of California at Irvine Machine Learning Repository.[20] The data set contains 103 data points for different concrete mixes. It has seven input variables in [kg/m^3] of concrete: cement, slag, fly ash, water, superplasticizer, coarse aggregate, and fine aggregate. It then has three output variables: slump [cm], flow [cm], and twenty-eight-day compressive strength [MPa].

11.2.1 Basic Concepts of Machine Learning

Machine Learning comprises two main families of techniques: (1) unsupervised learning and (2) supervised learning. This is probably the most fundamental concept to understand about ML. To make it easier, figure 11.14 shows the properties of each family.

Unsupervised learning consists of finding patterns in a data set when we are not sure whether any relationship exists. One of the most popular families of unsupervised learning

Example 11.4
Mining Patterns in the CST Data Set

To first investigate whether obvious relationships exist between the variables of the CST data set, (1) plot a graph of each possible relationship between the three dependent variables and the seven independent variables and (2) comment on the results.

Solution

(1) Essentially, we have to draw $3 \times 7 = 21$ graphs. Doing this by hand can be tedious, but it becomes quite simple with a short code. The figure below shows the 21 plots—it was made with the free matplotlib and seaborn libraries in Python.

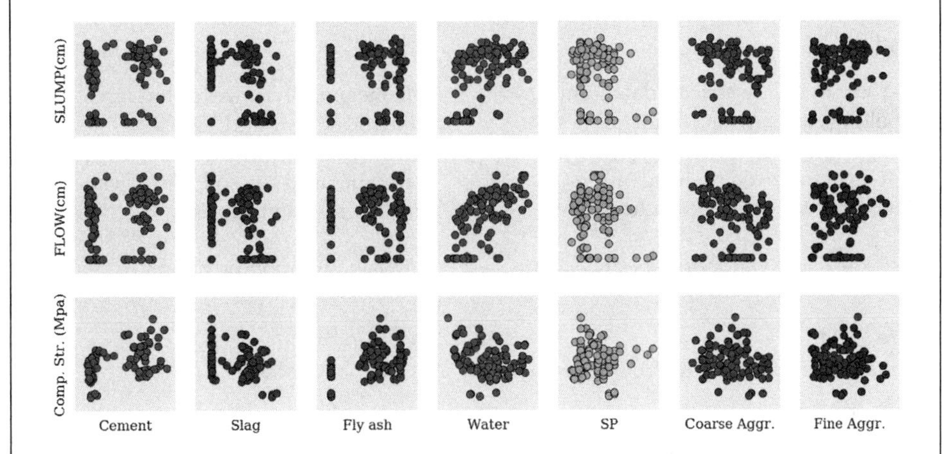

(2) From the figure, no obvious relationship can be identified between any of the dependent variables and the independent variables (i.e., we cannot draw any obvious straight lines, power laws, exponential functions, quadratic functions, and so on). This general observation confirms that we cannot use traditional statistics if we want to be able to predict the performance of a concrete mix, and as we are about to learn, this is where ML shines. We also see that the graphs for flow and slump are nearly identical, which suggests that both output variables are affected by some properties of a concrete mix. Results for compressive strength are a little different from the other two output variables.

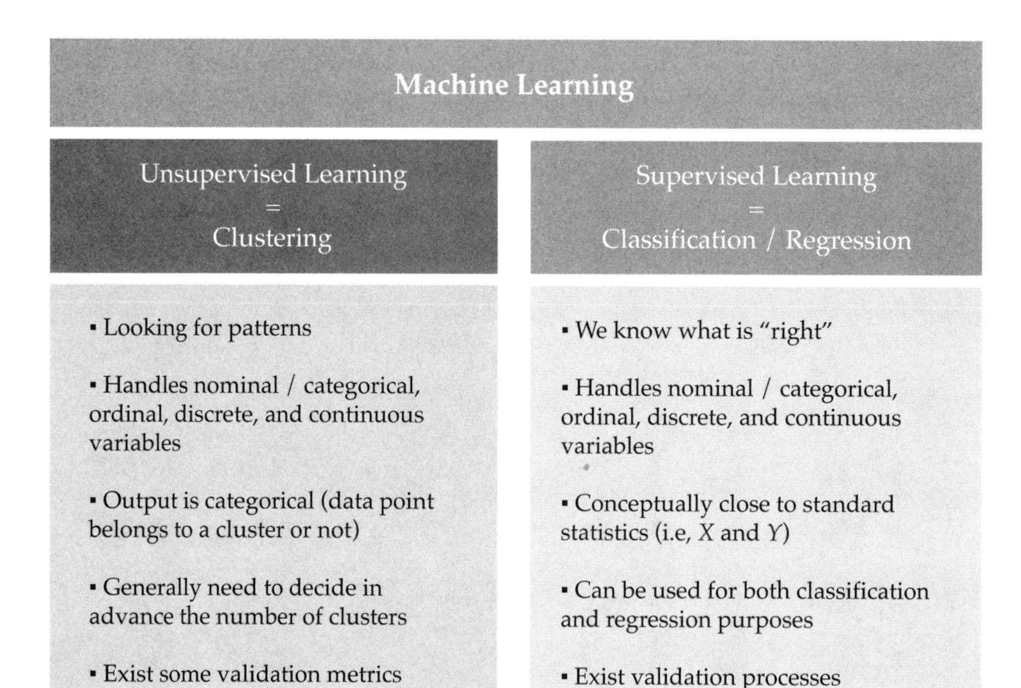

Figure 11.14
The main characteristics of Machine Learning.

techniques is *clustering*, in which we try to cluster (i.e., group) data points together based on their similarities. For example, we might find four main groups of points in a data set of 100 points. This type of technique can be used with any type of variable, but it outputs only categorical data—that is, it reports the cluster in which the data point belongs. We also have to specify the number of clusters that we think should exist. In practice, we simply try a bunch of them and identify the number of clusters that seems best. K-means clustering, as we will see later, is a typical unsupervised learning technique.

In *supervised learning*, we know that a "ground truth" exists (i.e., we have a dependent, *y*, variable), whether it is a *class* or a *continuous number*. Classes are discrete, like letter grades (i.e., A, B, C, D, F). As the name suggests, continuous numbers are continuous, like percentage grades (i.e., 95%, 82%, 75%). When we look for a class, we say we are dealing with a *classification* problem—we want to classify data into well-defined groups. It is a little like clustering, except that in this case we already know the groups that exist. In contrast, when we look for a continuous number, we say we are dealing with a *regression* problem. Here, regression has the same meaning as in statistics (e.g., simple linear regression), but we make full use of all the data. Decision tree learning is mostly used as a classification technique, and neural networks is used both as a classification and a regression technique. Both are supervised learning techniques.[21]

For both supervised and unsupervised learning, we need to *train* our algorithm. In chapter 7, we *calibrated* our logit model, here, we essentially do the same thing, but we say we *train* our model. For unsupervised learning, we essentially train our algorithm on all of our data points since we are looking for clusters of points. In supervised learning, however, we *train* our algorithm on some of the data points, and we *test* the trained algorithm on the remaining data points to measure its performance. For small data sets, we typically train the algorithm on about 80% of the data set, and we test the trained algorithm on the remaining 20% of the data. For larger data sets, we take a random (but fairly large) sample of data points, *train* the algorithm on this sample, and use different data points to *test* the trained algorithm.

To determine whether our trained algorithm is performing well, we calculate various statistics and analyze the results. One simple measure for classification is to count how many times the algorithm got it right—sometimes simply referred to as *accuracy*—thus giving us a percentage of times we were right. For regression, calculating the R^2 offers a good start, and many other measures, like mean square error (MSE), are sometimes used to train the algorithm in the first place. Moreover, we often like to further split the training set into multiple batches to train the algorithm multiple times and see whether the performance tends to stay stable or whether it varies constantly (which is generally a bad sign). This technique is called k-fold cross-validation (k for the number of batches we create). In contrast, unsupervised learning does not require any testing since we have nothing to validate (i.e., we have no ground truth), but we do have some metrics that can help us determine whether a pattern exists and the most accurate number of clusters.

Overall, objectively assessing the performance of a trained algorithm is far from trivial. In some cases, statistical indicators that are too good may actually suggest that the algorithm is overfitting the data used and may not perform well in the longer term. It is therefore desirable to test different techniques and to compare their performance. We then have to exercise our "scientific judgment" based on what we know about the data. A variety of techniques are available, but we will not discuss them here. What is certain is that one should always have a good understanding of the data being used to be able to assess the performance of a trained algorithm.

Here again, some coding tends to be required, although new ML tools come out frequently. Personally, at the time of this writing, I mostly used the free Python library Scikit-Learn (Pedregosa et al. 2011); it includes many ML techniques, is very easy to use, and is completely free. Nonetheless, many other libraries exist, including TensorFlow

(Abadi et al. 2015) for neural networks and deep learning, NLTK: The Natural Language Took Kit (Loper and Bird 2002) for text mining,[22] and many more.

We will now be introduced to our first ML technique: K-means clustering.

11.2.2 K-means Clustering

K-means clustering is an unsupervised learning technique that is all about clustering data points together based on their proximity to one another. We are not referring back to network science here. Instead, we need to calculate the actual (Euclidean) distance between two points on a graph. If we have data plotted on an x-y axis, then we calculate the distance between two data points. The goal is to group data points into separate clusters.

For example, figure 11.15 shows compressive strength versus cement content in the CST data set. Clearly, the figure shows no linear or simple nonlinear relationship, such as a scaling law. This does not mean that the data do not contain any relevant information, however. In fact, we can see at least two clusters of data points, roughly divided at 225 kg/m^3 of cement. This is typically where clustering algorithms can help us.

What we want is to determine which points are closer to one another. The first thing we must do is to determine the number of clusters we think exist. Starting with two

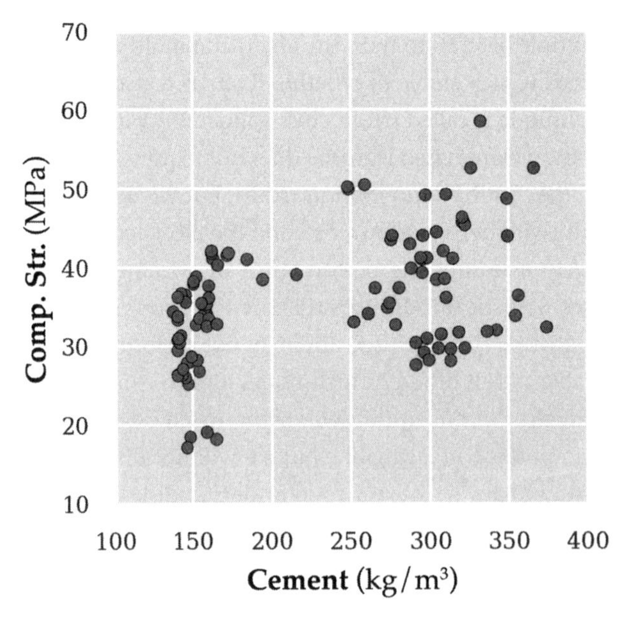

Figure 11.15
Compressive strength versus cement for CST data set.

clusters, K-means works as follows: (1) The algorithm picks two points at random and assigns them as cluster centers, and (2) it calculates the distance between every point and the two cluster centers and assigns the points to the cluster of the closest center. Distance is exactly what it sounds like. For two points A and B that are expressed by two variables x and y, the Euclidean distance between them is the following:[23]

$$dist(A,B) = \sqrt{(x_A - x_B)^2 + (y_A - y_B)^2} \tag{19}$$

Or more generally, when we have n variables x_i:

$$dist(A,B) = \sqrt{\sum_i^n (x_{A,i} - x_{B,i})^2} \tag{20}$$

Once the algorithm has figured out which data points are closest to the two randomly selected points, (3) it then finds the centroid of each of the two clusters. In K-means, the centroid is basically the center of the cluster (i.e., the average coordinates of all the points in the cluster), and it does not have to be an existing point. Once the centroids are calculated, (4) it does step (2) again. This is followed by step (3), step (4), and step (2) again, until all the points are clustered, and the clusters are stable (i.e., points are not assigned to different clusters).[24]

Mathematically, it essentially minimizes the sum of the distances between every point and their centroids, or in equation form:

$$\arg\min \sum_i \sum_j (\| x_i - c_j \|)^2 \tag{21}$$

where the double bars mean the Euclidean distance, and c_j is the centroid of cluster j.

Ideally, what we need afterward is some kind of metric that can tell us how well the clustering worked. After all, we not only have to find the right number of clusters that exists but we also have to choose the best one from the other clustering techniques that exist,[25] and this is not always obvious. It turns out that there is such a metric called the silhouette coefficient s. For a given point i in cluster C_i, its silhouette coefficient $s(i)$ is defined as:

$$s(i) = \frac{b(i) - a(i)}{\max[a(i), b(i)]} \tag{22}$$

where $b(i)$ is the minimum average distance between i and all points in any other cluster $C_{j \neq i}$ (i.e., the average distance to the points in the closest neighboring cluster), and $a(i)$ is the average distance between i and all points within its cluster C_i. Essentially, we want $b(i)$ to be large and $a(i)$ to be small, so $s(i)$ should be as close to 1 as possible. In equation form:

$$a(i) = \frac{\sum_{j \in C_i, i \neq j} dist(x_j, c_i)}{|C_i| - 1} \tag{23}$$

$$b(i) = \min_{C_j : 1 \leq j \leq k, j \neq i} \left\{ \frac{\sum_{j \in C_j} dist(x_j, c_i)}{|C_j|} \right\} \tag{24}$$

We can then calculate the average silhouette coefficient of a cluster by averaging the individual silhouette coefficients of all its points.

For classification purposes, an algorithm that is also based on distance exists, and it is called k-nearest neighbors (kNN). With kNN, the class of an unknown point takes the same class as its k neighbors. For example, with $k=3$, if two neighbors of a point belong to class A and one neighbor belongs to class B, kNN would assign the point to class A. To avoid ties as much as possible, we prefer to pick k as an odd number. When it comes to classification, however, other techniques may work best, and this is the expertise of our next ML technique.

11.2.3 Decision Tree Learning

Decision tree learning is a rule-based supervised learning technique—therefore we need a dependent variable y—and what the algorithm does is find the "best" rules to determine y. Here again, we will deal with tree networks, but this time we will be happy with the tree structure.

Decision tree learning works especially well as a classification technique—that is, when the dependent variable y is ordinal (e.g., small, medium, and large) or categorical (e.g., city name, or pass or fail). In the event that the dependent variable is continuous, we should try to make it categorical. An option is to look at whether a variable is lower or higher than the average (i.e., we create two classes: low and high); this is what we will do in example 11.6. This may not be enough, however. We can instead build classes manually (e.g., turn percentages to grades—more than 90% is A and so on) or use percentiles (i.e., every tenth percentile). If we really cannot divide our variable in a number of classes because we want to make an accurate prediction that needs to be continuous, then we need to use a regression technique such as neural networks, which we will cover in the next section. Although the general family of decision tree learning techniques can be used for regression purposes (e.g., gradient boosting machine), we will not cover it here except for one endnote.

Example 11.5

K-means Clustering of CST Data

Using the CST data set, (1) apply K-means clustering to compressive strength and cement content for two, three, four, and five clusters and (2) comment on the results.

Solution

(1) The figure shows four plots containing the results of applying a K-means algorithm for two, three, four, and five clusters on compressive strength versus cement content in the CST data.

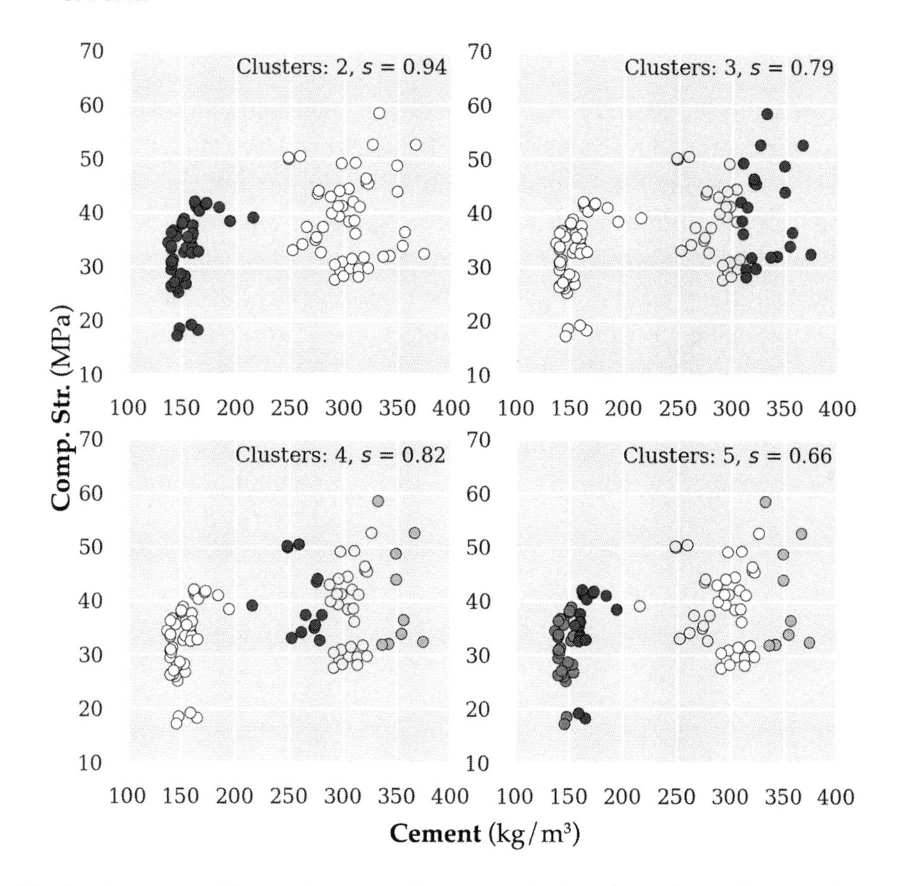

(2) The figure from (1) also shows the silhouette coefficient for each plot. For two, three, four, and five clusters, they are 0.94, 0.79, 0.82, and 0.75, respectively. This suggests that the K-means algorithm trained for two clusters performs best, which actually makes sense when we look at the figure.

Essentially, we aspire to split the independent variables *x's* based on their accuracy to determine the final class of a data point. We first take the most accurate variable and split it the best possible way. For example, let us say that whether it rains or not is the most accurate variable to determine the strength of concrete from examples 11.13 and 11.14. We would have a dummy variable—that is, 0 or 1 based on whether it rained the day we poured the concrete mix—and in the tree we would basically divide all the data points that we collected based on when it rained or when it did not rain.

Figure 11.16 gives us an example. Our first variable V_1 is whether it rained or not, and the tree tells us that if $V_1 \leq x_1$, we go to the left side of the tree, and if $V_1 > x_1$, we go to the right side of the tree. Because V_1 is a dummy variable, in this case x_1 will simply be 0.5.

Next, we go to the second level, and we do the same thing for each branch separately. Let us assume that for the first branch of the second level, $V_{2,1}$ is the cement content, and for the second branch of the second level, $V_{2,2}$ is the water content. Because these variables are continuous, we will need to "split" or "bin" them into two categories to again create two branches.[26] If the cement content is smaller than $x_{2,1}$, we take the left branch (the process is detailed in the next paragraph). In the end, we get the classes that we desire. In figure 11.16, four consecutive classes are shown (A, B, C, and D), but the results could easily be different, like A, B, A (again), and C. The most important factor is that at the end, we have a decision tree that has a series of specific rules that we can follow—hence the term *ruled-based* technique for decision tree learning—and when we arrive at the bottom of the tree, we have determined the class of our data points.

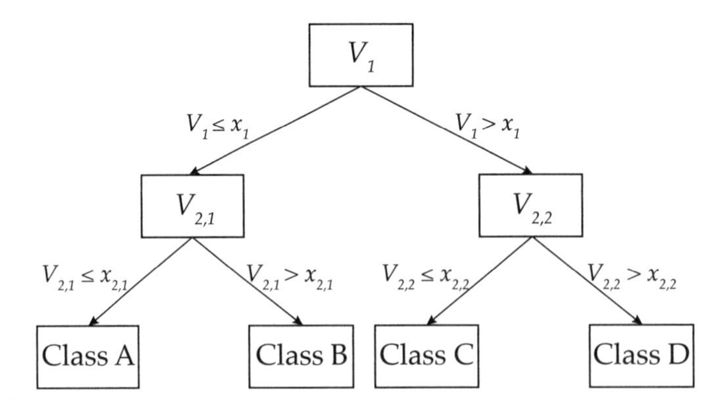

Figure 11.16
A sketch of the decision tree process.

The question now is how do we measure "accuracy" to determine which variable should be used first? Essentially, we want to select a variable that is "better" at predicting a class. For example, imagine we are trying to predict the income of people and we have two variables for them: level of education and city of birth. Clearly, level of education should be a better predictor of income than city of birth. In this case, we are more likely to use level of education in the first level and perhaps city of birth in the second level if it is found relevant. In technical terms, level of education contains more "information." We can also say that level of education is more "certain" or "pure."

To calculate the *information*, *certainty*, or *purity* of a variable, we mainly use two concepts.[27] We actually saw the first one, entropy, briefly in the transport chapter. In the information theoretic sense, entropy H_E is defined as:

$$H_E = -\sum_i p_i \log p_i \tag{25}$$

The second concept is Gini impurity H_G, which is defined as:

$$H_G = \sum_j p_j (1 - p_j) = 1 - \sum_j p_j^2 \tag{26}$$

Figure 11.17 shows how these two measures work. Entropy and Gini impurity are plotted for cases of two probabilities p and $(1-p)$. Think of it as a heads or tails game, except that the coin can have a bias. The normal case is when heads has a 50% chance of happening (i.e., $p=0.5$), and tails has a 50% chance of happening (i.e., $1-p=0.5$).

With a bias, however, heads can have only a 10% chance of happening (i.e., $p=0.1$), which means that tails now has a 90% chance of happening (i.e., $1-p=0.9$). If you

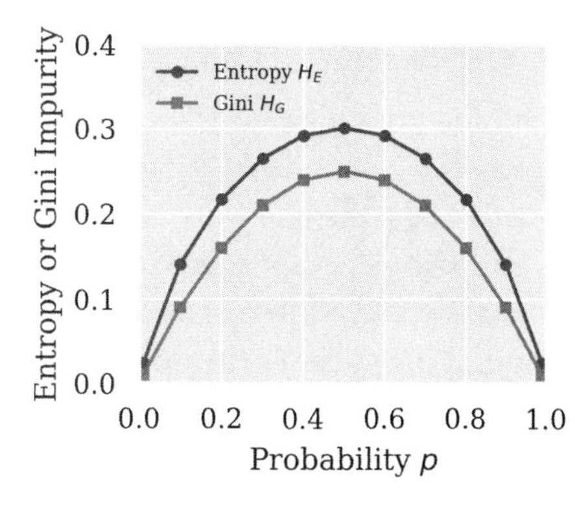

Figure 11.17
Entropy H_E and Gini impurity H_G.

were asked to bet on it, you would prefer to bet on the coin that has a bias, since you would have a 10/90 chance of being right as opposed to a 50/50 chance. Here, this is exactly what we mean by *information* or *purity*. When the coin has a bias, we are much more certain to get tails, and therefore we have more information, purity is higher, and both entropy and Gini impurity are lower. Figure 11.17 shows these for cases when heads have a 1, 10, 20, 30, 40, 50, 60, 70, 80, 90, and 99% chance of happening.

In the learning process of the decision tree, we calculate either entropy or Gini impurity for all variables, and the variable with the lowest value is selected—these are the "rules" learned by the algorithm. In practice, we actually go one step further. From equations (25) and (26), we can see that the number of probabilities (not just two, as in our example) will have an impact on the value of entropy and Gini impurity. We therefore need some kind of a normalizing factor to ensure we select the right variable. This is partly why we prefer to only split any variable, discrete or continuous, into two parts. Information G for each variable i is then calculated as:

$$G_i = \frac{n_{left}}{N} H_{i,left} + \frac{n_{right}}{N} H_{i,right} \tag{27}$$

where n_{left} and n_{right} are the number of data points on the left and right branch, respectively, N is the total number of data points, and H_i is the entropy or Gini impurity of the variable i applied either to all the points on the left or the right. We will not get into the details here, but equation (27) is part of a common decision tree technique called classification and regression tree (CART).

In the end, decision tree learning is extremely useful, but it is prone to overfitting because only one "perfect" tree is trained. To address this problem, we use what are called *ensemble* techniques that develop multiple trees. For example, instead of using the full 80% of the training data, we use less than 80% (say, 40%) but we train multiple trees on a different 40% of the entire 80%. Another possibility is to omit some of the independent variables and train multiple trees again. Essentially, we are adding "randomness" to the data set. The next time we try to predict a class, we "count" the number of votes for a class and pick the one with the most votes. For example, four trees may predict class A, and two trees may predict class B, and therefore the algorithm would return class A. As a whole, while one of these trees may not perform well, together they generally perform much better than one single "perfect" decision tree. Because multiple imperfect trees are trained, this technique is called *random forest*.[28]

This is all we are going to see about decision tree learning. What is fascinating about decision trees is that they work in a radically different way than conventional statistical techniques since we essentially try to determine some "rules" to follow. The concept

Example 11.6

Decision Tree Learning of CST Data

Using the CST data set, (1) apply decision tree learning to twenty-eight-day compressive strength by defining two classes: below or above average (the average was calculated to be 36.04 MPa; although this is a little simplistic for this case study, it will allow us to easily visualize the tree that will be trained) and report the results and (2) show and discuss the tree trained.

Solution

(1) The free Python library Scikit-Learn was used to train and test the decision tree. The algorithm was trained on 80% of the data and tested on the remaining 20%. Gini impurity was used for the training, and the tree trained was right 90.5% of the time. These results are actually quite high considering the relatively low number of data points.

(2) The figure below shows the tree that was trained in (1). The figure was made directly in Scikit-Learn. We can see that cement content is the first variable used, and the split is done for 159.5 kg/m^3 of concrete. Out of eighty-two data points (i.e., the 80% train set), forty data points have a cement content below 159.5 kg/m^3, and forty-two data points have a cement content above 159.5 kg/m^3. The Gini impurity value was 0.5. At the second level, water is used as the splitting variable on the left, and fly ash is used as the splitting variable on the right. Water is split at the value 174.45 kg/m^3, and fly ash is split at the value 117.5 kg/m^3. In the third level, we have the final results. From left to right, the leaves (which is what bottom boxes are called) predict classes: Below, Above, Above, and Below.

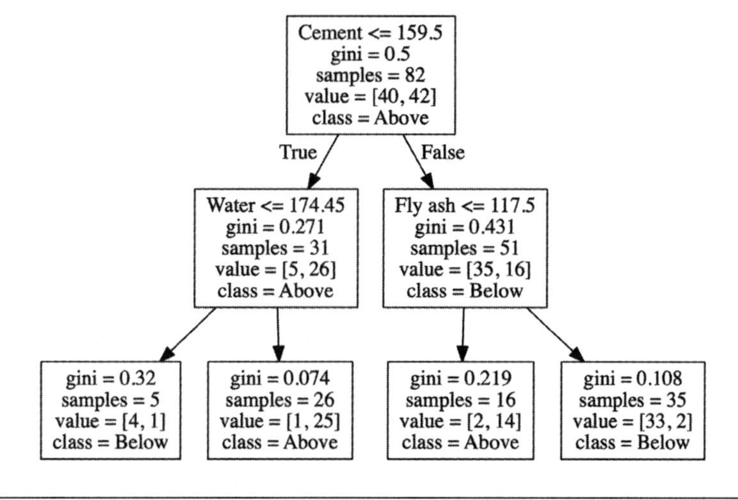

Example 11.6 (continued)

The hierarchy of the variables as well as the logic captured is also quite interesting and teaches us something about the impact of the variables present in the decision tree.

As a side technical note, the depth of the tree was limited to two (i.e., two levels) so that the figure below could be clear. This is why the Gini values in the leaves are not 0, and this is despite the fact that the problem was greatly simplified (i.e., predicting below or above). Decision trees are traditionally much larger. Moreover, we could have tried to predict every tenth percentile, for instance, as opposed to simply below or above, which would have generated an even larger tree (sometimes, the tree becomes so large that it is preferable to stop at a certain tree depth).

of taking variables one at a time and splitting them is extremely interesting. This is the power of algorithms in general, which are rarely used in conventional statistics. Another powerful algorithmic technique is artificial neural networks, which are featured in the next section.

11.2.4 Neural Networks

The last technique we will cover in this chapter, and in this book in fact, is one that has become extremely popular. Developed initially in the 1970s, the modern version of artificial neural networks, or neural networks or neural nets for short, is relatively old. It was not before the 2000s, however, that its use increased exponentially. The more evolved version of neural networks, called *deep learning*, is commonly used by large tech companies (like Google, Facebook, and Amazon). In fact, deep learning was at the core of the latest advances in artificial intelligence at the time of this writing. In this section we will learn the rudiments of how neural networks function, focusing on one specific learning process called backpropagation.[29]

As the name suggests, the mathematical concept of neural networks was inspired by the human neural network. The human brain is an extremely complex network of billions of neurons that transmit and share information with one another. The neurons are the cells where the information is processed and transformed before it is sent somewhere else. In particular, these transformations happen in the synapses, where information as an electric signal is converted to a chemical signal and then converted back to an electric signal.

For us, the independent variables x's provide information that will be transformed in the neurons to output a dependent variable y. Figure 11.18 shows a sketch of the main mechanism of neural networks. The input layer consists of all input variables (i.e.,

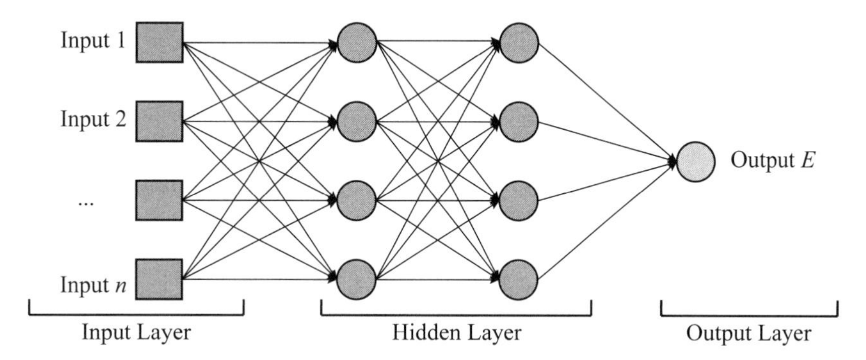

Figure 11.18
The mechanism of a neural network.

the independent variables). The hidden layer consists of the neurons that we will have to train. The figure shows two hidden layers, but we can have as many as we want. In fact, there are no well-established rules to determine the number of hidden layers and the number of neurons per hidden layer. Generally, for a similar performance, we prefer fewer layers and few neurons, since it makes the neural network simpler and more efficient.[30] Finally, the output layer consists of the output variables (i.e., the dependent variables); the figure shows only one output variable, but there can be multiple (e.g., one for each class).

To better understand what happens at each neuron, let us look at a simple example with one neuron and three input variables. Figure 11.19 shows how three input variables, x_1, x_2, and x_3, are assigned weights w_1, w_2, and w_3, which are used to calculate a variable y_1.

The variable y_1 also has a variable z_1, which we call a bias—but we can essentially think of it like the intercept b in a linear equation of the form: $y = ax + b$. The weights and the bias are initially totally random, usually between −1 and 1, and the training

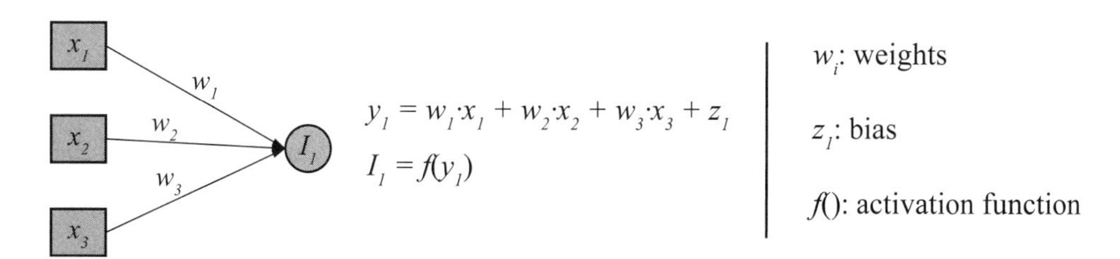

Figure 11.19
A closer look at one neuron.

process adapts their values to maximize the performance of the algorithm. More generally, y is defined as:

$$y_i = \sum_j w_j \cdot x_j + z_i \tag{28}$$

It is not y_1 that is passed on, however. We first need to transform it into I_1, which is generally defined as:

$$I_i = f(y_i) \tag{29}$$

The function f that transforms y_1 is called an activation function, and it is extremely important here. In fact, this is where the "magic" of neural networks happens, akin to synapses in human neural networks. Four types of activation functions are typically used: step, sigmoid (also called logistic[31]), tanh, and rectified linear unit (ReLU). Mathematically, they are defined as:

$$\text{Step: } f(y) = \begin{cases} -1 & \text{for } y < 0 \\ 1 & \text{for } y \geq 0 \end{cases} \tag{30}$$

$$\text{Sigmoid: } f(y) = \frac{1}{1+e^{-y}} \tag{31}$$

$$\text{Tanh: } f(y) = \tanh(y) \tag{32}$$

$$\text{ReLU: } f(y) = \begin{cases} 0 & \text{for } y < 0 \\ y & \text{for } y \leq 0 \end{cases} \tag{33}$$

This transformation process is important since it can completely change the "signal" sent by y_1. In fact, most activation functions are able to simulate any nonlinear relationship. This is incredible! Remember that scaling laws, for example, are expressed by power laws, which means we need to use the right statistical technique to fit a power law to our data. Here, we do not need to identify the type of linear or nonlinear relationship that exists. We simply feed the data to the neural network, and the activation function takes care of the nonlinearities. Figure 11.20 shows each of the four types of activation functions defined above.

The three first functions are actually a little limited because the values that come out of each node tend to be binary, either −1 or 1, or 0 or 1. As figure 11.20 shows, sigmoid and tanh can vary a little more for values around 0, and it is better to systematically standardize the input data for the neural network to perform better.[32] ReLU is quite different. When the x's are positive, the weights determine whether any information will be passed on. Although it may look a little crude and arbitrary at first, we must realize that for many nodes (e.g., 50), neural networks can really simulate complex relationships.

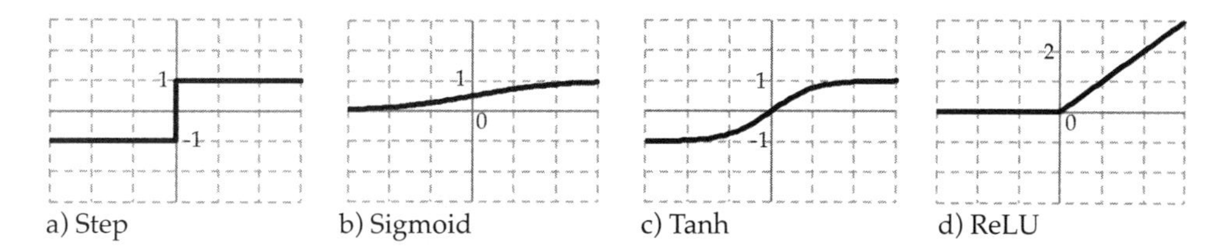

Figure 11.20
Four types of activation functions.

Once we have selected the number of neurons and hidden layers and the activation function, the backpropagation process applied to regression problems goes as follows: (1) assign random weights to calculate the *y*'s and *I*'s, (2) calculate the mean square error (MSE) between the training data and the data predicted by the neural network, (3) adjust the weights to decrease the MSE,[33] and run step (2) again until we have reached either a predefined number of iterations (e.g., 1,000 times) or a certain error threshold (e.g., less than 5%) error.

We can then use additional statistics to assess the performance of the trained neural network. The R^2 indicator (coefficient of determination) is fairly popular. For simple linear regressions, we can use the square of the Pearson correlation coefficient, but more generally (and used in Scikit-Learn for instance), R^2 is defined as:

$$R^2 = 1 - \frac{\sum (y_i - \hat{y}_i)^2}{\sum (y_i - \bar{y})^2} \tag{34}$$

where y_i is the actual value of a data point from the testing data set, $\hat{y}_i$ is the predicted value from the trained neural network, and $\bar{y}$ is the arithmetic mean of all the actual values. This R^2 can range from $-\infty$ (no correlation) to 1 (perfect correlation).

As mentioned before, neural networks can also be used for classification problems. The backpropagation technique is a little different, but it works essentially in the same way—one difference is that we now try to maximize *accuracy* instead of minimizing MSE. Because we focused on classification in decision tree learning, we focus solely on regression problems here.

Neural networks are therefore simple to use, although the internal mechanisms can be fairly complex. More importantly, they are extremely powerful. The main drawback is that it is difficult to qualitatively analyze the weights found since they are many, sometimes contradicting one another by neuron. Put differently, for linear relationships, the values of *a* and *b* tell us a lot about the relationship. Similarly, exponents for power laws tell us a lot about the nature of the scaling. For neural networks, we simply cannot do any of that, and this is really their main limitation. At the time of this writing, one of the best things we could do was to perform a sensitivity analysis—for instance, by switching one variable at a time by 5%, 10%, 50%, or even 100% or more and by plotting how the results change (interested readers should look up *partial dependence plots*).

This is it for neural networks. We should realize, however, that backpropagation is just one method to train a neural network (and not the best actually). Other popular, and sometimes better-performing, types of neural networks include generalized regression neural networks, probabilistic neural networks, convolutional neural networks, recurrent neural networks, and many more, but we will not discuss them here.

Overall, this is all we will see about ML. To learn more, Hastie, Tibshirani, and Friedman (2009) and Han, Kamber, and Pei (2011) offer great, easy-to-read resources.

By this point, we should all realize the extreme power of ML and its potential for urban engineering. Whether we are trying to predict power and water consumption or travel demand, ML can easily be used. Where ML might become our biggest asset, however, is to control smart devices so we consume less power and water, for example. Moreover, ML has a bright future in transport since it is heavily used to control autonomous vehicles. As more sensors are installed to monitor engineering systems, ML also offers incredible power to rethink how we design and operate urban infrastructure systems so we can become more sustainable.

11.3 Conclusion

In this chapter we talked about cities and the science that surrounds them. We first saw that making cities more sustainable is essential since more people now live in cities than in rural areas for the first time in the history of humanity. Dedicating an entire science to cities may therefore be very well worth it. Moreover, this science is fundamentally grounded within the realm of complexity science since cities are, after all, highly complex and adaptive systems.

This complexity of cities manifests in various ways, one of which is through scaling laws. These scaling laws tell us that some urban properties evolve in nonlinear ways as cities grow. This is paramount and may in fact hold the key for us. If we can first understand these scaling laws and control them to make sure that energy use and GHG emissions are lower in cities, then we will surely be on the right track. But this is obviously not an easy task to carry out. As importantly, urban infrastructure systems also affect one another in nonlinear ways, and this type of scaling may be present here as well, although little work had been done on this topic at the time of this writing.

Example 11.7

Neural Network of CST Data

Using the CST data set, (1) train and compare the performance of three neural networks, respectively, using sigmoid, tanh, and ReLU as the activation function (use one hidden layer with 50 nodes). (2) Using the best-performing activation function from (1), train and compare the performance of neural networks with one hidden layer and a number of neurons ranging from 1 to 100.

Solution

(1) Using Scikit-Learn, we can devise a small *for* loop to train three neural networks with one hidden layer and 50 nodes and record the performance (i.e., R^2) of each neural network. As recommended, we trained the neural networks on 80% of the data and tested it on the remaining 20%.

The results are displayed in the bar chart below. Sigmoid achieved a performance of 0.70, tanh achieved 0.73, and ReLU achieved 0.96. ReLU tends to perform better because it is not bounded between −1 and 1.

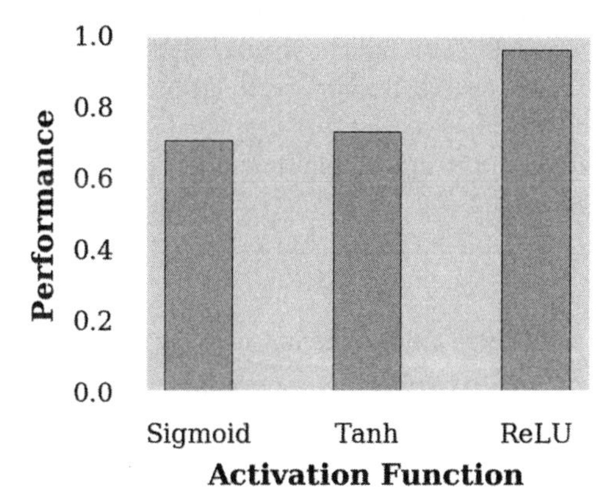

(2) Here again, we can devise a small for loop to record the performance (i.e., R^2) of our neural networks that ranges from one neuron in the hidden layer to 100. The results are displayed in the figure below.

(continued)

Example 11.7 (continued)

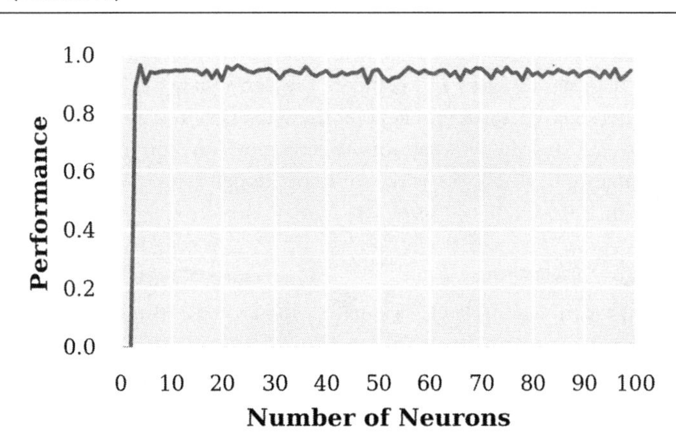

As we can see, the performance is initially poor when the neural network has few nodes. After about 5 nodes, it then quickly shoots up to about 95% all the way to 100 nodes. This is actually typical. We generally do not require many more nodes than the number of variables. We can also add extra hidden layers, but the performance is not likely to increase much further than the 95% that we see. In fact, we should be a bit careful since a higher performance might simply suggest that the neural network is overfitted to the data. As a general rule, there should be at least ten times more data points than the number of weights that have to be trained.

We also looked at the distribution of city sizes, learning about Zipf's law, which shows a power law relationship between city population and rank. Notably, this means that we will never have only large cities like Chicago or Beijing. The world contains a plethora of small cities that house billions of people, and the new solutions we develop should not only help large cities curb their carbon footprint but also help small cities. In fact, if scaling and power laws exist between cities, then perhaps our designs should also respect these laws. We then went over a few simple models to understand how city size is redistributed. These models are extremely simple, and yet they manifest the same kind of complex patterns that we observe in real cities. This shows us how complex cities are.

Subsequently, we ventured into the world of network science. Defined by their nodes and links, complex networks can be manifestations of complex systems, and studying their properties can therefore shed light into some of the properties of the complex systems themselves. In particular, we defined a series of properties that can be calculated, from network-wide properties such as the cyclomatic number and degree of connectivity to node and link-specific properties such as between centrality. We also quickly discussed the concept of multiplex—that is, network of networks—that seems to be particularly applicable in urban engineering.

The Science of Cities is still in its infancy, but it is emerging fast, and it will hopefully help us develop more sustainable infrastructure designs in cities around the world.

After the Science of Cities, we were introduced to ML. In particular, we saw that ML's big advantage is that the algorithms learn by themselves, and it is therefore easy to use ML in general. We learned about the two families of techniques in ML: unsupervised and supervised learning. Unsupervised learning is about finding patterns in a data set when we are not really sure whether something interesting exists or not. In contrast, supervised learning (or classification/regression) is closer to traditional statistics. We have some data points and a "ground truth" (i.e., a dependent variable), and we want to train a model that we can apply to other data points for which we do not have a ground truth. Moreover, depending on the type of data we are looking for—that is, categorical/ordinal versus continuous—we are dealing with classification or regression problems.

At the time of this writing, ML was not commonly used in engineering, despite the fact that its potential applications are numerous. With classification, for example, we can classify whether a project succeeds or not, and we can determine whether we should act or not (i.e., whether a bridge needs to be repaired). Even the level of service of a road (chapter 7) could be redefined since it is, after all, a classification problem.

We then saw three specific ML techniques: K-means (clustering), decision tree learning (classification), and neural networks (regression and classification). K-means uses Euclidean distance to cluster data points together in a predefined number of clusters. The silhouette coefficient can be used to assess whether the clusters found make sense or not.

Decision tree learning determines a hierarchy of variables, based on how accurate they are at predicting the final classification. Fundamentally, it works in a completely different way compared to traditional engineering techniques. Training a decision tree tends to be fast, and using a decision tree is even faster, thus it is perfect for making real-time decisions. Moreover, multiple trees can be trained to get a consensus of trees, and one powerful ensemble technique is random forest.

Neural networks are again very different, and they are directly inspired by our brain's neural network. Input variables feed into a hidden layer and are assigned weights and a bias. The information is then transformed by an activation function that can simulate any function. Neural networks are incredibly powerful, and it is therefore not surprising to know that they are used extensively and that they are evolving rapidly. At the time of this writing, deep learning (as a neural network with many layers) had become impressively powerful.

This concludes the chapter on the Science of Cities and ML. Just one last time, however, we can briefly go over all the material that we have covered. Most importantly, we can go over the big lessons that we have learned and that we should remember in all of our future projects. As importantly, in the conclusion we can try to list some of the work that we should tackle in the years to come and come up with a short checklist to implement in all future urban infrastructure projects.

Problem Set

11.1 From the United Nations website (i.e., World Urbanization Prospects) or any other urban population data set, (1) replot (or update if new data are available) figure 11.1 and (2) comment on the figure.

11.2 From the United Nations website (i.e., World Urbanization Prospects) or any other urban population data set, (1) download historical data on urban and rural populations for more developed and less developed countries in the world. Plot the data akin to figure 11.1 and comment. (2) From the same data set, select one country other than your own and plot the historical urban and rural populations.

11.3 In your own words, define and describe the Science of Cities.

11.4 In your own words, briefly describe the four properties of complex systems listed by Holland (2014).

11.5 From the four properties of complex systems listed by Holland (2014), select one and describe and illustrate it through several examples.

11.6 As an example of Monte Carlo analysis, using Microsoft Excel or any other commercial statistical software, (1) generate ten random values between 0 and 1 (i.e., essentially, assume a uniform distribution) and report the formula used. Similarly, (2) generate ten random values of a normal distribution with a mean of 170 and a standard deviation of 40 and report the formula used.

11.7 In your own words, define the concept of scaling law and describe the importance of the value of the scaling factor.

11.8 The table below shows the population, GMP, and road length for ten cities. Based on this data, determine whether (1) GMP and (2) road length follow a scaling law and if so, calculate the scaling factor and mention whether the relationship is sublinear, linear, or superlinear.

Population	GMP ($ billion)	Road length (miles)
505,682	57.60	590,000
654,785	67.24	707,000
2,568,793	383.78	2,467,000
1,284,739	170.96	1,537,000
999,273	121.85	975,000
3,948,675	530.25	4,082,000
823,746	91.54	793,000
728,348	79.39	709,000
928,376	117.66	1,043,000
4,968,547	745.96	4,779,000

11.9 The total length of water pipes and the GMP of five cities are listed in the table below. (1) Determine whether these two systems follow scaling laws and if so, whether the scaling is sub or superlinear. (2) A scaling law was found between the number of potholes N_p in a city and the total road length L (in meters) such that $Np = 3 \times 10^{-4} \cdot L^{0.82}$. Determine the number of potholes for a city with 100,000 km of roads and a city with 1,000,000 km of roads.

Population	Water pipes (km)	GMP ($100 million)
50,000	40	60
75,000	55	90
100,000	70	130
150,000	100	200
250,000	150	350

11.10 From the U.S. Census Bureau or any other legitimate data source, (1) select any variable other than population for U.S. cities and plot it versus population akin to figure 11.3 (even if no pattern can be discerned). (2) From the plot, determine whether a scaling law applies and if so, calculate the corresponding scaling factor.

11.11 In your own words, briefly describe the concept of Zipf's law applied to urban populations and describe the case of a pure Zipf's law.

11.12 (1) Determine whether the country with the city populations shown in the table below follows a pure Zipf's law. If yes, explain why, and if not, identify which city(ies) would have to be removed to obtain a pure Zipf's law. (2) The GDPs (in millions of dollars) of world countries in 2015 are fitted to a Zipf's law: $\mathrm{GDP}(r) = 45 \times 10^6 \cdot r^{-1.47}$.

Determine the GDPs of countries with rank 1, 10, and 100. Latvia, which was the 100th country in the list, had in reality a GDP of about $31 billion. Comment on the difference with your finding.

Population	Rank
100,000	1
70,000	2
50,000	3
35,000	4
25,000	5
20,000	6

11.13 From the U.S. Census Bureau or a similar source, (1) download the population data for all cities in a U.S. state of your choice or any country other than the United States, sort the population according to rank, and determine whether the data obey Zipf's law. (2) Comment on how you envision this plot to evolve in the future based on your idea of city growth in this specific state or country.

11.14 In your own words, describe the difference between the multiplicative and additive population models covered in the chapter.

11.15 Using Microsoft Excel or any commercial statistical software and starting with a population of 1,000, (1) apply equation (6) from the chapter for five different cases, varying λ between –0.01 and +0.01 for 500 time steps. Plot the results akin to figure 11.6. (2) Comment on whether or not you expected the results.

11.16 By doing your own research, describe the evolution of graph theory and network science since it was first invented by Euler in the eighteenth century.

11.17 By doing your own research from past studies, identify a network and describe how network science was used to learn the important properties of the network (do not limit yourself to engineering networks).

11.18 By doing your own research, identify any important property of networks not covered in the chapter and discuss it.

11.19 By doing your own research, describe what a scale-free network is and how it differs from a random network.

11.20 In your own words, describe what a multilayer network is and give an example of a multilayer network.

11.21 For the network below, (1) calculate N, L, S, μ, α_p, β, and γ_p. (2) Manually, calculate the network diameter Δ and determine the degree d, closeness C_C, normalized closeness C_{CN}, and betweenness centrality C_B of each node.

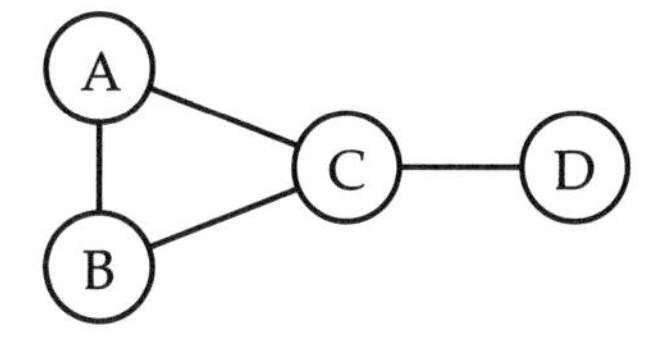

11.22 For the network below, (1) construct the adjacency matrix. Based on the adjacency matrix, (2) calculate N, L, S, μ, α_p, β, and γ_p. Using any software, (3) calculate the network diameter Δ and determine the degree d, normalized closeness C_{CN}, and betweenness centrality C_B of each node and (4) visualize the network in various layouts.

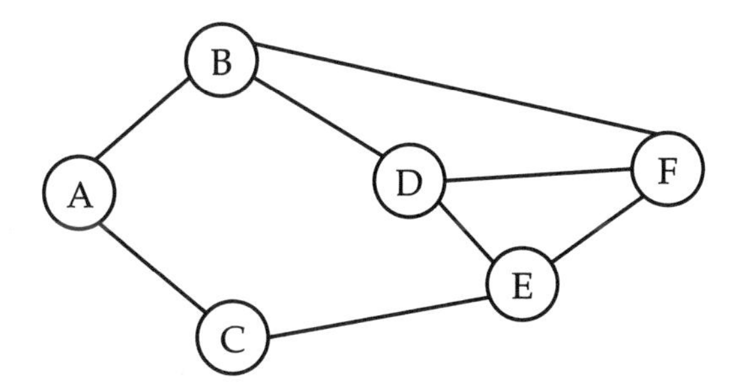

11.23 The network below shows the same network as problem 7.22 in chapter 7, with weights on each link. Using any software, calculate (1) the strength s of each node and (2) the link betweenness of each link both with and without the weights. (Note: at the time of this writing, Gephi cannot calculate link betweenness; an alternative is to use the NetworkX Python library).

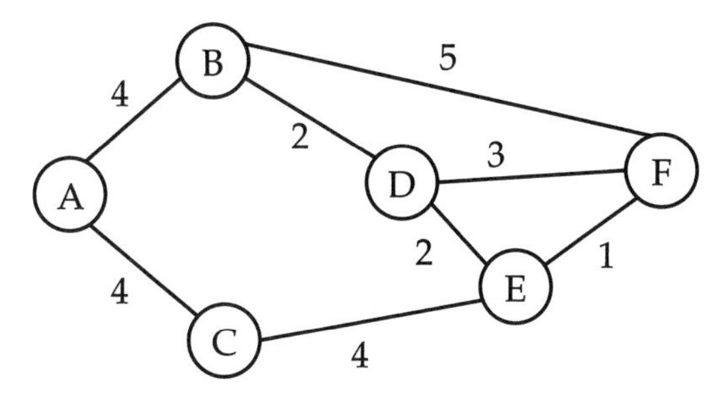

11.24 For the network below, (1) construct the weighted adjacency matrix. Based on the adjacency matrix, (2) calculate N, L, S, μ, α_p, β, and γ_p. Using any software, (3) calculate the network diameter Δ and determine the degree d, strength s, normalized closeness C_{CN}, and betweenness centrality C_B of each node and (4) the link betweenness of each link both with and without the weights. Moreover, similar to problem 7.39 in chapter 7, use the software to find the shortest path between nodes A to L both with and without the weights. (Note: it is recommended to use a more advanced software, such as the NetworkX Python library, to solve this problem).

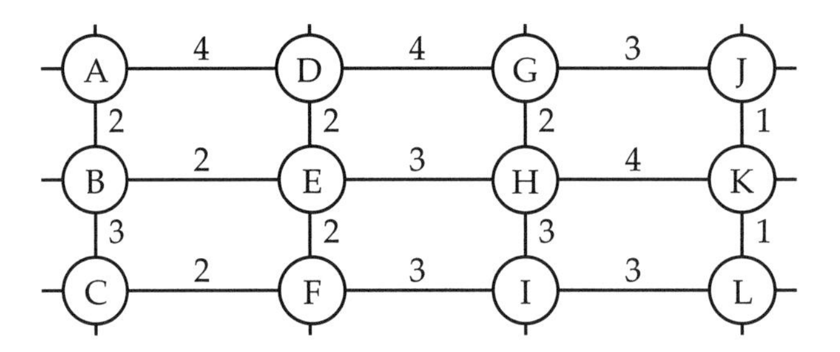

11.25 In your own words, describe how Machine Learning differs from traditional statistics.

11.26 In your own words, describe the difference between categorical, ordinal, discrete, and continuous variables.

11.27 In your own words, describe the difference between unsupervised and supervised learning.

11.28 In your own words, describe the difference between clustering, classification, and regression.

11.29 In your own words, describe how K-means clustering functions.

11.30 Calculate the Euclidean and Manhattan distance between the points in the table below with a point with coordinate (5,5).

x	y
2	9
7	6
4	1
9	4
5	7

11.31 For the data points in the table below, (1) select points A and B and calculate the distance between them and all other points (use equation (19)). (2) Based on the results, form two clusters, 1 and 2, and readjust the centroid of each cluster by either picking a point that is more centrally located in the cluster or by calculating the coordinates of the centroid (i.e., the mean coordinate of all points in the cluster) and calculate the distance between the two new centroids to all other points. (3) If the centroids need to be adjusted again, repeat the sequence from question (2) as many times as needed. (4) Once the final clusters have been identified, calculate the silhouette coefficient of the two clusters (equation (22)).

Point	x	y
A	1	2
B	9	9
C	3	3
D	2	5
E	1	4
F	5	8
G	5	5
H	1	8
I	9	6
J	7	7

11.32 Using any software (such as Scikit-Learn), select any data set and perform a clustering analysis using K-means clustering. Make sure to try several different number of clusters and calculate the silhouette coefficient each time.

11.33 Based on your own research, select any clustering technique other than K-means and describe it.

11.34 In your own words, describe how decision tree learning functions.

11.35 Based on your own research, describe how random forest functions.

11.36 Based on your own research, describe how gradient boosting machine functions.

11.37 The table below shows twenty values in the range [0, 49]. By dividing the data into bins of ten (i.e., 0–9, 10–19, 20–29, 30–39, 40–49), (1) count the number of values per bin and transform these counts into probabilities (i.e., between 0 and 1) and (2) calculate the Gini impurity and entropy of the data.

3	6	10	4	11	42	25	4	40	17	6	1	37	47	39	10	6	41	18	3

11.38 The first table below shows the results of a small survey of ten students. Each student was asked three questions about a class and had to answer with yes (Y) or no (N). Each was also asked whether she or he had passed or failed the course. (1) Fill out the second table below and calculate the Gini impurity and entropy based on the probabilities calculated. (2) Based on these results, determine which variable you would select first in a decision tree. (3) Without doing more calculations, estimate the variable that would be used in the second row of the decision tree.

Student	Attend classes?	Work at home?	Extra work?	Grade
1	Y	N	Y	Pass
2	Y	N	N	Fail
3	Y	Y	Y	Pass
4	Y	Y	N	Pass
5	Y	N	Y	Pass
6	N	Y	N	Fail
7	Y	N	N	Fail
8	Y	Y	N	Pass
9	N	Y	Y	Pass
10	N	N	N	Fail

			Attend classes?	Work at home?	Extra work?
Frequency	Y	Pass			
		Fail			
	N	Pass			
		Fail			
Probability	Y	Pass			
		Fail			
	N	Pass			
		Fail			
Gini impurity					
Entropy					

11.39 Using any software (such as Scikit-Learn), select any data set and solve a classification problem using decision tree learning. Make sure to use a portion of your data for training and a separate portion for testing.

11.40 In your own words, describe how neural networks functions.

11.41 In your own words, describe the role of the activation function in backpropagation neural networks.

11.42 The table below shows the grades of two students. The goal is to train a neural network to predict the grade for the final exam based on the grades from the homework, quiz, and midterm. The neural network has one hidden layer with one single neuron, and it uses the ReLU activation function. The initial weight for the neuron for the homework, quiz, and midterm are 0.22, 0.18, and 0.25, respectively, and the neuron has a bias of 30 (i.e., at the moment, the output of the neuron is determined as $0.22 \times \text{homework} + 0.18 \times \text{quiz} + 0.25 \times \text{midterm} + 30$). (1) Based on the current weights,[34] calculate the predicted grade for the final of the two students. (2) Calculate the squared error between the predicted final grade and the actual final grade. (3) Based on your knowledge of neural networks, estimate whether the weights for the three variables are likely to increase or decrease.

	Student 1	Student 2
Homework	92	90
Quiz	85	78
Midterm	89	81
Final	90	81

11.43 Using any software (such as Scikit-Learn), select any data set and solve a regression problem using a backpropagation neural network. Make sure to use a portion of your data for training and a separate portion for testing.

11.44 Based on your own research, select any neural network technique other than backpropagation neural networks and describe it.

Notes

1. According to the rough 2017 population.

2. We quickly mentioned him in the last chapter as one of the earlier people to look at cities from a biology viewpoint as opposed to a physics viewpoint, which is particularly relevant in the context of this paragraph.

3. There is even an article titled "Cities as Organisms" (Samaniego and Moses 2008).

4. Mitchell (2009) offers a great introduction to the field.

5. There is a great and famous publication titled "The Architecture of Complexity" by Simon (1962) that is worth the read.

6. It is completely free to download and use at https://ccl.northwestern.edu/netlogo/ (accessed March 27, 2019), and many tutorials are available to learn how to use it.

7. This statistical technique to find β is quite common, but we should be aware that some problems exist. Essentially, it is difficult to prove statistically that values of β roughly in the range 0.85–1.20 are statistically different than 1 (i.e., a linear relationship). But the concept of scaling remains phenomenal.

8. If interested to know more, see the short commentary by Louf and Barthelemy (2014).

9. The writings have been republished and can be found in von Thünen (1966).

10. That starts with linguistics and the number of words in James Joyce's *Ulysses*!

11. The data was generously shared by Matthieu Cristelli, the lead author of the article.

12. David Levinson (whom we saw in chapter 7 as a leading scholar in accessibility) has also produced extensive research on road networks (Levinson 2012; Xie and Levinson 2011), and I have personally done a lot of work on metro/subway (Derrible 2012; Derrible and Kennedy 2009, 2010) and infrastructure networks (Karduni, Kermanshah, and Derrible 2016; Peiravian and Derrible 2017).

13. In scale-free networks, few nodes possess many links, which can help transfer information efficiently, but this also makes the network more vulnerable since a targeted attack on these nodes can have a substantial impact on the performance of the network. This is the typical case in which the structure of the network has a large impact on how well it can perform.

14. It is also referred to as *geodesic distance*.

15. Available at https://gephi.org/ (accessed March 28, 2019).

16. Libraries are essential functions that someone else coded and that can be imported in our code for us to use. Libraries make the whole process significantly easier. In particular, they make it possible to do some fairly sophisticated analysis with just a few lines of code.

17. Available at https://en.wikipedia.org/wiki/Social_network_analysis_software (accessed March 28, 2019).

18. In 1997 IBM's Deep Blue computer beat Gary Kasparov, the best chess player in the world. Deep Blue was actually pretty "dumb." Every single move was coded manually, in a way that does not take too much memory. In contrast, in 2016 AlphaGo (Google's DeepMind algorithm) beat the best Go player in the world, Lee Sedol. AlphaGo had "learned" how to play Go by itself by being fed an impressive number of past games from which to learn. AlphaGo was therefore easier to program than Deep Blue and also much more powerful.

19. In computer science, these rules are called *heuristics*—again a term that sounds complicated, although it really is not. In fact, our two sustainability principles are actually heuristics.

20. The repository contains many great data sets that anyone can use for research or as examples when learning ML. It is available at http://archive.ics.uci.edu/ml/index.php (accessed March 28, 2019).

21. Technically, neural networks can also be used as an unsupervised learning technique, but we will not cover it here.

22. Technically, text mining is referred to as *natural language processing*.

23. Another distance is the Manhattan distance (i.e., no diagonals; we only follow the grid), which is defined as $|x_A - x_B| + |y_A - y_B|$.

24. The personal website of Andrey A. Shabalin at http://shabal.in/visuals/kmeans/2.html and the YouTube video at https://www.youtube.com/watch?v=BVFG7fd1H30 offer great visualizations of how K-means clustering works (accessed March 28, 2019).

25. Like K-means, some use distance, such as hierarchical clustering, and others use something else, like density, for instance, which is used by the DBSCAN algorithm, which stands for density-based spatial clustering of applications with noise.

26. Depending on the process used, some algorithms create more than two branches, but most do not. Plus, it really does not matter since we can successfully split a variable in multiple consecutive nodes. Small, medium, and large, for instance, can be split in three at one node, or they can be split in two at the first node (e.g., small + medium and large) and then again be split in two at the second node (e.g., small and medium).

27. This is where decision tree learning used for regression diverges. Instead of using entropy or Gini impurity, mean square error (MSE) is generally adopted, as are many ML techniques when used for regression. In the end we still have classes, but we have many of them, and we simply assign the value of the class to a data point. These types of decision trees are generally called regression trees, and they generally have a poor accuracy. Ensemble tree techniques such as gradient boosting machines, however, can perform extremely well on regression problems. They are based not on a single tree but rather on many trees that are built sequentially. We will not cover them here.

28. Random forest is part of the ensemble family of techniques referred to as *bagging*. As mentioned in note 27 and in the text, another powerful ensemble family is *boosting*, which includes gradient boosting machine. Gradient boosting machine tends to perform well for both classification and regression purposes, but the technical approach is different.

29. To demonstrate how a backpropagation neural network works without doing any coding, the company TensorFlow (which makes the Python library TensorFlow cited earlier in the text) made a useful online tool available at http://playground.tensorflow.org/ (accessed March 28, 2019).

30. In deep learning, neural networks are trained to handle significantly complex problems, and they tend to have many hidden layers that can be trained differently.

31. When using sigmoid, the neural network becomes a little like a traditional logistic regression (similar to the logit model), except that we have multiple neurons and potentially multiple hidden layers.

32. For example, by using the min-max standardization method so that all values are between 0 and 1.

33. We will not get into the mathematics of the process here, but it is relatively straightforward.

34. In practice, it is preferable to first standardize the data, but we do not here for simplicity. Moreover, we do not add another bias in the output layer, although it is common in practice.

References

Abadi, Martín, Ashish Agarwal, Paul Barham, Eugene Brevdo, Zhifeng Chen, Craig Citro, Greg S. Corrado, et al. 2015. "TensorFlow: Large-Scale Machine Learning on Heterogeneous Systems." http://download.tensorflow.org/paper/whitepaper2015.pdf.

Barabási, Albert-László. 2016. *Network Science*. Cambridge: Cambridge University Press.

Barthelemy, M. 2011. "Spatial Networks." *Physics Reports* 499(1–3): 1–101. https://doi.org/10.1016/j.physrep.2010.11.002.

———. 2016. *The Structure and Dynamics of Cities: Urban Data Analysis and Theoretical Modeling*. Cambridge: Cambridge University Press.

Barthelemy, M., P. Bordin, H. Berestycki, and M. Gribaudi. 2013. "Self-Organization versus Top-Down Planning in the Evolution of a City." *Scientific Reports* 3 (July): 2153.

Batty, Michael. 2013. *The New Science of Cities*. Cambridge, MA: MIT Press.

Bettencourt, Luís M. A. 2013. "The Origins of Scaling in Cities." *Science* 340(6139): 1438–1441. https://doi.org/10.1126/science.1235823.

Brand, Stewart. 2009. *Whole Earth Discipline: An Ecopragmatist Manifesto*. New York: Viking.

Bureau of Economic Analysis. 2016. *Gross Domestic Product by Metropolitan Area, 2015*. http://www.bea.gov/newsreleases/regional/gdp_metro/gdp_metro_newsrelease.htm.

Cristelli, Matthieu, Michael Batty, and Luciano Pietronero. 2012. "There Is More than a Power Law in Zipf." *Scientific Reports* 2 (November). https://doi.org/10.1038/srep00812.

Derrible, Sybil. 2012. "Network Centrality of Metro Systems." *PLoS ONE* 7(7): e40575. https://doi.org/10.1371/journal.pone.0040575.

Derrible, Sybil, and Christopher Kennedy. 2009. "Network Analysis of World Subway Systems Using Updated Graph Theory." *Transportation Research Record: Journal of the Transportation Research Board* 2112(1): 17–25. https://doi.org/10.3141/2112-03.

———. 2010. "The Complexity and Robustness of Metro Networks." *Physica A: Statistical Mechanics and Its Applications* 389(17): 3678–3691.

Euler, L. 1741. "Solutio Problematis Ad Geometriam Situs Pertinentis." *Commentarii Academie Scientiarum Imperialis Petropolitanae* 8: 128–140.

Federal Highway Administration. 2014. *Highway Statistics 2010*. https://www.fhwa.dot.gov/policy information/statistics/2010/hm71.cfm.

Han, J., M. Kamber, and J. Pei. 2011. *Data Mining: Concepts and Techniques*. Morgan Kaufmann Series in Data Management Systems. San Francisco, CA: Elsevier Science.

Hastie, Trevor, Robert Tibshirani, and J. H. Friedman. 2009. *The Elements of Statistical Learning: Data Mining, Inference, and Prediction*. 2nd ed. New York: Springer.

Holland, J. H. 2014. *Complexity: A Very Short Introduction*. Oxford: Oxford University Press.

Karduni, Alireza, Amirhassan Kermanshah, and Sybil Derrible. 2016. "A Protocol to Convert Spatial Polyline Data to Network Formats and Applications to World Urban Road Networks." *Scientific Data* 3 (June): 160046.

Levinson, David. 2012. "Network Structure and City Size." *PLoS ONE* 7(1): e29721. https://doi.org/10.1371/journal.pone.0029721.

Loper, Edward, and Steven Bird. 2002. "NLTK: The Natural Language Toolkit." In *Proceedings of the ACL-02 Workshop on Effective Tools and Methodologies for Teaching Natural Language Processing and Computational Linguistics*, vol. 1, 63–70. Stroudsburg, PA: Association for Computational Linguistics. https://doi.org/10.3115/1118108.1118117.

Louf, R., and M. Barthelemy. 2014. "Scaling: Lost in the Smog." *Environment and Planning B: Planning and Design* 41(5): 767–769.

Mitchell, Melanie. 2009. *Complexity: A Guided Tour*. New York: Oxford University Press.

Newman, M. 2010. *Networks: An Introduction*. Oxford: Oxford University Press.

Pedregosa, F., G. Varoquaux, A. Gramfort, V. Michel, B. Thirion, O. Grisel, M. Blondel et al. 2011. "Scikit-Learn: Machine Learning in Python." *Journal of Machine Learning Research* 12: 2825–2830.

Peiravian, Farideddin, and Sybil Derrible. 2017. "Multi-dimensional Geometric Complexity in Urban Transportation Systems." *Journal of Transport and Land Use* 10(1): 589–625.

Samaniego, Horacio, and Melanie E. Moses. 2008. "Cities as Organisms: Allometric Scaling of Urban Road Networks." *Journal of Transport and Land Use* 1(1): 21–39.

Simon, Herbert A. 1962. "The Architecture of Complexity." *Proceedings of the American Philosophical Society* 106: 467–482.

United Nations. 2018. *World Urbanization Prospects, the 2018 Revision*. New York: United Nations. https://esa.un.org/unpd/wup/CD-ROM/.

von Thünen, J. H. 1966. *Isolated State: An English Edition of Der Isolierte Staat*. Edited by P. G. Hall. Originally published in 1826. Jena, Germany: Pergamon.

Xie, Feng, and David M. Levinson. 2011. *Evolving Transportation Networks*. New York: Springer.

Yeh, I-Cheng. 2007. "Modeling Slump Flow of Concrete Using Second-Order Regressions and Artificial Neural Networks." *Cement and Concrete Composites* 29(6): 474–480. https://doi.org/10.1016/j.cemconcomp.2007.02.001.

Zipf, G. K. 1949. *Human Behavior and the Principle of Least Effort: An Introduction to Human Ecology*. Cambridge, MA: Addison-Wesley.

12 Conclusion

> We stand now where two roads diverge. But unlike the roads in Robert Frost's familiar poem, they are not equally fair. The road we have long been traveling is deceptively easy, a smooth superhighway on which we progress with great speed, but at its end lies disaster. The other fork of the road—the one "less traveled by"—offers our last, our only chance to reach a destination that assures the preservation of our earth.
>
> —Carson 1962, p. 278

The quote used to start this conclusion comes from Rachel Carson's seminal book *Silent Spring*, which was published in 1962. Her book is not about cities and infrastructure but about environmental science and the dramatic effects of spraying large amounts of chemical pesticides on crops in the United States in the 1940s and 1950s to get rid of invasive insect populations. Insects were indeed killed, but the birds and fish that ate the insects were also killed or contaminated, and they were then sometimes eaten by humans, thus having a direct impact on human health. In other cases, entire cities were sprayed with pesticides, directly affecting the residents of a city, especially children. Without knowing it, Carson did a network analysis by looking at the chain reaction caused by one seemingly minor and well-intentioned decision. In this instance, one small local decision generated large regional impacts.[1] Carson's work has notoriously contributed to the establishment of the U.S. Environmental Protection Agency (EPA) created by Richard Nixon, the former U.S. president, in 1970.[2]

The message of *Silent Spring* is strong, and it directly speaks to us as urban engineers. After all, well-intentioned decisions were made about the use of pesticides, but they turned out to have dramatic consequences that could have been avoided if the people in charge had looked at the whole system in the first place. Instead, the decisions made had an impact on the three elements of the triple bottom line of sustainability (people, planet, profit). Are we not in the same situation when we look at how we consume energy and resources and how we plan, design, engineer, and operate cities? Many people adopt a myopic view of the future by focusing on pressing issues individually, discarding the general context, but this can easily lead to the development of decisions that might have the opposite impact than that desired. Now is the time to change this practice, and we need to look at all urban engineering systems as a whole.

What should be clear now that we have reached the end of the book is that tremendous opportunities exist to better integrate our infrastructure systems. Even without using concepts of the Science of Cities and Machine Learning (ML), current infrastructure can be better integrated with relative ease.

The best example we have seen could be green infrastructure (e.g., rain gardens and bioswales), which can easily be integrated with transport infrastructure. Roads can, and in fact should, become more multifunctional. As done in Kuala Lumpur, tunnels used for traffic can be flooded during heavy rains. Moreover, water reservoirs at different elevations can be used as energy storage devices, as discussed in chapter 5. As seen in chapter 10, we can even argue that buildings contain a tremendous amount of embodied energy that is not used at the moment and that could be mined (chapter 9). This lack of integration directly stems from the separation of responsibilities to plan, design, engineer, and operate infrastructure systems in the twentieth century, as discussed in chapter 4.

Moreover, this integration does not necessarily have to happen on a large scale. We need to remember the picture from the house in Okinawa (Japan) in chapter 10. Power can be produced in a more distributed fashion, for example, by using solar panels and wind turbines. During the summer, water can be heated by placing water tanks in attics or on roofs. Air- and ground-source heat pumps can be used in the winter to heat buildings. All these solutions work, and they contribute to both principles of sustainability: demand goes down (e.g., water heated by the sun as opposed to a water heater, thus reducing electricity consumption) and supply is increased within reason (e.g., use of solar panels and air-/ground-source heat pumps).

These features of more integrated, decentralized, and even multifunctional systems offer the bases of sustainable infrastructure design, and they should be considered by architects, designers, and urban engineers. One of the reasons they have not become the norm in practice is because many of these concepts are new and were not taught in school. Moreover, we often underestimate the reluctance of people to change; changing one's practice adds risk, and risk is often not tolerated in professional firms. This is a particularly serious problem in civil engineering since we are constantly taught and told to use standards and codes. Safety is often used as an argument to support the use of codes, but with all the software packages we have access to, is this really justified? From conversations with practicing engineers, part of the reason seems to be litigation: firms are afraid to be sued if they do not follow the standards, but innovation never comes from standards.

Beyond the scope of urban engineering as discussed in this book, at least three other big and potentially paradigm-shifting changes in the years to come could substantially contribute to making cities more sustainable. Two are purely technological, and the third is organizational. Although they deserve an entire chapter to themselves, they will be briefly discussed in the next section. It will then be time to conclude this book and elaborate a four-step Urban Infrastructure Design (UID) process that can be applied to virtually all infrastructure projects throughout the world.

12.1 Three Paradigm-Shifting Changes

As already discussed, the first paradigm-shifting change is related to the increasing use of Information Technology (IT), which generally falls within the label of *Smart Cities*. This change was already happening at the time of this writing, and we can discuss it first.

The second paradigm-shifting change is about materials. Indeed, the materials we will use to build infrastructure in the future are likely going to be completely different than those being used at the time of this writing, including some that will literally be "living." We still have a long way to go before we get there, but there is no doubt that the use of new materials to build infrastructure will become increasingly important in the next decades.

The third paradigm-shifting change is organizational. The people who design infrastructure must communicate and exchange ideas with one another, and this is where the lessons of Christopher Alexander are relevant. Unlike the other two changes, this one is not technological, but it is nonetheless critical.

Only time will tell whether these three potential changes indeed have a paradigm-shifting impact on infrastructure design, or even if they happen. For now we can only speculate.

The big absence from this list is the Science of Cities. This is partly because, although the Science of Cities (or at least, elements of complexity) may be present in the three changes, it is quite uncertain at the moment how the Science of Cities will be used in practice for designing and operating infrastructure. We will discuss it a bit more in the final thoughts section. For now, let us start with Smart Cities.

12.1.1 Smart Cities

The term *Smart Cities*, or *Digital Cities*, has become a catchall term in the professional world. When companies refer to Smart Cities, they often include elaborate graphics with smiling people walking around, self-driving cars, ample greenery, and an absence of pollution. It seems that the types of figures shown in figure 12.1 have become stereotypical and even comical.[3] This is really too bad because, if well done and however it may be called in the future, Smart Cities can substantially contribute to urban engineering, sustainability, and even resilience.

We already talked about Smart Cities a little in chapter 5. To be precise, we talked about the smart grid. Although there is no commonly accepted definition of Smart Cities, most would recognize that it generally includes a large IT component. Put differently, infrastructure systems can be equipped with sensors to collect and provide data to better control infrastructure. This can contribute to using infrastructure more efficiently—but we recall from chapter 1 that efficiency tends to generate decreasing marginal returns—or it can help to radically change the way infrastructure systems are designed and operated. For example, the smart grid can be used to operate the current grid more efficiently or to design microgrids that are dramatically different. Other terms used and associated with Smart Cities are *Adaptive* or *Responsive Cities*. The goal is to enable the currently *passive* infrastructure systems to become more *active*, especially by better responding to real-time demand patterns. These elements fit particularly well with more integrated, decentralized, and multifunctional infrastructure.

There is therefore no doubt that when it comes to sustainability, cities can substantially benefit from IT. The form by which IT should be used is a little less certain. ML will play a big part, but we need to ensure that we look at the whole system. After all, using IT may actually increase energy consumption and increase interdependencies between systems instead of reducing them. For example, some studies suggest that autonomous vehicles may actually make traffic worse (Harper et al. 2016). At the time of this writing, initiatives that fit within the realm of Smart Cities have generally best contributed by providing more information to users. For example, I can see when my bus/train is coming using a smartphone app, which means I do not have to wait in the cold too long in the winter, and I am therefore more likely to keep riding transit. Moreover, I can use a smart thermostat to better set the temperature where I live or work—that is, making it not too hot in the winter and not too cold in the summer or turning it off or down when no one is around. Before long, however, IT will be used to operate entire infrastructure systems, and there is no better example than the autonomous vehicle.

For cities to become smart, we must first realize that the people who design infrastructure will have to learn how to work with IT as well. At the time of this writing, IT

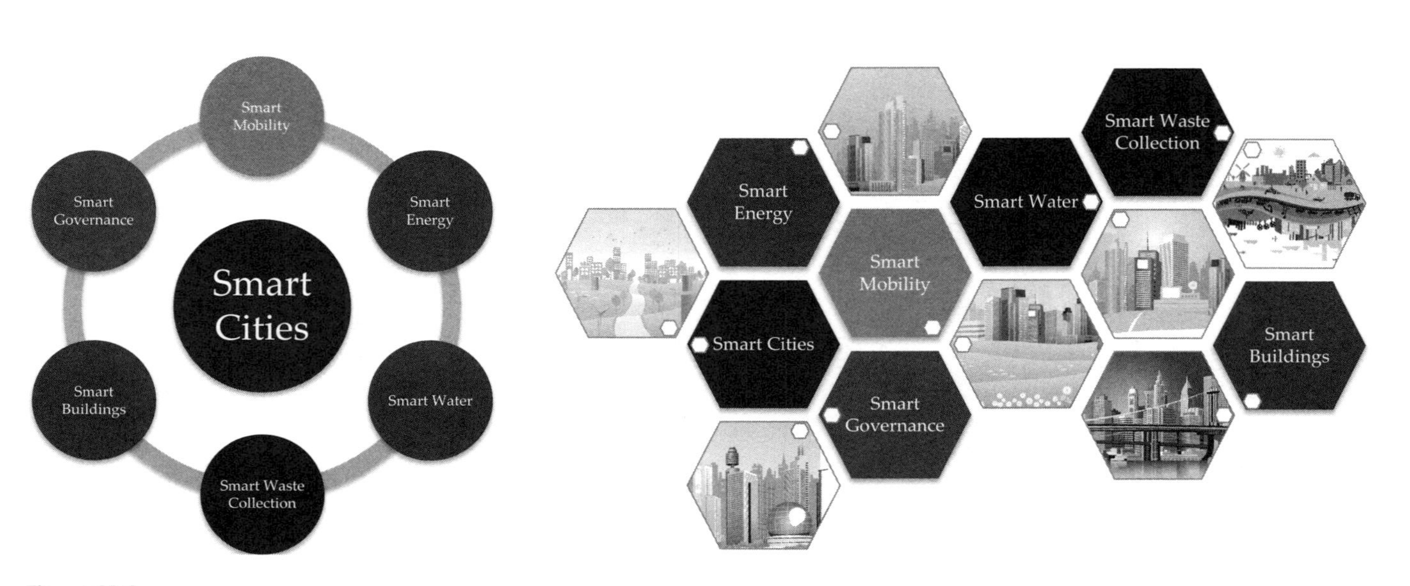

Figure 12.1
Typical Smart Cities figures.

had vastly remained a computer science discipline, but this needs to change. All urban engineers need to learn coding and the rudiments of database management and visualization. Indeed, regardless of how technically advanced Google and IBM are, urban engineers will be the ones redesigning infrastructure systems, and this can be done properly by IT-trained urban engineers or improperly by people who lack an understanding of the fundamental challenges of infrastructure design. Second, and as highlighted several times in this book, the traditional disciplinary walls must be brought down so that urban engineers can learn about all infrastructure systems in order to integrate them. As Frank Stella said, “Boundaries … are defining but not limiting.”

To enhance communication across disciplines, we will need to come up with quite literally a new vocabulary. In computer science, this is called an ontology. An ontology is a formal structure/schema to classify *objects* and *concepts*. A water conduit, for example, is an object that belongs to the concept of water distribution, which in turn belongs to the concept of water. Absolutely all elements of infrastructure systems can be classified by an ontology, and this should directly inform us as to how infrastructure systems can be better integrated.

It is therefore not too far-fetched, when we think about the future of cities, to imagine more remotely controlled and even autonomous infrastructure, from self-aware roads that send signals when they need to be repaired to floodgates that automatically adjust to the changing tide.

The potential of Smart Cities becomes even more prominent when coupled with self-healing materials, for example, that adjust their properties without the use of sensors, and this is the topic of the next section.

12.1.2 The Rise of New Materials

Where Smart Cities offer the potential to completely revisit how cities are operated, next-generation materials may dramatically alter how cities are built in the first place. While the field was still young at the time of this writing, new materials seem to possess tremendous opportunities for the future. These are sometimes referred to as *Smart Materials*, but it is doing them little justice to associate them with Smart Cities.

We must also realize that with peak oil (chapter 2), the use of petroleum-based materials should decrease substantially, and this includes all plastics. We are therefore in great need of alternative materials.

As an excellent reference, Miodownik (2015) offers an easy-to-read and short article titled “Materials for the 21st Century: What Will We Dream Up Next?” that was published in the prestigious journal *MRS Bulletin*. He notably spends an entire section discussing the role of materials for sustainable cities. Miodownik also references the excellent and incredibly relevant book *Sustainable Materials: With Both Eyes Open* by Allwood et al. (2012).

Naturally, many different types of materials possess fascinating properties relevant to infrastructure systems. For example, shape memory alloys recover their initial shapes when heated—a quick online search returns many entertaining videos. It does not take much effort to come up with applications for materials used in infrastructure systems that need to be responsive to temperature changes—for example, the difference in load-bearing capacity between warm and cold seasons.

Naturally, three-dimensional (3D) printing processes and materials offer great opportunities. They have been part of what is called *additive manufacturing* as a field of mechanical engineering, but applications in urban engineering are vast. From the custom printing of small infrastructure elements to the printing of entire infrastructure systems with complex and nonmachinable geometries, 3D printing has a lot to bring to infrastructure design and maintenance.

Moreover, two major families of relatively new materials are likely to have an important impact on the world and, accordingly, on urban engineering: metamaterials and bioinspired materials. Metamaterials are materials that possess properties not found in nature. In contrast, bioinspired materials either replicate what is found in nature or include living organisms, so the materials actually live and change shape over time.

In metamaterials, the term *meta* comes from Greek and essentially translates into "to go beyond." Metamaterials go beyond the natural properties of materials. Metamaterials are therefore purely artificial, which means they have to be synthesized. Up until the late 1990s and early 2000s, all materials were thought to have unalterable properties, including chromic (i.e., color), thermal, and even mechanical. What was found was that "the physical properties of metamaterials are not primarily dependent on the intrinsic properties of the chemical constituents, but rather on the internal, specific structures of metamaterials" (Liu and Zhang 2011). This means that materials can be custom made to have specific properties, which is a real game changer. These properties can vary widely. The most discussed properties deal with optics—that is, objects can be created to absorb and bend light in certain ways, and given the right material, some objects should even be able to stop emitting light altogether (yes, these objects would be invisible). Metamaterials can also be thermochromic, which means they change color based on the temperature. As Miodownik (2015) pointed out, can you imagine buildings that change color with the seasons? Perhaps more importantly for us, metamaterials can be given specific mechanical properties. For example, a metamaterial could behave differently based on pressure or temperature. Perhaps even more fascinating, it is possible to design metamaterials that expand under compression and shrink

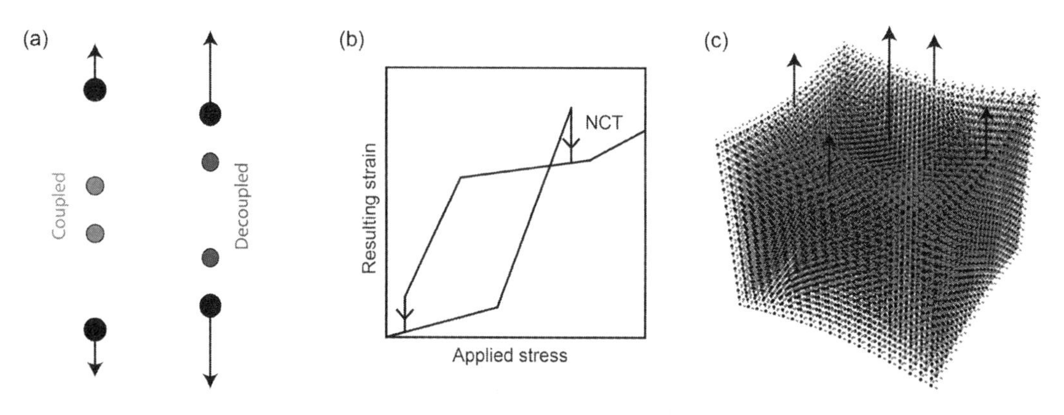

Figure 12.2
Metamaterial exhibiting negative compressibility transitions. (a) The constituent units possess a coupled configuration (*left*) for small applied tension and a decoupled configuration (*right*) for large applied tension, where the arrows show the applied tension. (b) Stress-strain projection of the hysteresis loop of the constituents. The resulting strain (length) counterintuitively decreases as the applied stress (tension) increases during the negative compressibility transition (NCT) from the coupled configuration to the decoupled configuration. (c) Metamaterial formed by a cubic lattice of constituents, where the color code is the same as in (a) and arrows indicate the applied stresses, which are larger in the center. The metamaterial contracts in the center as the applied stress increases during the NCT as a large number of constituents simultaneously decouple. The arrow lengths are exaggerated for visualization. Courtesy of Nicolaou Zachary and Adilson Motter.

under tension (Nicolaou and Motter 2012, 2013). This means that the more you pull on a material, the more it shrinks. Figure 12.2 illustrates the phenomenon (courtesy of the two authors just cited).

The second major family of new materials consists of bioinspired materials. In some form or another, we have all heard of them. To be precise, there are two types of bioinspired materials: biomimicking and living materials. As the name suggests, biomimicking materials (also called biomimicry and biomimetic) essentially try to copy nature. The most famous example is Velcro, with its tiny hooks and loops, which mimics how burs (i.e., from plants) can attach to animal hair; see figure 12.3. The list of biomimicking materials is actually quite long—it notably includes adhesive inspired by mussels, solar cells inspired by leaves, and soft/stiff materials that depend on water content and are inspired by sea cucumbers.[4]

In contrast, living materials literally contain living organisms. For example, materials can contain organisms that emit light. Imagine a wall that would light up at night without electricity. Perhaps more importantly for urban engineering, materials can be

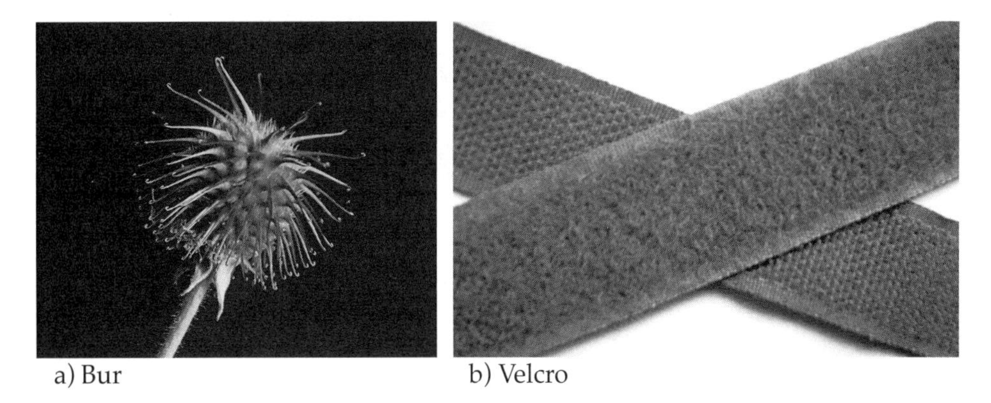

a) Bur b) Velcro

Figure 12.3
Velcro biomimicking a bur. Credit: (a) Zephyris; (b) Ryj.

self-healing. For example, aging infrastructure that shows cracks could repair itself automatically. More precisely, the living organisms in the material produce a bonding agent that fills up the cracks. This is a little like the human body, which repairs itself when injured (albeit not with living organisms but by human cells); even casts to treat broken bones are just there to limit movement and let the body do its work to reseal the break.

Figure 12.4 shows an example of how the process can work. The light gray medium is the original material (e.g., concrete), the large circles are healing agents, and the small circles are catalysts. Without cracks, the healing agent is passive and does not change. When a crack appears and starts to mix with the catalyst, the healing agent is activated and starts to fill the void. In this example, a separate catalyst was included, but depending on the healing agent selected, we can make it so that infiltrated water becomes the catalyst.

Similarly, we can think of self-assembling or growing material. Instead of shipping large amounts of materials for production, we could ship only the initial elements and let the material grow. This is particularly attractive for odd-shaped infrastructure elements that cannot be machined easily (this is a similar use to the 3D printing mentioned above). Put differently, can you imagine designing 200-year-old bridges that fix themselves? Pouring concrete or asphalt that simply heals itself when cracked (i.e., not more potholes)? The opportunities brought by these materials are truly paradigm shifting for urban engineering.

Despite the fact that humanity has been trying to understand, master, and exploit the properties of materials for our own benefit for at least the past 3,000 years, we are

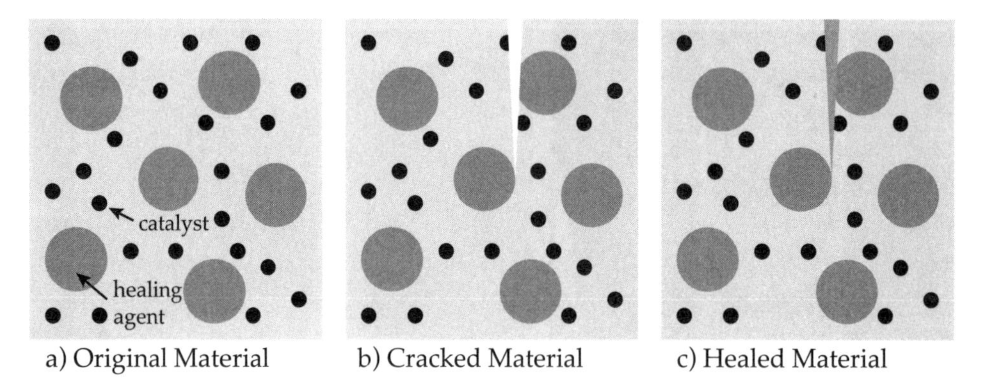

Figure 12.4
An example of the self-healing material process.

still at the beginning of the adventure. Wherever it may lead us, urban engineering will most certainly benefit from it.

After two technological paradigm-shifting changes, it is time to switch to a social paradigm-shifting change.

12.1.3 Organizational Change

As we learned in chapter 4, a chief architect and an engineer used to overlook the planning and design of all infrastructure systems. As infrastructure systems became increasingly complicated in the twentieth century, the responsibilities for designing and managing infrastructure systems were divvied up into separate departments, and these departments seldom communicate with one another. This entire process is quite logical. One of the best strategies to tackle a complex problem is the "divide and conquer" strategy.

This strategy is problematic when it comes to infrastructure, however, because there is little incentive for departments to work together. In fact, it is often quite the opposite. Those managing these departments are told to ensure that their departments function well. They are therefore responsible only for their departments, and coordinating projects across multiple departments simply adds risk and uncertainty. This is how rational stormwater engineers concluded that the best strategy to address flooding issues in paved cities was to build massive underground storage facilities, as opposed to talking to their colleagues in transport and working with them to make cities more permeable. At times, while individual decisions make sense at the local scale, when we look at the entire system, they do not. We can recall Don Norman's (2013) quote that we saw in chapter 1: "Engineers and businesspeople are trained to solve problems. Designers are trained to discover problems. A brilliant solution to the wrong problem can be worse than no solution at all: solve the correct problem" (p. 218). By containerizing responsibilities into departments that do not communicate with one another, we become great at solving the wrong problems.

In scientific and professional communities, we typically refer to this type of departmental organization as a *silo*. Just look at any organization's structure, and you will quickly find an organogram in the form of a tree, in the sense of Christopher Alexander. Figure 12.5 shows two examples of typical organograms taken in spring 2019 for the Illinois Department of Transportation and the Chicago Department of Water Management. From a cognitive point of view, these tree organograms make sense. After all, this is how the brain works, and by looking at these organograms, it is obvious who is responsible for what. Everyone knows, however, that informal relationships rapidly emerge, and information sharing does not follow the structure of the organogram.

So what is the solution? On the one hand, to integrate infrastructure systems we need to be able to make better decisions that transcend current departments' competencies, like it used to be before responsibilities were divvied up. On the other hand, the human brain has limits that we must acknowledge and that present a significant hurdle to designing organizations that communicate and share with one another. Professor Thomas Seager likes to say: "Rather than seeking to avoid conflict, work to build trust."[5] He is absolutely right, and we should find a way to provide institutional incentives to encourage cross-departmental communication and collaboration. Although there is no definitive answer about how to achieve this, we can think about ways and means through which new organizational structures could work.

The first thing we should do is recall the lesson from Minett that we saw at the end of chapter 10. In his 1975 paper, Minett offered a partial response to address the problem raised by Christopher Alexander. He notably argued for a supraplanning organization that would coordinate departments and agencies. Whether this type of agency would be sufficient is uncertain, but Minett also wrote something that resonates particularly well with our situation: "It is the links that create the system." We can interpret this quote as the need to create formal relationships between agencies and have people work together; without these relationships, we have only separate individual systems.

Figure 12.6 shows four options for a systemwide organizational structure. Option A is basically the status quo. A fairly powerful head, whether a company chief executive officer or a mayor, communicates with departments individually, and there is little knowledge sharing and no resource sharing across departments. We have elaborated quite heavily on this option and realize that it should be changed. Option B takes a radical approach and argues for a massive merger of all competencies. While this may be

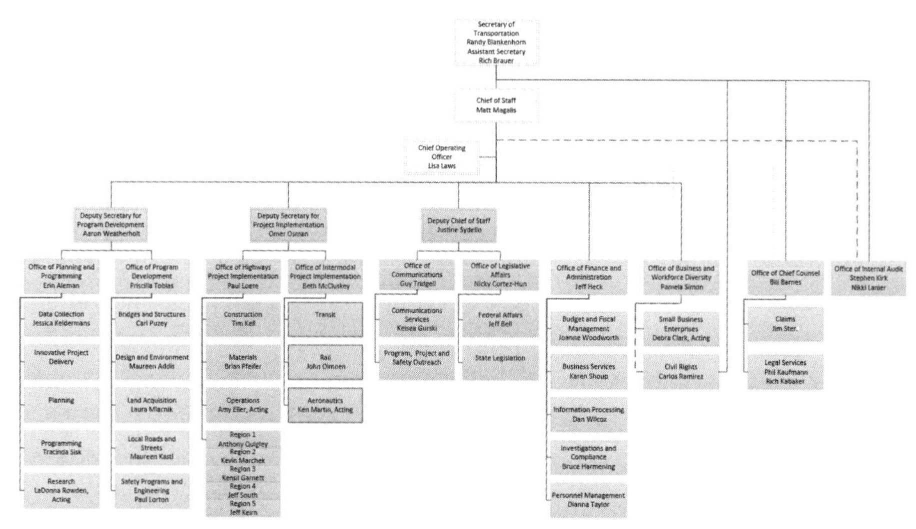

a) Illinois Department of Transportation

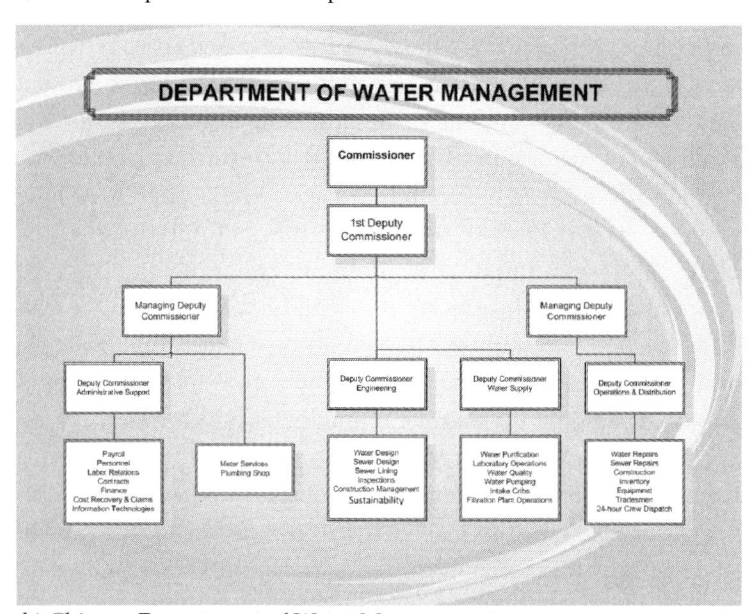

b) Chicago Department of Water Management

Figure 12.5

Two examples of an organogram. Courtesy of the Illinois Department of Transportation and the City of Chicago Department of Water Management.

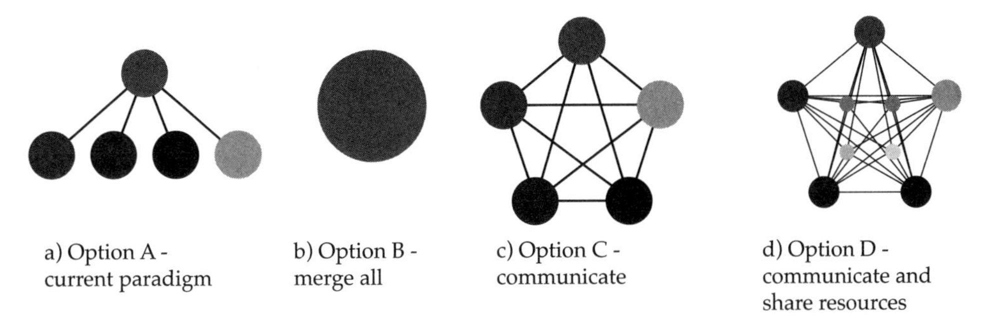

Figure 12.6
Four options for organizational structure.

appealing, we should remember that it used to be like that, and organizations were split because infrastructure systems became too complicated. Similar to Christopher Alexander favoring semilattices and Jane Jacobs favoring bottom-up initiatives, a gigantic organization operating all aspects of cities in a top-down manner is doomed to fail.

Option C starts to formally create relationships between departments so they communicate with one another, especially on projects that can overlap, whether functionally or geographically. Naturally, people can only work together if they speak the same language, which brings us back to the need to create a formal ontology of infrastructure systems from the Smart Cities section.

With option D, we can go even further and imagine services and resources that could be shared across multiple departments. Although this drawing is not as "clear" and easily understandable as the other three, it really is not that complicated. For example, database management and visualization services could (or perhaps even should) be handled by one single department to ensure that data can easily be accessed, shared, and visualized. We often overlook this aspect, but simple data format conflicts can stall entire projects for significant periods of time. Similarly, keeping a record of infrastructure assets in Geographic Information Systems format has become the norm for cities around the world (Balasubramani et al. 2017). This particularly makes sense for underground infrastructure systems since they occupy the same space (Derrible 2017). Moreover, other services like public relations, billing, and asset management could be maintained by one entity for all infrastructure systems. It may take time before we get there, but option D may be the best.

In the end, we must remember that the current way municipal, regional, and national departments are organized is fairly new. For example, the U.S. Department of Transportation was only created in 1966. It is therefore high time to try different organizational models. In fact, it is necessary if we aspire to successfully start integrating urban infrastructure systems.

Overall, these three paradigm-shifting changes have the potential to completely change the way infrastructure is planned, designed, engineered, and operated in cities. Moreover, they are as likely to completely change the engineering profession. These prospects are exciting, and it is now time for a few final thoughts and a four-step process to conclude this book.

12.2 Final Thoughts and the Four-Step Urban Infrastructure Design Process

As mentioned briefly in the previous section, the Science of Cities is missing from the three paradigm-shifting changes. This does not mean that it is not important, because it is, but we must realize that the Science of Cities is a young field. At the time of this writing, the Science of Cities had been mostly successful at discovering the patterns and the properties of cities, but to ever be useful in engineering and planning, it must be able to produce knowledge that can be translated into design. While this may not happen rapidly, engineering scholars around the world are adopting concepts of the Science of Cities. In particular, the Science of Cities is key to better understanding how humans perceive and design cities—remember Kevin Lynch's *The Image of the City*? In fact, it is only when we understand how humans perceive cities and how humans cognitively go through a design process that we can start designing cities that are both sustainable and livable.

This actually brings us to another point. Although we have not discussed it as a paradigm-shifting change, the general fields of design and psychology should possess a large role in urban engineering in the future. Presumably, we should be able to design infrastructure that incentivizes people to use it in a certain way. The term *nudge* has been famously used in the field of behavioral economics, resulting in a great book by Thaler and Sunstein (2009).[6] The goal here would therefore be to design infrastructure systems in such a way to promote behavioral change—for example, design electrical and water appliances that incentivize people to turn them off when not in use, design buildings to reduce crime and to promote social well-being, or design roads in certain ways to improve safety.[7]

Furthermore, we must remember the two sustainability principles: (1) controlling the demand and (2) increasing the supply within reason. As engineers, we often like to focus purely on supply, but we must remember that demand is key! Without developing policies, we can design systems that "shift" the demand. If we remember the three demand curves from chapter 10 (figure 10.1), one of our objectives is to reduce the peak—for example, by shifting the demand to other times during the day. Moreover, without focusing on efficiency, there are many ways to control demand, and we have seen quite a few in most chapters.

As general guidelines, we can follow a four-step urban infrastructure design (UID) process sequentially. Each time, an example is showcased to illustrate the goal of each step. The four UID processes are further explained and applied to infrastructure in Derrible (2019).

1. *Control the demand to reduce the need for new infrastructure.* More often than not, an insufficient capacity can be solved by reducing the demand rather than adding infrastructure to increase capacity. This does not mean that we should forbid people from consuming water/electricity or from traveling—especially in countries that need these resources—but it means that incentives should be put in place to reduce the overall demand for energy and resources.

   Example: Roads. In most high-income countries, effort should be put into reducing congestion before building new roads, such as by adopting travel demand–management strategies (and many such strategies exist).

2. *Integrate a needed service within current infrastructure.* If controlling the demand is not an option or if it is not sufficient, see what is in the area and whether the new infrastructure can be integrated within current infrastructure instead of directly seeking space to build infrastructure. Cities have a tremendous amount of embodied energy already. Seek to use it before building new.

   Example: Telecom lines. Remember that telecom lines are built next to railroad tracks to link two cities. Moreover, in many U.S. cities, telecom lines often use old Western Union conduits.

3. *Make new infrastructure multifunctional to provide for other infrastructure systems.* If new infrastructure needs to be built from scratch, see whether it can have multiple functions, following the colloquial saying "killing two birds with one stone." This includes adding functions relevant to different departments.

   Example: Green infrastructure. Every road project, both new and for maintenance, should include a green infrastructure element like rain gardens and bioswales.

4. *Design for specific interdependencies and decentralize infrastructure if possible.* Most new projects require the service of a different infrastructure system. Try to design these interdependencies to avoid putting pressure on existing systems, which can help reduce the demand and help decentralize infrastructure, thus contributing both to sustainability and resilience.

Example: Solar panels. When adding new sensors/monitoring devices on infrastructure systems (e.g., traffic signals, water gauges) add a solar panel or small wind turbine in addition to connecting the device to the power grid. This not only decreases the demand for electricity when the solar panel or wind turbine is generating electricity, it also makes the infrastructure system less dependent on the power grid.

The four-step UID process is extremely simple. It may not be applicable systematically depending on the context of the project or the standard of practice of a firm, but it can significantly help to design more sustainable infrastructure systems. Again, to learn more about this four-step process and as a quick recap to some of the concepts covered in this book, see "An Approach to Designing Sustainable Urban Infrastructure" (Derrible 2019).

To end, I would like to share my optimism for the future of humanity and the earth. People in the realm of sustainability are often negative and fearful of the future. Often, this is linked to the incredibly slow (and sometimes backward) pace of some of the world's leaders to take meaningful actions—I certainly will not argue that the pace is fast enough. However, I travel quite a lot and interact with faculty, students, industries, public officials, and a variety of people from different countries, and many recognize the problem we face as a society and are passionate about sustainability. At the same time, every year I have the privilege to teach future generations, and I see how dedicated students are to developing designs for a more sustainable world. All these interactions, experiences, and manifestations of a common goal toward designing a sustainable world make me optimistic about the future of our society.

Now it is time for us, as urban engineers, to use all the concepts, equations, findings, and lessons from this book, and collectively do the best we can and design sustainable urban infrastructure.

Problem Set

12.1 In your own words, describe and explain some of the main concepts you have learned from reading this book.

12.2 Select your favorite chapter from the book and discuss what you particularly liked about it and how it changed your viewpoint.

12.3 In your own words, describe some of the challenges that we, as a society, are facing when it comes to urban engineering and list some of the changes we should make to become more sustainable.

12.4 In your own words, describe a Smart City.

12.5 Based on your own research, select any biomimicking material and describe how it works and its potential to transform cities and urban engineering.

12.6 Based on your own research, select any metamaterial and describe how it works and its potential to transform cities and urban engineering.

12.7 In your own words, describe the current problems with the way organizations are structured and how responsibilities are divided and offer some ways to improve the situation.

12.8 By performing your own research, describe how the Science of Cities can be used in urban engineering.

12.9 By performing your own research, select an example of infrastructure, anywhere in the world, that was built to leverage human psychology (i.e., in the sense of a nudge) and explain why it succeeded or failed.

12.10 By performing your own research, select an example of infrastructure, anywhere in the world, that could have followed the four-point checklist but did not—for example, infrastructure that was built but was not required to be built or infrastructure that could have been made multifunctional.

12.11 In your own words, describe whether you are optimistic or pessimistic about the future of cities when it comes to sustainability.

Notes

1. This phenomenon is also well captured by the *butterfly effect*, which states that the flapping of a butterfly's wings can generate a hurricane on the other side of the planet. Remember that we actually saw it in the last chapter, as it is one of the four properties of a complex system listed by Holland (2014).

2. And we have used a large amount of data in this book that came from the EPA. Thank you EPA!

3. In fact, Microsoft's SmartArt feature was used to draw these two figures.

4. The Wikipedia page on "Biomimetics" is fairly good. Moreover, *Wired* magazine published an article in 2013 titled "Strange Biology Inspires the Best New Materials" that contains some examples of biomimicking materials. It is available at https://www.wired.com/2013/03/biomimetic-materials/ (accessed March 30, 2018).

5. Professor Seager is an incredibly creative and forward-thinking researcher. Look him up! At the time of this writing, he was a civil engineering faculty member at Arizona State University.

6. Richard Thaler, one of the two coauthors, actually received the 2017 Nobel Prize in Economics for his work.

7. At the time of this writing, a lot of work had already happened on this one already.

References

Allwood, Julian M., Jonathan M. Cullen, Mark A. Carruth, Daniel R. Cooper, Martin McBrien, Rachel L. Milford, Muiris C. Moynihan, and Alexandra C. H. Patel. 2012. *Sustainable Materials: With Both Eyes Open*. Cambridge: UIT Cambridge.

Balasubramani, Booma Sowkarthiga, Omar Belingheri, Eric S. Boria, Isabel F. Cruz, Sybil Derrible, and Michael D. Siciliano. 2017. "GUIDES: Geospatial Urban Infrastructure Data Engineering Solutions (Demo Paper)." In *SIGSPATIAL '17: Proceedings of the 25th ACM SIGSPATIAL International Conference on Advances in Geographic Information Systems*. New York: ACM. https://doi.org/10.1145/3139958.3139968.

Carson, Rachel. 1962. *Silent Spring*. Boston: Houghton Mifflin.

Derrible, Sybil. 2017. "Complexity in Future Cities: The Rise of Networked Infrastructure." *International Journal of Urban Sciences* 21 (supp. 1): 68–86. https://doi.org/10.1080/12265934.2016.1233075.

———. 2019. "An Approach to Designing Sustainable Urban Infrastructure." *MRS Energy & Sustainability* 5: E15. https://doi.org/10.1557/mre.2018.14.

Grima, Joseph N., and Roberto Caruana-Gauci. 2012. "Mechanical Metamaterials: Materials That Push Back." *Nature Materials* 11(7): 565–566. https://doi.org/10.1038/nmat3364.

Harper, Corey D., Chris T. Hendrickson, Sonia Mangones, and Constantine Samaras. 2016. "Estimating Potential Increases in Travel with Autonomous Vehicles for the Non-driving, Elderly and People with Travel-Restrictive Medical Conditions." *Transportation Research Part C: Emerging Technologies* 72 (November): 1–9. https://doi.org/10.1016/j.trc.2016.09.003.

Holland, J. H. 2014. *Complexity: A Very Short Introduction*. Oxford: Oxford University Press.

Liu, Yongmin, and Xiang Zhang. 2011. "Metamaterials: A New Frontier of Science and Technology." *Chemical Society Reviews* 40(5): 2494–2507. https://doi.org/10.1039/C0CS00184H.

Minett, John. 1975. "If the City Is Not a Tree, nor Is It a System." *Planning Outlook* 16(1–2): 4–18. https://doi.org/10.1080/00320717508711494.

Miodownik, Mark. 2015. "Materials for the 21st Century: What Will We Dream Up Next?" *MRS Bulletin* 40(12): 1188–1197.

Norman, D. 2013. *The Design of Everyday Things: Revised and Expanded Edition*. New York: Basic Books.

Nicolaou, Zachary G., and Adilson E. Motter. 2012. "Mechanical Metamaterials with Negative Compressibility Transitions." *Nature Materials* 11(7): 608–613. https://doi.org/10.1038/nmat3331.

———. 2013. "Longitudinal Inverted Compressibility in Super-Strained Metamaterials." *Journal of Statistical Physics* 151(6): 1162–1174. https://doi.org/10.1007/s10955-013-0742-8.

Thaler, Richard H., and Cass R. Sunstein. 2009. *Nudge: Improving Decisions about Health, Wealth, and Happiness*. Rev. and exp. ed. New York: Penguin Books.

Appendix

A. Tables

Table A.1
Standard urban engineering units.

Name	Unit	Symbol
General		
Time	Second	s
	Minute	min
	Hour	h
Temperature	Kelvin	K
	Celsius	C
Population		
People	Person	pers
Electricity		
Power	Watt	W
Energy	Watt-hour	Wh
Voltage	Volt	V
Current	Amp	A
Resistance	Ohm	Ω
Water		
Volume	Liter	L
Velocity	Meters per second	m/s
Flow rate	Cubic meters per second	m^3/s
Pressure	Pascal	Pa
	Meter	m

(continued)

Table A.1 (continued)

Name	Unit	Symbol
Transport		
Vehicles	Vehicle	veh
Trip	Trip	trip
Distance	Vehicle kilometers traveled	VKT
	Passenger kilometers traveled	PKT
Buildings		
Heat transfer	Watt	W
Heat transfer per unit area	Watt per square meter	W/m^2
Heat transfer per unit length	Watt per meter times kelvin	$W/(m{\cdot}K)$
Solid waste		
Weight	Metric ton	ton

Table A.2

British units to metric system conversion factors

Category	Conversion factor	British units
Mass		
	1 lb = 0.453 kg	lb: pound
	1 short ton = 0.907184 ton	short ton: U.S. short ton
	1 long ton = 1.016047 ton	long ton: U.S. long ton
Temperature		
	0 K = 273.15°C	
	F = 1.8°C + 32	F: Fahrenheit
Distance		
	1 in = 25.04 mm	in: inch
	1 ft = 0.3048 m	ft: feet
	1 yd = 0.9144 m	yd: yard
	1 mi = 1.609 km	mi: mile
Volume		
	$1\ m^3 = 1{,}000\ L$	
	1 gal = 3.78541 L	gal: gallon
	$1\ ft^3 = 0.02832\ m^3$	ft^3: cubic feet
	$1\ yd^3 = 0.76455\ m^3$	yd^3: yard

Table A.2 (continued)

Category	Conversion factor	British units
Pressure		
	1 psi = 6,894.76 Pa	psi: pound per square inch
Flow rates		
	1 gpm = 0.063 L/s	gpm: gallon per minute
Density		
	1 lb/ft^3 = 16.0184 kg/m^3	ft^3: cubic feet
	1 lb/yd^3 = 0.593276 kg/m^3	yd^3: cubic yard
	1 lb/gal = 119.826 kg/m^3	gal: gallon
Energy		
	1 cal = 1 kcal	cal: calorie
	1 cal = 4.187 kJ	
	1 cal = 1.16222 Wh	
	1 Btu = 1,055.056 J	Btu: British thermal unit
	1 Btu = 0.293071 Wh	
Energy density		
	1 Btu/lb = 0.64611 Wh/kg	
Power		
	1 hp = 745.7 W	hp: horsepower
Fuel economy		
	1 mpg = 0.4251 km/L	mpg: mile per gallon

Table A.3
Constants

Name	Symbol	Value	Units	Information
Electricity				
Planck's constant	h	6.63×10^{-34}	J.s	
Water				
Water density	ρ	998	kg/m^3	At 20°C
Gravitational acceleration	g	9.81	m/s^2	
Specific weight	$\gamma = \rho \cdot g$	9.79	kN/m^3	At 20°C
Water viscosity	μ	1.002×10^{-3}	kg/m·s	At 20°C
Kinematic viscosity	$\nu = \mu/\rho$	1.003×10^{-6}	m^2/s	At 20°C
Transport emissions				
Passenger car[a]		215	veh-km	Technically not a constant, but used by the EPA for nationwide carbon accounting
Light-duty truck[b]		297	veh-km	
Medium & heavy-duty truck		914	veh-km	
Motorcycle		120	veh-km	
Bus		35	per-km	
Transit rail		74	per-km	
Regional rail		101	per-km	
Intercity rail (i.e., Amtrak)		88	per-km	
Buildings				
Stefan-Boltzmann constant	σ	5.67×10^{-8}	W/m^2K^4	

[a]Includes passenger cars, vans, SUVs, and small pickup trucks.
[b]Includes full-size pickup trucks, full-size vans, and extended-length SUVs.

Table A.4

U.S. states consumption values used in this book

	Electricity					
	Grid emission factor[a]	*Consumption*[b]	Gas[c]		Water[d]	
State	*kg CO_2e/MWh*	*MWh/year*	*MWh/year*	*m^3/year*	*L/day*	*gal/day*
AK	421.85	7.08	2.20	207.96	374.44	99
AL	416.16	14.57	8.17	774.11	278.06	73
AR	509.21	12.99	6.09	576.67	336.62	89
AZ	425.06	12.36	10.35	980.05	547.03	145
CA	205.96	6.57	7.92	750.49	324.33	86
CO	670.40	8.33	4.25	402.99	465.90	123
CT	227.77	8.53	3.52	332.99	133.91	35
DC	219.11	9.65	3.90	369.60	252.36	67
DE	403.40	11.36	5.15	488.16	302.91	80
FL	466.74	13.48	14.25	1,349.51	313.28	83
GA	456.98	13.65	4.62	438.01	283.91	75
HI	695.32	6.06	15.13	1,433.58	545.31	144
IA	455.36	10.37	4.47	423.18	246.29	65
ID	85.85	11.44	4.58	434.27	697.43	184
IL	370.16	8.79	3.01	285.39	302.25	80
IN	827.78	11.70	4.11	389.56	287.62	76
KS	546.16	10.78	4.78	452.96	249.76	66
KY	892.72	13.46	5.03	476.95	257.20	68
LA	400.14	14.88	8.54	808.95	435.28	115
MA	375.36	7.18	3.99	377.54	212.96	56
MD	462.42	11.94	4.42	418.50	272.73	72
ME	157.68	6.56	3.81	360.89	190.98	50
MI	501.69	8.01	3.29	311.86	254.90	67
MN	462.80	9.17	3.83	362.65	219.26	58
MO	770.65	12.49	4.71	445.92	336.20	89
MS	427.76	14.44	6.51	616.57	378.54	100
MT	571.57	9.76	4.26	403.91	404.27	107
NC	395.84	13.21	5.70	540.12	266.11	70
ND	759.70	12.55	4.26	403.37	302.53	80
NE	585.47	11.68	4.84	458.52	294.86	78
NH	143.70	7.24	4.55	431.17	226.84	60
NJ	253.94	8.29	3.82	361.74	302.92	80
NM	717.87	7.57	5.35	506.59	307.90	81

(continued)

Table A.4 (continued)

	Electricity					
	Grid emission factor[a]	*Consumption*[b]	Gas[c]		Water[d]	
State	*kg CO_2e/MWh*	*MWh/year*	*MWh/year*	*m^3/year*	*L/day*	*gal/day*
NV	349.97	11.10	6.42	607.81	478.09	126
NY	211.32	7.14	3.25	307.84	269.57	71
OH	669.21	10.69	3.89	368.16	237.89	63
OK	475.51	13.12	5.55	525.46	268.17	71
OR	139.33	10.88	5.52	523.18	402.96	106
PA	390.12	10.10	3.83	362.34	216.18	57
RI	395.40	7.03	4.15	393.48	223.55	59
SC	287.04	13.85	7.11	673.86	378.54	100
SD	234.31	11.78	4.74	448.69	211.83	56
TN	452.88	14.85	5.76	545.93	303.41	80
TX	478.37	13.87	7.80	738.46	311.09	82
UT	743.17	8.99	4.24	401.18	638.47	169
VA	371.49	13.44	3.76	356.07	303.10	80
VT	30.31	6.59	4.68	443.75	166.29	44
WA	85.25	11.46	4.52	427.79	391.21	103
WI	633.45	8.19	4.34	411.23	207.97	55
WV	902.92	13.22	4.14	392.34	302.42	80
WY	925.78	10.20	4.10	388.04	589.77	156
US	455.42	10.77	4.71	445.80	311.42	82

Sources:

[a]U.S. Environmental Protection Agency 2016 values, https://www.epa.gov/energy/egrid (accessed August 18, 2018). Used in chapter 5.

[b]U.S. Energy Information Agency 2016 values, https://www.eia.gov/electricity/data/state/ (accessed August 18, 2018). Used in chapter 5.

[c]U.S. Energy Information Agency 2016 values, https://www.eia.gov/naturalgas/data.php (accessed August 18, 2018). Not used in the book.

[d]U.S. Geological Survey 2015 values, https://pubs.er.usgs.gov/publication/cir1441 (accessed August 18, 2018). Used in chapter 6.

B. Moody Diagram

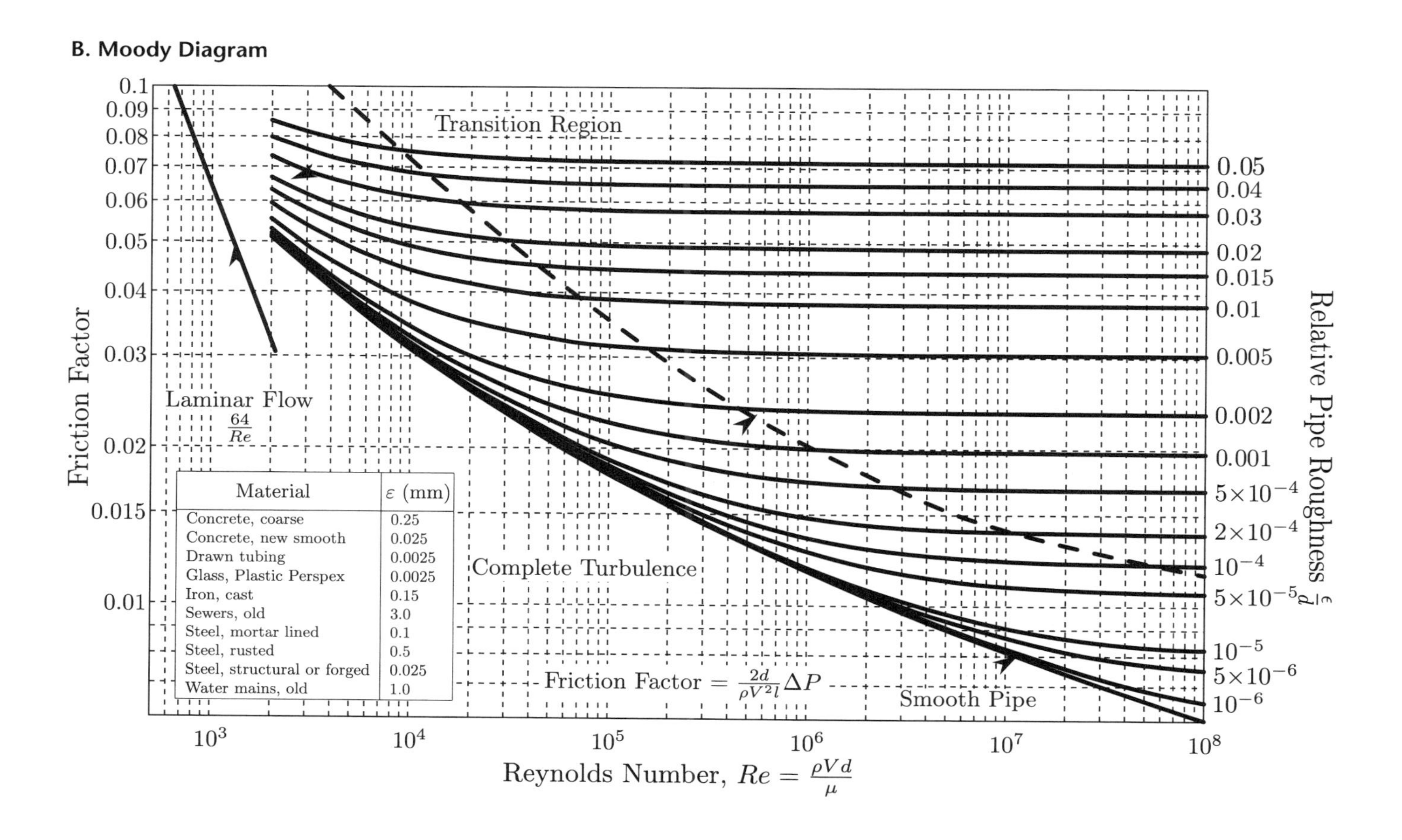

Material	ε (mm)
Concrete, coarse	0.25
Concrete, new smooth	0.025
Drawn tubing	0.0025
Glass, Plastic Perspex	0.0025
Iron, cast	0.15
Sewers, old	3.0
Steel, mortar lined	0.1
Steel, rusted	0.5
Steel, structural or forged	0.025
Water mains, old	1.0

C. Level-of-Service Diagram

Metric System

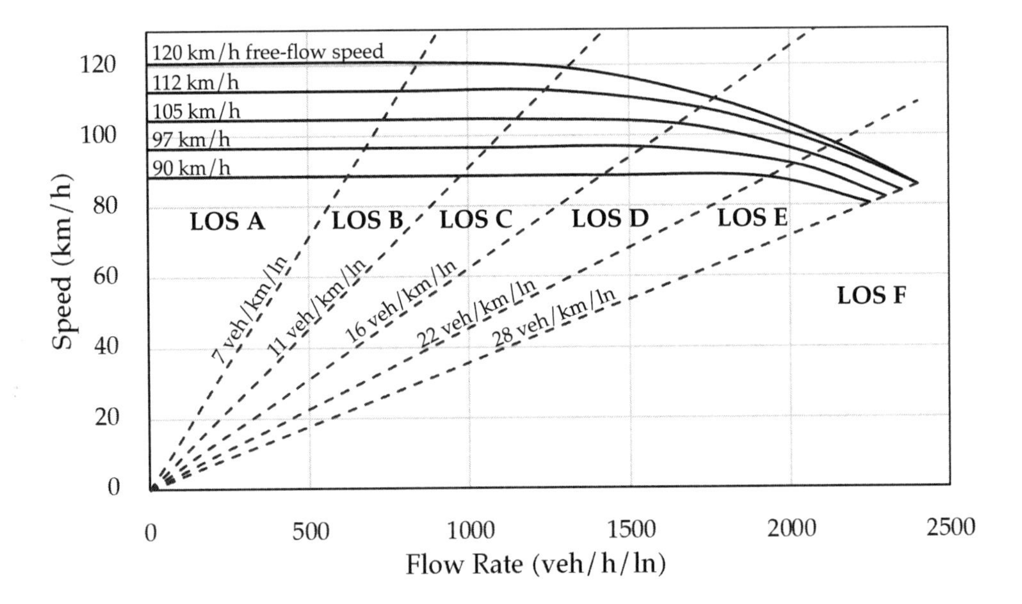

Free-flow speed	120 km/h		112 km/h		105 km/h		97 km/h		90 km/h	
LOS	*q*	*v*	*q*	*v*	*q*	*v*	*q*	*v*	*q*	*v*
A	825	120.7	770	112.65	710	104.61	660	96.56	605	88.51
B	1,330	118.8	1,260	112.65	1,170	104.61	1,080	96.56	990	88.51
C	1,775	109.9	1,735	107.34	1,665	103.00	1,560	96.56	1,430	88.51
D	2,130	98.0	2,110	97.04	2,060	94.63	2,000	91.89	1,915	88.03
E	2,400	85.8	2,400	85.78	2,350	84.01	2,300	82.24	2,250	80.47

Note: q is in veh/km/ln, and v is in km/h.

Imperial System

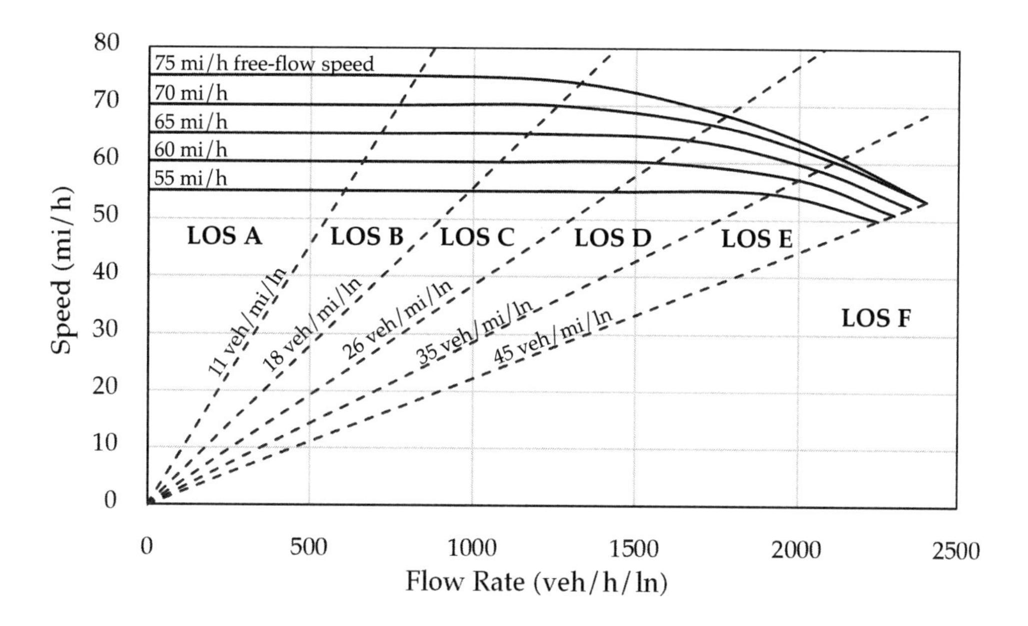

Free-flow speed	75 mph		70 mph		65 mph		60 mph		55 mph	
LOS	q	v	q	v	q	v	q	v	q	v
A	825	75.0	770	70.0	710	65.0	660	60.0	605	55.0
B	1,330	73.8	1,260	70.0	1,170	65.0	1,080	60.0	990	55.0
C	1,775	68.3	1,735	66.7	1,665	64.0	1,560	60.0	1,430	55.0
D	2,130	60.9	2,110	60.3	2,060	58.8	2,000	57.1	1,915	54.7
E	2,400	53.3	2,400	53.3	2,350	52.2	2,300	51.1	2,250	50.0

Note: q is in veh/mi/ln, and v is in mph.

D. Equation Sheet

Chapter 1: Introduction

No equations.

Chapter 2: Sustainability

$$P - C \geq 0 \quad (1)$$

$$\frac{dP}{dt} - \frac{dC}{dt} \geq 0 \quad (2)$$

$$I = P \cdot A \cdot T \quad (3)$$

$$E = P \cdot A \cdot T \cdot \varepsilon \quad (4)$$

Chapter 3: Population

$$p_n = p_{n-1} r = p_0 r^n \quad (1)$$

$$S_n^p = \frac{p_0(1 - r^{n+1})}{1 - r} \quad (2)$$

$$a_n = a_{n-1} + d = a_0 + n \cdot d \quad (3)$$

$$S_n^a = \frac{(n+1)}{2}(a_0 + a_n) \quad (4)$$

$$\frac{dP}{dt} = k_1 P \quad (5)$$

$$\ln(P) = k_1 t + c \quad (6)$$

$$P = c' e^{k_1 t} \quad (7)$$

$$P = P_0 e^{k_1 t} \quad (8)$$

$$k_1 = \frac{\Delta P}{\Delta t} \times \frac{1}{P} \quad (9)$$

$$\frac{dP}{dt} = k_2 \quad (10)$$

$$P = k_2 t + c \quad (11)$$

$$P = k_2 t + P_0 \quad (12)$$

$$k_2 = \frac{\Delta P}{\Delta t} \quad (13)$$

$$\frac{dP}{dt} = k_3(P_{sat} - P) \tag{14}$$

$$P = P_{sat} - (P_{sat} - P_0)e^{-k_3 t} \tag{15}$$

$$k_3 = \frac{\Delta P}{\Delta t} \times \frac{1}{(P_{sat} - P)} \tag{16}$$

$$\frac{dP}{dt} = k_4 P\left(1 - \frac{P}{K}\right) \tag{17}$$

$$P(t) = \frac{P_{sat}}{1 + a \cdot e^{bt}} \tag{18}$$

$$P_{sat} = \frac{2P_0P_1P_2 - P_1^2(P_0 + P_2)}{P_0P_2 - P_1^2} \tag{19}$$

$$a = \frac{P_{sat} - P_0}{P_0} \tag{20}$$

$$b = \frac{1}{\Delta t}\ln\left[\frac{P_0(P_{sat} - P_1)}{P_1(P_{sat} - P_0)}\right] \tag{21}$$

$$P(t+1) = P(t) + B(t+1) - D(t+1) + M(t+1) \tag{22}$$

$$P_i(t+1) = P_{i-1}(t)S_{i-1}(t) \tag{23}$$

$$P_n(t+1) = P_{n-1}(t)S_{n-1}(t) + P_n(t)S_n(t) \tag{24}$$

$$P(t+1) = \sum_{i=2}^{n-1} P_{i-1}(t)S_{i-1}(t) \tag{25}$$

$$B(t+1) = \sum_{i=1}^{n-1} P_i(t) \times \Delta t \times \frac{1}{2}(F_i(t) + F_{i+1}(t)) \tag{26}$$

$$M(t+1) = P(t+1) - P(t) - B(t+1) + D(t+1) \tag{27}$$

Chapter 4: Urban Planning

No equations.

Chapter 5: Electricity

$$V = R \cdot I \tag{1}$$

$$V_C = V_S - RI \tag{2}$$

$$P = V \cdot I \tag{3}$$

$$E = P \cdot t \tag{4}$$

$$V_S = V_1 + V_2 \tag{5}$$

$$i_S \cdot R_{eq} = i_1 \cdot R_1 + i_2 \cdot R_2 \tag{6}$$

$$R_{eq} = R_1 + R_2 \tag{7}$$

$$R_{eq} = \sum R_i \tag{8}$$

$$i_S = i_1 + i_2 \tag{9}$$

$$\frac{V_S}{R_{eq}} = \frac{V_1}{R_1} + \frac{V_2}{R_2} \tag{10}$$

$$R_{eq} = \frac{1}{\frac{1}{R_1} + \frac{1}{R_2}} \tag{11}$$

$$R_{eq} = \frac{1}{\sum \frac{1}{R_i}} \tag{12}$$

$$V(t) = 170 \cdot \sin(2\pi \cdot 60 \cdot t) \tag{13}$$

$$V_C = \frac{P}{I} - RI \tag{14}$$

$$P = \frac{1}{2} \rho A v^3 \tag{15}$$

$$E = h\upsilon \tag{16}$$

$$E_p = \eta \cdot E_s \cdot A \tag{17}$$

Chapter 6: Water

$$T = \frac{n+1}{m} \tag{1}$$

$$i = \frac{a}{(b+t)^c} \tag{2}$$

$$\textit{total energy} = \textit{hydraulic} + \textit{kinetic} + \textit{potential} \tag{3}$$

$$P + \frac{1}{2} \rho V^2 + \rho g z = \text{constant} \tag{4}$$

$$H = \frac{P}{\gamma} + \frac{1}{2} \frac{V^2}{g} + z \tag{5}$$

$$H_L = f \cdot \frac{L}{D} \cdot \frac{V^2}{2g} \tag{6}$$

$$\frac{\varepsilon}{D} \tag{7}$$

$$\mathrm{Re} = \frac{\rho V D}{\mu} = \frac{VD}{\nu} \tag{8}$$

$$H_1 = H_2 + H_L \tag{9}$$

$$H_1 + H_P = H_2 + H_L \tag{10}$$

$$P_w = \gamma \cdot Q \cdot H_P \tag{11}$$

$$P_s = \frac{P_w}{\eta} = \frac{\gamma \cdot Q \cdot H_P}{\eta} \tag{12}$$

$$H_L = K \cdot Q^2 \tag{13}$$

$$K = f \cdot \frac{L}{D} \cdot \frac{1}{2gA^2} \tag{14}$$

$$K_{series} = K_1 + K_2 + K_3 = \sum_i K_i \tag{15}$$

$$K_{BC} = \frac{1}{\left(\frac{1}{\sqrt{K_B}} + \frac{1}{\sqrt{K_C}} \right)^2} \tag{16}$$

$$K_{parallel} = \frac{1}{\left(\sum_i \frac{1}{\sqrt{K_i}} \right)^2} \tag{17}$$

$$Q = \frac{C_m}{n} A R^{2/3} S_f^{1/2} \tag{18}$$

$$E_s = y + \frac{V^2}{2g} \tag{19}$$

$$Fr = \frac{V}{\sqrt{gD}} \tag{20}$$

$$Q = KA \frac{\phi_1 - \phi_2}{L} \tag{21}$$

$$K = k\frac{\gamma}{\mu} \tag{22}$$

$$q = \frac{Q}{A} = K\frac{\phi_1 - \phi_2}{L} \tag{23}$$

$$v = \frac{q}{n_e} \tag{24}$$

$$Q_u = \frac{\pi K\left(\phi_2^2 - \phi_1^2\right)}{\ln\left(\frac{r_2}{r_1}\right)} \tag{25}$$

$$Q_c = \frac{2\pi Km(\phi_2 - \phi_1)}{\ln\left(\frac{r_2}{r_1}\right)} \tag{26}$$

$$Q = C \cdot i \cdot A \tag{27}$$

$$P = Q + I_a + F \tag{28}$$

$$\frac{F}{S} = \frac{Q}{P - I_a} \tag{29}$$

$$Q = \frac{(P - I_a)^2}{P - I_a + S} \tag{30}$$

$$Q = \frac{(P - 0.2S)^2}{P + 0.8S},\ P > 0.2S \tag{31}$$

$$S = 25.04\left(\frac{1000}{CN} - 10\right) \tag{32}$$

Chapter 7: Transport

$$q = \frac{n}{t} \tag{1}$$

$$\bar{v}_t = \frac{\sum v_i}{n} \tag{2}$$

$$\bar{v}_s = \frac{l}{\bar{t}} \tag{3}$$

$$\bar{t} = \frac{1}{n}\sum t_i \tag{4}$$

$$\bar{v}_s = \frac{l}{\frac{1}{n}\sum t_i} = \frac{1}{\frac{1}{n}\sum\left[\frac{1}{(l/t_i)}\right]} \tag{5}$$

$$k = \frac{n}{l} \tag{6}$$

$$q = vk \tag{7}$$

$$v = v_f\left(1 - \frac{k}{k_j}\right) \tag{8}$$

$$q = v_f\left(k - \frac{k^2}{k_j}\right) \tag{9}$$

$$q = k_j\left(v - \frac{v^2}{v_f}\right) \tag{10}$$

$$\frac{dq}{dk} = v_f\left(1 - \frac{2k}{k_j}\right) = 0 \tag{11}$$

$$k_{cap} = \frac{k_j}{2} \tag{12}$$

$$v_{cap} = \frac{v_f}{2} \tag{13}$$

$$q_{cap} = \frac{v_f k_j}{4} \tag{14}$$

$$q_p = v_p k_p \tag{15}$$

$$q_p = \frac{v_p}{s_p} \tag{16}$$

$$T = T_{o,1} + T_{o,2} + t_{t,1} + t_{t,2} = 2(T_o + t_t) \tag{17}$$

$$T = 2(T_o + 0.15 \cdot T_o) = 2.3 \cdot T_o \tag{18}$$

$$T_o = \frac{L}{V_o} \tag{19}$$

$$T_o = t_m + t_s + t_i \tag{20}$$

$$t_a = \frac{V_c}{a} \tag{21}$$

$$t_d = \frac{V_c}{d} \tag{22}$$

$$A_{shaded} = \frac{1}{2}V_c \cdot t_a = \frac{1}{2}V_c \cdot \frac{V_c}{a} = \frac{V_c^2}{2a} \tag{23}$$

$$L = V_c \cdot t_m - \frac{V_c^2}{2a} - \frac{V_c^2}{2d} \tag{24}$$

$$t_m = \frac{L}{V_c} + \frac{V_c}{2}\left(\frac{1}{a} + \frac{1}{d}\right) \tag{25}$$

$$t_m = \frac{L}{V_c} + (n_s - 1)\frac{V_c}{2}\left(\frac{1}{a} + \frac{1}{d}\right) \tag{26}$$

$$t_m = \sqrt{\frac{2(a+d)l}{a \cdot d}} \tag{27}$$

$$t_s = n_s \cdot t_{avg} \approx 20n_s \tag{28}$$

$$t_i = n_c \cdot \frac{t_{traffic}}{4} \tag{29}$$

$$n_{tr} = w \cdot T \cdot f = w \cdot \frac{T}{h} \tag{30}$$

$$t_{wk} = \frac{d_{wk}}{v_{wk}} \tag{31}$$

$$S = \frac{L}{n_s - 1} \tag{32}$$

$$t_w = \begin{cases} \frac{h}{2}, & \text{when } h \leq 10 \text{ min} \\ \sqrt{h}, & \text{when } h > 10 \text{ min} \end{cases} \tag{33}$$

$$PKT = N_o \cdot VKT \tag{34}$$

$$E = P \cdot A \cdot T \cdot \varepsilon \tag{35}$$

$$E = \sum_m S_m \cdot P \cdot A \cdot T \cdot \varepsilon \tag{36}$$

$$E = \sum_m S_m \cdot P \cdot A \cdot (T \cdot \varepsilon)_m \tag{37}$$

$$O_i = \sum_j T_{ij} \tag{38}$$

$$D_i = \sum_j T_{ij} \tag{39}$$

$$T = \sum_i \sum_j T_{ij} = \sum_i O_i = \sum_j D_j \tag{40}$$

$$A_{im} = \sum_j E_j \cdot f\left(c_{ijm}\right) \tag{41}$$

$$T_{ij} = KO_iD_jf_{ij} \tag{42}$$

$$K = \frac{T}{\sum_{ij} O_iD_jf_{ij}} \tag{43}$$

$$K_i = \frac{1}{\sum_j D_jf_{ij}} \tag{44}$$

$$K_j = \frac{1}{\sum_i O_jf_{ij}} \tag{45}$$

$$T_{ij} = K_iK_jO_iD_jf_{ij} \tag{46}$$

$$H = -\sum_i \sum_j p_{ij} \log p_{ij} \text{ with} \sum_i \sum_j p_{ij} = 1 \tag{47}$$

$$U_{im} = \beta_1 X_{1m} + \beta_2 X_{2m} + \cdots + \beta_n X_{nm} = \sum_i \beta_i X_{im} \tag{48}$$

$$P_{im} = \frac{U_{im}}{\sum_m U_{im}} \tag{49}$$

$$U_{im} = \sum_i \beta_i X_{im} + \varepsilon_{im} = V_{im} + \varepsilon_{im} \tag{50}$$

$$P_{i1} = P\ (U_{i1} \geq U_{i2}) \tag{51}$$

$$P_{i1} = P\ (V_{i1} - V_{i2} \geq \varepsilon_{i2} - \varepsilon_{i1}) \tag{52}$$

$$P_{im} = \frac{e^{V_{im}}}{\sum_m e^{V_{im}}} \tag{53}$$

$$A_i = \ln\left(\sum_m e^{V_{im}}\right) \tag{54}$$

$$t_i = t_{i,f}\left[1 + a\left(\frac{d_i}{c_i}\right)^b\right] \tag{55}$$

Chapter 8: Buildings

$$M = E + C + R \tag{1}$$

$$Q^{\bullet} = -kA\frac{dT}{dx} \tag{2}$$

$$Q^{\bullet} = -kA\frac{\Delta T}{\Delta x} = kA\frac{T_1 - T_2}{\Delta x} \tag{3}$$

$$q^{\bullet} = k\frac{T_1 - T_2}{\Delta x} = \frac{T_1 - T_2}{\Delta x / k} \tag{4}$$

$$R = \frac{\Delta x}{k} \tag{5}$$

$$q^{\bullet} = \frac{\Delta T}{R_1 + R_2 + \cdots + R_n} \tag{6}$$

$$R_{eff} = \sum_{i=1}^{n} R_i \tag{7}$$

$$U = \frac{1}{R} = \frac{k}{\Delta x} \tag{8}$$

$$q^{\bullet} = U_{eff} \cdot \Delta T \tag{9}$$

$$Q^{\bullet} = A \cdot U_{eff} \cdot \Delta T \tag{10}$$

$$Q^{\bullet}_{total} = \sum_{j=1}^{m} Q^{\bullet}_{m} \tag{11}$$

$$U_{eq} = \frac{1}{A}\sum_{i=1}^{m} A_i U_i = \frac{1}{A}\sum_{i=1}^{m}\frac{A_i}{R_i} \tag{12}$$

$$Q^{\bullet} = h_c \cdot A \cdot \Delta T \tag{13}$$

$$q^{\bullet} = h_c \cdot \Delta T \tag{14}$$

$$R = \frac{1}{h_c} \tag{15}$$

$$L^3 \Delta T < 63 \tag{16}$$

$$vL < 1.4 \tag{17}$$

Type	Laminar	Turbulent
Free flow		
Warm horizontal facing up/cold horizontal facing down	$h_c = 1.32\left(\frac{\Delta T}{L}\right)^{1/4}$ (18)	$h_c = 1.52\ (\Delta T)^{1/3}$ (19)
Warm horizontal facing down/cold horizontal facing up	$h_c = 0.59\left(\frac{\Delta T}{L}\right)^{1/4}$ (20)	
Tilted surface. Applies for angles between 30° and 90°.	$h_c = 1.42\left(\frac{\Delta T \sin\beta}{L}\right)^{1/4}$ (21)	$h_c = 1.31\ (\Delta T \sin\beta)^{1/3}$ (22)
Forced flow		
Applies to all surfaces. *L* is the length of the surface in the direction of the flow.	$h_c = 2.0\left(\frac{v}{L}\right)^{1/2}$ (23)	$h_c = 6.2\left(\frac{v^4}{L}\right)^{1/5}$ (24)

$$E_b = \sigma T^4 \tag{25}$$

$$\varepsilon = \frac{E}{E_b} \tag{26}$$

$$\alpha + \tau + \rho = 1 \tag{27}$$

$$\dot{Q}_{12} = \frac{A\sigma}{1/\varepsilon_1 + 1/\varepsilon_2 - 1}(T_1^4 - T_2^4) \tag{28}$$

$$\dot{q}_{12} = \frac{\sigma}{1/\varepsilon_1 + 1/\varepsilon_2 - 1}(T_1^4 - T_2^4) \tag{29}$$

$$h_r = \frac{\dot{q}_{12}}{T_1 - T_2} \tag{30}$$

$$h_r = \frac{4\sigma T^3}{1/\varepsilon_1 + 1/\varepsilon_2 - 1} \tag{31}$$

$$\dot{Q} = h_r \cdot A \cdot \Delta T \tag{32}$$

$$\dot{q} = h_r \cdot \Delta T \tag{33}$$

$$\dot{q} = F \cdot h_r \cdot \Delta T \tag{34}$$

$$dF_{1,2} = \frac{\cos\theta_1 \cdot \cos\theta_2}{\pi S_{1,2}^2} dA_2 \tag{35}$$

$$F_{1,2}A_1 = F_{2,1}A_2 \tag{36}$$

$$\dot{q}_{solar} = \alpha \cdot I_{solar} \tag{37}$$

$$Q^{\bullet}_{sys} = U_{sys} \cdot \Delta T = \frac{\Delta T}{R_{sys}} \tag{38}$$

$$R_{sys} = \frac{1}{U_{sys}} = \frac{1}{h_i} + R_{layer} + \frac{1}{h_o} \tag{39}$$

$$Q^{\bullet}_{w} = U_w \cdot A_w \cdot \Delta T \tag{40}$$

$$V^{\bullet} = A \cdot c \cdot \Delta p^n \tag{41}$$

$$V^{\bullet} = a \cdot V \tag{42}$$

$$Q^{\bullet}_{air} = \rho \cdot c_p \cdot V^{\bullet} \cdot \Delta T \tag{43}$$

$$U_{air} = \rho \cdot c_p \cdot V^{\bullet} \tag{44}$$

$$D = \sum_{days} (T_o - T_{ref}) \tag{45}$$

$$E = U_{sys} \cdot A \cdot D \tag{46}$$

$$C = \frac{V}{A} \tag{47}$$

$$RC = \frac{C}{C_{ref}} \tag{48}$$

$$S = \frac{L}{W} \tag{49}$$

Chapter 9: Solid Waste

$$M = 100 \cdot \left(\frac{W_{wet} - W_{dry}}{W_{wet}} \right) \tag{1}$$

$$S_c = l \tag{2}$$

$$S_c = \left(\frac{l + w}{2} \right) \tag{3}$$

$$S_c = \left(\frac{l + w + h}{3} \right) \tag{4}$$

$$S_c = (l \times w)^{1/2} \tag{5}$$

$$S_c = (l \times w \times h)^{1/3} \tag{6}$$

$$K = Cd^2 \frac{\gamma}{\mu} = k \frac{\gamma}{\mu} \tag{7}$$

$$E_{dry} = E_{asdiscarded}\left(\frac{100}{100-M}\right) \quad (8)$$

$$E_{recoverable} = E_{asdiscarded}\left(\frac{100}{100-M-Ash}\right) \quad (9)$$

$$E_{Dulong} = 94C + 394\left(H_2 - \frac{1}{8}O_2\right) + 6N + 26S \quad (10)$$

$$\text{organic matter} + O_2 + \text{nutrients} \rightarrow \text{new cells} + \text{humus} + CO_2 + H_2O + NH_3 + SO_2^{2-} + \text{heat} \quad (11)$$

$$\text{organic matter} + H_2O + \text{nutrients} \rightarrow \text{new cells} + \text{humus} + CO_2 + CH_4 + NH_3 + H_2S + \text{heat} \quad (12)$$

Chapter 10: Urban Metabolism and Infrastructure Integration

$$S_{i,m} = P \cdot A_{i,m} \cdot M_{i,m} \quad (1)$$

$$S_{i,m} = D \cdot A_{i,m} \cdot M_{i,m} \quad (2)$$

$$Y_{i,m} = S_{i,m} \cdot T_{i,m} \quad (3)$$

$$S = \sum_i \sum_m S_{i,m} \quad (4)$$

$$Y = \sum_i \sum_m Y_{i,m} \quad (5)$$

$$\frac{dS_m}{dt} = I_m(t) - O_m(t) \quad (6)$$

$$\Delta S_m^{t_2-t_1} = [I_m(t_2) - I_m(t_1)] - [O_m(t_2) - O_m(t_1)] \quad (7)$$

$$I_F + P_F + I_{W,Kit} = O_{F,FW} + O_{F,Met} + O_{F,S} \quad (8)$$

$$I_E = I_{E,\ buildings} + I_{E,\ transport} + I_{E,\ industry} + I_{E,\ construction} + I_{E,\ water\ pumping} + I_{E,\ waste} \quad (9)$$

$$I_{E,\ buildings} = I_{E,\ heating} + I_{E,\ cooling} + I_{E,\ lights\ and\ appl.} + I_{E,\ water\ heating}. \quad (10)$$

$$I_{i,\ heating} = \sum_i P \cdot A_i \cdot HDD \cdot T_{i,\ heating} \quad (11)$$

$$I_{i,\ cooling} = \sum_i P \cdot A_i \cdot CDD \cdot T_{i,\ cooling} \quad (12)$$

$$I_{E,S} + I_{E,F} + I_{E,I} = O_{E,I} + O_{E,G} + O_{E,E} \quad (13)$$

$$Q_W = Q_{W,D} + Q_{W,L} \quad (14)$$

$$Q_{W,D} = Q_{W,D,\ base} + CDD \cdot T_{W,\ cooling} \quad (15)$$

$$T_{W,\,cooling} = \frac{Q_{W,\,max\ daily} - Q_{W,avg\ daily}}{CDD} \tag{16}$$

$$Q_{W,L} = L \cdot r_{leak} \tag{17}$$

$$Q_{W,L} = D \cdot \rho_W \cdot r_{leak} \tag{18}$$

$$Q_{WWT} = Q_{WW,D} + Q_{WW,R} + Q_{Inf} \tag{19}$$

$$I_{W,\,precip} + I_{W,\,pipe} + I_{W,\,sw} + I_{W,\,gw} = O_{W,\,evap} + O_{W,\,out} + \Delta S_W \tag{20}$$

Chapter 11: Science of Cities and Machine Learning

$$x_{n+1} = r \cdot x_n(1 - x_n) \tag{1}$$

$$Y = \alpha X^{\beta} \tag{2}$$

$$\log(Y) = \beta \log(X) + \log(\alpha) \tag{3}$$

$$P = \frac{P_0}{r^{\beta}} \tag{4}$$

$$P \cdot r = \text{constant} \tag{5}$$

$$P(t+1) = (1+\lambda)P(t) \tag{6}$$

$$\left.\begin{array}{l} P_i(t+1) = P_i(t) + P_0 \\ P_j(t+1) = P_j(t) - P_0 \end{array}\right\} \tag{7}$$

$$\mu = L - N + S \tag{8}$$

$$\alpha = \frac{L - N + S}{\frac{1}{2}N(N-1) - (N-S)} \tag{9}$$

$$\alpha_p = \frac{L - N + S}{2N - 6 + S} \tag{10}$$

$$\beta = \frac{L}{N} \tag{11}$$

$$\gamma = \frac{L}{\frac{1}{2}N(N-1)} \tag{12}$$

$$\gamma_p = \frac{L}{3N - 6} \tag{13}$$

$$A_{ij} = \begin{cases} 1 & \text{if nodes } i \text{ and } j \text{ are connected} \\ 0 & \end{cases} \tag{14}$$

$$d_i = \sum_j A_{ij} \tag{15}$$

$$C_{Ci} = \frac{1}{\sum_j p(i,j)} \tag{16}$$

$$C_{CNi} = \frac{N-1}{\sum_j p(i,j)} \tag{17}$$

$$C_{Bi} = \sum_{j,k} \frac{p_{jk}(i)}{p_{jk}} \tag{18}$$

$$dist(A, B) = \sqrt{(x_A - x_B)^2 + (y_A - y_B)^2} \tag{19}$$

$$dist(A, B) = \sqrt{\sum_i^n (x_{A,i} - x_{B,i})^2} \tag{20}$$

$$\arg\min \sum_i \sum_j (\| x_i - c_j \|)^2 \tag{21}$$

$$s(i) = \frac{b(i) - a(i)}{\max[a(i), b(i)]} \tag{22}$$

$$a(i) = \frac{\sum_{j \in C_i, i \neq j} dist\ (x_j, c_i)}{|C_i| - 1} \tag{23}$$

$$b(i) = \min_{C_j : 1 \leq j \leq k, j \neq i} \left\{ \frac{\sum_{j \in C_j} dist(x_j, c_i)}{|C_j|} \right\} \tag{24}$$

$$H_E = -\sum_i p_i \log p_i \tag{25}$$

$$H_G = \sum_j p_j (1 - p_j) = 1 - \sum_j p_j^2 \tag{26}$$

$$G_i = \frac{n_{left}}{N} H_{i,left} + \frac{n_{right}}{N} H_{i,right} \tag{27}$$

$$y_i = \sum_j w_j \cdot x_j + z_i \tag{28}$$

$$I_i = f(y_i) \tag{29}$$

$$\text{Step: } f(y) = \begin{cases} -1 & \text{for } y < 0 \\ 1 & \text{for } y \geq 0 \end{cases} \tag{30}$$

$$\text{Sigmoid: } f(y) = \frac{1}{1+e^{-y}} \tag{31}$$

$$\text{Tanh: } f(y) = \tanh(y) \tag{32}$$

$$\text{ReLU: } f(y) = \begin{cases} 0 & \text{for } y < 0 \\ y & \text{for } y \leq 0 \end{cases} \tag{33}$$

$$R^2 = 1 - \frac{\sum (y_i - \hat{y}_i)^2}{\sum (y_i - \bar{y})^2} \tag{34}$$

Chapter 12: Conclusion

No equations.

Index